Springer Proceedings in Earth and Environmental Sciences

The series Springer Proceedings in Earth and Environmental Sciences publishes proceedings from scholarly meetings and workshops on all topics related to Environmental and Earth Sciences and related sciences. This series constitutes a comprehensive up-to-date source of reference on a field or subfield of relevance in Earth and Environmental Sciences. In addition to an overall evaluation of the interest, scientific quality, and timeliness of each proposal at the hands of the publisher, individual contributions are all refereed to the high quality standards of leading journals in the field. Thus, this series provides the research community with well-edited, authoritative reports on developments in the most exciting areas of environmental sciences, earth sciences and related fields.

More information about this series at http://www.springer.com/series/16067

V. I. Karev · Dmitry Klimov ·
Konstantin Pokazeev
Editors

Physical and Mathematical Modeling of Earth and Environment Processes (2018)

4th International Scientific School for Young Scientists, Ishlinskii Institute for Problems in Mechanics of Russian Academy of Sciences

Springer

Editors
V. I. Karev
Institute for Problems in Mechanics Russian Academy of Sciences (IPMech RAS)
Moscow, Russia

Dmitry Klimov
Institute for Problems in Mechanics Russian Academy of Sciences (IPMech RAS)
Moscow, Russia

Konstantin Pokazeev
Faculty of Physics
Lomonosov Moscow State University
Moscow, Russia

ISSN 2524-342X ISSN 2524-3438 (electronic)
Springer Proceedings in Earth and Environmental Sciences
ISBN 978-3-030-11532-6 ISBN 978-3-030-11533-3 (eBook)
https://doi.org/10.1007/978-3-030-11533-3

Library of Congress Control Number: 2019930995

This Springer imprint is published by the registered company Springer Nature Switzerland AG
The registered company address is: Gewerbestrasse 11, 6330 Cham, Switzerland

Preface

This book, entitled *Physical and Mathematical Modeling of Earth and Environment Processes (2018)—4th International Scientific School for Young Scientists, Ishlinskii Institute for Problems in Mechanics of Russian Academy of Sciences*, is the result of a collaborative work on the frame of the 4th International Scientific School for Young Scientists held at the Ishlinskii Institute for Problems in Mechanics of the Russian Academy of Sciences on October 24–26, 2018. This book includes results of the theoretical and experimental studies of processes in the atmosphere, oceans, the lithosphere, and their interaction; environmental issues; problems of human impact on the environment; and methods of geophysical research.

The conference articles included in book describe the studies on the dynamics of natural systems: geosphere, hydrosphere, atmosphere—and their interactions, the human contribution to naturally occurring processes, laboratory modeling of processes in various environments, and testing of newly developed physical and mathematical models.

A number of articles in the book are devoted to issues related to the oil and gas production and storage. Special mention should be made the works developing the geomechanical approach to solving these problems. Oil and gas remain the main sources of energy on the planet for the coming decades. But large and accessible oil and gas fields have been depleted; therefore, new sources of hydrocarbon raw materials are needed. These may be shale oil and gas, Arctic hydrocarbon reserves, as well as oil and gas from deep horizons (more than 5.5 km). New technologies are required to develop them. At the same time, environmental problems are becoming increasingly acute. Therefore, the need arises to create environmentally friendly technologies. The foundation for the creation of such technologies can be a geomechanical approach based on the use of a huge reserve of elastic energy stored in a rock mass by controlling the stress–strain state of the formation. The book presents new results of experimental and theoretical modeling of the processes of deformation, destruction, and filtration in rocks in connection with the creation of the scientific foundations of new technologies for the production of hydrocarbons. The investigation of the dependence of well stability and permeability of rocks on the stress–strain state in conditions of high rock pressure is represented also.

Our book

- Enriches the understanding of the geophysical processes taking place in various environments (lithosphere, hydrosphere, atmosphere), including anthropogenic, and promotes to the intensification of their development.
- Includes the results of theoretical and experimental studies on the development of the geomechanical approach to creating new technologies of hydrocarbon production based on controlling the stress–strain state in the reservoir.
- Contains the results of recent research in the field of interactions of the lithosphere, atmosphere, and hydrosphere of various scales, energy exchange of the atmosphere and the ocean, including under anthropogenic influences.
- Proposes new physical and mathematical models of processes occurring in the environment, both natural and man-made, as well as the improvement of existing ones.
- Develops methods of solving problems of stability oil and gas wells in the process of their construction and operation, and an increase of the productivity of hydrocarbon production, including on the ocean shelf, thereby raising the efficiency of training specialists in oil reservoir physics.
- Presents modern methods of research and modeling of various processes in environment, means and ways of monitoring of natural systems, methods of research and forecasting of natural and man-made disasters, and ways to eliminate their consequences.
- Contains additional material for specialists working in the oil and gas industry to expand, improve, and disseminate new acknowledgments in this field.

Program Committee of the School, which included the leading scientists on the scientific directions of the School, has conducted peer review of the papers submitted to the book and produced a competitive selection. 121 papers were submitted. As a result of double-blind peer review, 48 papers were selected.

Moscow, Russia

V. I. Karev
Dmitry Klimov
Konstantin Pokazeev

Contents

The Influence of Anticyclonic Movement Over the Sea of Azov on Variations of Maximum Instantaneous Current Speed in the Kerch Strait During 1948–2017 Ice Seasons

A. V. Kholoptsev[1], T. Ya. Shul'ga[2(✉)], O. Ye. Shchodro[2], and S. A. Podporin[3]

[1] N.N. Zubov State Oceanographic Institute, Sevastopol Branch, Sovietskaya 61, 299011 Sevastopol, Russia

[2] Marine Hydrophysical Institute, Kapitanskaya 2, 299011 Sevastopol, Russia
shulgaty@mail.ru

[3] Sevastopol State University, Universitetskaya 33, 299053 Sevastopol, Russia

Abstract. The maximum absolute value of instantaneous speed increase of the Azov Current in the Kerch Strait, caused by anticyclones moving in southward directions with parameters recorded during 1948–2017 ice seasons, is estimated by use of Princeton Ocean $3D$ hydrodynamic model of general circulation in combination with hydraulic methods of channel calculation. Probability distributions of characteristics of such anticyclones, as well as modern tendencies of their frequency variation are estimated for winters of different severity on the Sea of Azov.

Keywords: Kerch Strait · Modeling · Instantaneous current speed · Atmospheric pressure · Frequency · Anticyclone · Hydrodynamic modeling · Tendency

1 Introduction

Unexpected intensification of a current in any narrow strait poses significant danger to ships navigating its waters. During ice seasons, it may also endanger hydrotechnical installations located there. Therefore, estimation of maximum values of instantaneous current speed increase in a strait, caused by certain natural factors, is of significant importance for oceanographers and those involved in operation and design of transport complexes.

The solution of this problem may also be of considerable interest, as far as factors with relatively high frequency and worldwide presence are concerned. Among them are variations in distribution of atmospheric pressure and wind velocity components over a body of water caused by rapid movement of an

V. I. Karev et al. (Eds.): *Physical and Mathematical Modeling of Earth and Environment Processes (2018)*, SPEES, pp. 1–14, 2019.
https://doi.org/10.1007/978-3-030-11533-3_1

anticyclone over it, which, in turn, leads to a water level rise at the entrance of the strait and an anemobaric wave propagation along it [1,2]. The influence of these phenomena on variations in instantaneous current speed in the strait is the more noticeable, the higher the average increase rate of average daily atmospheric pressure at its entrance is (hereinafter atmospheric pressure increase is referred to as API). Among other affecting factors are the relief of the strait seabed and the shape of its coastlines [3,4]. The absolute maximum of instantaneous speed increase (ISI) of the ensuing current is usually attained at the strait's cross-section with the minimal area.

Estimation of ISI levels of a current that can be caused by phenomena under consideration is most practically significant for straits with high intensity of marine traffic and/or ongoing hydrotechnical construction operations. In Russia, one of such straits is the Kerch Strait, where the construction of an automobile-railway bridge connecting the Crimean Peninsula with the mainland regions of Russia has recently been completed. During severe and moderate winters, the Azov Current transfers ice along the Kerch Strait from the Sea of Azov to the Black Sea. The impact of the ice on the supports of the previous bridge, built in the middle of the 20th century, resulted in their failure and a consequent bridge collapse. Therefore, the issue of whether a similar catastrophe can happen again in the future is far from being meaningless.

During severe and moderate winters, an ice season in the Kerch Strait begins in mid-December and ends mostly by the end of March. In the Sea of Azov, individual ice formations can also be met in April. During mild winters, ice in the strait is practically never met [5,6]. The last few winters in the region have been mild, which, of course, cannot guarantee that moderate or severe winters will not happen again in the future and the bridge supports will not have to face perceptible ice attacks. Possible consequences of such attacks, everything else being equal, are the more dangerous, the greater the absolute value of the instantaneous speed of the Azov Current carrying the drifting ice along the strait is.

During 1955–2014 ice seasons, expeditions held by the Sevastopol Branch of N.N. Zubov State Oceanographic Institute (SB SOIN) periodically monitored average speed variations of the existing current in the Kerch Strait, including its cross-section that adjoins the island of Tuzla (hereinafter referred to as the Tuzla cross-section). The average speed values of the Azov Current, as measured at the specified fairway cross-section, were normally found within 0.14–0.18 m/s. Yet, higher average speed values of 0.7–0.83 m/s were recorded on the stretch of the strait corresponding to its northern narrow part (where the ferry route from port Crimea to port Caucasus lies) [5]. Variations of absolute values of instantaneous current speed in the Tuzla cross-section were, however, not monitored at that time, although it is obvious that their peak levels could not but exceed the above-mentioned average ones.

The Azov Current intensification, if it falls on an ice season, also leads to an increase in the amount of ice transferred from the Sea of Azov to the northern part of the strait per unit time. Therefore, if such a phenomenon is to take place during a severe or moderate winter, the result will be a significant increase of ice extent in the water area in question. Due to the relief features of the Kerch

Strait seabed at its part where the bridge is located, the Azov Current transfers ice from the Sea of Azov to the Black Sea mainly along the fairway (between the bridge supports). The greater the value of the absolute velocity of ice carried by the Azov Current is, the stronger the mechanical action produced by this ice will be on ships passing the strait, as well as on the bridge supports [5–8]. Therefore, an estimation of ISI of the Azov Current at the part of the fairway where the bridge is located will be required to assess feasibility of constructing special ice protection facilities for the bridge supports.

One of the most appropriate approaches to such an estimation is based on results of mathematical modeling carried out with help of the three-dimensional nonlinear Princeton Ocean Model (*POM*) [9]. As shown in [10], this model, after having been adapted to the area of the Sea of Azov [8], enabled to demonstrate a significant role of the resonance mechanism in the occurrence of extremely high amplitudes of wind setup-setdown oscillations and seiches generated by anticyclones moving over the water in southward directions at speeds equal to the speed of propagation of a free long wave in this water area [11]. API parameters in these anticyclones, as well as characteristics of the ensuing tides, surge phenomena oscillations, and seiches, are comprehensively studied in [12]. As shown in [13], the field of the current velocities in the Kerch Strait is most significantly affected by anticyclones moving over the Sea of Azov in southward directions.

On the coast of the Sea of Azov, there is a number of Russian and Ukrainian weather stations, which perform routine monitoring of variations of basic meteorological parameters, including atmospheric pressure [14]. An important source of information on distribution of average daily atmospheric pressure reduced to sea level for particular days is NCEP/NCAR reanalysis, which contains data for the period from 1.01.1948 to 31.12.2017 [15]. The information presented therein is uniformly precise and corresponds to any nodes of the coordinate grid with a grid step of 2.5°. The grid includes nodes located between the 15 and 45E meridians, as well as 40 and 80N parallels (the Azov–Black Sea region, including the Kerch Strait, lies within this range).

However, the Azov Current ISI caused by southward movement of anticyclones over the Sea of Azov during ice seasons have not yet been estimated for any part of the Kerch Strait fairway. Neither have features of interannual frequency variability of API exceeding a certain level been considered. These features thus cannot be adequately accounted for when developing strategies to enhance the navigation safety in the strait and to ensure safe operation of the existing bridge.

The goal of this study includes estimation of the Azov Current ISI that may be attained in the Tuzla cross-section during ice seasons when anticyclones move south-ward over the Sea of Azov and identification of probability distribution of API values and modern tendencies of their frequency variation. To achieve the stated goal, the following problems were addressed:

- Estimation of distribution of the Azov Current ISI values at the Kerch Strait northern entrance, as well as differences between water levels at its northern and southern entrances during southward movement of anticyclones over the Sea of Azov.

- Estimation of ISI values that can be attained in the Tuzla cross-section.
- Finding probability distribution of particular API values that were attained when anticyclones were moving southward over the Sea of Azov during 1948–2017 ice seasons for winters of different severity and identification of modern tendencies of interannual variations in frequency of the API considered.

2 Materials and Methods

To address the stated problems, factual material for a fairly long period was required to estimate average daily atmospheric pressure at any point of the Sea of Azov and the Kerch Strait. NCEP/NCAR reanalysis was considered as a possible source of such information. However, consistency of the results presented therein with the actual observation data collected in this region had not been previously verified. Therefore, in order to ensure that the reanalysis could indeed provide trustworthy factual material to solve the problems in question, we compared its results with the information obtained from routine observations of atmospheric pressure variations for the period from 01.01.1972 to 31.03.1988. The specified information was collected from observation sites located in Opasnoye, Mysovoye, and Taman and can be found in the archives of SB SOIN.

To obtain the verification, the reanalysis data corresponding to each observation site were interpolated by means of Delaunay triangulation technique [16]. The results were then compared with the mean daily atmospheric pressure values derived from the observed data, which enabled to determine their relative error. Also, for each observation site we estimated relevance of statistical correlation between the time dependences of the interpolation results and the actual atmospheric pressure values using the Student's criterion [17]. In this way, validity of the factual material and its applicability for solution of the problem under consideration were confirmed.

The first problem was addressed using mathematical modeling. The Azov Current ISI and the water surface level at the northern cross-section of the Kerch strait were estimated with help of the *POM* hydrodynamic model, which is based on solution of the equations of the viscous fluid turbulent motion in a hydrostatic approximation [11]. The estimation was carried out for cases when there was movement of anticyclones over the Kerch Strait. Parameterization of the vertical viscosity index and the turbulent diffusivity was performed in accordance with the Mellor-Yamada differential model [18]. The horizontal viscosity index was derived by use of the subgrid viscosity model [19]. The projections of tangential wind stresses were expressed through the values of wind velocity at standard meteorological altitude. The allowance was made for the coefficient of the sea surface aerodynamic drag [20]. The normal velocity component at the bottom of the strait was assumed to be zero; the near-bottom tangential stresses were assumed to be related to the velocity logarithmically. Adhesion conditions at the lateral boundaries were considered to be fulfilled. The speed of the Azov Current at the starting moment of time was also assumed to be zero. Integration steps for time and spatial coordinates were picked in accordance with the stability criterion for barotropic waves [21].

Information on bathymetry and coastline shapes of the Sea of Azov, obtained by digitization of hydrographic maps, was used as an input for modeling purposes. Spatial resolution of the model grid connecting the extreme points of the water area in question (the Sea of Azov, the Kerch Strait, and part of the Black Sea within 44.501°–48°N and 34°–40.021°E) along horizontal coordinates was 1 km, the number of vertical calculation levels was 11.

Variations of distributions of wind velocity components over the sea surface were investigated with help of a mesoscale numerical atmospheric *SKIRON/Eta* model [22]. The prediction results obtained using this model were in a form of a 72-h forecast of meteorological parameters for the Azov–Black Sea and the Mediterranean basin. Atmospheric model data, available after 6 h on a grid with 0.1° resolution for both latitude and longitude, were interpolated on the computational grid of the Azov Basin.

When calculating ISI of the Azov Current in the fairway of the Kerch strait at the Tuzla cross-section, we assumed the entire strait to be a channel, in which a unidirectional flow was created by difference in water levels between its northern and southern entrances. The calculations were performed using a technique, based on numerical solution of the Chezy equation and the related formulas 1:

$$\begin{gathered} v = C\sqrt{RI}, Q = v\omega, C = R^y/n, R = \omega/\chi, \\ y = -0.13 + 2.5\sqrt{n} - 0.75\left(\sqrt{n} - 0.1\right)\sqrt{R} \cong 0.16, \end{gathered} \tag{1}$$

where v, Q, C, n and ω are velocities, flows, Chezy coefficients, roughness coefficients, clear flow areas, and hydraulic radii for the entire flow section under steady flow conditions; I – hydraulic slope (is constant over the entire length of the channel), χ – wetted perimeter of the channel. The equation was solved using a trial-and-error method. For more accurate determination of velocities within the plane hydraulics problem (represented at the horizontal plane), the shape of the channel cross-sections was also accounted for by use of the approximate method, proposed by Velikanov [23–25].

The unsteady flow motion was calculated by solving the set of Saint-Venant non-linear equations numerically. The set included the equation of dynamics (2) and the continuity equation (3):

$$\frac{1}{A}\frac{\partial Q}{\partial t} + \frac{1}{A}\frac{\partial}{\partial x}\left(\frac{Q^2}{A}\right) + g\frac{\partial y}{\partial x} - g\left(S_0 - S_f\right) = 0. \tag{2}$$

$$\frac{\partial Q}{\partial x} + \frac{\partial A}{\partial t} = 0. \tag{3}$$

In Eqs. 2 and 3 standard notation is used.

To obtain a numerical solution of the above equation set, the method of splitting by physical processes and state variables was applied [23]. To evaluate the Azov Current ISI at the Tuzla cross-section, we used ISI distributions obtained for the northern section of the Kerch Strait, as well as the differences in levels between the northern section and the average Black Sea level at the southern strait entrance. The areas of the above-mentioned cross-sections were also accounted for. As is shown in [26], the cross-sectional area of the Kerch Strait at its northern part (port Crimea – port Caucasus) is 26 km^2, whereas the same of the Tuzla cross-section (cape Ak–Burun—the Island of Tuzla) amounts to 12 km^2.

As far as the second problem was concerned, its most challenging part was identification of all anticyclones that occurred during the months of 1948–2017 ice seasons and could possibly cause the phenomena under consideration. To estimate the average daily atmospheric pressure over the Sea of Azov, we picked 100 points of its water area using the Monte Carlo method [27]. For each point, the appropriate reanalysis results were obtained through Delaunay triangulation [15]. The results for all the points, corresponding to the same day, were then averaged.

For identification of the anticyclones that drifted southward over the Sea of Azov towards the northern entrance of the Kerch Strait, we used synoptic charts compiled using the reanalysis results that were obtained through Delaunay triangulation. For this particular European region, the charts in question enabled to obtain distributions of the values of surface mean daily atmospheric pressure, reduced to sea level, for each day of API occurrence, as well as for the day before it. When identifying the anticyclones that were able to increase instantaneous current speed in the Kerch Strait, we compared the charts, forming each such pair. Since only anticyclones moving over the sea from the north were of interest, a certain API was taken into account and used in further analysis only if the position of the corresponding anticyclone, as plotted on the chart, tended to change southward, which followed from the analysis of charts for two consecutive days.

The severity of winters in the Kerch Strait was estimated using the method described in [28]. For each ice season and each month of it, the total number of anticyclones and the related API values were determined. Frequency values of the events under consideration were derived as ratios of the above mentioned parameters to the total number of days in a particular month of an ice season, averaged in a 30-year-long sliding window. For each ice season and each month of it, maximum API values were determined. The estimates of the indicators obtained in such a way enabled to form their time series, which were then used for estimation of API probability distributions and identification of modern tendencies of their frequency variation.

The angular coefficient value of the linear trend of a corresponding time series was used as a quantitative characteristic of the modern tendency of variations in API frequency and their maximum values for a particular month of an ice period [17]. The angular coefficient values were estimated for the period from 2005 to 2017.

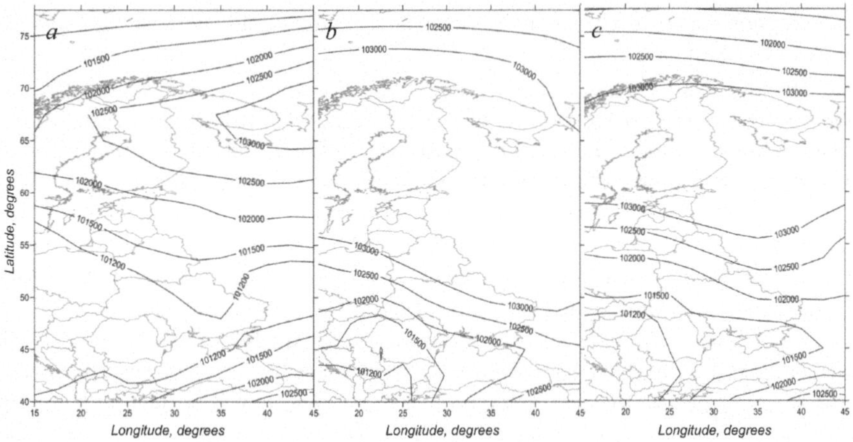

Fig. 1. Distributions of average daily values of ground-level atmospheric pressure over Eastern Europe, reduced to sea level: 5.12.2015 (a); 6.12.2015 (b); 7.12.2015 (c).

3 Analysis and Discussion

The results of the factual material analysis have revealed that, during 1948–2017 ice seasons, the highest API values for the phenomena under consideration took place on December 6, 2015 (18.7 hPa). Figure 1 shows distributions of average daily values of the surface atmospheric pressure reduced to sea level for the period December 5–7, 2015. As follows from Fig. 1a (05.12.2015), a vast area of low atmospheric pressure was present over the Sea of Azov (the average daily pressure in the region was approximately 1012 hPa). At the same time, a high-pressure area, which was a part of the Arctic anticyclone, was located over Scandinavia, regions of the North-Western Federal District of Russia, and the Barents Sea. Yet, a subtropical anticyclone, which was not associated with the Arctic one, rested above Turkey and the Caucasus. As is evident from Fig. 1b, on 06.12.2015 an invasion of the Arctic air occurred. This led to formation of a submeridional belt of increased atmospheric pressure that linked the above-mentioned anticyclones with each other.

As a result, penetration of the Arctic air into European regions of Russia situated in temperate latitudes caused an atmospheric pressure rise. On 06.12.2015 it almost reached 1015 hPa over the Sea of Azov. From Fig. 1 it is clear that on 07.12.2015 the invasion of the Arctic air continued, causing significant expansion of the above mentioned high pressure belt. At the same time, atmospheric pressure over the Sea of Azov increased further by almost 9 hPa, so this event was regarded as an API. Comparing Fig. 1b and c, we see that the isobars shown there were moving over the water area in question from the northeast towards the Kerch Strait northern entrance. Hence, a conclusion was made that this anticyclonic situation was capable of generating significant changes in the instantaneous current speed in the strait and, in particular, could increase the speed of the Azov Current.

Moreover, from Fig. 1c we see that when this process was developing, the atmospheric pressure over the Sea of Azov was significantly higher compared to the Black Sea area near the southern entrance to the Kerch Strait, which (owing to the law of communicating vessels) also contributed to further intensification of the Azov Current. The variation features of atmospheric pressure distribution over the Azov Region that emerged during the API that happened on February 7, 2017, were as well typical for other API events that occurred when anticyclones were moving southward over the region during 1948–2017 ice seasons.

In accordance with the above methodology, hydrodynamic modeling of the wind velocity field and API-induced currents in the Sea of Azov was performed by use of atmospheric forcing obtained from the *SKIRON/Eta* data [22]. Having analyzed the dynamics of change of atmospheric pressure maximum values over the Azov Sea area (the information was obtained from *SKIRON/Eta* data), we could see that API situations were accompanied by atmospheric pressure rise. For instance, in the first ten-day period of December 2015, atmospheric pressure changed from its minimum (1009.9 hPa) to its maximum (1038.9 hPa), the average increase rate being 9.5 hPa per day.

The estimation results have shown that the field of the wind velocity vector over the sea was characterized by considerable spatial inhomogeneity (both in its absolute value and in direction) during the process under consideration. The movement of the anticyclone over the region led to variations in distribution of wind velocity components, which on 06.12.2015 caused a significant rise of the Azov Sea level at the Kerch Strait northern entrance due to a positive water surge.

The above was established using the *POM* model, which also enabled to estimate instantaneous speed values of the Azov Current at various points of the Kerch Strait northern entrance on 06.12.2015. The computations were performed for both the surface and the near-bottom layers of the water body in question, as well as for time moments differing by 3 min, which enabled to estimate the Azov Current ISI. We have made similar estimates for other episodes of anticyclonic movements that occurred in the region during 1948–2017 ice seasons. As an example, ISI values of the Azov Current for two of such episodes that occurred on 06.12.2015 and 24.02.2017 are considered. The estimation was made for 5 points (P_1–P_5) located equidistantly at the northern entrance to the Kerch Strait at latitude 39.33°N (point P_3 was exactly in the fairway). The results obtained for the specified points and date ranges are presented in Table 1.

According to the computation results, in each episode under consideration instantaneous speed values of the Azov Current at different points of the northern entrance to the Kerch Strait fell on the same day. As follows from Table 1, the highest values of the current speed were attained in the middle of the strait (point P_3). In particular, the absolute maximum of this indicator on the surface was 85 cm/s, whereas the same averaged over the depth was 64 cm/s on 06.12.2015. That date was also noted for the greatest for the whole study period absolute

Table 1. Absolute maximum values of instantaneous speed (cm/s) of the Azov Current averaged over the vertical and over the sea surface layer at the points of the Kerch Strait northern entrance, as identified in December 2015 and February 2017 during API.

Points	1–31 December, 2015		1–28 February, 2017	
	Average	Surface velocity	Average	Surface velocity
P_1	37	45	35	50
P_2	41	48	37	44
P_3	64	85	56	76
P_4	47	54	45	48
P_5	50	53	53	61

Table 2. Differences between the levels of the Sea of Azov at the northern entrance to the Kerch Strait and the average level of the Black Sea at the southern cross-section of the strait at the moments when the Azov Current was at its maximum.

Time	P_1	P_2	P_3	P_4	P_5
06.12.15 22:00	26	7	17	2	21
24.02.17 15:45	24	7	13	1	22

maximum value of the average increase rate of atmospheric pressure among all API events considered (1870 Pa). The same characteristic for January was attained in 1987 (1650 Pa), for February—in 1979 (1750 Pa), and for March—in 2006 (1660 Pa). Differences between the levels of the Sea of Azov at points P_1–P_5 compared to average levels at the southern entrance to the Kerch Strait in December 2015 (41 cm) and February 2017 (36 cm) are presented in Table 2.

As follows from Table 2, on days when the Azov Current was at its maximum, the greatest differences in the levels of the Sea of Azov (26 cm and 24 cm), as compared to the average level of the Black Sea, in both cases were observed at point P_1 located in the western part of the Kerch Strait northern entrance.

Similar research into all cases of API occurrence over the Sea of Azov has confirmed that the absolute maximum of the current speed at the points of the Kerch Strait northern entrance was attained on 6.12.2015. Therefore, the sought-for values of the Azov Current ISI at the Tuzla cross-section were determined using the simulation results corresponding exactly to this date. To obtain them, the following was considered: (i) the Azov Current ISI values at points P_1–P_5, both at the surface and at depths of 3.5 m and 10 m; (ii) differences between the level of the Sea of Azov at points P_1–P_5 and the average Black Sea level at the Kerch strait southern cross-section; (iii) the strait's cross-sectional areas corresponding to both its northern cross-section and the Tuzla one. It has been established that in case of synoptic situations similar to the one that happened on 6.12.2015, the Azov Current speed at the Tuzla cross-section can reach as much as 0.85 m/s. Anticyclones moving in southward directions over Eastern Europe and coming over the Sea of Azov can cause increases of the Azov Current surface speed at the Tuzla cross-section as much as 0.54 m/s. Such a substantial and unexpected current intensification is likely to pose significant risks to ships passing under the bridge, as well as to endanger integrity of the bridge supports during ice seasons. Consequently, estimation of frequency of such events, as well as tendencies of its variation remains of considerable interest.

Similar estimations were performed for winters of different severity in the Azov region. It has been found that, during 1948–2017, severe winters in the Kerch Strait area occurred only before 1987. During the last 30 years, winters in this region have been either mild or moderate [7]. At the same time, the number of moderate winters was 12, while mild winters numbered 18 (for the previous thirty years, these numbers were, respectively, 15 and 11, while 4 winters were considered severe). In the modern period (2005–2017), there were only 4 moderate winters, all the rest being mild. As it can be seen, the frequency of mild winters has increased.

Identification of all API-events for the period from 1948 to 2017 has allowed to estimate their frequency, particularly for cases when the average rate of atmospheric pressure increase during these events exceeded a certain level. Such estimates were made for each ice season for both moderate and mild winters. It has been found that during the considered period, irrespective of winter's severity, for API events with the average atmospheric pressure change rate not exceeding 500 Pa/day, API frequency had a tendency to decrease during ice seasons. On the contrary, for API events with this parameter exceeding 1000 Pa/day, it tended to increase. Moreover, during ice seasons corresponding to mild winters, the frequency of any API-events also tended to increase. Since 1985, this process has been showing significant acceleration. For moderate winters, decrease in the frequency of API-events with the average atmospheric pressure change rate not exceeding 500 Pa/day during the ice season has been established. In case of API-events, in which this parameter did not exceed 1000 Pa/day, API frequency remained unchanged.

It can thus be seen that during the period from 1966 to 2017, the frequency of API-events, capable of causing most perceptible sudden increases in the speed of the Azov Current in the Kerch Strait, did not diminish. During the above period, the probability distribution densities of API-events, corresponding to winters of different severity, changed. This is evidenced by Fig. 2, which demonstrates comparison of the related histograms for mild and moderate winters.

As follows from Fig. 2a, during mild winters, API-events with the average atmospheric pressure increase rates (600 Pa/day) have been and remain the most probable. At the same time, their probability in the modern period has increased. Figure 2b shows that during moderate winters, the most probable API will be those with the lowest values of average atmospheric pressure increase rates. At the same time, probability of API-events with this parameter exceeding 1000 Pa has increased for both mild and moderate winters. Hence, the problem of accounting for these processes during months of ice seasons remains relevant for those involved in steering and control of vessels navigating the Kerch Strait.

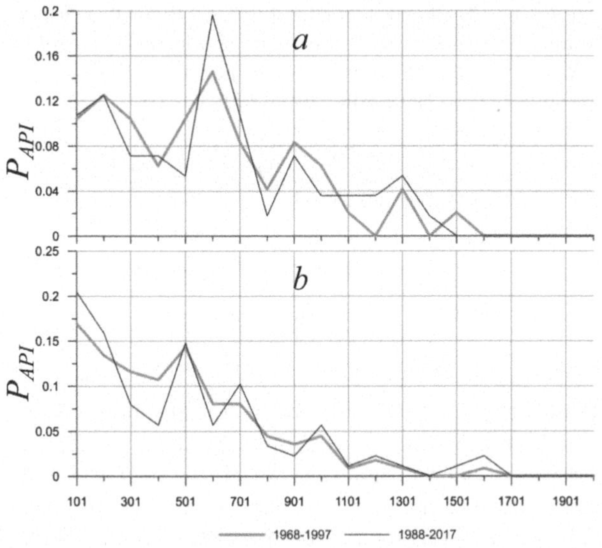

Fig. 2. Histograms of probability distribution densities of API-events (hPa) corresponding to: mild winters (a), moderate winters (b).

The values of the linear trend angular coefficients (they serve as a quantitative characteristic of interannual variation tendency of maximum API values for a particular month), as well as corresponding values of differences in mean atmospheric pressure over the Sea of Azov and the Black Sea area adjacent to the Kerch Strait southern entrance are presented in Table 3.

As follows from Table 3, for the period from 2005 to 2017 the values of the linear trend angular coefficients, which describe interannual changes in maximum API levels, are maximal in December and minimal in March. At the same time, in December, January, and February API tends to increase, whereas in March there is a tendency for its decrease. As far as changes in the difference in mean daily atmospheric pressure over the Sea of Azov and the Black Sea area adjacent to the Kerch Strait southern entrance are concerned, in December and January of 2005–2017 there were tendencies for their increase (most notable in December). The values of the same parameter in February and March, on the contrary, tended to decrease (most rapidly in March). This allows to assume that peak monthly values of the instantaneous current speed at various cross-sections of the northern part of the Kerch Strait in the present period tend to increase in December and January, whereas in February and March they tend to decrease.

From the results obtained, we can conclude that navigation risks that can be posed by sudden increases in instantaneous speed of the Azov Current at various cross-sections of the Kerch Strait in case of API-events tend to increase in the present period. Therefore, improving methods of their prediction with consideration of local microcirculation processes is a relevant and timely issue for meteorologists and ocean-ographers.

Table 3. The values of the linear trend angular coefficients describing interannual changes in maximum API levels (A); differences in mean daily atmospheric pressure over the Sea of Azov and the Black Sea area adjacent to the Kerch Strait southern entrance (B) corresponding to the period from 2005 to 2017.

Month	A (Pa/year)	B (Pa/year)
December	69.61264	31.3956
January	45.69231	18.43407
February	6.414835	- 4.46978
March	−85.989	−27.2802

The risks of damage that can be done to the bridge supports by ice formations drifting along the Kerch Strait rely not only on their speed, but also on individual characteristics of such formations, which, in turn, depend on severity of a particular winter. Despite the fact that for the last few decades severe winters have not been observed in the region, whereas the number of mild winters has significantly exceeded the number of moderate ones, preservation of this course of events is not guaranteed in the future. Although the risk of damage to the bridge supports by ice formations drifting along the Kerch Strait has by now become lower compared to the mid-20th century, the frequency of API occurrences in December and January has a tendency to increase. Consequently, consideration of the related risks seems appropriate. One must also bear in mind that possible economic losses ensuing from the bridge damage may greatly exceed the cost of building ice protection facilities for the bridge supports.

4 Conclusion

The main results of this study can be formulated as follows.

(1) During ice seasons, anticyclones moving in southward directions and coming over the Sea of Azov are capable of causing instantaneous speed increases of the Azov Current at the Tuzla cross-section as much as 1.2 m/s.
(2) The frequency of the anticyclones capable of creating greatest ISI of the Azov Current in the Kerch Strait has increased since 1948. This is true for both mild and moderate winters.
(3) In the present period, during months of ice seasons, the frequency of API attains its maximum in December and drops to a minimum in March.
(4) The issues related to safe navigation in the Kerch Strait, as well as to protection of the existing hydrotechnical installations during ice seasons, remain relevant.

Acknowledgments. The work was carried out within the framework of the State Project No. 0827–2018–0004 "Complex interdisciplinary studies of Oceanological processes that determine the functioning and evolution of ecosystems of the Black and Azov seas" (the code "Coastal Research").

References

1. Tengt, M.H., Wu, T.Y.: Evolution of long water waves in variable channels. Fluid Mech. **266**, 303–317 (1994)
2. Kurkin, A.A.: Nonlinear and Nonstationary Dynamics of Long Waves in the Coastal Zone, p. 330. N. Novgorod State Technical University, N. Novgorod (2005). (in Russian)
3. Chang, H.K., Lioul, C.: Long wave reflection from submerged trapezoidal breakwaters. Ocean Eng. **34**(1), 185–191 (2007)
4. Bautista, E.G., Mendez, F., Bautista, O., Mora, A.: Propagation of shallow water waves in an open parabolic channel using the WKB perturbation technique. Appl. Ocean Res. **33**(3), 186–192 (2011)
5. Dyakov, N.N., Tymoshenko, T.Y., Belogudov A.A., Gorbach, S.B.: Atlas of Ice of the Azov and Black Seas, p. 219. OSI-Gidrofizika, Sevastopol (2016). (in Russian)
6. Hydrometeorological conditions of the seas of Ukraine. The Azov Sea, vol. I, p. 402. Gidrofizika, Sevastopol (2009). (in Russian)
7. Hydrometeorology and Hydrochemistry of the USSR Seas. The Azov Sea, vol. V, p. 236. Gidrometeoizdat, St. Petersburg (1991). (in Russian)
8. Oceanographic Atlas of the Black and Azov seas, p. 356. Ukrmorcartografia, Kiev (2009). (in Ukrainian)
9. Blumberg, A.F., Mellor, G.L.: A description of three dimensional coastal ocean circulation model. In: Heaps, N. (ed.) Three-Dimensional Coastal Ocean Models, vol. 4, pp. 1–16. American Geophysical Union, Washington, D.C. (1987)
10. Cherkesov, L.V., Shul'ga, T.Ya.: Investigation of the effect of the baric formations' parameters on free and forced oscillations of the level and flow in the Sea of Azov. Phys. Oceanogr. **4**, 14–26 (2016). https://doi.org/10.22449/1573-160X-2016-4-12-24. (in English)
11. Cherkesov, L.V., Ivanov, V.A., Khartiev, S.M.: Introduction into Hydrodynamics and Wave Theory, p. 264. Gidrometeoizdat, St. Petersburg (1992). (in Russian)
12. Cherkesov, L.V., Shul'ga, T.Ya.: Numerical analysis of the influence of parameters of atmospheric fronts to the free and forced oscillations level and inflows into the Azov Sea. In: Proceedings of the SOIN, vol. 218, pp. 120–137 (2017). (in Russian)
13. Cherkesov, L.V., Shul'ga, T.Ya.: Numerical analysis of the effect of speed and direction of the duration of action of the wind on the circulation of the Azov Sea water with and without water exchange through the Kerch Strait. Oceanology **58**(1), 19–27 (2018)
14. Dubinin, M., Pospelov, I.: Sources of meteorological data on the territory of the Russian Federation by stations. http://gis-lab.info/qa/meteo-station-sources.html. Accessed 15 Aug 2018
15. Reanalysis of average daily values of atmospheric pressure. https://www.esrl.noaa.gov/psd/data/gridded/data.ncep.reanalysis.pressure.html. Accessed 15 Aug 2018
16. Skvortsov, A.V.: Delaunay Triangulation and Its Application, p. 28. Tomsk State University, Tomsk (2002). (in Russian)
17. Kobzar, A.I.: Applied Mathematical Statistics. A Handbook for Engineers and Researchers, p. 816. Fizmatlit, Moscow (2006). (in Russian)
18. Mellor, G.L., Yamada, T.: Development of a turbulence closure model for geophysical fluid problems. Rev. Geophys. Space Phys. **20**, 851–875 (1982)
19. Smagorinsky, J.: General circulation experiments with primitive equations, Part I, The basic experiment. Mon. Weath. Rev. **91**, 99–164 (1963)

20. Wannawong, W., Wongwises, U., Vongvisessomjai, S.: Mathematical modeling of storm surge in three dimensional primitive equations. Int. J. Math. Comput. Phys. Electr. Comput. Eng. **5**(6), 797–806 (2011)
21. Courant, R., Friedrichs, K.O., Lewy, H.: On the partial difference equations of mathematical physics. IBM J. Res. Dev. **11**(2), 215–234 (1967)
22. Kallos, G., Nickovic, S., et al.: The regional weather forecasting system SKIRON and its capability for forecasting dust uptake and transport. In: Proceedings of the WMO Conference on Dust Storms, Damascus, p. 9 (1997)
23. Savenko, V.Ya.: Mathematical models and methods for calculating quasi-three-dimensional free surface flows, p. 188. Technics, Kiev (1995). (in Russian)
24. Shchodro, O.E.: Construction of the planned pattern of the flow and spatial deformations of the channel of arbitrary shape. Hydro-energetics Ukraine **3**, 36–39 (2010). (in Russian)
25. Mishuev, A.V., Sladkevich, M.S.: Investigation of flow parameters in zones of sharp channel width changes during passage of the displacement wave. In: Methods for Studying and Calculating of Waves, Their Effects on Hydraulic Structures and Banks/Materials of Conferences and Meetings on Hydraulic Engineering, Leningrad, pp. 178–183 (1982). (in Russian)
26. Matishov, G.G., Chikin, A.L.: Study of wind currents in the Kerch Strait applying mathematical modeling. Bull. SRC RAS **8**(2), 27–32 (2012). (in Russian)
27. Fishman, G.S.: Monte Carlo Concepts, Algorithms, and Applications, p. 698. Springer, Heidelberg (1996)
28. Sirotkina, A.I.: To the question of assessing the severity of winters in the southern seas. In: Proceedings of NIIAC, vol. 8, pp. 62–69 (1959). (in Russian)

Numerical Simulations of the Black Sea Hydrophysical Fields Below the Main Pycnocline: Validation by ARGO Data

N. V. Markova(✉), O. A. Dymova, and S. G. Demyshev

Marine Hydrophysical Institute RAS, Sevastopol, Russia
n.v.markova@mail.ru

Abstract. Modeling of hydrophysical fields of the Black Sea for the one year period (2011) is carried out using the MHI z-coordinate nonlinear model with a spatial resolution of 1.6 km. Comparison of the simulation results with ARGO floats data shows a quite satisfactory agreement between the model and the measured parameters below the main pycnocline up to the maximum profiling depth of the floats equal to 1500 m in 2011. The greatest differences for temperature and salinity are in the seasonal thermocline, which is reproduced by the model a few meters deeper in comparison with the data of measurements. In most cases the shift of the thermocline reconstructed is the main reason of the discrepancy between the model and in-situ data in the upper layer. It is shown, that in the layer below 300 m, the circulation features are simulated in a quite sufficient agreement with ARGO data and allow us to describe the field of the deep Black Sea currents.

Keywords: Black Sea · Modeling · Circulation · Deep currents · Temperature · Salinity · ARGO

1 Introduction

The number of recent publications focused on the Black Sea deep circulation [1–4] is rather limited. The most completely reconstruction of the three-dimensional hydrophysical field structure below the Black Sea main pycnocline is carried out by numerical modeling [2–4]. Modern numerical models successfully reproduce the dynamics of the upper sea layer, especially in experiments involving data assimilation of some measured hydrophysical parameters. The prevalent quantity of in-situ measurements is concentrated in the upper layer (depth from 0 to 200–300 m), and satellite remote sensing involves the operative obtaining of information only about the state of the ocean surface. So, as a rule, numerical models are well validated according to field measurements and remote sensing data mainly in the upper sea layer.

At the same time, the deep-water measurements are quite rare. To fill this data gap, in the Black Sea since the beginning of the 2000s such measurements are carried out by ARGO floats [5]. An ARGO float drifts at its parking depth and performs the profiling of the sea water parameters when it ascends to the surface for data transfer via satellite

V. I. Karev et al. (Eds.): *Physical and Mathematical Modeling of Earth and Environment Processes (2018)*, SPEES, pp. 15–21, 2019.
https://doi.org/10.1007/978-3-030-11533-3_2

or descends back. For the Black Sea, in open access the ARGO data are available since 2005 [5].

In this paper, we compare results of the prognostic numerical experiment on modeling the Black Sea parameters in 2011 and data on the temperature and salinity obtained by ARGO floats during the same period.

2 Model and ARGO Data

The calculation is performed by using of nonlinear z-coordinate ocean hydrodynamics model of the Marine Hydrophysical Institute (MHI) [6]. The bathymetry from MHI Oceanography Database [7] is applied. The SKIRON [8] reanalysis products such as the wind stress, heat fluxes, precipitation and evaporation for 2011 are used as boundary conditions on the sea surface. The sea surface temperature is assimilated daily. The rivers runoff and water exchange through the straits are also taken into account. Sea level, temperature, salinity, and horizontal current velocity are set at the initial moment from MHI reanalysis [9].

The model is implemented on a C-grid, the horizontal resolution is 1.6 km, and 27 z-level is defined vertically. Modification of the MHI model with a spatial resolution of 5 km was early validated in the frame of the MyOcean project [10].

All boundary and initial fields are interpolated linearly into the grid nodes. Horizontal viscosity and diffusion are described by biharmonic operators with constant coefficients of 10^{16} $cm^4 \cdot s^{-1}$ and $5 \cdot 10^{16}$ $cm^4 \cdot s^{-1}$, respectively. The coefficients of vertical viscosity and diffusion are computed using the Mellor-Yamada turbulence close model [11]. The equations of the model [6] are integrated for the period from January 1 till December 31, 2011, and three-dimensional hydrophysical fields for each day of 2011 are obtained.

For comparison with the simulation results, the data from 6 floats drifted at different parking depths and programmed to vertical profiling from horizons of 500–1500 m up to the surface are taken. The ID numbers of the floats, operating periods under consideration and number of profiles are presented in Table 1.

Table 1. ARGO floats in the Black Sea in 2011: operating period under consideration and number of valid profiles.

Float ID	Time range	Profiling depth, db	Number of profiles
7900465	02.01.11–28.12.11	500	70
7900466	02.01.11–28.12.11	500	65
1901200	05.01.11–31.12.11	700/1500	72
6900803	19.03.11–29.12.11	700/1500	55
6900804	19.03.11–29.12.11	700/1500	58
6900805	19.03.11–29.12.11	700/1500	57

So, we considered 377 profiles of both temperature (T) and salinity (S) and processed several decade thousands measurements. Observations covered the abyssal part of the sea and are performed in all seasons.

3 Comparison Results

The deviations of the model temperature and salinity from those measured during profiling are calculated. The spatial resolution of the model is high enough to reduce the interpolation error when the model data are determined at the profiling point. All data are divided according to the character sea layers: subsurface (0–5 m), upper mix (5–30 m), cold intermediate (30–100 m), and main pycnocline (100–300 m). The subpycnocline layer is considered as divided in two parts: of 300–800 m and 800–1500 m depth. The horizon of 800 m is chosen due to the increase in the currents velocities detected at the depth of 800–1000 m [4], and the horizon of 1500 m is the maximum of the ARGO profiling depth in 2011 (see Table 1).

Then the mean values of deviations for each float are averaged and reduced to the model horizons. The results of comparative analysis are shown in Table 2. It is obtained that the maximum of temperature deviations is reached in the layer of 5–30 m and equal to 0.39 °C, and salinity one – in the 30–100 m layer and equal to 0.33‰. Below the horizon of 300 m, the deviation of model values from measured ones does not exceed 0.03 °C and 0.09‰, for temperature and salinity, respectively. So, for the horizons under the main pycnocline, the model results and measured data are agreed enough.

Table 2. The mean and RMS deviations of model temperature and salinity from those measured with ARGO floats.

Depth, m	Temperature, °C		Salinity, ‰	
	Mean	RMS	Mean	RMS
0–5	0.167	0.182	–0.154	0.177
5–30	0.206	0.386	–0.164	0.172
30–100	0.200	0.242	0.007	0.326
100–300	0.107	0.112	0.239	0.231
300–800	–0.009	0.013	–0.072	0.091
800–1500	–0.026	0.027	0.001	0.007

The vertical profiles of T- and S-deviations for all floats trajectories are considered. It is shown that the maximal deviations are observed in summer period. The differences in temperature can reach several degrees in the upper mix layer. Figure 1 illustrates an example of mean and average T- and S-deviation profiles for the float 6900803. The data are used for the period March 19 – December 29, 2011 – from the float deployment till the last station before the end of 2011.

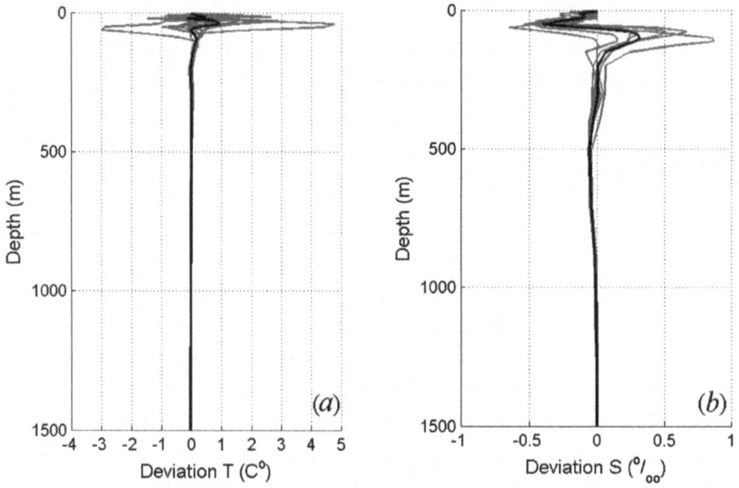

Fig. 1. Vertical profiles of the mean-track (black line) and deviations (grey lines) along the float 6900803 trajectory: a – temperature; b – salinity.

There are at least two possible reasons of these deviations in the upper layer. Firstly, the intensive mesoscale eddies causing an upwelling or downwelling are formed in summer [12], and it could lead to significant density variations. Such vortex structures could not be reconstructed by the model with the sufficient accuracy. Secondly, in the model results, the shift of the seasonal thermocline down several meters can be associated with an inaccurate description of the vertical redistribution of the heat flux, which leads to a deepening of the lower boundary of the upper mixed layer. These effects should certainly be eliminated, but also at the present time, the model shows quite satisfactory agreement in the subpycnocline layers under consideration. As seen in Fig. 1, the deviations are quite small on the horizons of more than 300 m. As an example, the deviations of T and S at lower horizons are shown in Fig. 2 for the same float 6900803 along its track in 2011. It is obvious that, for both temperature and salinity, the deviations are quite small and in some cases are comparable to the accuracy of the sensor measurements sated on Argo floats (0.002 °C and 0.005‰ correspondently).

In comparison to amount of temperature and salinity measurement data, the database of currents in the Black Sea is the most deficient [1, 13]. Moreover, the data of deep current soundings are in the greatest shortage. Measurements of currents by the ARGO floats are not carried out. Based on the displacement of the ARGO floats, the lagrangian velocities on their parking depth are just estimated from 2 to 6 $\text{cm} \cdot \text{s}^{-1}$ [14, 15]. So, numerical model output data are often used as the basic information on the state of the deep sea. The velocity field under the main pycnocline, obtained in accordance with the temperature and salinity verified, believed to be close enough to the actual picture of the currents.

The analysis of the subpycnocline currents derived from modeling results is carried out. The currents velocity maps for all model horizons are built. It is obtained that the

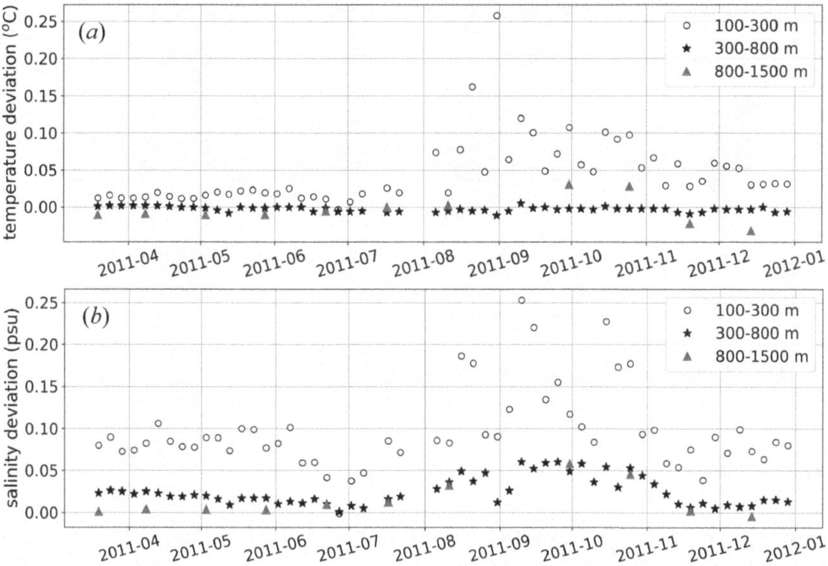

Fig. 2. Along-track temperature (a) and salinity (b) deviations between simulated and measured by the float 6900803.

basin-scale circulation is generally cyclonic. The qualitative structure of the circulation, corresponding to the surface one, is saved up to the depth of 500 m. The seasonal intensification of currents is observed in winter and autumn. The average velocities in the lower layers are in an order weaker than the surface ones. The mean values are about 2–5 cm • s^{-1}, and the instantaneous ones can reach 15 cm • s^{-1}. The structure of the velocity field in the layer of 500–900 m is unordered and consists of numerous eddies and currents. Several steady eddies with dimensions over 100 km and orbital velocity of about 10 cm • s^{-1} are located in the central and eastern parts of the basin throughout the year.

Some features of the currents field not typical for the general cyclonic circulation scheme are found at a depth of about 1000 m and more. Anticyclonic flows within the velocity of 3–10 cm • s^{-1} are detected in different parts of the Black Sea along the continental slope. These currents are about 10 km wide; their lifetime ranges from one up to several weeks, and they are found in all seasons. For areas characterized by the anticyclonic deep currents, normal to the shore sections of along-shore velocity are drawn. As an example, cross-sections normal to the shore near the Bulgarian (28°0′ E; 42°0′ N) and the Caucasian (39°12′ E; 43°54′ N) coasts in August 2011 are shown in Fig. 3. It is obvious that in these areas currents change their direction at horizons of 700–1100 m. Positive velocities correspond to the cyclonic currents, and negative values refer to the anticyclonic ones. Note that the velocity modules near continental slope at the depths of 300–500 m and 900–1100 m are close.

The modeling results are agreed with the data of numerical experiments on the reconstruction of climatic circulation of the Black Sea [16], field measurements [1] and prognostic numerical experiments [4].

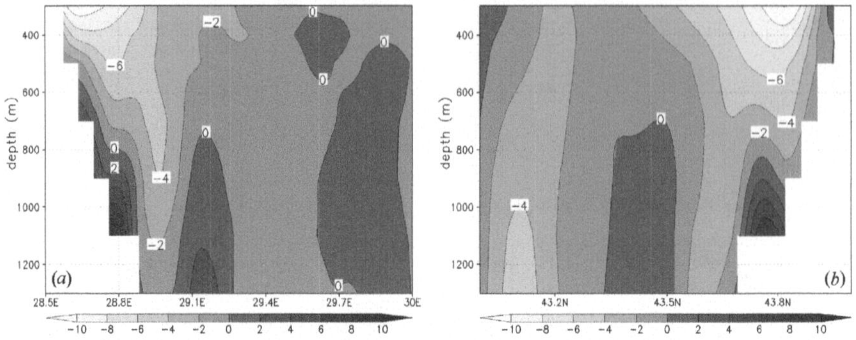

Fig. 3. The along-shore component of the velocity (cm · s^{-1}) on the normal to the shore sections: a – at point (28°0′ E; 42°0′ N) and b – at point (39°12′ E; 43°54′ N).

4 Discussion

We compare the model temperature and salinity of the Black Sea with ARGO measurements in 2011. It is shown that the greatest deviations are obtained in the upper layer of the sea. This is primarily due to the fact that the model reproduces the seasonal thermocline some deeper than the one obtained from the measurement data.

As well, it was found that in the deep layers, the differences between the model and the measured data are in order less. At the horizons below the main pycnocline, the deviations of the calculated temperature and salinity from the measured ones are quite small (of the order of 10^{-2} °C and 10^{-2}–10^{-3} ‰, respectively). So, we admit the temperature and salinity to be correct. The minimum errors are reached at depths of 500 m and more. This fact allow us to consider the model field of deep currents and to reveal its features such as eddies and narrow along-slope anticyclonic currents at horizons of 700–1100 m. In this regard, the present version of the model is seems to be suitable for investigation of the deep Black Sea. However, for the full realistic picture of the circulation, the parameterization refinement of the heat vertical redistribution in the upper sea layer is obviously required. We also hope for increase the number of measurements of the velocity of deep currents in order to be able to verify the velocity field as well.

Acknowledgments. The numerical experiment for 2011 was carried out with the support of RFBR (grant № 18-05-00353 A). The data comparison was carried out within the framework of the State assignment (theme № 0827-2018-0002).

References

1. Ostrovskii, A.G., Zatsepin, A.G., Soloviev, V.A., et al.: Autonomous system for vertical profiling of the marine environment at a moored station. Oceanology **53**(2), 233–242 (2013)
2. Arkhipkin, V.S., Kosarev, A.N., Gippius, F.N., Migali, D.I.: Seasonal variations of climatic fields of temperature, salinity and water circulation in the Black and Caspian seas. Moscow University Bulletin. Series 5: Geography **5**, 33–44 (2013)

3. Lukyanova, A.N., et al: The Black Sea deep-water circulation research by results of numerical modeling and in-situ data: INMRAS model numerical experiment. In: Ecological Safety of Coastal and Shelf Zones of the Sea, vol. 3, pp. 9–14 (2016). (in Russian)
4. Demyshev, S.G., Dymova, O.A., Markova, N.V., et al.: Numerical experiments on modeling of the Black Sea deep currents. Phys. Oceanogr. **2**, 38–50 (2016)
5. USGODAE ARGO Page. http://usgodae.org/argo/argo.html. Accessed 17 July 2018
6. Demyshev, S.G., Korotaev, G.K.: Numerical energy-balanced model of the baroclinic ocean currents on a C-grid. In: Numerical Models and Results of Calibration Calculations of Currents in the Atlantic Ocean, pp. 163–231. INM RAS, Moscow (1992). (in Russian)
7. Belokopytov, V.N., Khaliulin, A.K., Godin, E.A., et al.: Information products to study environmental threats and dangerous phenomena in the Black, Azov and Caspian seas. In: NATO Science for Peace and Security. Series C: Environmental Security, pp. 91–104 (2008)
8. NonHydrostatic SKIRON/Eta Modelling System Page. http://forecast.uoa.gr/forecastnewinfo.php. Accessed 18 July 2018
9. Korotaev, G.K., Knysh, V.V., Lishaev, P.N., et al.: Reanalysis of seasonal and interannual variability of Black Sea fields for 1993–2012. Izvestiya Atmos. Oceanic Phys. **52**(4), 418–430 (2016)
10. Demyshev, S., Knysh, V., Korotaev, G., Kubryakov, A., Mizyuk, A.: The MyOcean Black Sea from a scientific point of view. Mercator Ocean Q. Newsl. **39**, 16–24 (2010). http://marine.copernicus.eu/wp-content/uploads/2016/06/r63_9_quarterly_letter-_issue_39.pdf. Accessed 31 July 2018
11. Mellor, G.L., Yamada, T.: Development of a turbulence close model for geophysical fluid problems. Rev. Geophys. Space Phys. **20**, 851–875 (1982)
12. Kubryakov, A.A., Stanichny, S.V.: Seasonal and interannual variability of the Black Sea eddies and its dependence on characteristics of the large-scale circulation. Deep-Sea Res. Part I Oceanogr. Res. Pap. **97**, 80–91 (2015)
13. Markova, N.V, Plastun, T.V.: Investigation of deep-water circulation of the Black Sea with the use of contact measurements. In: Proceedings of the II All-Russian Scientific Conference of Young Scientists on Complex studies of the World Ocean, pp. 169–170. IO RAS, Moscow (2017)
14. Korotaev, G., Oguz, T., Riser, S.: Intermediate and deep currents of the Black Sea obtained from autonomous profiling floats. Deep-Sea Res. II **53**(17–19), 1901–1910 (2006)
15. Markova, N.V., Bagaev, A.V.: Velocities of the Black Sea deep currents estimated from the profiling drifters Argo data. Phys. Oceanogr. **3**, 23–35 (2016)
16. Demyshev, S.G., Ivanov, V.A., Markova, N.V.: Analysis of the Black Sea climatic fields below the main pycnocline obtained on the basis of assimilation of the archival data on temperature and salinity in the numerical hydrodynamic model. Phys. Oceanogr. **19**(1), 1–12 (2009)

Internal Waves Kinematic Characteristics in the Central Atlantic in 2014–2015

K. S. Grigorenko[1(✉)] and S. M. Khartiev[2]

[1] Southern Scientific Center RAS, 41 Chekhova Street, Rostov on Don 344006, Russia
klim_grig@mail.ru
[2] Southern Federal University, 105/42 Bolshaya Sadovaya Street, Rostov-on-Don 344006, Russia

Abstract. Internal waves generation hydrology conditions based on CTD-profiling data, received in the 39-th and the 40-th cruise of research vessel "Akademik Sergey Vavilov", are investigated. The Atlantic ocean water masses thermochaline structure was considered in the area of investigations, geographical demarcation based on hydrology conditions differences was conducted. It was found out that salinity is smaller, and temperature is bigger in the surface layer in the area of Doldrums and Bogdanov fracture zones, picnocline is deeper, its thickness is bigger, therefore another hydrodynamics mode of internal waves in the area of this fracture zones was generated in comparison with stations, situated in the area of Vema fracture zone.

Keywords: Internal waves · The Central Atlantic · Abyssal flows · Kinematic characteristics of internal waves

1 Introduction

It is known that the thermohaline conveyor of the World Ocean forms the climate of the Earth. The Mid-Atlantic Ridge consists of a huge number of separate ridges and troughs, which form transform faults, over which Antarctic waters are spread [15]. Complex and variable hydrological conditions form a special dynamic regime of the water area, multidirectional ocean currents are stratified according to density, density transition zones are fixed at the interfaces between them and internal waves are formed [5, 6, 9, 10], which, in their turn, transfer a huge amount of energy of the ocean. An all-encompassing work, including the results of expeditionary investigations of internal waves in almost all the waters of the World Ocean, and in particular on the equatorial latitudes of the Atlantic, was published in 2018 [1]. Temperature and flow measurements at buoy stations used in the monograph obtained during the international expeditions POLIMODE, PIGAP, SICAR and the Southern Ocean.

This paper is devoted to the study of the kinematic characteristics and the modal composition of internal waves in the Vema fracture zone region. The CTD-profiling data obtained in the 39th and 40th voyages of the research vessel "Academik Sergei Vavilov" (Fig. 1) are used in the work. Profiling was carried out using a deep water sampling complex SBE-32 with a CTD-probe Sea-Bird 19plus mounted on it. The data obtained was processed using the software of the probe manufacturing company SBE Data Processing.

V. I. Karev et al. (Eds.): *Physical and Mathematical Modeling of Earth and Environment Processes (2018)*, SPEES, pp. 22–32, 2019.
https://doi.org/10.1007/978-3-030-11533-3_3

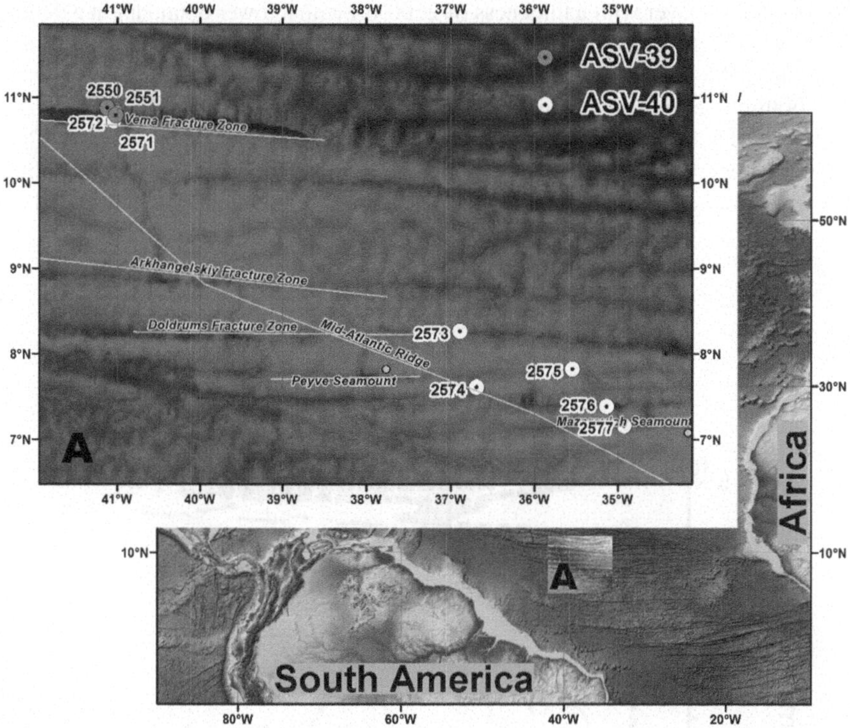

Fig. 1. Expedition stations location scheme from the 39-th and 40-th scientific cruise of "Akademik Sergei Vavilov" research vessel in 2014–2015.

2 Hydrology of the Atlantic Ocean in the Area of Vema and Doldrums Fracture Zones

The section of the Mid-Atlantic Ridge in the Vema and Doldrums fault regions is very important in the context of investigations of the distribution of bottom Antarctic waters to the north, according to studies [2, 3], the transfer of Antarctic waters is 0.1–0.7 Sv to the east.

The upper layer at station 2569 to a depth of 150 m is a surface water, with temperature values, at this depth, 14 °C, salinity 35.5 PSU, density 1027.5 kg/m^3 (Fig. 2). The main transition zone separates the surface waters from intermediate Antarctic waters, the temperature decreases, values near 6 °C are observed, the salinity in this water mass is reduced to a minimum of 34.7 PSU, at the horizon of 800–900 m, after which increases to 35 PSU, the density, at the lower boundary of the layer increases to 1031 kg/m^3. On the horizon of 1400–1500 m there is a boundary between the layers of the Antarctic intermediate water and the North Atlantic deep water mass, which occupies a layer with a thickness of 2500 m, with a lower boundary at a depth of 4000 m, where temperatures are approximately 2.5 °C, salinity of 34.9 PSU, density of 1046 kg/m^3. The bottom horizons are occupied by Antarctic bottom water, with values at the bottom: temperature approximately 1.8 °C, salinity - 34.8 PSU, density of 1048 kg/m^3. To analyze the thermohaline structure differences of the Atlantic Ocean the upper layers, with a depth of up to 1500 m, with significant temperature, salinity, and density gradients in the investigated polygon, were subsequently considered.

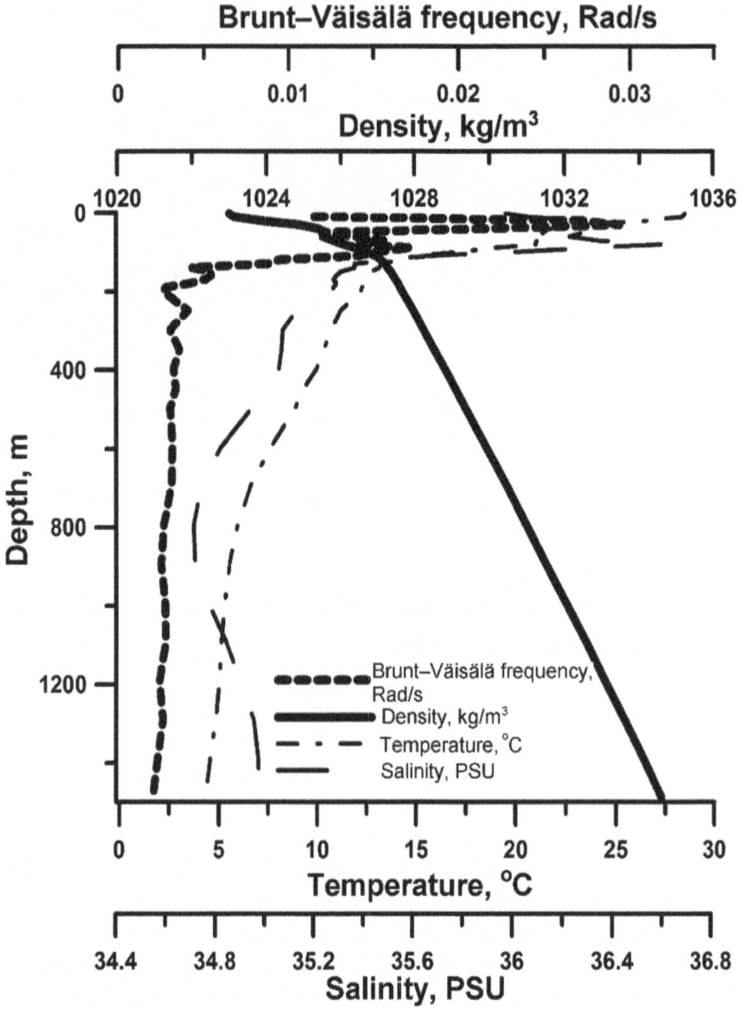

Fig. 2. Vertical distribution of temperature, salinity and density at the station 2569.

Stations 2573-2577 are located much southward of stations 2569–2572, in other hydrological conditions. The surface layer at these stations is more desiccated (33.6

PSU at st.2575 and 35.8 at station 2572) and is heated by about 1 °C, respectively, the surface water density is less (1021 kg/m^3 at station 2575 and 1022.9 at station 2572), due to which the density difference in the main transition zone increases. Pycnocline at the "southern" stations is located deeper, its lower boundary at stations 2569–2572 is observed on the horizons of 170–180 m, whereas at stations 2573–2577 - approximately 200 m. The core of the Antarctic intermediate waters is located at a depth of 800 m, the temperature on this horizon at stations 2569–2572 is 6.6–6.8 °C, at stations 2573–2577 it is 6.1–6.4 °C. Salinity is 34.7 PSU at stations 2569-2572 and 34.6 at stations 2573–2577. The density at all stations was 1030.9 kg/m^3. With depth the features of the thermohaline structure of stations 2569–2577, associated with their different geographical location can be traced less noticeably. The temperature at the horizon of 1500 m varies in the interval 4.3–4.5 °C, the salinity is 34.96–34.98 PSU, the density is 1034.6 kg/m^3.

3 The Mathematical Statement of the Problem

$$\frac{\partial \mathbf{V}}{\partial t} = -\frac{1}{\rho_0}\nabla p + \mathbf{F}, div\mathbf{V} = 0, \tag{1}$$

$$\frac{\partial p}{\partial t} + \frac{d\rho_0}{dz} v_z = 0 \tag{2}$$

$$v_z = 0 \text{ when } z = -1; v_z = 0 \text{ when } z = 0, \tag{3}$$

where the dimensionless values are related to the dimensional values as:

$$(\bar{x}, \bar{y}, \bar{z}, k) = (x, y, z, \bar{k})\ H, \quad (\bar{t}, \sigma, f) = \sqrt{H/g}(t, \bar{\sigma}, \bar{f}), \tag{4}$$

Here $\mathbf{V} = \{v_x, v_y, v_z\}$ - velocity vector of wave perturbations. $\mathbf{F} = \{fv_y, -fv_x, -\rho/\rho_0\}$ - body forces vector, p, ρ - wave perturbations of pressure and density, respectively; $\bar{f} = 2\Omega \sin\varphi$ - Coriolis parameter; Ω - angular velocity of rotation of the Earth; ϕ - the latitude; g - acceleration of gravity; $\bar{\sigma}$ - wave frequency; $\bar{k}$ - horizontal wave number.

The solution of the homogeneous boundary value problem (1)–(3) can be represented in the form:

$$[v_x, v_y, v_z, p, \rho] = [U(z), V(z), W(z), P(z), R(z)] \exp(i(kx - \sigma t)) \tag{5}$$

Substituting the expression (5) into Eqs. (1), (2) and the boundary conditions (3), we obtain for a given frequency σ the eigenvalue problem of the following form (the prime denotes the derivatives with respect to z).

$$W'' - N^2(z)W' - k^2 \frac{\sigma^2 - N^2(z)}{\sigma^2 - f^2} W = 0 \tag{6}$$

$$W = 0 \text{ when } z = -1 \tag{7}$$

Where $N^2(z) = -\frac{\rho_0'}{\rho_0}$ - square of Vaisala-Brent's dimensionless frequency. In [4–6] there is a review of papers devoted to the solution of the spectral boundary-value problem (6)–(7) with a detailed description of the schemes of used analytic-numerical methods. In this paper, the homogeneous boundary value problem (6)–(7) is solved numerically, using the finite element method, in the mathematical package Flex PDE [7, 8].

Let us formulate the mathematical formulation of the problem in the presence of horizontal density diffusion. In the hydrodynamic Eqs. (1), we assume that the $f = 0$, boundary conditions (3) remain unchanged, and the incompressibility Eq. (2) is represented in the form [1, 11, 13]:

$$\frac{\partial \rho_1}{\partial t} + v_z \frac{\partial \rho_0}{\partial z} = \delta \left(\frac{\partial^2 \rho_1}{\partial x^2} + \frac{\partial^2 \rho_1}{\partial y^2} \right) \tag{8}$$

Where $\delta = D/(H\sqrt{gH})$, D - coefficient of horizontal density diffusion [9]. Substituting (5) into (1), (3), (8), after separating the variables, we obtain an ordinary differential equation of second order with boundary conditions.

$$W'' - N^2(z)W' - k^2 \frac{\omega^2 - N^2(z)}{\omega^2} W = 0 \tag{9}$$

$$W = 0 \text{ when } z = -1; W = 0 \text{ when } z = 0. \tag{10}$$

Here:

$$\omega^2 = i\sigma(k^2\delta - i\sigma) \tag{11}$$

It is easy to verify that problem (6), (7) coincides with the problem (9), (10) for $f = 0$ and $\sigma = \omega$. In a similar formulation, the solution of the problem was considered in [6].

Assuming the wave number k to be given, we determine, using the finite element method, the eigenvalues of the parameter ω for which the homogeneous boundary value problem (9)–(10) has a nontrivial solution. We obtain the dispersion relations $\omega = \omega_n(k)$, where n is the number of the mode (n = 1, 2, 3…). Substituting $\omega = \omega_n(k)$ in (11), we have:

$$\omega_n^2(k) = i\sigma(k^2\delta - i\sigma) \tag{12}$$

We rewrite the last inequality in the form:

$$\sigma^2 + ik\delta\sigma - \omega_n^2 = 0, n = 1, 2\ldots \quad (13)$$

Solve (13) with respect to σ, we find:

$$\sigma_{1,2}^n = i\alpha k^2 \pm \sqrt{\omega_n^2(k) - \alpha^2 k^4} \quad (14)$$

Here it is denoted: n = 1, 2, 3…, $\alpha = \frac{\delta}{2}$.
Let us consider the case when

$$\omega_n^2(k) > \alpha^2 k^4 \quad (15)$$

Substituting the expression (14) (5), we obtain:

$$v_{zn}^{\pm} = W_n(z)e^{-\alpha k^2 t} e^{i(kx \pm \sqrt{\omega_n^2(k) - \alpha^2 k^4})t} \quad (16)$$

Selecting the real part in (16), we find:

$$\mathrm{Re} v_{zn}^{\pm} = W_n(z)e^{-\alpha k^2 t} \cos(kx \pm \sqrt{\omega_n^2(k) - \alpha^2 k^4 t}) \quad (17)$$

As far as $\omega_n(k) \to 0$, when $n \to \infty$, the inequality (15) is violated, starting with some number n_0 ($n > n_0$). Thus, when condition (15) is satisfied, there are a finite number n_0 of progressive internal waves (17) that are damped with the same dimensionless decrement:

$$d = \frac{\delta}{2} k^2 = \frac{Dk^2}{2H\sqrt{gH}} \quad (18)$$

Consider the situation that arises when $n > n_0$, and the following inequality holds:

$$\omega_n^2(k) < \alpha^2 k^4 \quad (19)$$

The expression (14) in this case takes the form:

$$\sigma_{1,2}^n = i(\alpha k^2 \pm \sqrt{\alpha^2 k^4 - \omega_n^2(k)}) \quad (20)$$

Substituting (20) into (5), we have:

$$v_{zn}^{\pm} = W_n(z)e^{-(\alpha k \pm \sqrt{\alpha^2 k^4 - \omega_n^2(k)})t} e^{ikx} \quad (21)$$

Selecting the real part from (21), we find:

$$\mathrm{Re}v_{zn}^{\pm} = W_n(z)e^{-(\alpha k^2 \pm \sqrt{\alpha^2 k^4 - \omega_n^2(k)})t}\cos kx \tag{22}$$

Formula (22) describes the standing internal waves for each n-th mode, which corresponds to two decrements:

$$d_1 = \alpha k^2 + \sqrt{\alpha^2 k^4 - \omega_n^2(k)} \tag{23}$$

$$d_2 = \alpha k^2 - \sqrt{\alpha^2 k^4 - \omega_n^2(k)} \tag{24}$$

The first corresponds to a fast, aperiodic mode of attenuation, the second to a slow one. In both cases $n > n_0$, $n \to \infty$. There are infinitely many aperiodic modes.

4 Internal Waves Kinematic Characteristics

Figure 3 shows the dispersion curves of internal waves calculated from the density stratification data at station 2569. The areas of convergence of the dispersion curves of various modes are seen at different dimensionless frequencies: the 1 st and 2 nd modes when $\sigma = 0,14$; 3-rd, 4-th and 5-th modes when $\sigma = 0,135$; and the 2 nd and 3 rd modes at $\sigma = 0,15$. In the surroundings of these frequencies, there are regions of fast change of group velocity of internal waves and concentration of the dispersion curves of various modes. There are so-called "resonance zones" are the anomalous regions of

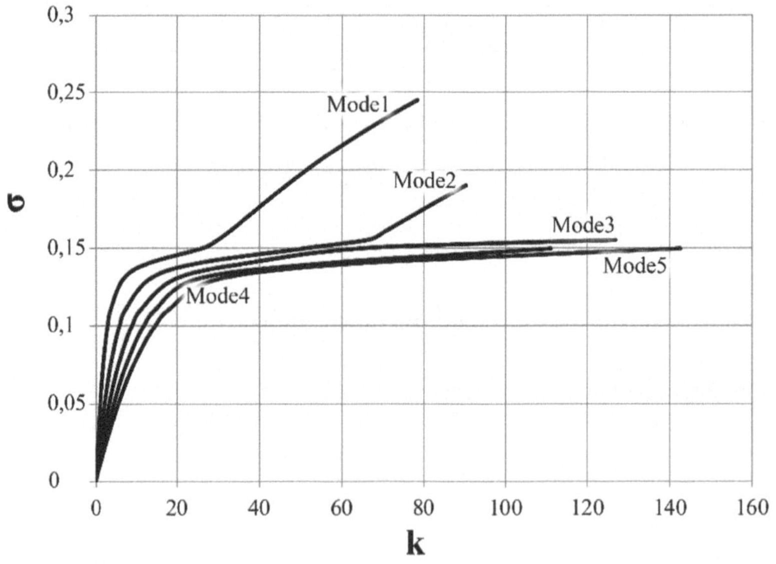

Fig. 3. Dispersion curves of internal waves in a dimensionless form at station 2569.

the spectrum arising in the surroundings of the critical frequencies for which some eigenvalue of a single solitary waveguide coincides with some other value [9–11]. The difference in the stratification of the density on stations 2569-2572 and stations 2573-2577 leads to significant differences in the nature of the dispersion curves.

At “southern” stations 2573–2577 the thickness of the transition zone is more (100–120 m, vs. 50–80 m at stations 2569–2572), therefore, despite the fact that these stations are not characterized by maximum density gradients and maximum Väisälä-Brent frequency values, internal waves are more high-frequency.

The first and the second modes increase rapidly, so with the wave number k = 0.015 1/m, as shown in Table 1, the first modes of stations 2973–2977 have frequencies greater than 0.014 rad/s, and the second modes exceed 0.0076 rad/s. The third mode at stations 2969–2971 increases parallel to the fourth and fifth, in the high-frequency part of the spectrum, continuously in the region of fast change in the group velocity of internal waves $(d\sigma/dk)$ [10–12]. At stations 2972–2977, the dispersion curve of the third mode at a defined frequency leaves this region (0.0072 Rad/s, 0.0072 Rad/s, 0.0073 Rad/s, 0.0074 Rad/s, 0.0072 Rad/s for stations 2973, 2974, 2975, 2976 and 2977, respectively), and increases parallel to the higher-frequency second and first modes.

The wave motions of the internal waves of the diurnal and semidiurnal periods involve the entire thickness of the ocean from the surface to the bottom, with increasing frequency, the waveguide becomes smaller, and the velocity maxima tend to the depth of the density transition zone with the maximum gradient, respectively, the power and density difference of the main pycnocline, The local density transition zones deeper than the main pycnocline affect to the formation of modes of internal waves at a wave number k = 0.015 1/m. The influence of density stratification on the vertical component of the speed of internal waves of the diurnal and semi-diurnal period was not revealed. When considering the influence of the horizontal component of the internal waves of the diurnal and semi-diurnal period, a difference is visible and the effect of various hydrological conditions is noticeable. At stations where the main pycnocline has a large thickness, where, respectively, the higher modes have higher frequencies with the same lengths, the horizontal component of the velocity is amplified in the main layer of the density jump, reaching the maximum amplitudes. For k = 0.015 1/m, the effect of vertical density stratification is noticeable for both components and for all modes under consideration. At the “southern” stations the thickness of the waveguide becomes larger, the wave motions of the lowest considered mode, the fifth, propagate almost to the entire depth of the ocean (article 2577).

At different stations, respectively, with different thermohaline structure, there is a noticeable difference in attenuation modes. In particular, at $\sigma = 0,015$ and at $\alpha = 9,6 \times 10^{-6}$ on station 2569, from the periodic to aperiodic regime, the 5-th mode passes, at station 2570-3, 4, 5, at station 2571-3, 4, 5, at station 2,572 - only 5, on stations 2573-4 and 5, at stations 2574, 2575 and 2576 - all modes have a periodic mode of attenuation, at station 2577 - 4 and 5 modes go into aperiodic mode of attenuation.

Consider seasonal changes in hydrological conditions at the polygon in an area of the Vema fracture zone and trace their influence on the kinematic characteristics of internal waves and their damping regimes.

Figure 4 shows the vertical stratification of temperature, salinity and density, as well as the distribution of the Väisälä-Brent frequency to a depth of 1500 m at station 2551, carried out in the Vima fault area on October 8, 2014.

The thermohaline structure at station 2551 is similar to the thermohaline structure at stations 2569-2571: the upper quasihomogeneous layer is thin, the maximum density gradients are observed already at 40 m horizon, the main pycnocline is also relatively thin, with a lower boundary of about 150 m. The Väisälä-Brent frequency profile has one pronounced maximum corresponding to the main pycnocline.

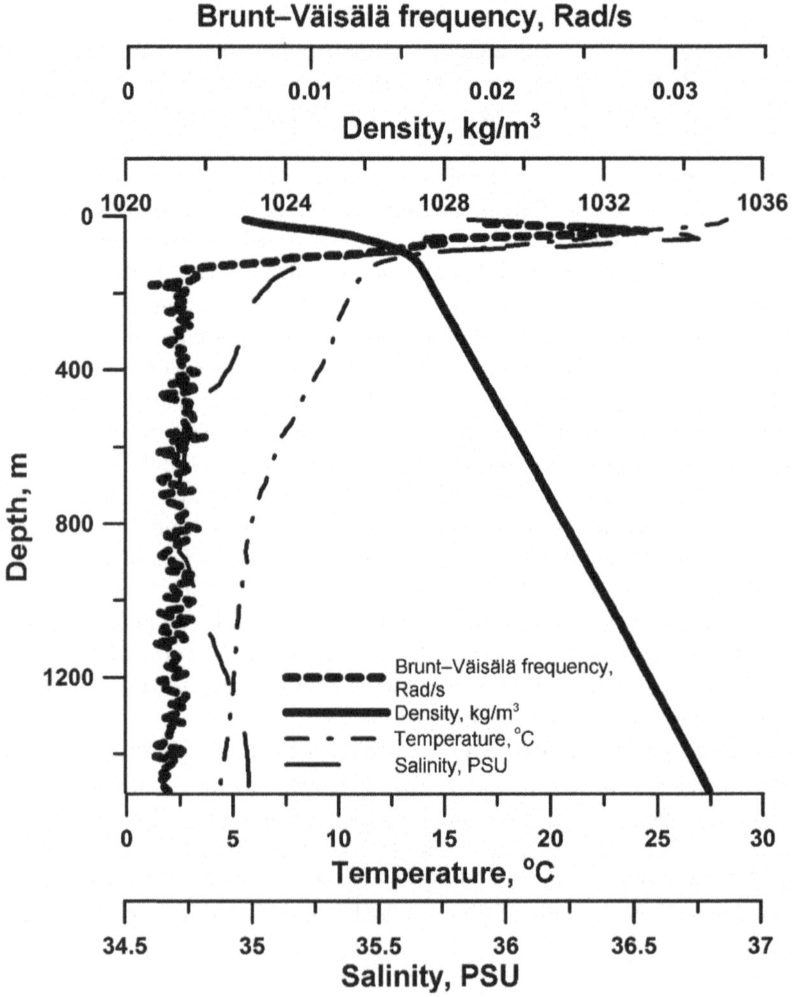

Fig. 4. Vertical distribution of temperature, salinity and density at the station 2551.

Dispersion curves at station 2551 are also similar to the curves at stations 2569-2571, which have similar regions of resonance zones, as a result of which, at frequency, the first, second and third modes of internal waves have a periodic damping mode, and the fourth and fifth are aperiodic.

5 Results and Discussion

The hydrological and climatic regimes of the equatorial regions of the Earth are stable for a period of several years, however, small differences in the thermohaline structure in 2014 and 2015 are still traced: in 2014, there was observed density stratification in which internal waves had smaller frequencies with the same lengths.

The influence of the seasonal variation of the vertical density structure in the Central Atlantic on the dissipation of internal waves was investigated. It is shown that the seasonal variability of stratification can quantitatively (the magnitude of time decrements) and qualitatively (periodic and aperiodic regimes) change the decay of the modal composition of internal waves.

Acknowledgments. The project was carried out within the RFBR project № 17-08-00085 “Processes in the deep channels connecting the East and West Atlantic basins”, Research project “Present-day situation and long-term variability of the coastal ecosystems of the southern seas of Russia” State registration CITIS № 01201363187 and SFedU Internal Project-07/2017-4 “Development of methodological bases and recommendations for integrated management of the coastal zone of the Azov Sea in conditions of growth of dangerous exogenous processes, recreational stress, climate variability”.

References

1. Morozov, E.: Oceanic Internal Tides: Observations, Analysis and Modeling. A Global View. Springer, Berlin (2018)
2. Demidov, A.N., Dobrolyubov, S.A., Morozov, E.G., Tarakanov, R.Yu.: Transport of bottom waters through the vema fracture zone in the mid-atlantic ridge. Dokl. Earth Sci. **416**(3), 1–5 (2007)
3. Morozov, E.G., Demidov, A.N., Tarakanov, R.Yu.: Transport of antarctic waters in the deep channels of the Atlantic Ocean. Dokl. Earth Sci. **422**(6), 815–818 (2008)
4. Krauss V.: Internal waves. 272 p. Hydrometeoizdat. Leningrad (1968)
5. Monin, A.S. (ed.): Physics of the Ocean. Ocean Hydrodynamics, vol. 2, 456 p. Nauka, Moscow (1978)
6. Cherkesov, L.V., Ivanov, V.A., Khartiev, S.M.: Introduction into Hydrodynamics and Wave Theory, 264 p. Gidrometeoizdat. St. Petersburg (1992)
7. Soloviev, A.N., Khartiev, S.M., Solovieva, A.A., Grigorenko, K.S., Matishov, D.G.: Internal gravity and acoustical waves parameters analysis based on “the climatic Atlas of the Arctic Regions seas 2004”. SSC RAS Bull. **6**(1), 24–32 (2010)
8. Khartiev, S.M., Morozov, E.G., Grigorenko, K.S., Matishov, D.G., Soloviev, A.N., Solovieva, A.A.: Internal waves in a stratified sea with layers of density inversions. Dokl. Earth Sci. **448**(1), 92–96 (2013)

9. Bukatov, A.V., Cherkesov, L.V.: Waves in an Inhomogeneous Sea, 224 p. Naukova Dumka, Kiev (1983)
10. Goncharov, V.V.: On some features of internal waves in the ocean. Tsunami and internal waves. MHI AS USSR. Sevastopol, pp. 87–96 (1976)
11. Grigorenko, K.S., Khartiev, S.M., Solovieva, A.A., Ermoshkin, A.V.: Kinematic characteristics of internal waves in the Central Atlantic investigation according to CTD-profiling data. Ecol. Bull. Res. Cent. Black Sea Econ. Coop. (BSEC) (1), 41–50 (2015)
12. Khartiev, S.M., Cherkesov, L.V.: Investigation of the influence of the vertical density structure on the dissipation of internal waves. Mar. Hydrophysical Investig. (3), 7–15 (1980)

Modeling the Black Sea Deep Circulation with ERA-Interim Forcing in Summer 2013

O. A. Dymova(✉), N. A. Miklashevskaya, and N. V. Markova

Marine Hydrophysical Institute of RAS, Sevastopol, Russian Federation
olgadym@yahoo.com

Abstract. The results of a prognostic numerical experiment on simulation of the Black Sea circulation are given for warm period of 2013. The MHI hydrophysical ocean model and ERA-Interim atmospheric forcing are used for the modeling. Comparison of observed and simulated temperature and salinity is carried out. A satisfactory agreement between measured and simulated data is obtained. The greatest attention is paid to the deep Black Sea circulation structure. Hydrophysical fields at the depths below the main pycnocline are studied in detail. It is confirmed that the field of deep currents contains vortex formations and flows that qualitatively and quantitatively differ from the surface ones. There are a number of vortexes that form not at the sea surface but near the low boundary of the main pycnocline (at depths of 150–300 m) and propagate down to the bottom without weakening. As well, in summer 2013, quasi-periodic narrow deep currents propagating anticyclonically are generated in some regions along the Black Sea continental slope.

Keywords: Black Sea · Modeling · Deep circulation · Currents · Vortexes · Temperature · Salinity · Measurements

1 Introduction

A clear understanding of the nature of the Black Sea deep-water circulation and its mesoscale structure has not been established yet. In total, it is known that a general structure of the deep-water circulation is not identical with the surface one. The cyclonic Rim Current and gyres in the eastern and western parts of the sea characterized for the surface layer are not supported there. Dynamical structures of lower scales are predominating, and the currents velocities are much weaker than ones on the surface. The descriptions of some separate dynamical structures (vortexes, currents) are presented mainly in the works based on the results of scientific expeditions [1, 2].

The complete structure and spatial-temporal variability of the deep hydrophysical fields in the Black Sea are studied insufficiently, and on some details, there are disagreements in opinions of the researchers. At present, most authors hold a view that the general direction of deep circulation is cyclonic. However, in some papers on the calculations of deep-sea currents and theoretical study based on measurements data, a number of conclusions are drawn about the probability of the existence of a reverse circulation relative to the surface one. For example, in [3, 4] the possibility of countercurrents at the depths of the main pycnocline is shown. At the same time, in [5, 6],

V. I. Karev et al. (Eds.): *Physical and Mathematical Modeling of Earth and Environment Processes (2018)*, SPEES, pp. 33–40, 2019.
https://doi.org/10.1007/978-3-030-11533-3_4

the presence of such currents is not supported by the analysis of the trajectories of the profiling floats.

Detailing the dynamics of the Black Sea at great depths will allow not only to contribute to theoretical concepts but also to expand the understanding of hydrophysical processes in the marine environment as a whole. So, our research focuses on the study of the Black Sea deep-sea circulation, and it is performed on the basis of numerical modeling. Nowadays, numerical modeling is the most accessible and informative method of obtaining data on three-dimensional hydrophysical fields and allows us to reconstruct their structure with good discreteness in space and time.

The numerical experiment to simulate the deep-water circulation of the Black Sea is provided using the model of the Marine Hydrophysical Institute (MHI) [7]. ERA-Interim reanalysis data [8], which include heat, moisture and wind velocity, are used as atmospheric forcing. The correction of the calculated temperature field by assimilation of the sea surface temperature or by any other techniques is not performed in the experiment.

The modeling is carried out for the period from May to September 2013. In summer, the Rim Current weakens and intensively meanders, and the probability of emergence of specific dynamical features such us countercurrents, etc., increases. As well, data of deep-water observations of temperature and salinity south-west of the Crimean peninsula (cruise of RV «Professor Vodyanitsky» №76CTD) and the data of the profiling up to the depth of 2000 m conducted by nine Argo floats [9], are available for the period mentioned. The map of the stations where the considered profiles have been executed is shown in Fig. 1.

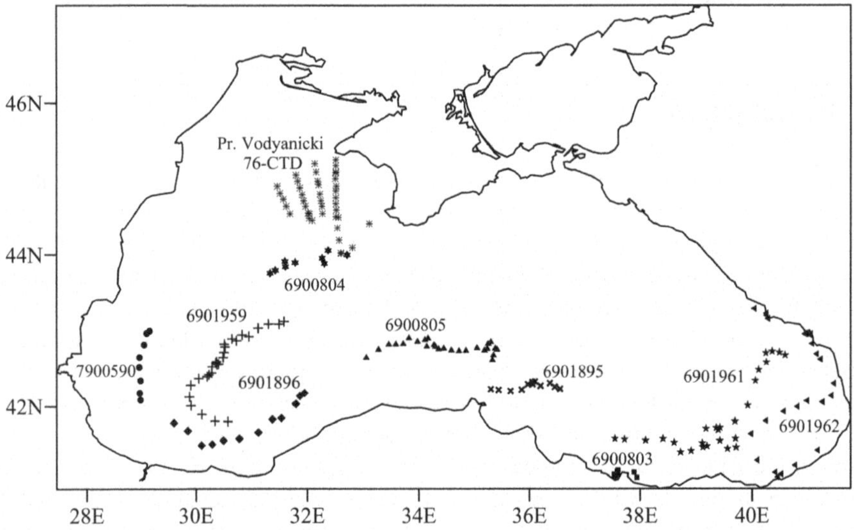

Fig. 1. Map of deep-water stations executed during 76CTD cruise of RV «Professor Vodyanitsky» (September 19–23, 2013) and Argo floats (May–September 2013). Numbers and symbols indicate ID and trajectories of Argo floats.

2 Comparison Measured and Simulated Data

The temperature and salinity simulated are compared with the relevant data of observations. In total, 193 profiles are considered (42 of them – the data of the RV «Professor Vodyanitsky» 76CTD cruise, and 151 are obtained by Argo floats). To quantify the results of the numerical experiment, the mean and standard deviations of the model temperature and salinity from the measured ones are calculated. The values of temperature and salinity deviations are presented in Table 1. For the surface layer 0–5 m the expeditional observations are available only, and the mean deviation is equal to 4.7 °C for temperature and 0.106‰ for salinity. For the lower horizons, according to the results of all observations, the maximum differences between modeled and measured data are in the layer of 5–30 m for temperature (0.94 °C) and in the layer of 30–100 m for salinity (0.308‰). At the depths from 300 to 1500 m the modeled and measured data are the closest. Here the mean deviations are at about ten times less than in the upper layers and do not exceed 0.031 °C for temperature and 0.014‰ for salinity. Thus, the quite satisfactory quality of the model reconstruction of the Black Sea deep thermohaline structure for the considered summer period is established.

Table 1. The mean and RMS deviations of model temperature and salinity from measurement data in summer 2013

Depth (m)	Temperature, °C		Salinity, ‰	
	Mean	RMS	Mean	RMS
0–5	–4.7	1.0366	0.106	0.0975
5–30	0.94	0.9518	0.087	0.129
30–100	0.87	0.1721	0.308	0.1476
100–300	0.169	0.0669	0.256	0.1445
300–800	0.005	0.0203	–0.008	0.0596
800–1500	–0.031	0.0423	0.014	0.0197

3 Structure of the Deep-Water Circulation

The situation with the lack of velocity measurements at the great depths is unfortunately typical. Thus, measurements of the deep currents during expeditions on research vessels are rare. The Argo floats do not measure currents velocity constructively. So, the deep circulation features analysis for the period from May to September 2013 in our work is based on the simulation results. A quite low error in modeling the deep-water temperature and salinity allows us to assume a good quality of the field of deep currents, which agrees with them. Examples of the currents fields on the 300 and 1700 m horizons for the end of June and the beginning of August, 2013, are shown in Figs. 2 and 3, respectively.

According to the modeling results, the Rim Current propagates over the continental slope as the uniform stream in the considered warm period of 2013. In the surface layer, its maximum velocities exceed 40 cm • s^{-1}, and the average ones are equal to

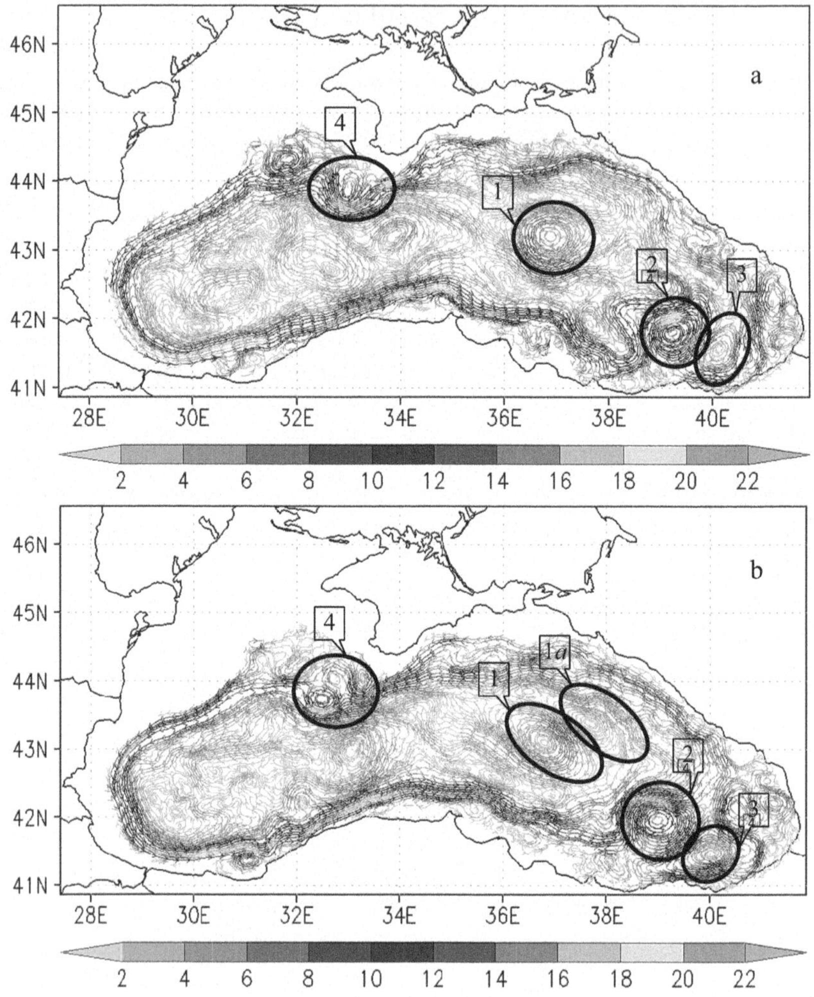

Fig. 2. Currents field at the level of 300 m on 26 June (a) and 8 August (b) 2013.

20–25 cm • s^{-1}. The Rim Current meandering leads to the forming of mainly anticyclonic mesoscale vortexes to the right of the mainstream. These eddies eventually move in accordance with the direction of the current. The main regions of their formation are the central part of the Anatolian coast and the Sukhumi area (near the Cape Pitsunda).

The Rim Current is clearly traced up to the horizon of 500 m. At this depth, at the end of May in the western and southern parts of the sea, its width is about 35 km, and velocities are of 8–10 cm • s^{-1}. At the northern and eastern boundaries of the basin, the current is weaker and its width is about 20–22 km with characteristic velocities of 6–8 cm • s^{-1}. In these regions, an episodic decrease in the current velocity down to 4 cm • s^{-1} is revealed.

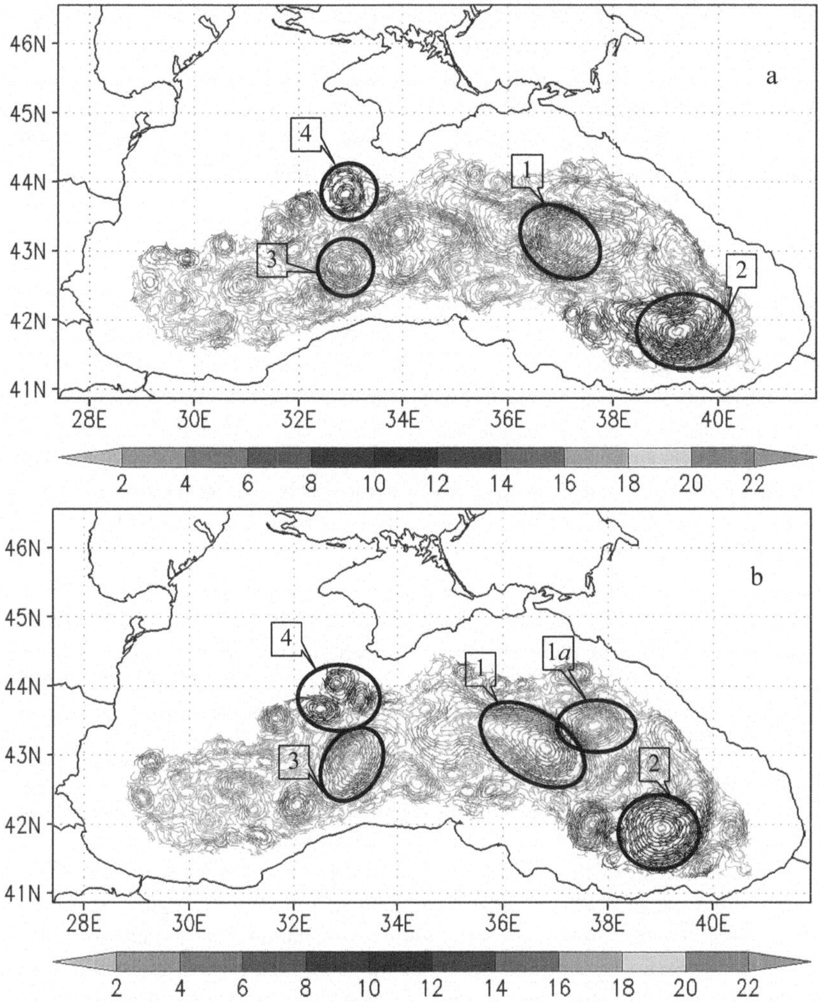

Fig. 3. Currents field at the level of 1700 m on 26 June (a) and 8 August (b) 2013.

An anticyclonic eddy with a diameter of approximately 100 km is traced in the middle of the eastern part of the sea (see in Figs. 2 and 3, region surrounded by the circle «1»). Its center has been shifted from 38° E in May to 36° E in early September. For 3.5 months period, the eddy stretches in the zonal direction and shifts to the West. It reaches the most developed stage by the end of July and keeps the size and velocity until the end of August. This eddy is visualized clearly not on the surface of the sea, but only from the horizon of 150 m and down to the horizon of 2100 m. At depths more than 900 m, its dimension is about of 200 km that exceeds the size of its upper part in 1.5–2 times. At horizons less than 900 m, the velocity does not exceed 8 cm • s^{-1}, but deeper it increases up to 10–12 cm • s^{-1}. At the end of the experiment, in September, at horizons less than 500 m, this anticyclone dissipates. In deeper layers it is transformed into a system of two anticyclonic eddies.

From the second decade of July, toward the North-East of the above mentioned anticyclone on the horizons of 200 m and more, the cyclonic eddy begins to form (see Figs. 2b and 3b, region marked «1a»). It moves half a degree to the South by the second decade of August. The cyclone gradually increases in size. It reaches the anticyclone size by the twentieth of August, and exceeds it by the beginning of September. On the lower horizons, due to orographic features, the dimension of this cyclone is smaller than in the overlying layers.

At the end of May – the beginning of June, in a south-eastern part of the Black Sea (near Trabzon, between 39° and 40° E), a steady anticyclonic circulation is formed starting from the depth of 150 m (Figs. 2 and 3, region marked «2»). It is intensified and shifted to the West (up to 38° E) by September. It can be traced to the horizon of 1700 m. The orbital velocity of the eddy is 10–12 cm • s^{-1} and does not vary with depth. This eddy is similar to the surface anticyclones which are formed at the central part of the Black Sea Anatolian coast and are moved by the Rim Current to the East. However, the deep anticyclone has a more stable localization compared to these surface vortexes.

Due to peculiarities of the topography and meandering of the Rim Current, a cyclonic eddy is formed in the layer down to 500 m to the right of the anticyclone considered (Fig. 2, region marked «3»). Its size is about 100 km. The center of this cyclone is placed in the area of 41.5° N, 40° E. The evolution of this eddy in the layer of the main pycnocline at the depths of 50–150 m leads to the shifting of east-located Batumi anticyclone to the North by 1° along the meridian.

In the western part of the sea, since the depth of 500 m and down to the bottom, there is a steady cyclonic vortex during all time of the integration. Its characteristic velocities are of 4–6 cm • s^{-1} on the horizons of 500–900 m, and 4–8 cm • s^{-1} in the lower layers. Diameter of this vortex is about 100 km. During its existence, the vortex is split and then merged with neighboring small eddies (Fig. 3, region marked «3»).

Main feature of the above considered vortexes is that they are traced from the lower boundary of the main pycnocline down to the bottom, keeping their structure with depth. At the same time, their maximum velocities even increase in the deep layers (by about 25–50%). These vortexes are formed and developed in the abyssal part of the Black Sea, in contrast to the coastal anticyclonic ones that are formed between the Rim Current and the shore [10].

In the region of 44° N, 32° E, the anticyclonic eddy with velocities of 15–20 cm • s^{-1} (Sevastopol anticyclone) is generated periodically. At the depth of 300 m, its center is displaced on a half-degree to the South relatively to the position on the sea surface (Fig. 2a, region marked «4»). Starting from a horizon of 200 m and deeper, the structure of the velocity field in this region changes. In the second half of July a system of three small vortexes (cyclonic and two anticyclonic) is formed here (Figs. 2b and 3b, marked area «4»). It can be traced down to the bottom (1700 m). The velocities in the cyclone reach 6–8 cm • s^{-1}, and in anticyclones 10–14 cm • s^{-1}, increasing near the bottom to 14–16 cm • s^{-1}.

The narrow anticyclonic currents propagating along the continental slope are observed at the depths about 1000 m. They are of the most intense at late May–June in the north-eastern part of the sea, from the southern coast of Crimea to the Caucasus coast near Sukhumi. As well, they exist along the coasts of Turkey and Bulgaria. The width of the currents does not exceed 5 km, and the average velocity is 6–8 cm • s^{-1}. Since the end of June, it decreases to 4 cm • s^{-1}. The current along the Caucasian coast is quasi-periodic, it appears for 3–4 weeks, and then disappears for a period of 4 up to 10 days.

4 Conclusion

The comparison of the simulated temperature and salinity with observed ones showed us a quite good agreement at the depths below the main Black Sea pycnocline. In the model currents field, the existence of local vortex structures of different signs of vorticity is confirmed. It was found, that the most part of the deep-water vortexes is formed not in the surface layer, but at depths just under the permanent pycnocline and propagates down to the sea bottom. Orbital velocities of vortexes are about 5–10 cm • s^{-1} and do not decrease significantly with depth. During the summer, these vortexes migrate within 1–1.5° of both the longitude and latitude relative to the site of their formation. As well, the narrow anticyclonic currents along the continental slope at the depths of about 1000 m are revealed.

Acknowledgments. The numerical experiment was carried out within the framework of the State assignment theme No. 0827-2018-0003. The data comparison was carried out within the framework of the State assignment theme No. 0827-2018-0002.

References

1. Oguz, T., Latun, V.S., Latif, M.A., et al.: Circulation in the surface and intermediate layers of the Black Sea. Deep-Sea Res. **40**(8), 1597–1612 (1993)
2. Morozov, A.N., Lemeshko, E.M.: Experience of the acoustic Doppler measuring instrument of currents (ADCP) use in the Black Sea. Phys. Oceanogr. **16**(4), 216–233 (2006)
3. Bulgakov, N.P., Bulgakov, S.N.: Manifestation of the Black Sea countercurrent in the seawater density and hydrostatic pressure [Proyavlenie Protivotecheniya v Chernom more v polyah plotnosti vody i gidrostaticheskogo davleniya]. Morskoy gidrofizicheskiy zhurnal **4**, 63–76 (1995). (in Russian)
4. Bulgakov, S.N., Korotaev, G.K., Whitehead, J.A.: The role of buoyancy fluxes in the formation of a large-scale circulation and stratification of sea water: 1, 2. Izvestiya Atmos. Oceanic Phys. **32**(4), 548–564 (1996)
5. Korotaev, G., Oguz, T., Riser, S.: Intermediate and deep currents of the Black Sea obtained from autonomous profiling floats. Deep-Sea Res. II **53**(17–19), 1901–1910 (2006)
6. Markova, N.V., Bagaev, A.V.: Velocities of the Black Sea deep currents estimated from the profiling drifters Argo data. Phys. Oceanogr. **3**, 23–35 (2016)
7. Demyshev, S.G., Korotaev, G.K.: Numerical energy–balanced model of the baroclinic ocean currents on a C–grid. In: Numerical Models and Results of Calibration Calculations of Currents in the Atlantic Ocean, pp. 163–231. INM RAS, Moscow (1992). (in Russian)

8. Dee, D.P., Uppala, S.M., Simmons, A.J., et al.: The ERA-Interim reanalysis: configuration and performance of the data assimilation system. Q. J. R. Meteorol. Soc. **137**(656), 553–597 (2011)
9. Argo Data Management. http://www.argodatamgt.org/. Accessed 29 June 2018
10. Titov, V.B.: Morphometric parameters and hydrophysical characteristics of nearshore anticyclonic eddies in the Black Sea. Russ. Meteorol. Hydrol. **4**, 52–57 (2002)

Determination of the Concentration of Contaminants at the Outlet of the Underwater Source Based on the Variation Algorithm

Sergey Vladimirovich Kochergin(✉) and Vladimir Vladimirovich Fomin

Marine Hydrophysical Institute RAS, Sevastopol 299029, Russian Federation
vskocher@gmail.com

Abstract. A variation algorithm for determining the concentration of the contaminating admixture at the outlet of the underwater source C_p is proposed. It is based on a minimization of the quadratic functional, which is determined by the residual values of the measured and calculated values of the pollution concentration and the solution of the corresponding problem in variations. The algorithm for determining the C_p tested in the framework of numerical three-dimensional hydrothermodynamic model of the spread of sewage in the coastal zone of the Crimean Peninsula, adjacent to the city of Sevastopol. Numerical experiments were carried out to restore the C_p for different number of measurements and different levels of random noise in the measurements. A measurement scheme was used in which a moving sensor consistently measured the vertical profile of pollution concentration at fifteen stations. It is shown that without taking into account the random noise in the initial data, the C_p parameter is restored accurately for any number and combination of stations. When random noise is taken into account, parameter C_p recovery occurs when the most informative stations are used. The proposed algorithm can be useful in the interpretation and planning of natural experiments on the study of wastewater distribution in coastal waters.

Keywords: Numerical modeling · Problem in variations · Functional minimization · Parameters identification · Data assimilation

1 Introduction

The environmental situation in coastal waters is influenced by various factors, including the release of pollutants from underwater reservoirs. Knowledge of reliable information about the situation in the area of releases is important for management decisions. The generation of such information is most often based on the use of clock and remote sensing methods of observations [1–3], as well as the use of mathematical modeling methods [4–6].

In mathematical modeling there is a problem of identification of model parameters on the available real information. The methods of parameter identification are most

V. I. Karev et al. (Eds.): *Physical and Mathematical Modeling of Earth and Environment Processes (2018)*, SPEES, pp. 41–50, 2019.
https://doi.org/10.1007/978-3-030-11533-3_5

often based on the minimization of the quality functional, which characterize the deviations of the model estimates of the concentration of the mixture from the measurement data. Approaches to minimization of such functionals can be different. One of the effective ways to find the optimal parameter is the linearization method [7, 8].

The aim of this work is to build and verify an algorithm for identifying the input parameters of the pollutant transport model. As a first step for solving problems of this type, we consider a numerical hydrothermodynamic model [9], in which the identification parameter is the concentration of pollution at the outlet of the underwater source (hereinafter C_p). It should be noted that the identification of the C_p parameter is a practically important problem, since it is usually not possible to obtain accurate C_p values for a number of reasons.

2 Dynamic Model

To calculate the flow fields, admixture concentration, temperature and salt content of sea water in the area of underwater discharge, a nonlinear baroclinic model described in [9–11] is used. The model is based on the system of three-dimensional equations of ocean dynamics in the Boussinesq and hydrostatics approximation. The system uses a dimensionless vertical coordinate $\sigma = (z - \eta)/D$, $D = h + \eta$ – dynamic depth, h – bottom relief, η – sea level.

The numerical algorithm for solving the system of equations is based on the division of the problem into barotropic and baroclinic modes and the use of explicit schemes for horizontal coordinates and implicit schemes for coordinates. The coefficients of horizontal turbulent diffusion is calculated with the formula Smagorinsky [12]. Vertical turbulent viscosity and diffusion coefficients are determined using the Mailor-Yamada model [9, 13]. TVD [14, 15] schemes are used for the approximation of advective operators in model equations. These schemes have the property of monotonicity, which is a necessary condition for adequate modeling of fields with large spatial gradients. A detailed description of the numerical algorithm of the model is given in [10].

The problem is solved at a time interval $[0, t_0]$ in a rectangular region $\Omega = \{0 \le x \le L; 0 \le y \le L; -1 \le \sigma \le 0\}$ with liquid lateral boundaries and a free surface (Fig. 1). At the initial time $t = 0$ the following conditions are set:

$$u = U_0, v = w = 0, \eta = 0, T = T_0(\sigma), S = S_0(\sigma), C = 0. \quad (1)$$

Here u, v, w – are the components of the flow velocity along the axes, x, y, σ respectively U_0 – he depth constant velocity of the background flow; T, S – temperature and salinity of water; $T_0(\sigma)$ – background temperature distribution; $S_0(\sigma)$ – background salinity distribution; C – concentration of the contaminant.

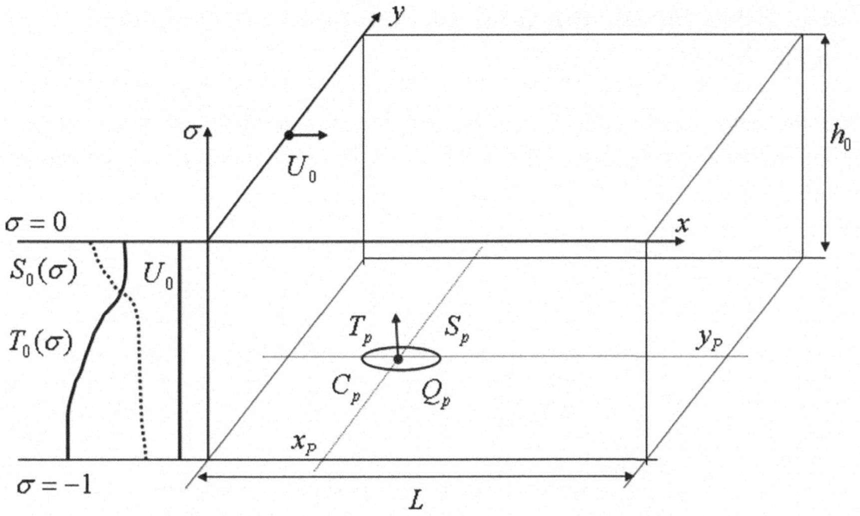

Fig. 1. Scheme of the computational domain and input model.

At $t > 0$ at the bottom of the pool ($\sigma = -1$) in the local area Ω_p the mass source having the following parameters begins to operate in the local area: w_p – the rate of water flow from the source; T_p, S_p – the temperature and salinity of the flowing water; C_p – the concentration of the admixture at the outlet of the source.

The field of admixture concentration C in the region Ω_p is described by a three-dimensional transport-diffusion equation:

$$\frac{\partial}{\partial t}(DC) + \Lambda C = \frac{\partial}{\partial x}\left(A_C \frac{\partial C}{\partial x}\right) + \frac{\partial}{\partial y}\left(A_C \frac{\partial C}{\partial y}\right) + \frac{\partial}{\partial \sigma}\left(\frac{K_C}{D}\frac{\partial C}{\partial \sigma}\right), \tag{2}$$

where $\Lambda C = \frac{\partial}{\partial x}(DuC) + \frac{\partial}{\partial y}(DvC) + \frac{\partial}{\partial \sigma}(wC)$ – is the transfer operator; A_C and K_C – are the coefficients of horizontal and vertical turbulent diffusion coefficients.

On the free surface ($\sigma = 0$) and out-of-source at the bottom $\left(\sigma = -1; (x,y) \notin \Omega_p\right)$ boundary conditions have the form

$$\frac{K_C}{D}\frac{\partial C}{\partial \sigma} = 0, \tag{3}$$

The boundary conditions at the bottom in the area of the source $\left(\sigma = -1; (x,y) \in \Omega_p\right)$ are written in the form [11]:

$$\frac{K_C}{D}\frac{\partial C}{\partial \sigma} = w_p\left(C - C_p\right) \tag{4}$$

3 Algorithm for Identification of Pollution Concentration in the Source

When choosing C_p as a parameter to be identified, the problem can be solved on the basis of the linearization method [7]. Let $V = \frac{\delta C}{\delta C_p}$ – the variation function C relative to the unknown parameter C_p. Following [16], we set in accordance to (2)–(4) the problem in variations:

$$\frac{\partial}{\partial t}(DV) + \Lambda V = \frac{\partial}{\partial x}\left(A_C \frac{\partial V}{\partial x}\right) + \frac{\partial}{\partial y}\left(A_C \frac{\partial V}{\partial y}\right) + \frac{\partial}{\partial \sigma}\left(\frac{K_C}{D}\frac{\partial V}{\partial \sigma}\right) \quad (5)$$

$$\frac{K_C}{D}\frac{\partial V}{\partial \sigma} = 0 \quad at \quad \sigma = 0 \quad (6)$$

$$\frac{K_C}{D}\frac{\partial V}{\partial \sigma} = 0 \quad at \quad \sigma = -1 \quad and \quad (x, y) \notin \Omega_p \quad (7)$$

$$\frac{K_C}{D}\frac{\partial V}{\partial \sigma} = w_p(V - 1) \quad at \quad \sigma = -1 \quad and \quad (x, y) \in \Omega_p \quad (8)$$

Let the problem (5)–(8) be solved on the time interval $[0, t_0]$ in the region Ω. Then, to solve the problem of C_p parameter identification, it is required to find the minimum of the following functional:

$$J = \frac{1}{2}\langle P(RC - C_{obs}), P(RC - C_{obs})\rangle \quad (9)$$

where $\langle a, b\rangle = \int_0^{t_0}\int\int\int_\Omega ab d\Omega dt$ – the scalar product; C_{obs} – the measured values of C at the given points of the domain Ω at given points in time; R – a projection operator in the points of observations; P – the fill operator the residuals field with zeros in the absence of measurement data.

Let's imagine a variable in the form:

$$C = \bar{C} + V\left(C_p - C_p^*\right), \quad (10)$$

where $\bar{C}$ – some assessment of the concentration of pollution, and C_p^* – its true value, to be determined. After substitution (10) in (9) we get:

$$J = \frac{1}{2}\left\langle P\left(R\left(\bar{C} + V\left(C_p - C_p^*\right)\right) - C_{obs}\right), P\left(R\left(\bar{C} + V\left(C_p - C_p^*\right)\right) - C_{obs}\right)\right\rangle \quad (11)$$

Further, for simplicity, omit the entry line $\bar{C}$. From the condition $\frac{\partial J}{\partial C_p} = 0$, finally have:

$$C_p^* = C_p + \frac{\langle P(RC - C_{obs}), PRV \rangle}{\langle PRV, PRV \rangle} \tag{12}$$

The formula (12) is also valid for the case when the measurements come, for example, only from the sea surface. The problem of constructing the quality functional (10) is solved by the appropriate choice of projection operators P and R.

Thus, the C_p parameter identification algorithm consists of the following:

- the problem (1)–(4) is solved with some initial given C_p;
- forecast residuals are calculated and the functional (9) is constructed;
- the problem in variations (5)–(8) is solved;
- evaluation is made C_p^{n+1} by the formula:

$$C_p^{n+1} = C_p^n + \frac{\langle P(RC - C_{obs}), PRV \rangle}{\langle PRV, PRV \rangle}.$$

These steps are repeated until the convergence criterion is met:

$$\left| C_p^{n+1} - C_p^n \right| < \varepsilon,$$

where ε – is the specified accuracy of the C_p parameter definition.

4 Numerical Experiments and Discussion of Results

Model calculations were carried out for the area of the Blue Bay of the Sevastopol city, where the main city wastewater is located. The estimated area had a horizontal size L is 2 km and a depth h_0 is 30 m (Fig. 1). A rectangular grid with a step d is 20 m horizontally and a discreteness of 1 m vertically was used. The integration step time was 5 s. On the western border of the computational domain ($x = 0$) was set conditions of the form (1). On other liquid boundaries, smooth continuation conditions were used for all variables: $\partial\varphi/\partial x = 0$ for $x = L$; $\partial\varphi/\partial y = 0$ for $y = 0$ and $y = L$.

The center of the source of pollution was located at the point with coordinates: $x_p = 600$ m; $y_p = 1000$ m. Horizontal size of the source along the x, y axes was equal to the step of the calculated grid d. It was assumed that the water flowing from the source is zero salinity $(S_p = 0)$, and its temperature is equal to the ambient temperature, i.e. $T_p = T_0(-1)$. As T_0 and S_0 was obtained from the sensing in the area of underwater production, performed May 20, 2016 based on data from ADCP measurements in the area of the Heraklean Peninsula [17], the value of U_0 was taken to be equal to 0.05 m/s.

At first, a "reference" calculation of dynamic characteristics and pollution concentration caused by the action of the source with a "true" value C_p^* is 0,05 kg/m^3 was carried out. The total integration time was 6 h. For the first hour of model time, the water flow at the source Q_p is linearly increased from 0 to 1.4 m^3/s and then remained unchanged. The rate of water flow from the source was estimated by the formula $w_p = Q_p/d^2$.

The spatial distribution of the field of contaminated water on a fixed level $\sigma = -0,3166$ ($z = -9,5$ m) for t = 6 h is shown in Fig. 2. There are isolines of relative concentration $c = 100\% \cdot C/C_p^*$. Dimensionless horizontal coordinates are of the form: $X = x/2d$; $Y = y/2d$. On Fig. 3. for t = 6 h the contours of the relative concentration at the section along the axis $X(Y = 25)$ are shown.

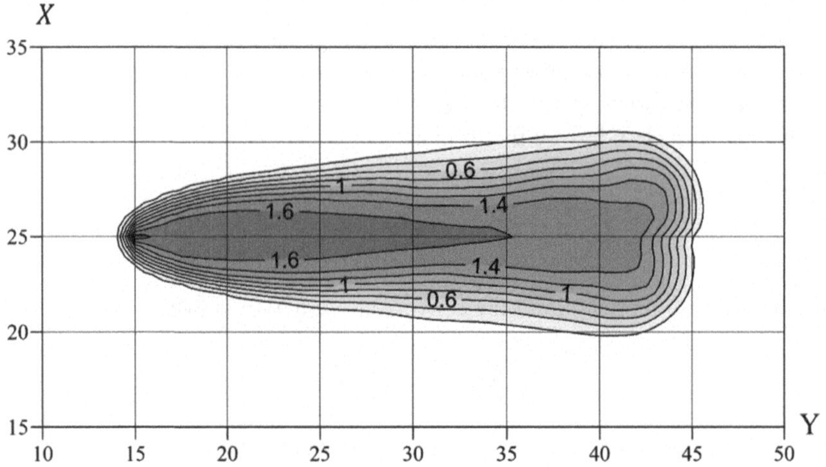

Fig. 2. Horizontal distribution of the relative concentration of the mixture C (%) in the area of the underwater source at $\sigma = -0,3166$ ($z = -9,5$ m).

The figures show that the field of polluted water consists of a plum and a jet stretched in the direction of the background flow and concentrated in the density jump layer (8–10 m). Detailed description of the influence of density stratification of sea water and the speed of the background flow on admixture this field given in [10, 11].

In the process of "reference" to calculate the value of C was accumulated in accordance with the scheme of measurements is shown in Fig. 4 as a numbered set of points. It was assumed that a moving sensor consistently measures the vertical profile C from the free surface to a depth of $z = 25$ m ($\sigma_B = -0,8333$) at stations 1–15. Measurements begin at time $t_1 = 2$ h at point $(X_1, Y_1) = (25, 20)$ and end at time $t_{15} = 5,5$ h at point $(X_{15}, Y_{15}) = (35, 30)$. The time resolution of measurements is 15 min. The specified measurement scheme is chosen from the considerations that in field studies the identification of contaminated waters is usually carried out by a series of vertical probes perpendicular to the direction of the jets [3].

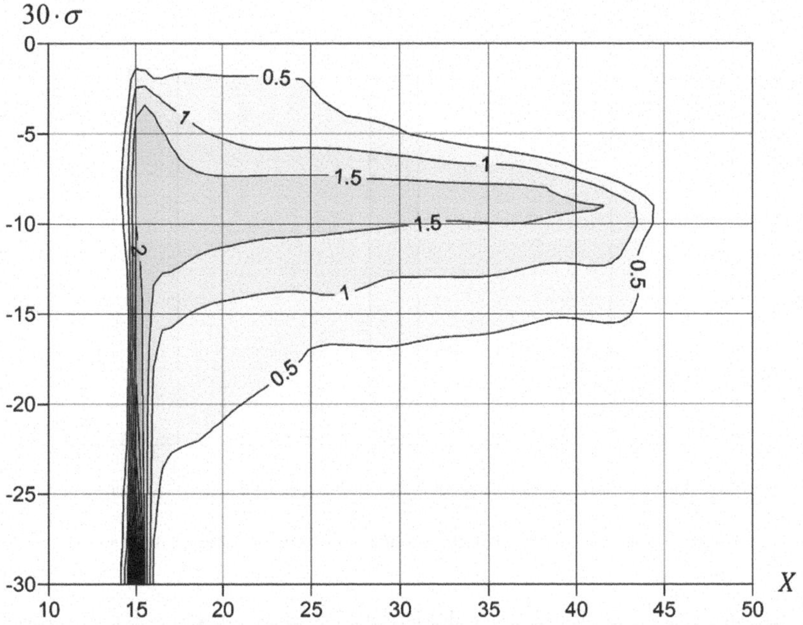

Fig. 3. Vertical structure of the field of relative concentration of C (%) in the area of the underwater source at $Y = 25$.

Functionality $I = \frac{1}{C_p^*}\int_{\sigma_B}^{0} C(\sigma)d(\sigma)$ characterizes the level of informativeness of measurements. Value I for each of the stations are given in Fig. 5. As you can see, the highest values of the I (stations 3, 8, 13) correspond to the hit of the sensor on the axis of the polluted water jet, and the lowest values of the I (stations 1, 2, 5, 6, 11, 12, 15) – cases where the sensor "slipped" the axial portion of the jet.

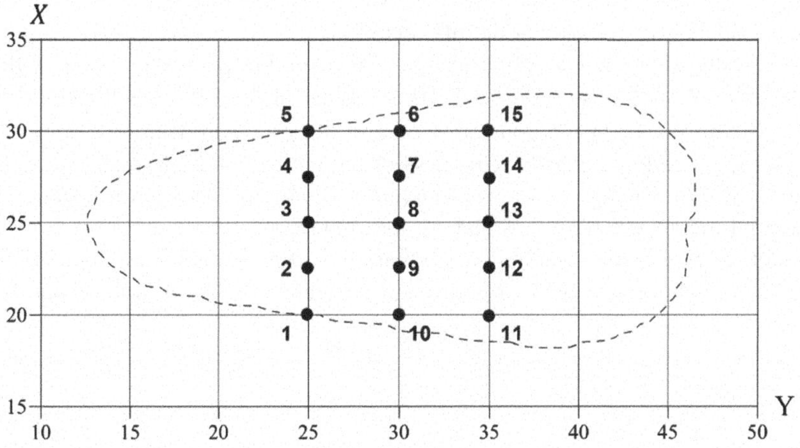

Fig. 4. Scheme of measurement of the admixture concentration field in the area of the underwater source. The position of the stations marked by points.

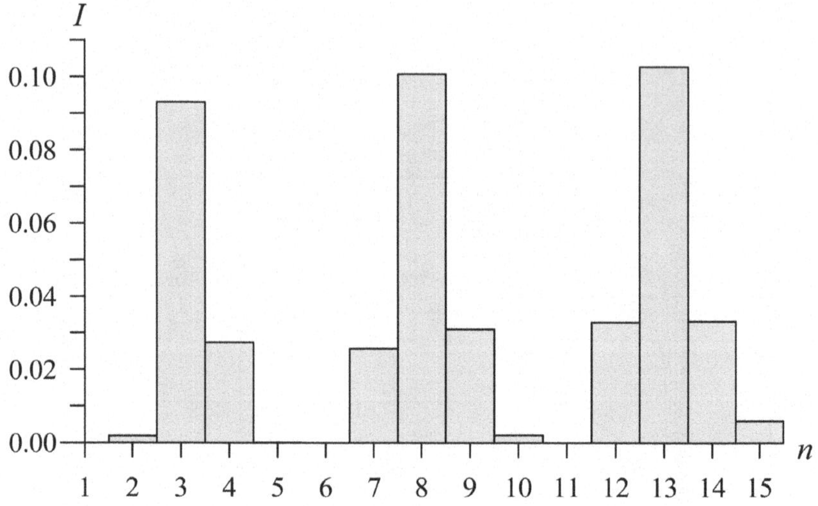

Fig. 5. Dependence of the functional I value on the station number n.

In parallel with the "reference" calculation, the problem was solved in variations (6)–(9). At its decision vertical profiles in at stations 1–15 in the moments of time $t_1 - t_{15}$ were kept. Since V is independent of the parameter C_p, the problem (6)–(9) was solved once.

Then a series of numerical experiments on testing the above algorithm of parameter C_p identification was performed. Numerical experiments were carried out with a different number of stations. In addition, the "measured" vertical profiles $C(\sigma)$ were supplemented with random noise $\xi \in [-r/2, r/2]$, where $\xi = r(\delta - 1/2)$; r – is a given noise level; $\delta \in [0, 1]$ – a normally distributed random variable. The initial approximation was set to $C_p = 0, 1\ \mathrm{kg/m^3}$. It is established that 2, 3 iterations are necessary for convergence of the iteration process. Moreover, the main decrease in the residual functional (10) occurs already at the first iteration.

The results of numerical experiments to recover the parameter C_p or different combinations of stations at $0 < r \leq C_p^*/10$ are given in the table. As the comparison shows, without taking into account the random noise (experiments 1, 5 and 9), the parameter C_p is restored exactly regardless of the number and informativeness of the stations used. When random noise is taken into account, the best recovery occurs when the most informative stations 3, 8 and 13 are used.

Results of restoration of pollution concentration at the output from the underwater source C_p at different number of stations and different noise level in the input data

№	Station	r, kg/m^3	Noise level $100\% \cdot r / C_p^*$, %	Recovering C_p, kg/m^3	Error recovery $100\% \cdot C_p / C_p^*$, %
1	1–15	0,0000	0,0	0,5000	0,0
2	1–15	0,0125	2,5	0,5276	5,5
3	1–15	0,0250	5,0	0,6487	29,7
4	1–15	0,0500	10,0	0,9670	93,4
5	3, 8, 13	0,0000	0,0	0,5000	0,0
6	3, 8, 13	0,0125	2,5	0,4791	4,2
7	3, 8, 13	0,0250	5,0	0,5814	16,3
8	3, 8, 13	0,0500	10,0	0,8432	68,6
9	1, 2, 5, 6, 11, 12, 15	0,0000	0,0	0,5000	0,0
10	1, 2, 5, 6, 11, 12, 15	0,0125	2,5	0,5836	16,7
11	1, 2, 5, 6, 11, 12, 15	0,0250	5,0	0,6932	38,6
12	1, 2, 5, 6, 11, 12, 15	0,0500	10,0	1,0986	119,7

5 Conclusions

On the basis of the analysis of the numerical experiments, correct work of the linearization algorithm is shown for the identification of the selected input parameter of the problem. The algorithm has a good iterative convergence of the process, which allows to quickly assess the concentration in the source of pollution. The algorithm can be used to solve a wide range of environmental problems, as well as to interpret and plan full-scale experiments on the study of wastewater distribution in coastal waters.

The work was carried out within the framework of the state task on the topic 0827-2018-0004 "Complex interdisciplinary studies of Oceanological processes that determine the functioning and evolution of ecosystems of the coastal zones of the Black and Azov seas" ("Coastal studies") and supported by grant RFFI 18-45-920035 p_a.

References

1. Bondur, V.G., Grebenyuk, Yu.V: Remote indication of anthropogenic influence on marine environment caused by depth waste water plum: modeling, experiment. Issledovanie Zemli iz kosmosa **6**, 49–67 (2001)
2. Bondur, V.: Complex satellite monitoring of coastal water areas. In: 31st International Symposium on Remote Sensing of Environment, St. Petersburg, June 20–24, p. 7 (2005)
3. Bondur, V.G.: Satellite monitoring and mathematical modelling of deep runoff turbulent jets in coastal water areas. In: Waste Water, Evaluation and Management. http://www.intechopen.com

4. Blumberg, A., Ji, Z., Ziegler, C.: Modeling outfall plume behavior using far field circulation model. J. Hydraul. Eng. **122**(11), 610–616 (1996)
5. Zhang, X., Adams, E.: Prediction of near field plume characteristics using far field circulation model. Ibid **125**(3), 233–241 (1999)
6. Bondur, V.G., ZHurbas, V.M., Grebenyuk, Yu.V.: Mathematical modeling of turbulent jets of deep-water sewage discharge into coastal basins. Oceanology **46**(6), 757–771 (2006)
7. Alifanov, O.M., et al.: Extreme Methods for Solving Ill-Posed Problems, Nauka, Moscow, 286 p. (1988)
8. Gorsky, V.G.: Planning of Kinetic Experiments. Nauka, Moscow, 240 p. (1984)
9. Ivanov, V.A., Fomin, V.V.: Numerical modelling of propagation of underwater runoff in the coast of Heraclean Peninsula. Phys. Oceanogr. **6**, 82–95 (2016)
10. Ivanov, V.A., Fomin. V.V.: Mathematical Modelling of the Dynamic Processing in the Sea-Land Zone. ECOSI-Gidrophisika, Sevastopol, 363 p. (2008)
11. Bondur, V.G., Ivanov, V.A., Fomin, V.V.: Features of propagation of contaminating water from an underwater out falling a stratified environment of coastal water. In: Izvestiya RAN, Fizika atmosfery i okeana, vol. 54, no. 4, pp. 453–461 (2018)
12. Smagorinsky, J.: General circulation experiments with primitive equations. I. The basic experiment. Mon. Weather Rev. **91**(3), 99–164 (1963)
13. Mellor, G.L., Yamada, T.: Development of a turbulence closure model for geo-physical fluid problems. Rev. Geophys. Space Phys. **20**(4), 851–875 (1982)
14. Harten, A.: On a class of high resolution total-variation-stable finite-difference schemes. J. Numer. Anal. **21**(1), 1–23 (1984)
15. Fomin, V.V.: Application of TVD schemes to numerical modeling of frontal salinity zones in a shallow sea. Meteorologiya i gidrologiya **2**, 59–68 (2006)
16. Kochergin, V.S., Kochergin, S.V.: Variational algoritm for identifying power pulsed source pollution. In: Ecologicheskiy vestnik nauchnyh centrov CHernomorskogo ehkonomicheskogo sotrudnichestva, no 3, pp. 62–72 (2017)
17. Morozov, A.N., Ivanov, V.A., Shutov, S.A., et al.: Spatial structure of currents near Heraclean Peninsula according ADCP-observations in 2015. In: Ecologicheskaya bezopasnost' pribrezhnoj i shel'fovoj zon morya, MGI, Sevastopol, no. 1, pp. 73–79 (2016)

The Role of Wind Forcing in the Interannual Variability of Upper Layer Heat Balance Components in the North Atlantic

A. B. Polonsky(✉) and P. A. Sukhonos

Institute of Natural and Technical Systems, Lenina Street, 28, Sevastopol 299011, Russia
pasukhonis@mail.ru

Abstract. Large-scale patterns of wind stress and wind stress curl (WSC) over the North Atlantic undergo significant changes at the interannual-to-decadal scales. These changes also manifest in variability of circulation and thermal state of the upper mixed layer (UML). This paper discusses the influence of wind forcing on interannual variability of the UML heat balance in the North Atlantic in January. It is shown that interannual variability of vertical heat advection at the UML base is largely due to the WSC anomalies over the North Atlantic, excepting the vicinity of western boundary currents, Labrador Sea and tropical latitudes (0–15° N). Changes in the intensity of the UML horizontal advective heat transport in the Subtropical gyre are related to the wind field pattern in the mid-Atlantic with a lag of ~2 years. In the years with anomalously high anticyclonic WSC values over the Subtropical Gyre, the vertical heat advection at the base of the UML intensifies. At the same time the anticyclonic WSC area decreases in meridional dimensions, the excessive oceanic heat release and enhanced UML cooling occur in the North Atlantic. Simultaneously the cyclonic WSC areas increase in the tropical and subpolar latitudes. The North Equatorial Current and the Azores Current also noticeably intensify. Alternatively, in years with low anticyclonic WSC values over the Subtropical Gyre, vertical heat advection at the base of the UML weaken, the anticyclonic WSC area increases in meridional dimensions, moderate oceanic heat release and moderate UML cooling occur in the North Atlantic. Simultaneously the cyclonic WSC areas decrease in the tropical and subpolar latitudes. The North Equatorial Current and the Azores Current also noticeably weaken.

Keywords: Wind stress curl · Heat balance · Interannual variability · Upper mixed layer · North Atlantic

1 Introduction

Upper ocean circulation in the North Atlantic plays an important role in heat transport between low and high latitudes. Thus, it is important to study the spatial-temporal variability of the currents field as well as wind field, which is one of the main causes of the ocean currents. The Subtropical Gyre (STG) and the Subpolar Gyre (SPG) are the most prominent features of the North Atlantic large-scale circulation. In general, spatial

V. I. Karev et al. (Eds.): *Physical and Mathematical Modeling of Earth and Environment Processes (2018)*, SPEES, pp. 51–62, 2019.
https://doi.org/10.1007/978-3-030-11533-3_6

structure of large-scale atmospheric circulation determines the location of oceanic gyres. The STG/SPG is situated in the area of anticyclonic/cyclonic wind stress curl (WSC). Distribution of the WSC, in its turn, is determined by the position and spatial structure of prevailing winds – the Northeastern trade wind in the tropics (which is the most intense in the north Equatorial Atlantic) and the westerlies in the middle latitudes weakened to the subpolar latitudes [1].

Interannual-to-decadal variations of the WSC cause changes in large-scale ocean circulation and manifest themselves in sea level variations and the depth of thermocline fluctuations in the STG [2], meridional shifts of the STG [3] and changes in position of the axis of the North Atlantic Current [4]. Low-frequency WSC variability can directly affect ocean temperature, thermal fronts and geostrophic advection in the Northwest Atlantic [5, 6]. Direct relationship was found between the WSC and ocean currents variability in the North Atlantic on interannual scales: currents in the Labrador Sea [7], the Florida current [8], the Gulf Stream [9], the Norwegian Current [10], the West Greenland Current [11]. Thus, the relevance of studying both the WSC interannual variations and the response of large-scale ocean circulation and thermal state to these changes is practically assured.

Model studies with low spatial resolution showed that generation of sea surface temperature anomalies on the interannual scale is determined by the gyre response to the wind impact, especially in the areas with high horizontal gradients of sea surface temperature [12, 13]. The authors of paper [14] found that 70–80% of the Atlantic meridional overturning circulation (AMOC) variability in the eddy-permitting ocean model could be reconstructed from surface forcing. The authors of other paper [15] revealed that the observed AMOC anomaly is largely driven by the response of the upper-mid ocean transports and Ekman layer to variability in atmospheric boundary conditions. However, some authors suggested that the observed AMOC anomaly could has been a random event not directly linked to atmospheric boundary forcing [16].

Wind forcing also plays an important role on longer time scales primarily affecting the interannual variability through the excitation of Rossby waves in the central Atlantic, which propagate westward and interact with the western boundary. The planetary waves are also modulating the decadal time-scale oceanic response to buoyancy forcing [17]. The dynamic response of the upper ocean layer to fluctuations of a large-scale wind field in the North Atlantic leads to formation of decadal thermal anomalies with opposite sign between the STG and SPG [18]. Therefore, evaluation of the impact of interannual variability of the surface momentum fluxes on circulation and thermal state of the upper ocean is important for understanding the mechanisms of climatic variability in the North Atlantic. At the same time, quantitative evaluation of influence of wind variation on the interannual-to-decadal variability of upper ocean heat balance components requires further clarification.

The purpose of this paper is to estimate the role of varying wind forcing in formation of interannual anomalies of heat balance components within the upper mixed layer (UML) in the North Atlantic in January. The estimation is based on homogeneous and long-term data sets.

2 Data and Methods

We used monthly data on the North Atlantic (0–70° N, 8–80° W) from the ocean reanalyses ORA-S3 for 1959–2011 [19] and GODAS for 1980–2016 [20]. In order to calculate the mixed layer depth in the ORA-S3 reanalysis, a scheme based on the semi-empirical theory of turbulence was applied. The main idea of the scheme is to calculate the Richardson number according to [21]. The mixed layer depth is assumed to be equal to the depth at which the Richardson number reaches a critical value. The mixed layer depth in the GODAS reanalysis is defined as a depth where the temperature deviation from the surface temperature is less than 0.8 °C. Monthly data on the wind stress, latent and sensible heat fluxes, balances of shortwave and longwave radiation on the ocean surface with a spatial resolution of $1° \times 1°$ were taken from the Coordinated Ocean Research Experiments version 2 (COREv2) for 1949–2006 [22]. According to the set of heat flux components, the net surface heat fluxes were calculated for each month. The positive direction of *x/y* axis is eastward/northward, while the positive direction of the *z* axis is upwards. The origin of the coordinate system lies at the undisturbed surface of the ocean.

The calculation procedures of the WSC and heat balance terms are described in detail in [23, 24], respectively. Heat balance terms were estimated for each month for the entire period. Further analysis is made for January. This is the month when the westerlies and trade winds are the most intense and the WSC modulus is the highest over the North Atlantic, except the tropical zone [23]. To analyze the relationship between the variability of wind forcing and the heat balance components on interannual scale, correlation analysis was used. Interannual variability of the average (over the entire STG and its western and central-eastern parts) WSC was analyzed too. This region is located between the coastlines in the North Atlantic (at the east and the west) and between 15° N and 40° N. The chosen coordinates for the northern and southern boundaries are stipulated by the peculiarities of meridional shifting of the anticyclonic WSC area at the intra-annual scale [23].

To determine the dependence of large-scale WSC pattern in the North Atlantic versus the high and low values (exceeding ± 1 standard deviation) of the anticyclonic WSC over the STG region, the composite method was applied. To calculate the composites, the years (1966, 1971, 1972, 1974, 1990, 2002, 2009) with high and the years (1963, 1969, 1979, 1981, 1983, 1992, 2005) with low anticyclonic WSC values over the STG region were chosen, which is about 15% of the total sample size. Then the data for each grid node were averaged for all Januaries, separately for years with high and low anticyclonic WSC values. The obtained average values (composites) for two samples were compared with each other. The composites of wind stress, the UML horizontal current velocity, the net surface heat fluxes, partial derivative of the UML temperature and vertical heat advection at the base of the UML were also calculated for these years. Evaluation of statistical significance for the composites was reduced to assess the significance of difference in the mean of two samples at each grid node: the Student's t-test was used.

3 Results

Consider the relationship between the large-scale WSC and vertical heat advection at the base of the UML. Vertical velocity at the base of the Ekman layer associated with the WSC. Interannual anomalies of vertical heat advection over most of the North Atlantic are almost entirely formed by the anomalies of the vertical velocity at the base of the UML. Correlations between the interannual anomalies of the WSC and vertical heat advection are positive at most of North Atlantic (Fig. 1). Low correlation coefficients are due to the discrepancy between the thickness of the Ekman layer and the mixed layer depth, which for the middle latitudes may differ by an order of magnitude in the winter season. The WSC interannual intensification or weakening over the North Atlantic leads to change in vertical movement rate at the base of the UML and, as a result, alter the vertical heat advection. However, exceptions are tropical latitudes, western boundary currents, the East Greenland Current and most of the Labrador Sea where the impact of the local WSC variability on the vertical heat advection is not important. The influence of the Ekman vertical velocity on vertical heat advection in the Canary upwelling and the Portuguese upwelling is also small. In this case, the upwelling is caused by the alongshore component of the wind-stress due to intense and relatively stable trade winds, but not due to the local WSC.

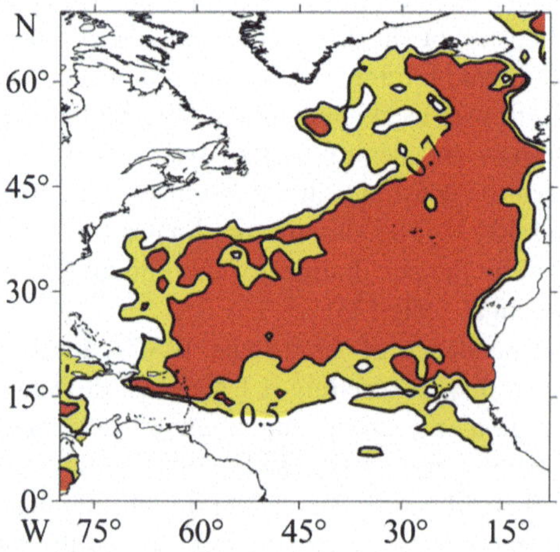

Fig. 1. Correlation coefficients between the wind stress curl and temperature advection caused by anomalous vertical velocity at the base of the UML in the North Atlantic. Correlations were calculated for January 1959–2011. Areas with the values of correlation coefficients above 0.5 are marked yellow and above 0.7 – red.

Figure 2a shows the time series of the UML horizontal heat advection modulus averaged in the latitudinal band 15–40° N according to the ORA-S3 and GODAS January data. These time series agree well. The calculated magnitude of correlation coefficient for 1980–2011 between the time series equals to 0.55. Figure 2b shows the WSC interannual variations over the STG (15–40° N) according to the ORA-S3 data,

GODAS data and COREv2 data for January. An overall good correspondence between these three time series is easily discernible. The cross-spectrum of time series of the UML horizontal heat advection modulus and the anticyclonic WSC has significant spectral peaks at 2–3 and 6–7 years. The oscillations of these characteristics are highly coherent at these periods. The coherence function reaches values greater than 0.8 and 0.7 at the considered scales.

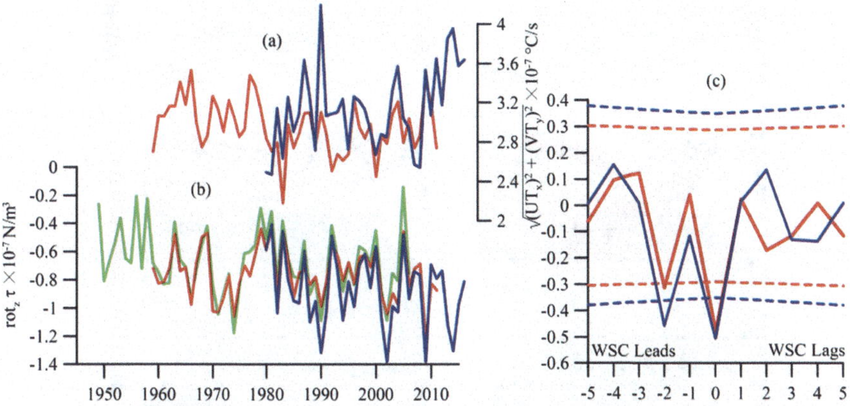

Fig. 2. Interannual variability of the horizontal heat advection modulus (× 10^{-7} °C/s) (a) and wind stress curl (× 10^{-7} N/m^3) (b) over the Subtropical Gyre (in the latitudinal band 15–40° N) for January from COREv2 data (green), ORA-S3 data (red) and GODAS data (blue). Lead-lag correlations between these time series (c). Dashed lines in (c) specify the confidence level of 95%.

Figure 2c shows the lead-lag correlation function between interannual variations of the UML horizontal heat advection modulus and the WSC over the STG region. There is a significant negative synchronous relationship between these time series at the 95% confidence level. In addition, there are significant negative peak with two-year leading of the WSC. It indicates that intensification/weakening of the anticyclonic WSC over the STG region (negative/positive anomaly) is accompanied by an increase/decrease in the UML advective heat transport in the middle latitudes (positive/negative anomaly) with a lag of two years. Thus, the upper-ocean advective heat transport in the central and eastern parts of the STG adjusts quasi-synchronously to the large-scale WSC variations in the mid-latitudes. However, the adjustment of UML horizontal heat advection to interannual WSC variability, which takes place in the western part of the STG, has a lag of two years. The delay in the Gulf Stream reaction to the WSC changes in the North Atlantic is about two years, which confirms the results of paper [9].

The composites of the WSC, wind stress, UML currents velocity, net surface heat fluxes, partial derivative of the UML temperature and vertical heat advection at the base of the UML clearly point to the fact that the atmospheric heat and momentum fluxes pattern, the upper ocean circulation and thermal state in the North Atlantic undergo significant interrelated changes during years with high and low values of the anticyclonic WSC over the STG region (Figs. 3 and 4).

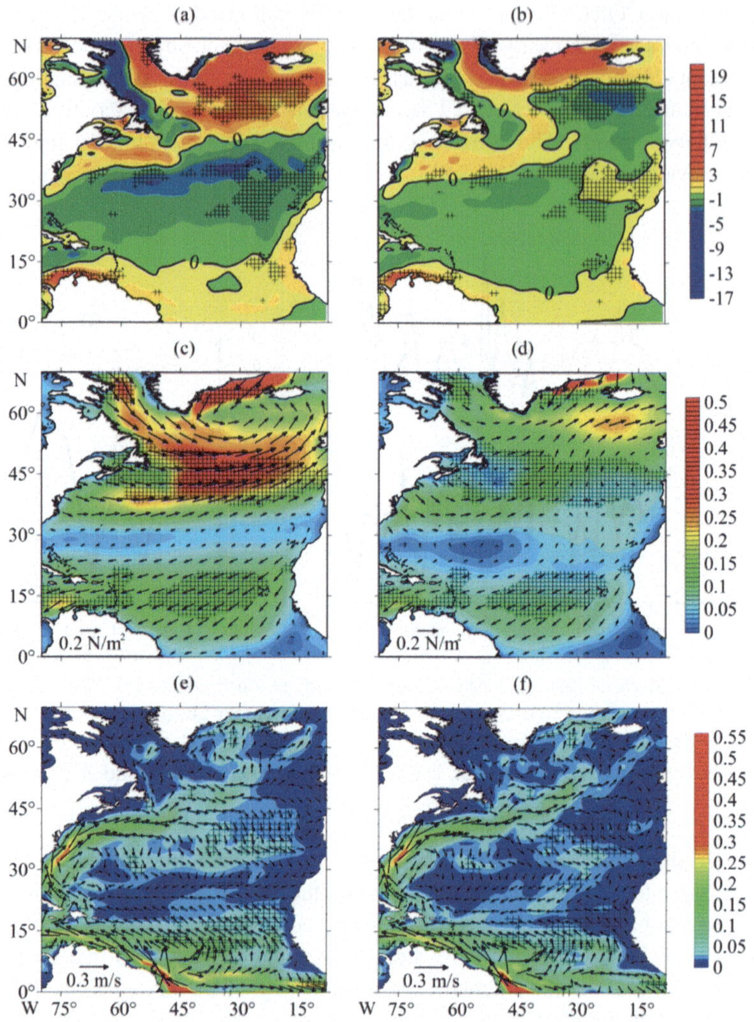

Fig. 3. Wind stress curl (× 10^{-7} N/m^3) (a, b), wind stress (N/m^2) (c, d) and UML currents velocity (m/s) (e, f) during years with high (a, c, e) and low (b, d, f) wind stress curl values over Subtropical Gyre region (in the latitudinal band 15–40° N) for January. The pluses specify the grid nodes where the composite difference is significant at the 95% confidence level.

The anticyclonic WSC area in the northern tropics and the subtropical zone intensifies and becomes narrower during years with high absolute values of the anticyclonic WSC over the STG region. Note that intensification of the subtropical oceanic gyre entails its radial contraction that was first mentioned by Iselin [25]. The WSC maximum is located over the Azores and corresponds to the Azores high. On the contrary, the anticyclonic WSC area is significantly weaker in the WSC magnitude and

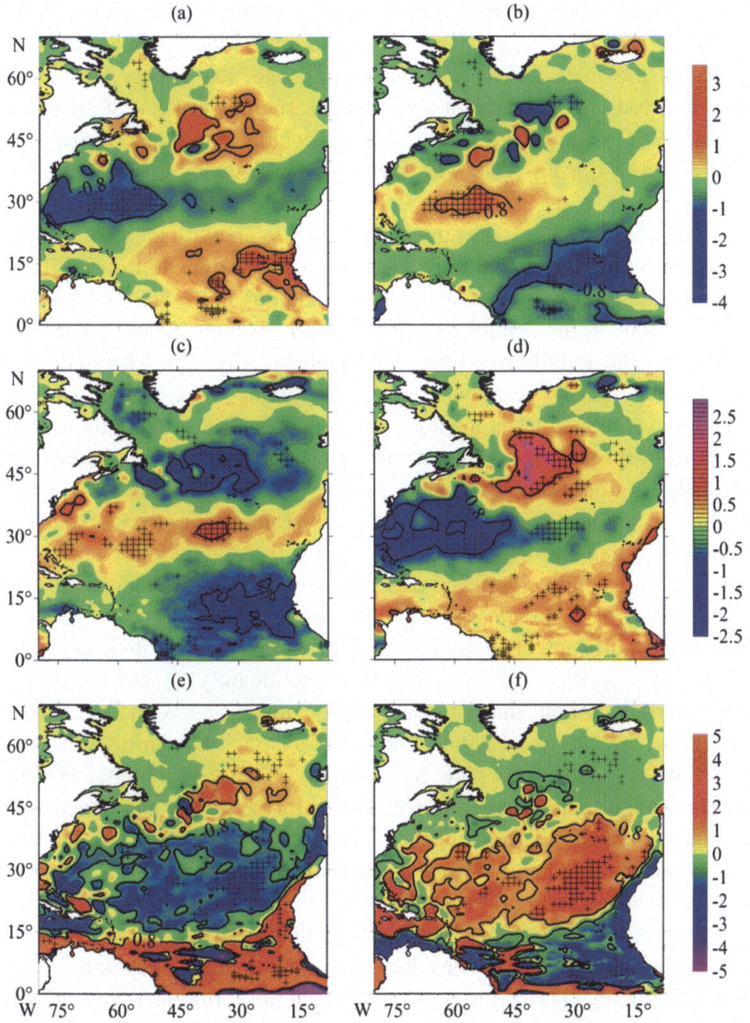

Fig. 4. Composite anomalies of heat fluxes at the sea surface Q_0 normalized by $\rho_0 C_P H$ ($\times\ 10^{-7}$ °C/s) (a, b), partial derivative of the UML temperature T_t ($\times\ 10^{-7}$ °C/s) (c, d) and vertical heat advection at the base of the UML ($\times\ 10^{-7}$ °C/s) (e, f) during years with high (a, c, e) and low (b, d, f) wind stress curl values over the Subtropical Gyre region (in the latitudinal band 15–40° N) for January. The isolines correspond to values $\pm 0.8 \times 10^{-7}$ °C/s. Pluses specify the grid nodes where the composite difference is significant at the 95% confidence level.

expands in the southern (up to 10° N) and northern (up to 60° N) directions during years with low values of the anticyclonic WSC over the STG region. The WSC maximum is located over the Bermuda and corresponds to the Bermuda high. In addition to that, the anticyclonic area "moves away" from the African continent during these years. The cyclonic WSC intensifies in the tropical and subpolar latitudes and the areas with cyclonic WSC increase during years with high anticyclonic WSC values over the STG region. Significant (95% level) differences on the WSC composites are distinguished in the Inter-Tropical Convergence Zone, the Azores high and the Icelandic low. In these areas, there is a significant variation of the WSC magnitude or a change of the WSC nature from anticyclonic to cyclonic (because of displacement of the zero WSC lines) during years with high and low values of the anticyclonic WSC over the STG region.

Both trade winds and westerlies intensify during years with high anticyclonic WSC values over the STG region. These winds have a zonal direction during these years. Moreover, the highest values of the wind stress modulus in the areas of the trade winds and westerlies focus in the central part of the North Atlantic. These values are almost twice as high as the long-term average for January (Fig. 3c). The trade winds and westerlies weaken during years with low anticyclonic WSC values over the STG region. The wind stress values in the trade winds do not differ essentially from the long-term average values for January. The maximum wind stress values in the westerlies are observed in the northeastern part of the North Atlantic. The westerlies are directed predominantly in the northeastern direction. This feature leads to an expansion of the anticyclonic WSC area in the eastern part of the North Atlantic to the north – up to 60° N (Fig. 3b, d). Statistically significant differences of wind stress composites take place mainly in the areas of the trade winds and the westerlies.

Statistically significant changes of the UML current velocities are also visible (Fig. 3 e, f). The current velocities increase significantly in the North Equatorial Current, the Azores current, the East Greenland Current, the West Greenland Current and in the Gulf Stream recirculation zone (30° N 55° W) during years with high anticyclonic WSC values over the STG region. These years also correspond to intensification in the trade winds and westerlies. In the above-mentioned regions the UML horizontal temperature gradients are small in magnitude and vary little. Thus, for these regions the wind-induced change in current velocities should lead to significant changes in the UML horizontal heat advection.

The increased ocean heat loss (the net heat flux positive anomaly) occurs in the areas of the trade winds and westerlies during years with high anticyclonic WSC values over the STG region. The decreased ocean heat loss (the net heat flux negative anomaly) occurs in the vicinity of 30° N. The net heat flux anomalies opposite in sign are formed during years with low anticyclonic WSC values over the STG region (Fig. 4 a, b). The structure of composites of partial derivative of the UML temperature anomalies is compatible with the composites of net heat fluxes anomalies in January. Negative anomaly of partial derivative of the UML temperature occurs in the tropics and high latitudes. It means that UML cooling intensifies during years with increased wind circulation over the STG region. Positive anomaly of partial derivative of the UML temperature occurs in the central part of the STG. It means less intensive UML cooling in winter. Anomalies of partial derivative of the UML temperature with opposite sign form during years with low anticyclonic WSC values over the STG region (Fig. 4 c, d). Statistically significant differences of composites of interannual anomalies of net heat fluxes and partial derivative of the UML temperature take place mainly in the vicinity of the Guiana Current, the central and eastern part of the Inter-Tropical Convergence Zone, the central part of the STG and the westerlies area. In these areas, anomalies with opposite sign appear during years with high and low anticyclonic WSC values over the STG region.

Figure 4 e, f shows the pattern of vertical heat advection anomalies composites at the base of the UML during years with high and low anticyclonic WSC values over the STG region. Despite the noisiness of the anomalous structure, one can distinguish some large-scale features. The magnitude of vertical heat advection anomalies is high in the tropics and subtropics. In these regions, the contribution of vertical heat advection to the total UML heat budget is significant. Anomalies of vertical heat advection are small in high latitudes (where the mixed layer depth is large). In these regions, the contribution of vertical heat advection to the total UML heat budget is small. The positive anomaly of vertical heat advection occurs in the vicinity of the Inter-Tropical Convergence Zone, and the negative anomaly – in the STG region during years with high anticyclonic WSC values over the STG region. The opposite in sign vertical heat advection anomalies occurs during years with low anticyclonic WSC values over the STG region. In general, such anomaly pattern corresponds to the spatial distribution of the correlation values shown in Fig. 1.

Thus, the increased heat loss from the ocean to the atmosphere and increased UML cooling take place during years with more intensive wind circulation in tropical and subpolar latitudes in January. On the contrary, heat loss from the ocean to the atmosphere and the UML cooling decrease in these years in the central part of the STG. The vertical heat advection at the base of the UML also intensifies in the North Atlantic.

4 Discussion and Conclusion

The results of the fundamental study [26] showed that the rapid year-to-year sea surface temperature variability results from the ocean–atmosphere heat fluxes that are forced by changing winds. In the STG this variability can arise as a response to long-term atmospheric forcing, specifically the changes associated with the strength and location of the subtropical high. This fact fully conforms to our results.

The described changes in the WSC spatial structure over the North Atlantic can be explained by the features of large-scale circulation processes in the ocean-atmosphere system [27]. The mechanism of ocean impact on the atmosphere in the North Atlantic, which is confirmed in this paper, is characterized by a positive feedback.

The cyclonic WSC in the tropics and subpolar latitudes and anticyclonic WSC in the subtropics simultaneously increase during years with high values of the anticyclonic WSC over STG region. Intensification of oceanic and atmospheric circulation causes an excessive upper ocean cooling due to additional heat release from the ocean into the atmosphere. Vertical heat transport at the base of the UML and turbulent mixing of the surface waters concurrently become more intensive. Intensification of convective processes in the Labrador Sea (increase in the heat fluxes at the base of the UML) and the increase of zonal circulation in the atmosphere also occur. Moreover, it agrees with the fact about the quasi-synchronous intensification of the Azores high and the Icelandic low [28], at the same time it does not contradict the condition of absolute vorticity conservation in the atmosphere. Intensification of zonal circulation should be accompanied with the emergence of additional instability and synoptic activity increase in the atmosphere. In addition, intensification of large-scale wind field and the WSC and, as a consequence, the increase in surface currents velocities lead to the delayed intensification of nonlocal processes in the upper ocean, such as advective heat transports and horizontal eddy diffusion of heat. The timescale response of western boundary currents to the wind forcing lasts up to several years. In turn, interannual anomalies of the net surface heat fluxes, which are mostly induced by varying winds, contribute to formation of opposite (in sign) anomalies of the UML temperature partial derivative in the STG and SPG.

The results obtained in this paper are the following:

The interannual fluctuations of large-scale WSC have a significant effect on variability of vertical heat advection at the base of the UML at most of the North Atlantic. The 50–60% vertical heat advection interannual variability is caused by the WSC variability on the extended part of the North Atlantic in winter. Exceptions are the latitudinal band 0–15° N, the vicinity of intensive currents, Labrador Sea and the Canary upwelling.

Intensification of atmospheric circulation synchronously causes increased ocean circulation and excessive UML cooling. The common lag of thermal oceanic response (advective heat transport) to the atmospheric forcing (anticyclonic WSC) is two years for the western part of the Subtropical Gyre.

The large-scale patterns of wind stress and the WSC undergo significant changes at the interannual-to-decadal scale. Variations in the magnitude of the WSC over the STG region are not only caused by an increase or decrease in wind speed, but also changes in the wind field structure. These changes also reflect in the upper ocean circulation, the net heat fluxes anomalies and the UML temperature partial derivative. The large-scale WSC field intensification (weakening), the meridional dimensions decrease (increase) of anticyclonic WSC area, the excessive (moderate) ocean heat release and excessive (moderate) UML cooling occur in the North Atlantic during years with high (low) anticyclonic WSC values over the subtropical latitudes in January. Meanwhile, the cyclonic WSC areas, the ocean heat loss and the UML cooling increases (decreases) in the tropical and subpolar latitudes. There is also a significant increase (decrease) in the UML currents velocities and vertical heat transports at the base of the UML during these years in the North Atlantic.

The results obtained in the paper confirm that interannual variability of atmospheric momentum fluxes plays an important role in formation of circulation and thermal state of upper ocean layer in the North Atlantic.

Acknowledgements. The research was partially supported by grant No. 17-35-50028 of the Russian Foundation for Basic Research (RFBR).

References

1. Stommel, H.M.: The Gulf Stream: A Physical and Dynamical Description, p. 248. University of California Press, Berkeley and Los Angeles (1958)
2. Sturges, W., Hong, B.G., Clarke, A.J.: Decadal wind forcing of the North Atlantic subtropical gyre. J. Phys. Oceanogr. **28**(4), 659–668 (1998)

3. Häkkinen, S., Rhines, P.B.: Shifting surface currents in the northern North Atlantic Ocean. J. Geophys. Res. **114**(C04005) (2009). https://doi.org/10.1029/2008JC004883
4. White, M.A., Heywood, K.J.: Seasonal and interannual changes in the North Atlantic subpolar gyre from Geosat and TOPEX/POSEIDON altimetry. J. Geophys. Res. **100**(C12), 24931–24941 (1995). https://doi.org/10.1029/95JC02123
5. Dong, S., Kelly, K.A.: Heat budget in the Gulf Stream region: the importance of heat storage and advection. J. Phys. Oceanogr. **34**(5), 1214–1231 (2004)
6. Peterson, I., Greenan, B., Gilbert, D., Hebert, D.: Variability and wind forcing of ocean temperature and thermal fronts in the Slope Water region of the Northwest Atlantic. J. Geophys. Res. Oceans **122**(9), 7325–7343 (2017). https://doi.org/10.1002/2017JC012788
7. Han, G., Tang, C.L.: Interannual variations of volume transport in the western Labrador Sea based on TOPEX/Poseidon and WOCE data. J. Phys. Oceanogr. **31**(1), 199–211 (2001)
8. DiNezio, P.N., Gramer, L.J., Johns, W.E., Meinen, C.S., Baringer, M.O.: Observed interannual variability of the Florida Current: wind forcing and the North Atlantic Oscillation. J. Phys. Oceanogr. **39**(3), 721–736 (2009). https://doi.org/10.1175/2008JPO4001.1
9. Dzhiganshin, G.F., Polonsky, A.B.: Low-frequency variations of the Gulf Stream transport: description and mechanisms. Phys. Oceanogr. **19**(3), 151–169 (2009). https://doi.org/10.1007/s11110-009-9047-5
10. Orvik, K.A., Skagseth, O.: The impact of the wind stress curl in the North Atlantic on the Atlantic inflow to the Norwegian Sea toward the Arctic. Geophys. Res. Lett. **30**(17), 1884 (2003). https://doi.org/10.1029/2003GL017932
11. Spall, M.A., Pickart, R.S.: Wind-driven recirculations and exchange in the Labrador and Irminger Seas. J. Phys. Oceanogr. **33**(8), 1829–1845 (2003). https://doi.org/10.1175/2384.1
12. Luksch, U.: Simulation of North Atlantic low-frequency SST variability. J. Clim. **9**(9), 2083–2092 (1996)
13. Czaja, A., Marshall, J.: Observations of atmosphere-ocean coupling in the North Atlantic. Quart. J. Roy. Meteor. Soc. **127**(576), 1893–1916 (2001). https://doi.org/10.1002/qj.49712757603
14. Hirschi, J.J.-M., Blaker, A.T., Sinha, B., Coward, A.C., de Cuevas, B.A., Alderson, S.G., Madec, G.: Chaotic variability of the meridional overturning circulation on subannual to interannual timescales. Ocean Sci. **9**(5), 805–823 (2013). https://doi.org/10.5194/os-9-805-2013
15. Roberts, C.D., Waters, J., Peterson, K.A., Palmer, M.D., McCarthy, G.D., Frajka-Williams, E., Haines, K., Lea, D.J., Martin, M.J., Storkey, D., Blockley, E.W., Zuo, H.: Atmosphere drives recent interannual variability of the Atlantic meridional overturning circulation at 26.5 N. Geophys. Res. Lett. **40**(19), 5164–5170 (2013). https://doi.org/10.1002/grl.50930
16. Thomas, M.D., Zhai, X.: Eddy-induced variability of the meridional overturning circulation in a model of the North Atlantic. Geophys. Res. Lett. **40**(11), 2742–2747 (2013). https://doi.org/10.1002/grl.50532
17. Polo, I., Robson, J., Sutton, R., Balmaseda, M.A.: The importance of wind and buoyancy forcing for the boundary density variations and the geostrophic component of the AMOC at 26 N. J. Phys. Oceanogr. **44**(9), 2387–2408 (2014). https://doi.org/10.1175/JPO-D-13-0264.1
18. Williams, R.G., Roussenov, V., Smith, D., Lozier, M.S.: Decadal evolution of ocean thermal anomalies in the North Atlantic: the effects of Ekman, overturning, and horizontal transport. J. Clim. **27**(2), 698–719 (2014). https://doi.org/10.1175/JCLI-D-12-00234.1
19. Balmaseda, M.A., Vidard, A., Anderson, D.L.T.: The ECMWF ocean analysis system: ORA-S3. Mon. Wea. Rev. **136**(8), 3018–3034 (2008). https://doi.org/10.1175/2008MWR2433.1

20. Behringer, D.W., Xue, Y.: Evaluation of the global ocean data assimilation system at NCEP: the Pacific Ocean. In: Proceedings Eighth Symposium on Integrated Observing and Assimilation Systems for Atmosphere, Oceans, and Land Surface, Seattle, WA, pp. 11–15. American Meteor Society (2004). http://ams.confex.com/ams/pdfpapers/70720.pdf
21. Pacanowski, R.C., Philander, S.G.H.: Parameterization of vertical mixing in numerical models of tropical oceans. J. Phys. Oceanogr. **11**(11), 1443–1451 (1981)
22. Large, W.G., Yeager, S.G.: The global climatology of an interannually varying air-sea flux data set. Clim. Dyn. **33**(2), 341–364 (2009). https://doi.org/10.1007/s00382-008-0441-3
23. Polonsky, A.B., Sukhonos, P.A.: Variability of the wind stress, the field of currents, wind stress curl and vorticity of surface currents in the North Atlantic. Izvestiya Rossiiskoi Akademii Nauk, Seriya Geograficheskaya **1**, 62–73 (2017). https://doi.org/10.15356/0373-2444-2017-1-62-73
24. Polonsky, A.B., Sukhonos, P.A.: Interannual variations in the components of heat budget in the upper layer of the North Atlantic in different seasons. Izv. Atmos. Oceanic Phys. **53**(4), 459–466 (2017). https://doi.org/10.1134/S0001433817040107
25. Iselin, C.O'D.: Preliminary report on long-period variations in the transport of the Gulf Stream System. Papers in Phys. Oceanogr. and Meteor. **8**(1), 1–40 (1940). https://doi.org/10.1575/1912/1048
26. Bjerknes, J.: Atlantic air-sea interaction. Adv. Geophys. **10**, 1–82 (1964). https://doi.org/10.1016/S0065-2687(08)60005-9
27. Polonskii, A.B.: The Role of the Ocean in Climate Changes, 183 p. Naukova Dumka, Kiev (2008)
28. Mächel, H., Kapala, A., Flohn, H.: Behaviour of the centres of action above the Atlantic since 1881. Part I: Characteristics of seasonal and interannual variability. Int. J. Climatol. **18** (1), 1–22 (1998)

Numerical Study of Haline Convection in a Porous Domain with Application for Geothermal Systems

E. B. Soboleva(✉)

Ishlinsky Institute for Problems in Mechanics RAS, Moscow, Russia
soboleva@ipmnet.ru

Abstract. We study stochastic haline convection in a porous domain in a context of dissolved admixture transport problems in geothermal reservoirs. An isotropic homogeneous rectangular porous domain divided by a horizontal interior layer at lower porosity and permeability is considered. The upper boundary is assumed to be an admixture source leading to admixture diffusion into a volume and a development of gravity-driven haline convection. Numerical simulations of convective flows and mass transfer based on the continuity, Darcy's and admixture transport equations are performed. An influence of properties of interior layer on the convection structure and dynamic characteristics is investigated.

Keywords: Haline convection · Porous medium · Geophysical flows · Geothermal reservoir · Numerical simulation

1 Introduction

Many fields of human activity such as ecology, environmental management, mining industry need an information on admixture transport in groundwater. Salts dissolved in water move in soils and rocks in the Earth because of different reasons. Usually, the admixture transport is associated with forced or natural modes of convection and the last one is typically caused by the temperature and salinity gradients leading in the gravity field to a development of motions [1, 2]. In special cases, a growth of instability and motions arisen can be independent of gravity, for example, under the capillary pressure gradient [3–5] which, however, is not considered in the present study. Our study is devoted to natural gravity-driven convection associated with an admixture transport in a geothermal reservoir; the geothermal system is an alternative energy source using deep heat of Earth [6]. Similar problems on convective admixture transport are solved in the context of carbon capture and sequestration technologies [7, 8]. Groundwater with dissolved admixture is involved in thermal gravity-driven convection under the geothermal gradient resulted from the temperature rising in the direction to the hot Earth's core; this gradient is normally about 20–30K per each kilometer [9]. However, if salinity gradients are sufficiently high, haline (or as is also called, density-driven) convection can be more intensive in a comparison with thermal convection. Estimations show that, for a domain with the vertical size of 10 m, haline

V. I. Karev et al. (Eds.): *Physical and Mathematical Modeling of Earth and Environment Processes (2018)*, SPEES, pp. 63–74, 2019.
https://doi.org/10.1007/978-3-030-11533-3_7

convection is dominant starting with concentration differences in the domain of about 6.10^{-4} [10]; parameters of water correspond to the real temperature and pressure in geothermal systems being of 500K and 150 atm, respectively. The concentration differences exceeding 6.10^{-4} are observed in many real cases in which only haline convection can be taken into consideration disregarding thermal convection.

In this paper, we investigate numerically haline convection in a porous rectangular reservoir initially filled with pure motionless water. The constant admixture concentration is held at the upper boundary modeling a close location of salt source. An admixture diffuses into a volume forming a dense brine near the boundary which tries to fall so it gives rise to a development of haline convection. A domain consists of homogeneous isotropic porous medium divided by an interior horizontal layer at lower porosity and permeability. An effect of interior layer on the convection structure and dynamic characteristics is investigated. Here, we continue our study of haline convection in heterogeneous domains starting with [10, 11]. We consider wide interior layers, convective motions inside them and analyze spectrums of convection structures.

2 Mathematical Model and Numerical Method

We consider a homogeneous porous domain divided by a horizontal interior layer at different porosity and permeability (see Fig. 1). Initially, the domain is filled with pure motionless water. The constant admixture concentration is held at the upper boundary providing the admixture to diffuse into the volume. The solute density is lineally related to the admixture concentration so the dense brine is formed near the upper boundary. This brine tries to fall under the gravity force giving rise to a development of haline convective flows and mass transfer.

Haline convection in a porous domain is described by the governing equations, which are the continuity, Darcy's and admixture transport equations. The governing equations can be written in a dimensionless form as follows

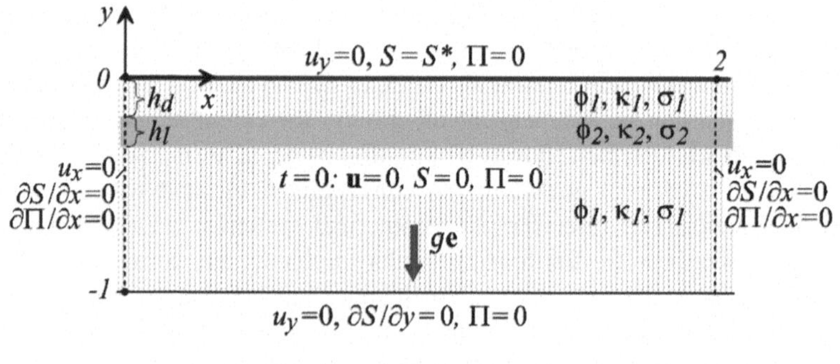

Fig. 1. Problem under study.

$$\nabla \cdot \mathbf{u} = 0 \quad (1)$$

$$\mathbf{u} = -Rd\, \kappa(\nabla \Pi - S \cdot \mathbf{e}) \quad (2)$$

$$\phi \frac{\partial S}{\partial t} + \mathbf{u} \cdot \nabla S = \nabla(\sigma\, \phi \nabla S) \quad (3)$$

The equations include $\mathbf{u} = \mathbf{U}\, H/D_c$, $\Pi = (P - P_{in})/(\alpha\, \rho_c^{sat} gH)$, and $S = \rho_c/\rho_c^{sat}$ to be the dimensionless filtration velocity, pressure, and admixture density, respectively. The velocity has the components: $\mathbf{u} = (u_x,\, u_y)$. Here, $\mathbf{U}$, P, P_{in}, ρ_c, ρ_c^{sat}, H, D_c, α, and g are the physical filtration velocity, actual and initial values of pressure, actual and maximal (corresponding to saturated brine) values of admixture density, height of domain, coefficient of molecular diffusion, salinity expansion coefficient, and module of gravity acceleration. The geometrical size and time are normalized by H and H^2/D_c, respectively. The initial and boundary conditions are given in Fig. 1.

Dimensionless parameters of the system are the Rayleigh-Darcy number Rd, relative permeability κ and coefficient of diffusion σ, as well as the porosity ϕ and the characteristic density S^*. From the definition

$$Rd = \frac{\alpha \rho_c^{sat} g k_1 H}{\mu_s D_c}, \kappa = k/k_1 \sigma = D/D_c \quad (4)$$

Subscripts "1" and "2" denote the main medium and interior layer, respectively. Superscripts "*" and "*sat*" denote the upper boundary and saturated brine. The conditions at the boundaries of interior layer are

$$\Pi_2 = \Pi_1, S_2 = S_1, u_{x2} = u_{x1}\kappa_2, u_{y2} = u_{y1}, w_{x2} = w_{x1}\kappa_2 \frac{\phi_1}{\phi_2}, w_{y2} = w_{y1}\frac{\phi_1}{\phi_2} \quad (5)$$

Here, w_x and w_y are the components of the fluid velocity $\mathbf{w} = (w_x,\, w_y) = \mathbf{u}/\phi$. Notice, a flow direction has to change when flow lines cross the interface that is resulted from the conditions (5).

Haline convection is simulated at parameters corresponding to real geological conditions: ρ_0 = 842.5 kg m^{-3}, μ_s = 1.212.10^{-4} Pa s, c^{sat} = 0.3417, D_c = 3.10^{-9} m^2 s^{-1}, α = 0.815. We put ϕ_1 = 0.2, k_1 = 1.0.10^{-13} m^2, D_1 = 1.58.10^{-9} m^2 s^{-1}, ϕ_2 = 0.1, k_2 = 1.0.10^{-14} m^2, D_2 = 1.19.10^{-9} m^2 s^{-1}. The height of domain is H = 25 m, the distance from the upper boundary to the interior layer is H_d = 1.25 m. The layer width H_l varies. The appropriate dimensionless quantities are Rd = 2.19.10^4, κ_1 = 1, σ_1 = 0.525, κ_2 = 0.1, σ_2 = 0.398, h_d = 0.05. The dimensionless layer width is h_l = 0; 0.032; 0.11. We put admixture concentration at the top c^* = 0.04 that is equivalent to the dimensionless density $S^* = \frac{(1-\alpha c^{sat})c*}{c^{sat}(1-\alpha c*)} = 0.0873$. The aspect ratio of considered domain equals to 2.

Simulations are performed with the use of 2D finite-difference numerical code employed during several years to the problems of underground haline convection [10, 12–14]. The complicated mathematical model and last version of the code including tests are described in [8]. To obtain accurate numerical results we use very fine space grids up to 5000×2500.

3 Results and Discussion

An admixture diffuses from the upper boundary to the volume so brine locating nearby is denser then pure water under this brine. Brine is unstable in the gravity field and tries to fall that leads to developing haline gravity-driven convection which is stochastic at considered parameters. The evolution of convective flows in a domain containing an interior layer at the width H_l = 0.8 m (or h_l = 0.032 in dimensionless form) is shown in Fig. 2. The interior layer is displayed in Figs. 2 and 4 with gray color. In the initial stage, brine fingers travelling from the upper boundary and filling the space above the interior layer completely occur (Fig. 2(a)). The layer passes an admixture difficultly because the permeability is lowered ten times comparing with a main medium. When an admixture reaches the layer, the maximal fluid velocity $|\mathbf{W}_{max}|$ characterizing the intensity of motions falls comparing with the case of no layer; the fluid velocity is defined as $\mathbf{W} = \mathbf{U}/\phi$. This trend is exhibited in Fig. 3: the curve 2 deviates from the curve 1 corresponding to the case of no layer and goes to bottom starting from t $\approx$ 400 days. In the time period t $\approx$ 400–800 days, an admixture is transported throughout the layer by forced mode of convection. The motions of fluid in the layer are downward everywhere and the velocity is lowered due to the conservation law: no fluid comes to the space above the interior layer. Later, we observe crucial changing in the convection pattern that is resulted from convection in the layer to transform to natural. Some pure water starts to rise from the space below the layer, cross the layer and fill the space above the layer leaved by brine. Rising fingers of pure water split brine above the layer into a set of roughly periodic "brine balls" (Figs. 2(b–d)). This stage is characterized by an intensification of motions as an inflow of pure water makes up for an outflow of brine. As visible in Fig. 3, the velocity $|\mathbf{W}_{max}|$ increases starting from t $\approx$ 800 days and even exceeds the appropriate value in a domain with no layer after t $\approx$ 2000 days. The convection pattern in a domain as a whole is rearranged with time as well. We see small brine fingers travelling below the interior layer in the initial stage (Fig. 2(a)) which later merge and are located less often (Fig. 2(b)). To the time t $\approx$ 2550 days and later, the brine fingers resemble jets taking off the layer (Fig. 2(c, d)).

For a comparison, a domain with a wider interior layer at H_l = 2.75 m (or h_l = 0.11 in dimensionless form) is considered. Convection patterns are shown in Fig. 4. The main features of convection evolution is the same as at H_l = 0.8 m discussed early. Initially, an admixture fills the space above the layer completely that is associated with a forced mode of convection in the layer (Fig. 4(a)). Later, fluid motions here are rearranged, some extended zones of pure water appear (Fig. 4(b, c)) so that pure water

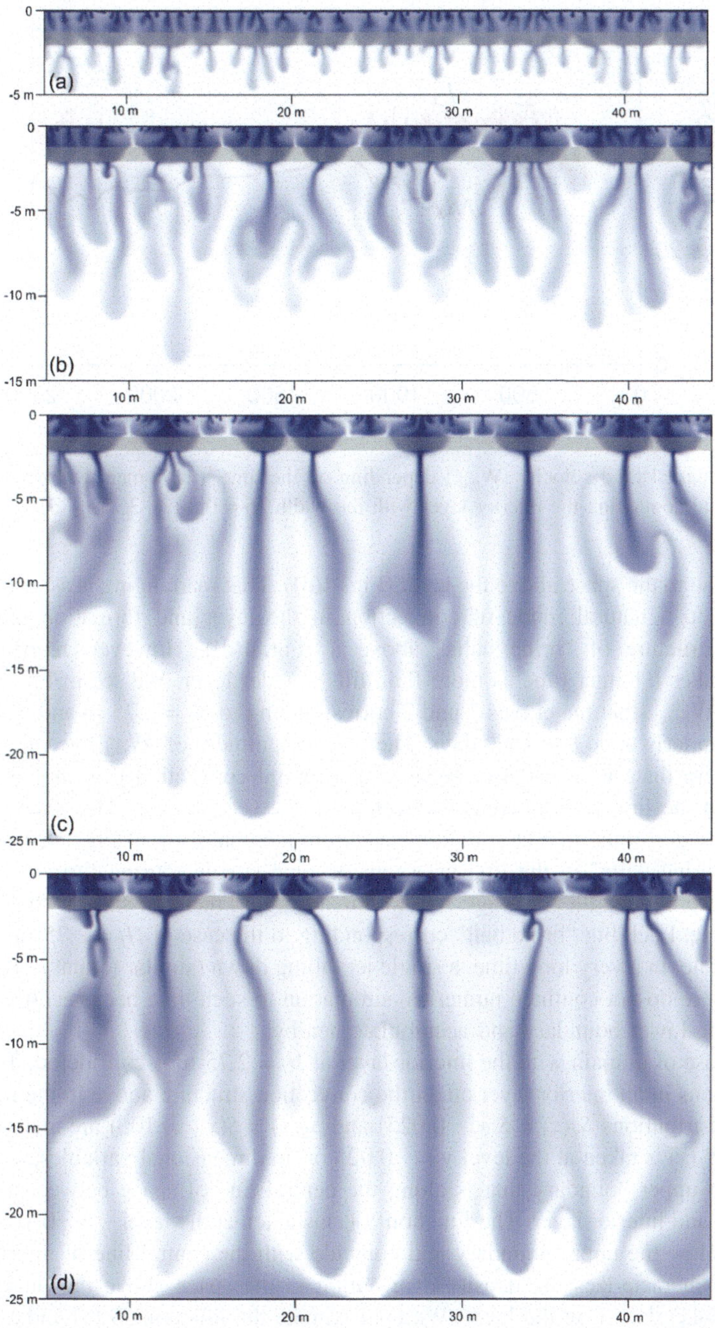

Fig. 2. Concentration field in a domain containing an interior layer at the width $H_l = 0.8$ m at times t = 675 (a), 1500 (b), 2550 (c), 3150 days (d).

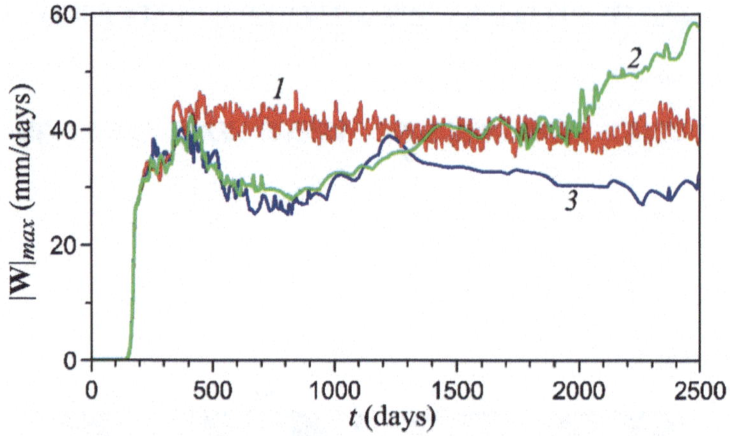

Fig. 3. Maximal fluid velocity $|\mathbf{W}_{max}|$ depending on the time in a domain without an interior layer (*1*) and containing the interior layer with the width H_l = 0.8 (*2*), 2.75 m (*3*).

starts to rise to the space above the layer (Fig. 4(d)). That rearranging is associated with transition to a natural mode of convection in the layer and formation of roughly periodic structure of "brine balls" above it (Fig. 4(d)). However, rearranging of motions takes a longer time as more fluid fills a wider layer. We see in Fig. 3 for the velocity $|\mathbf{W}_{max}|$, that the curves 3 and 2 corresponding to H_l = 2.75 m and H_l = 0.8 m behave similarly up to t ≈ 1300 days. Then, the magnitudes of $|\mathbf{W}_{max}|$ become lower at H_l = 2.75 m than at H_l = 0.8 m because the natural convection does not manage to develop in the first case whereas it develops in the second case. The scale of "brine balls" is bigger in the first case that is clear from a comparison of Figs. 4(d) and 2(d). Maybe, "brine balls" in the first case will become smaller during further evolution, however we believe they average size to remain bigger then the size of "balls" in the second case. Each big "brine ball" corresponding to the case of H_l = 2.75 m, perhaps, will generate, in a very long time, a single jet falling down (similar to that in Fig. 2(d)), however, we do not continue numerical simulation to such time because an admixture reaches the lower boundary and accumulate nearby.

The case of domain with the interior layer at H_l = 2.75 m is considered. To exhibit how motions in an interior layer effect the convection structure above it, the horizontal density distributions $S_1(x) = S(x, -0.025)$ and $S_2(x) = S(x, -0.100)$ are analyzed. The function $S_1(x)$ is taken at the level $y = -0.025$ or, in dimensional variables, at 0.625 m below the upper boundary that is along the central line of space between the upper boundary and interior layer. The function $S_2(x)$ is taken at the level $y = -0.100$ to be at 2.50 m below the upper boundary and coincide with the central line of interior layer. So, $S_1(x)$ characterizes the density distribution in the space above the interior layer whereas $S_2(x)$ does it in the layer. We approximate the functions $S_1(x)$ and $S_2(x)$ with

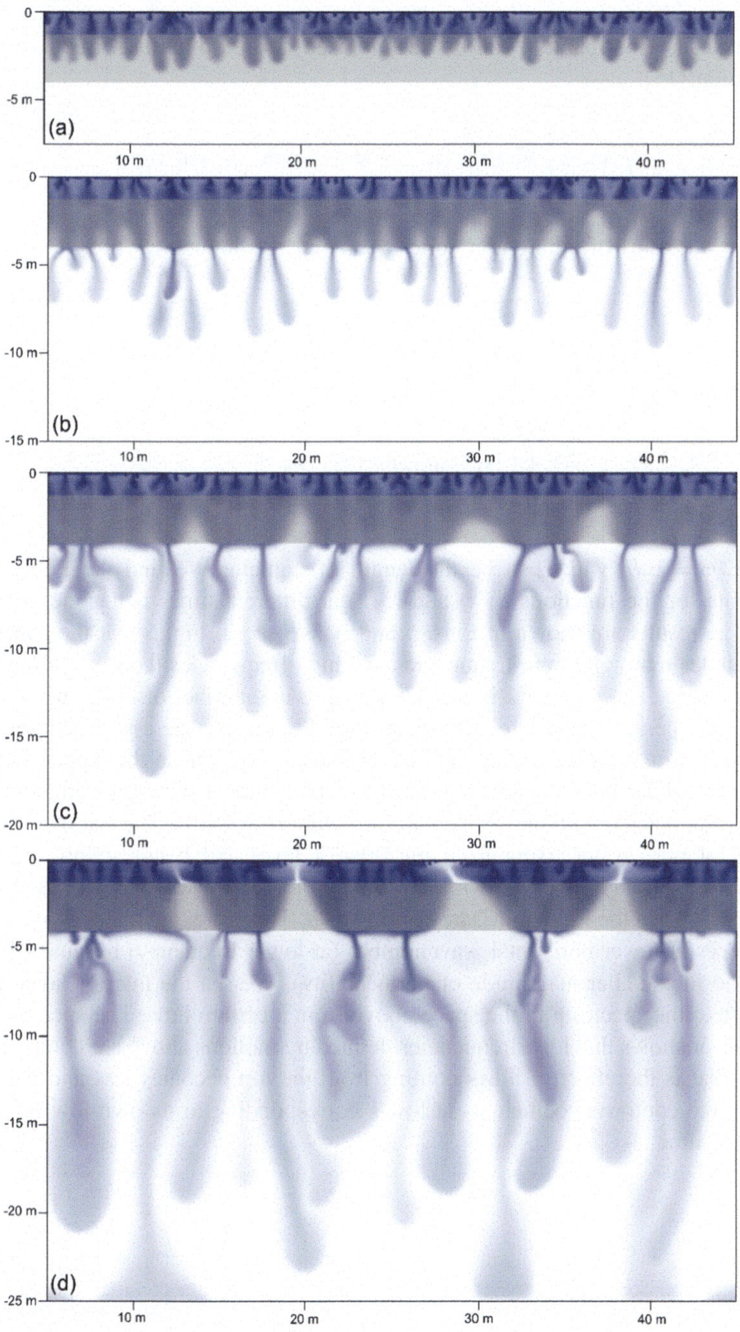

Fig. 4. Concentration field in a domain containing an interior layer with the width $H_l = 2.75$ m at times $t = 675$ (a), 1275 (b), 1950 (c), 3150 days (d).

the Fourier series to find the frequencies of periodic space structures. Let the function $f(x)$ be known. This function can be decomposed into the Fourier series according to the equation [15]

$$f(x) = \frac{1}{2}a_0 + \sum_{k=1}^{\infty} a_k \cos(k\omega x) + \sum_{k=1}^{\infty} b_k \sin(k\omega x) \tag{6}$$

The coefficients of the Fourier series a_0, a_k, b_k are defined as follows

$$\begin{aligned} a_0 &= \frac{2}{L}\int_{-L/2}^{L/2} f(\xi)\,d\xi, a_k = \frac{2}{L}\int_{-L/2}^{L/2} f(\xi)\cos(k\omega\xi)d\xi, \\ b_k &= \frac{2}{L}\int_{-L/2}^{L/2} f(\xi)\sin(k\omega\xi)d\xi \end{aligned} \tag{7}$$

Here, $\omega = 2\pi/L$ with L to be the length of domain. We find the Fourier series coefficients for the functions $S_I(x)$ and $S_2(x)$ at different times being the same as in Fig. 4. The results are shown in Figs. 5 and 6. We see in Fig. 5 that the spectrum of $S_I(x)$ includes many different frequencies and high frequencies (corresponding to high k) are available. The number k is roughly equal to the number of repeating fragments, which form a periodic structure. Therefore, high frequencies are associated with small-scale space structures generated by the boundary layer near the upper boundary. However, at a long time, the coefficients of low frequency at $k \approx 4$ is dominant (Fig. 5 (d)) that is associated with the "brine ball" structure (Fig. 4(d)). The last one is induced by flows of pure water rising from the interior layer and being at low frequencies generated by this layer. As known, the average horizontal wavenumber is decreased with decreasing in the Rayleigh-Darcy number Rd. Therefore, the interior layer induces disturbances at lower horizontal wavenumber (at lower frequency) because of smaller Rd (due to the smaller magnitude of permeability). We see the interior layer starts to contribute to the spectrum in the space above from a certain time. The spectrum in the interior layer shows that high frequencies defined by motions above are damped in time (Fig. 6), that is the effect of flows coming from the top becomes less and less significant. On the contrary, coefficients at low k corresponding to its own low frequencies increase in time.

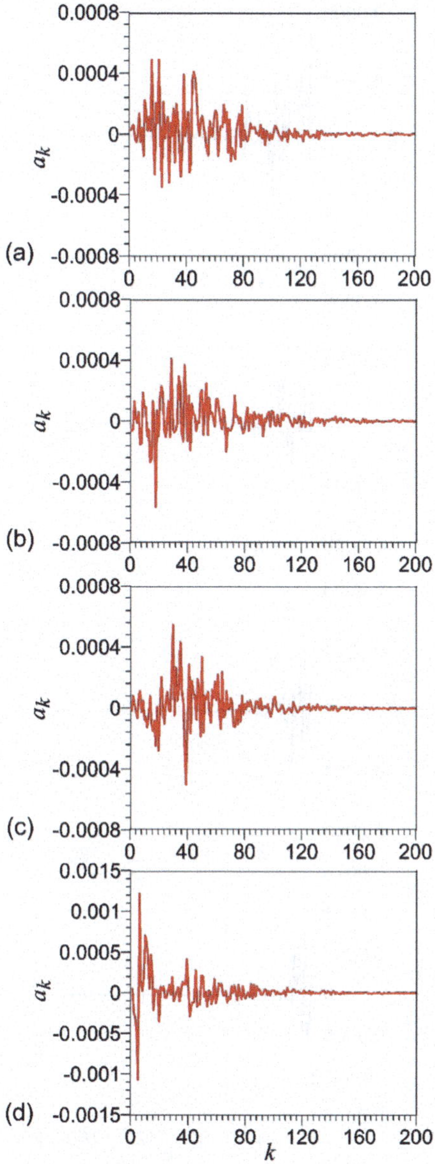

Fig. 5. Coefficients a_k (left) and b_k (right) of the Fourier series of the function $S_I(x)$ in the main medium at the level $y = -0.625$ m at times $t = 675$ (a), 1275 (b), 1950 (c), 3150 days (d).

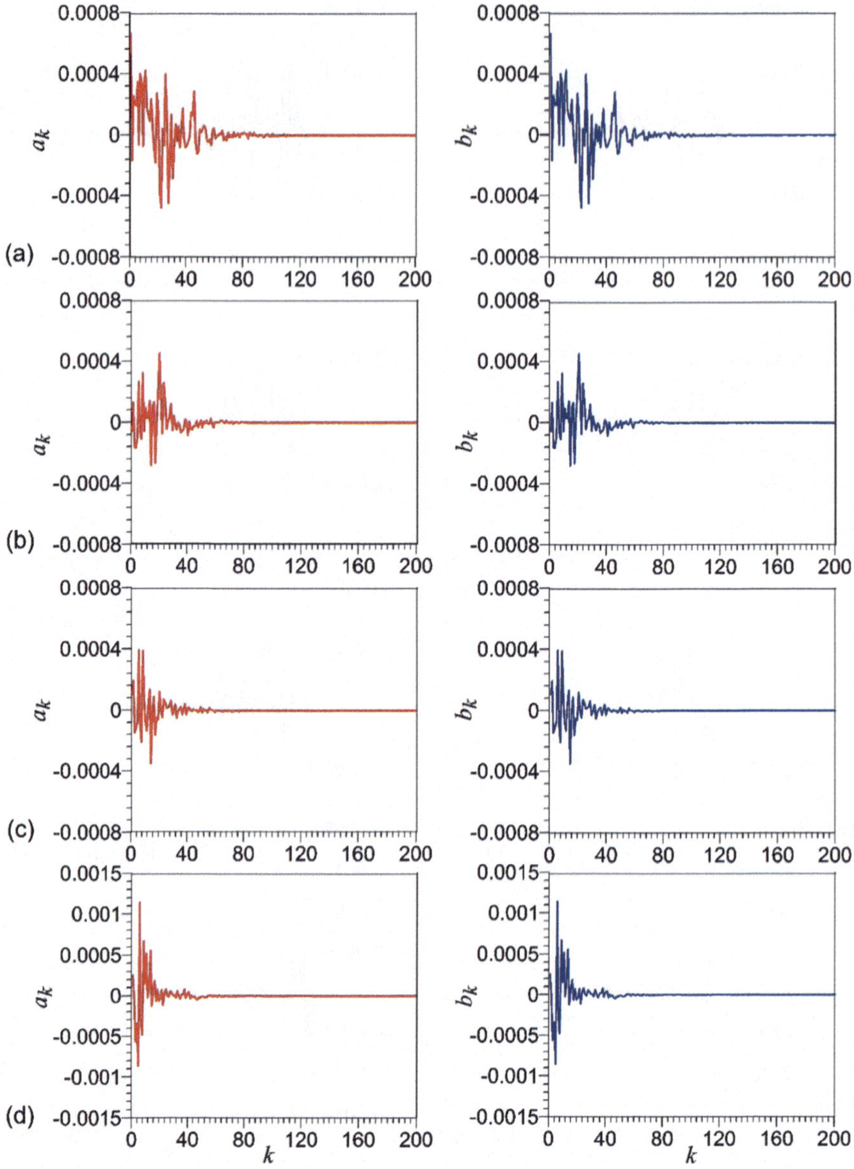

Fig. 6. Coefficients a_k (left) and b_k (right) of the Fourier series of the function $S_2(x)$ in the interior layer at the level $y = -2.50$ m at times $t = 675$ (a), 1275 (b), 1950 (c), 3150 days (d).

4 Conclusion

Stochastic haline convection in a rectangular porous domain containing a horizontal layer at lower porosity and permeability was investigated numerically. Convection is driven by more dense brine appearing near the upper boundary due to admixture diffusion and becoming unstable in the gravity field. As found, the interior layer is able to pass brine by different mode of convection. Initially, fluid here is involved in forced convection: the vertical component of velocity is directed downward or equal to zero. During time, the motion is rearranged and the upward velocity develops as well so forced convection transforms into natural convection. This dynamic stage is characterized by some fingers of pure water which rise from the layer to the space above and by flows intensified. Water fingers split brine between the upper boundary and interior layer into roughly periodic "brine balls". An influence of layer on convection structure above is observed in this stage and our spectral analysis of density distributions confirms this conclusion.

Acknowledgments. The author would like to thank G.G. Tsypkin for many fruitful discussions. This work has been supported by the Russian Science Foundation (grant № 16-11-10195).

References

1. Nield, D.A., Bejan, A.: Convection in Porous Media, 3rd edn. Springer, New York (2006)
2. Bear, J., Cheng, A.H.-D.: Modeling Groundwater Flow and Contaminant Transport. Springer, New York (2010)
3. Shargatov, V.A.: Instability of a liquid-vapor phase transition front in inhomogeneous wettable porous media. Fluid Dyn. **52**(1), 146–157 (2017). https://doi.org/10.1134/s0015462817010148
4. Tsypkin, G.G.: Stability of the evaporation and condensation surfaces in a porous medium. Fluid Dyn. **52**(6), 777–785 (2017). https://doi.org/10.1134/S0015462817060118
5. Shargatov, V.A., Tsypkin, G.G., Bogdanova, YuA: Filtration-flow fragmentation in medium with capillary-pressure gradient. Dokl. Phys. **63**(5), 199–2002 (2018). https://doi.org/10.1134/s1028335818050038
6. McKibbin, R.: Modeling heat and mass transport in geothermal systems. In: Vafai, K. (ed.) Handbook of Porous Media, chap. 7, pp. 545–572. Taylor & Francis, Boca Raton (2005)
7. Farajzadeh, R., Samili, H., Zitha, P.L.J., Bruining, H.: Numerical simulation of density-driven natural convection in porous media with application for CO_2 injection projects. Int. J. Heat Mass Transf. **50**, 5054–5064 (2007). https://doi.org/10.1016/j.ijheatmasstransfer.2007.08.019
8. Huppert, H.E., Neufeld, J.A.: The fluid mechanisms of carbone sequestration. Ann. Rev. Fluid Mech. **46**, 255–272 (2014). https://doi.org/10.1146/annurev-fluid-011212-140627
9. Anderson, L.: New Theory of the Earth. Cambrige University Press, Cambrige (2007)
10. Soboleva, E.: Numerical simulation of haline convection in geothermal reservoirs. J. Phys.: Conf. Ser. **891**, Paper 012105 (2017). https://doi.org/10.1088/1742-6596/891/1/012105
11. Soboleva, E.: Density-driven convection in an inhomogeneous geothermal reservoir. Int. J. Heat Mass Transf. **127**(C), 784–798 (2018). https://doi.org/10.1016/j.ijheatmasstransfer.2018.08.019

12. Soboleva, E.B., Tsypkin, G.G.: Numerical simulation of convective flows in a soil during evaporation of water containing a dissolved admixture. Fluid Dyn. **49**(5), 634–644 (2014). https://doi.org/10.1134/S001546281405010X
13. Soboleva, E.B., Tsypkin, G.G.: Regimes of haline convection during the evaporation of groundwater containing a dissolved admixture. Fluid Dyn. **51**(3), 364–371 (2016). https://doi.org/10.1134/S001546281603008X
14. Soboleva, E.: Numerical investigations of haline-convective flows of saline groundwater. J. Phys.: Conf. Ser. **891**, Paper 012104 (2017). https://doi.org/10.1088/1742-6596/891/1/012104
15. Korn, G.A., Korn, T.M.: Mathematical Handbook for Scientists and Engineers: Definitions, Theorems, and Formulas for Reference and Review, 2nd edn. McGraw-Hill Book Company, New York (1968)

Experimental and Analytical Study of Submerged Jet

T. O. Chaplina[1,2(✉)], V. I. Karev[2], and E. V. Stepanova[2]

[1] Faculty of Physics, Lomonosov Moscow State University, Moscow, Russia
chaplina_to@inbox.ru

[2] A. Yu. Ishlinsky Institute for Problems in Mechanics RAS, Moscow, Russia

Abstract. Experimental studies of submerged jets with various parameters (density, viscosity, salinity, temperature) are carried out, and the velocity of the jet propagation is estimated. A theoretical model for the description of jet motion is proposed. A comparison was made of the values of the jet outflow velocity calculated and estimated from experimental data, which showed a satisfactory coincidence. All obtained results are consistently reproduced with preservation of both functional dependencies and numerical values.

Keywords: Submerged jets · Experimental and analytical model · Visualization · Shadow method · Moire effect · Jet discharge rate

1 Introduction

Submerged jets and shear flows are often found in nature and various technical applications, such as mixing, burning, generation of sound disturbances, etc. Due to the small number of determining parameters, the submerged jet of incompressible fluid is a classical object of theoretical studies and is of interest for modeling in the experiments, which are of both general scientific and applied importance. Increased interest to the topic is associated with rapid development of a number of scientific fields, including geophysical hydrodynamics, chemical technology, and nuclear power engineering. The study of the submerged jet is also relevant for solving environmental problems associated with the release of petroleum products into the ocean when damaged oil-producing equipment and other pipeline structures. Currently widely used experimental methods for the study of jet streams [1–3]. Experiments, as a rule, are carried out in laboratory conditions, where the dimensions of the studied region of propagation of a jet stream are limited to several meters. In real conditions, in particular, in the case of the release of oil pollution, the scale is larger by several orders of magnitude – up to tens of kilometers. Therefore, it is necessary to take into account the relativity of the obtained experimental results.

The purpose of this work was to study the characteristics of the submerged jet under various experimental conditions, as well as to analyze the results using the theoretical model.

V. I. Karev et al. (Eds.): *Physical and Mathematical Modeling of Earth and Environment Processes (2018)*, SPEES, pp. 75–82, 2019.
https://doi.org/10.1007/978-3-030-11533-3_8

2 Jet Classification and Their Main Characteristics

A jet is a form of fluid flow in which a fluid (gas) flows in a medium (gas, liquid, plasma) with parameters that differ from the jet (velocity, temperature, density, etc.). Jet streams are extremely common and diverse – from a jet flowing from a rocket engine nozzle to jet streams in the atmosphere and the ocean [4]. A classification of jet flows is given in [5]. It is carried out according to the following characteristic properties: cross-sectional shape (flat, round, square, etc.), the direction of the flow velocity at the nozzle section (axial, fan, swirling jets), the state of aggregation of the jet substance (liquid, gas, plasma, two jets), the Mach number at the nozzle section (subsonic jets (M < 1), supersonic jets (M > 1)), the direction of the flow velocity in the medium relative to the jet, the phase state of the jet and the medium, the flow regime in the jet boundary layer (laminar, turbulent).

Interest in the study of the submerged jet has a nearly century-long history [6–8]. Pioneering work was published in 1937 [6]. To determine the velocity field at a distance significantly exceeding the diameter of the nozzle, the method of suspended particles was used. If we assume that the jet flows from a point fictitious source located inside the tip, then the measured velocity profiles are consistent with the Schlichting self-similar solution [9]. The authors of the article proposed a method for calculating the position of a fictitious source, based on the condition of equality of the measured kinetic energy of the flow in the initial cross section of the jet and the energy calculated by the analytical formula from the definition.

In the classical hydrodynamics Landau found a rare case of the exact solution of the Navier–Stokes equations, namely the problem of a submerged jet. Considering the emergence of turbulence, Landau proposed a new approach to this problem. In 1944, an article [7] was published with the exact solution of the problem of a submerged jet in an infinite liquid medium in the absence of a gravitational field. The problem was solved in spherical coordinates r, θ, φ. From the continuity equations and Navier-Stokes taking into account axial symmetry around the polar axis $\left(v_{\varphi}=0\right)$, and preserving the total flow of jet impulse through a closed surface as a result of transformations, the solution was obtained:

1. Spatial distribution of jet velocity

$$v_r = \frac{1}{r}F(\theta), v_\theta = \frac{1}{r}f(\theta) \tag{1}$$

where $f(\theta) = -\frac{2\nu \sin\theta}{A-\cos\theta}$, $F(\theta) = 2\nu\left\{\frac{A^2-1}{(A-\cos\theta)^2} - 1\right\}$, A – constant.

2. Pressure distribution

$$p - p_0 = -\frac{4\rho\nu^2}{r^2}\frac{A\cos\theta - 1}{(A-\cos\theta)^2} \tag{2}$$

where the constant A is related to the total flow of momentum in the jet P by the relation

$$P = 16\pi v^2 \rho A \left\{ 1 + \frac{4}{3(A^2 - 1)} - \frac{A}{2} \ln \frac{A+1}{A-1} \right\} \tag{3}$$

3 Experimental Studies of the Submerged Jet

In this work, we investigated a submerged jet of water of various density, obtained by changing the salinity, and also studied the distribution of jets of water-immiscible liquids (oil and gasoline). Two methods of jet visualization used: the shadow method (Fig. 1) and the method based on the Moire effect.

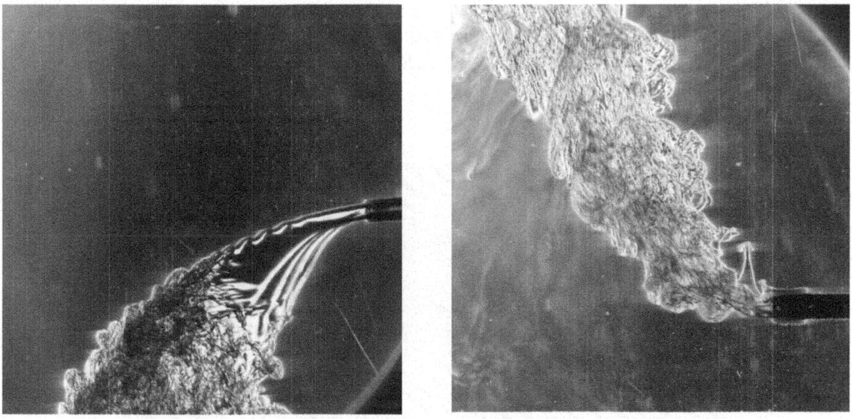

a b

Fig. 1. Shadow visualization of the submerged jet: **a** more saline water (ρ = 1015 kg/m^3) of the same temperature, **b** slowly pours into salt water (t = 20 °C, ρ = 1010 kg/m^3) water of the same salinity slowly flows in, heated to t = 30 °C.

The shadow method allows to obtain an image of a jet with a high resolution, but only in the initial part (Fig. 1, a, b), while the method based on the Moire effect has a much larger area of measurement. The experiments are carried out at different positions of the reservoir, corresponding to different speeds and different angles of the outflowing fluid (Figs. 2 and 3).

As a result of processing the experimental data, the positions of the central line of the jet were obtained at different values of the initial velocity. The characteristic dependence is presented in Fig. 4, where the density of the base fluid is 1020 kg/m^3, the density of the inflowing fluid is 1050 kg/m^3. With an increase in the initial velocity, the central lines turn out to be less bended. The change in the position of the central line of the jet as a function of the angle of inclination of the initial velocity to the horizon are showed in Fig. 5 The degree of steepness of the line is the greater, the greater the angle of inclination of the flowing fluid.

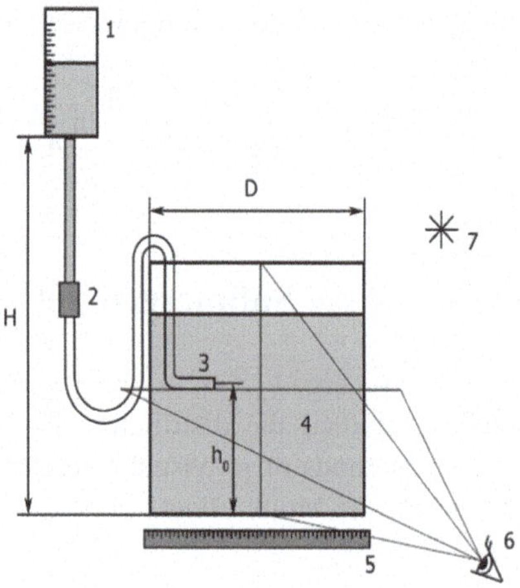

Fig. 2. The experimental setup: 1 – the tank with liquid, 2 – valve, 3 – a tube (nozzle) of small radius (2.1 mm), 4 – a vessel with a base fluid (S_0 density, kg/m^3), 5 – line, 6 – photo/ video camera, 7 – light source.

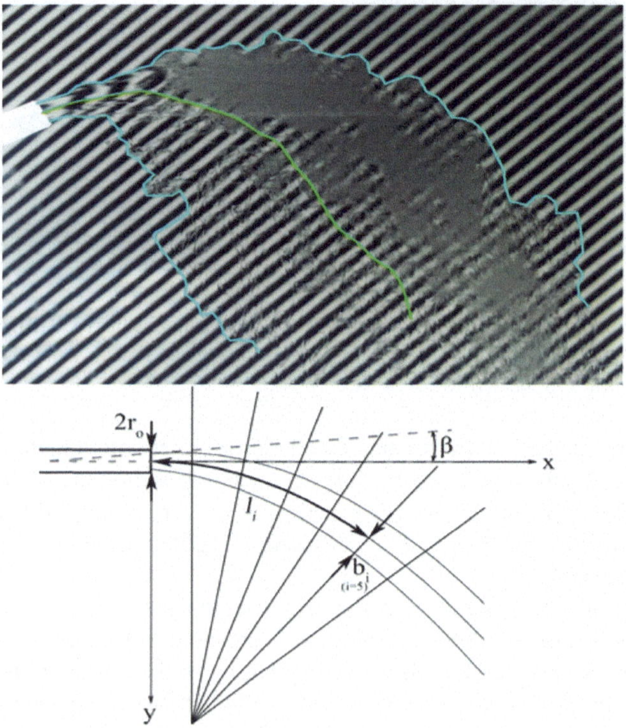

Fig. 3. The image and the scheme of the jet (r_0 is the nozzle radius (2.7 mm), l_i – the distance from the nozzle along the center line of the jet, b_i – the half width of the jet on the i-th division, β – the angle of the jet solution, v_0 – the initial velocity.

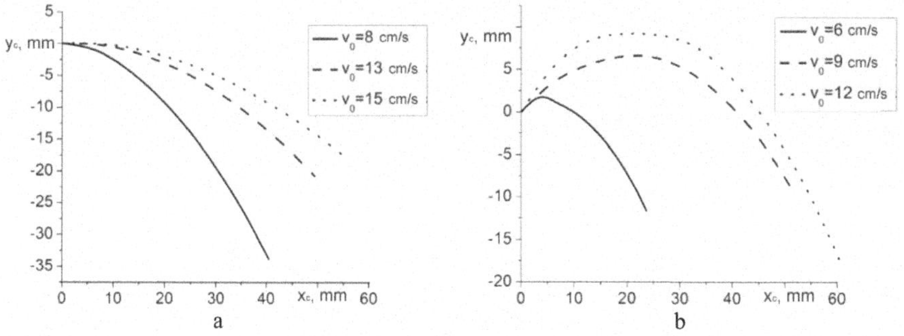

Fig. 4. The central lines of the jet (ρ_0 = 1020 kg/m^3, ρ = 1050 kg/m^3) at various flow rates. The angle of inclination of the velocity of the flow to the horizon: **a** 0°; **b** 30 °.

The positions of the central line of the jet, depending on its density with the same position of the tank H = 70 cm, as well as the dependence of the half width of the jet on the distance from the nozzle along the central line of the jet are shown in Figs. 6 and 7.

Experiments were carried out for the same difference in the densities of the base and flowing fluids with a change in the density of the base fluid (Fig. 8).

When describing the dynamics of a submerged jet, the paper considers an analogy with the motion of a body thrown at an angle to the horizon. The equation of the trajectory of motion of a body thrown from a point (x0, y0) horizontally with a speed v_0 in a field with a vertical acceleration ξ, is as follows:

$$y - y_0 = \frac{\xi}{2v_0^2}(x - x_0)^2 \tag{4}$$

From here we express the initial speed through the coefficient C:

$$C = \frac{\xi}{2v_0^2}, v_0 = \sqrt{\frac{\xi}{2C}} \tag{5}$$

ρ_0 – ambient fluid density, ρ – density of fluid in the jet, V – particle volume, m – mass. The particle is acted upon by gravity mg and Archimedes force F_A. According to the second law of Newton:

$$F_A - mg = m\xi \tag{6}$$

Then $\rho_0 gV - \rho Vg = \rho V\xi$, $\xi = g\frac{\rho_0 - \rho}{\rho}$

For liquid outflow velocity finally is obtain

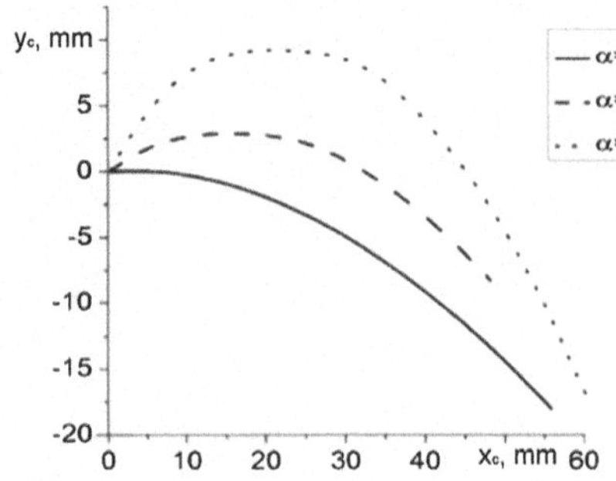

Fig. 5. The central line of the jet, depending on the angle between the direction of the initial velocity and the horizon (ρ_0 = 1020 kg/m^3, ρ = 1050 kg/m^3).

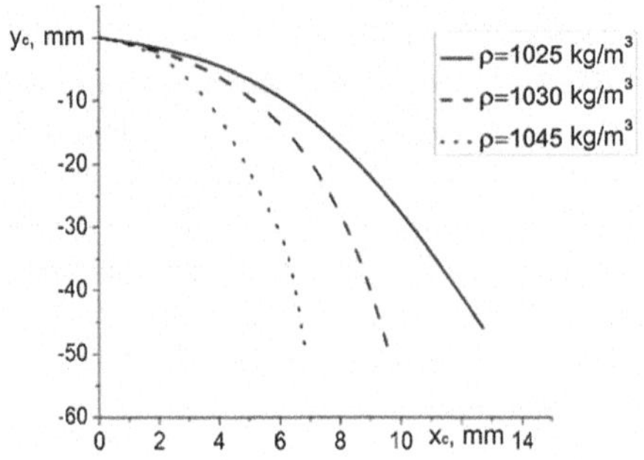

Fig. 6. The position of the central line of the jet, depending on its density. The density of the base fluid is 1020 kg/m^3.

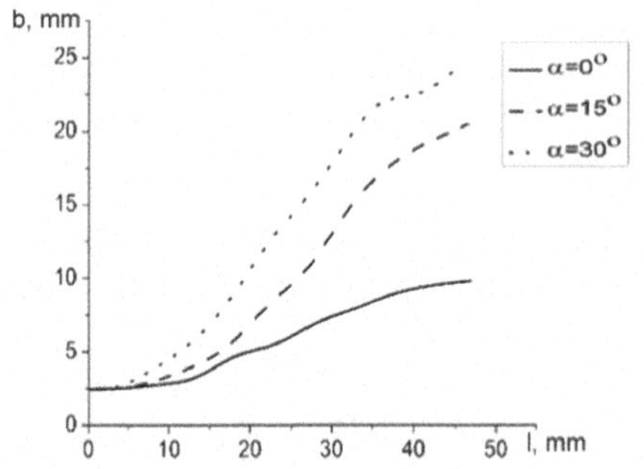

Fig. 7. The dependence of the half-width of the jet on the distance from the nozzle along the central line of the jet. The density of the base fluid is 1000 kg/m^3, the density of the inflowing liquid is 1030 kg/m^3.

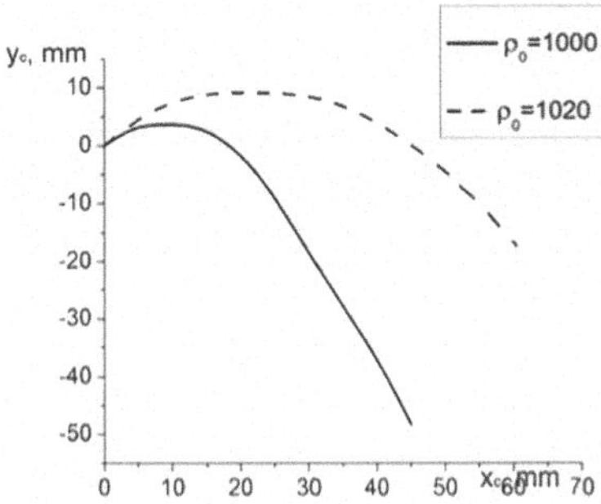

Fig. 8. The position of the central line of the jet at different densities of the base fluid (ρ = 1030 kg/m^3, ρ_0 = 1000 kg/m3; ρ = 1050 kg/ m3, $\rho0$ = 1020 kg/m^3).

$$v_0 = \sqrt{\frac{g(\rho_0 - \rho)}{2C\rho}} \tag{7}$$

The Reynolds number is determined by the ratio

$$Re = \frac{\rho v_0 d}{\mu} \tag{8}$$

where d – nozzle diameter, μ – dynamic viscosity of the flowing fluid.

The calculated parameters for a series of experiments are shown in Table 1.

Table 1. Submerged jet parameters

ρ, kg/m^3	ρ_0, kg/m^3	H, cm	α^0	C	β^0	$v_{0analit}$, mm/s	v_{0exp}, mm/s	a	$l_{нач}$, mm	Re
1005	1020	100	0	0.008	11.04	95.26	97.03	0.06	32	514
1010	1020	100	0	0.005	15.59	96.68	94.92	0.08	22	524
1010	1000	100	0	0.004	14.19	116.25	114.30	0.07	24	630
1010	1000	100	15	0.006	21.80	89.40	133.94	0.12	15	485
1010	1000	100	30	0.008	29.20	75.64	101.54	0.16	11	410
1030	1000	100	0	0.007	14.23	146.39	143.63	0.07	24	809
1030	1000	100	15	0.016	19.28	93.43	113.71	0.10	18	517
1030	1000	100	30	0.025	30.36	75.72	110.26	0.17	10	419
1050	1020	100	0	0.008	11.31	129.02	114.72	0.06	31	727
1050	1020	100	15	0.013	21.80	104.99	105.06	0.12	15	592
1050	1020	100	30	0.016	21.80	92.62	103.76	0.12	15	522

4 Conclusions

A comparison was made of the values of the jet outflow velocity calculated and estimated from experimental data, which showed a satisfactory coincidence. All obtained results are consistently reproduced with preservation of both functional dependencies and numerical values.

Acknowledgement. This work is supported by Russian Foundation for Basic Research (Grant 18-01-00116).

References

1. Wen, Q., et al.: Structure analysis of a low reynolds number turbulent submerged jet interacting with a free surface. J. Fluids Eng. **136**(10), 101–104 (2014)
2. Sodjavi, K., et al.: PIV and electrodiffusion diagnostics of flow field, wall shear stress and mass transfer beneath three round submerged impinging jets. Exp. Thermal Fluid Sci. **70**, 417–436 (2016)
3. Shao, D., et al.: Flow patterns and mixing characteristics of horizontal buoyant jets at low and moderate Reynolds numbers. Int. J. Heat Mass Transf. **105**, 831–846 (2017)
4. Prohorov, A.M., et al.: Physical Encyclopedic Dictionary. Soviet Encyclopedia (1984). (in Russian)
5. Matveev, S.G., et al.: Calculation of the characteristics of jet streams of gaseous media in GTE combustion chambers benefit. Samara State Aerospace UN-t, Samara (2011). (in Russian)
6. da C Andrade, E.N., Tsien, L.C.: The velocity-distribution in a liquid-into-liquid jet. Proc. Phys. Soc. **49**(4), 381–391 (1937)
7. Landau, L.D.: On one new exact solution of the Navier-Stokes equations. Dokl. Acad. Sci. USSR. **43**(7), 299–301 (1944). (in Russian)
8. Sobey, R.J., Johnston, A.J., Keane, R.D.: Horizontal round buoyant jet in shallow water. J. Hydraul. Eng. **114**(8), 910–929 (1988)
9. Kirchhoff, G.: Mechanics. M.: USSR Academy of Sciences, 404 p. (1962)

Internal Waves over the Continental Shelf of the Heracles Peninsula: Modeling and Observation

Vitalii Ivanov, T. Ya. Shul'ga(✉), Andrei Bagaev, Tatiana Plastun, Alesia Medvedeva, Ludmila Verzhevskaia, and Irina Svisheva

Marine Hydrophysical Institute, Kapitanskaya 2, 299011 Sevastopol, Russia
shulgaty@mail.ru

Abstract. The paper considers the internal waves on the shelf of the Crimean coast, their structure and dynamics at the periphery of the Black Sea basin. Manifestations of internal waves were studied by numerical experiments and by processing of remote sensing data from high-resolution data. Maps of the vertical velocity of the first three modes of internal waves on the shelf of the Crimean coast were constructed. The phase velocity of the internal waves in the deep part of the sea for the first mode at the given profiles varies from 2.6 to 5 ms^{-1}; for the waves of the second mode—1.1–2.3 ms^{-1}; for the waves of the third mode—0.7–1.4 ms^{-1}. Wavelengths detected from the satellite data had a mean value of 0.4 km. The longest waves, around 0.8 km, were observed between Sevastopol and Yevpatoria. During June, 2017 the waves of 2nd and 3rd modes were observed in the area of study the most frequently.

Keywords: Black Sea · Phase velocity · Dispersion relations · Mode structure · Remote sensing · Intensive internal waves · Coastal upwelling · Topographic effects

1 Introduction

The paper investigates the appearance and propagation of internal waves at the shelf of the Black Sea, near the Heracles Peninsula. This complex phenomenon was addressed in a set of papers [1–5]. In present study, we apply numerical experiments in order to investigate the dynamics of internal waves and the conditions for their development. We studied the conditions when intense internal waves appear on the periphery of the Rim Current off the shelf of the Crimean coast. For the first time, highres images of the sea surface near the Heracles Peninsula were used. They provided a unique opportunity to assess parameters of internal waves. The results allowed us to establish theoretical spatio-temporal characteristics of internal waves that correspond to shallow pycnocline.

V. I. Karev et al. (Eds.): *Physical and Mathematical Modeling of Earth and Environment Processes (2018)*, SPEES, pp. 83–92, 2019.
https://doi.org/10.1007/978-3-030-11533-3_9

2 Study Area

In some areas of the Crimean shelf irregularity of the coastline/bathymetry and the Rim Current passing near the shore create specific conditions for the internal waves generation (see Fig. 1). One of such areas is to the North of the Heracles Peninsula—in Kalamita Bay, between the cities of Sevastopol and Yevpatoria. The shelf here is wide and the isobath of 150 m is located up to 60 km away from the coastline. Another area is to the South–East of the Heracles Peninsula, between the cities of Sevastopol and Yalta. The shelf here is convex and narrow (just 30 km wide). Between these two areas the shelf is even more narrow, down to 10 or even 6 km wide.

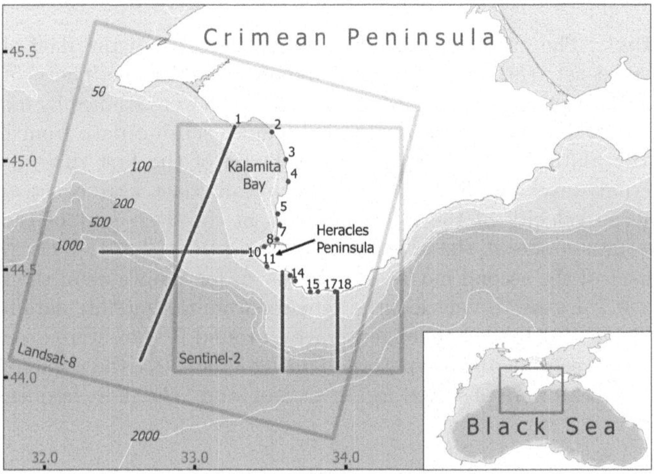

Fig. 1. The bathymetry of the Black Sea near the Crimean Peninsula. Thin white contours represent 20–, 50–, 100–, 200–, 500–, 1000–, and 2000–m isobaths. Dark-grey numbered lines depict some transects considered for the numerical experiments (for the rest only starting points and their numbers are plotted). Horizontal axis units are degrees East, vertical axis—degrees North. Light–grey rectangles outline the areas of satellite data availability. In the inset location of the area of interest is shown.

3 Methodology

In our work, we tackled the main problems stage by stage, as shown in the flowchart (Fig. 2). At the first stage, we detected and evaluated the characteristics of the surface manifestations of internal waves. At the second stage, we obtained theoretical estimates of the internal wave parameters based on the development and implementation of the algorithm for solving the full boundary value problem [6]. At the final stage, the results of theoretical and numerical studies were compared with the data obtained by remote sensing methods.

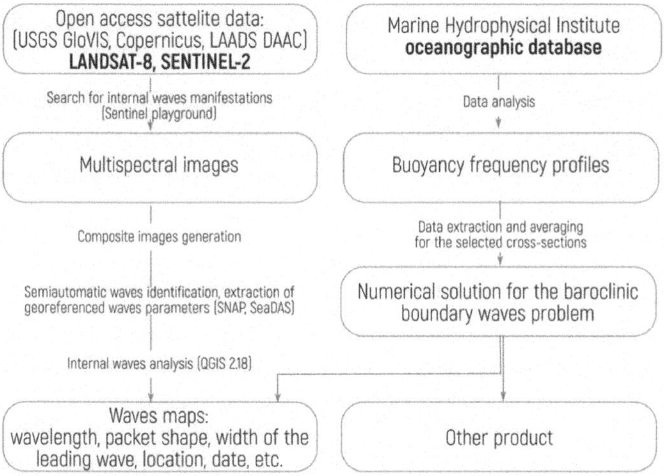

Fig. 2. Schematic flowchart of the major steps in deriving the results of the present study.

4 Data and Methods

4.1 Deriving Vertical Density Structure

Decadal temperature and salinity fields at standard levels (in the 0–1500 m layer) in the Black Sea were reconstructed using in-situ data and satellite measurements which had been collected from the 1950–es up to nowadays [7]. These fields were used to investigate the vertical density structure in the vicinity of the slope–shelf area at the Black Sea coast of the Crimea.

The following contact and remote sensing data on the Black Sea were used:

(A) Research Vessel measurements on temperature and salinity from 1951 to 2008 and temperature measurements by drifters from 2001 to 2017, from the Bank of Oceanographic Data of Marine Hydrophysical Institute.

(B) Temperature and salinity measured by deepsea buoy–profilers at horizons from 4 to 1500 m (565 profiles for 2005–2008).

(C) Monthly averages of nighttime satellite measurements of SST at a 4×4 km grid for 1985–2007 [7].

Hydrological fields were constructed on the basis of the data grouped over decades with a five–year overlap. Combining the data over the decades ensured their uniform spatial distribution on most horizons, which allowed horizontal interpolation. An array of temperature and salinity climate data of the Black Sea contains 12 files with mean monthly fields. Each file consists of thirty–six 141×88 arrays. The data is available at the following depths: 0, 10, 20, 30, 40, 50, 75, 100, 120, 150, 200, 300, 500, 600, 800, 1000, 1200, 1500, 1800, 2000 m. Horizontal longitudinal resolution was $1/10°$, latitudinal—$1/16°$ in the region of 27.8°–41.8°E, 40.875°–46.3125°W. Temperature and salinity data were selected for each of the 18 transects, shown at Fig. 1, for 4 months, from May to August

(144 files). The seawater density (ρ, kg·m^3) was calculated from the International Equation of State (UNESCO) [8]. Profiles of buoyancy frequency $N(z)$ defined as $N^2(z) = -\frac{g}{\rho(z)}\frac{\partial\rho(z)}{\partial z}$ were estimated for all density profiles. Here, $\rho(z)$ is the density, z is the vertical coordinate (positive upward), and g is the acceleration due to gravity. Both $\rho(z)$ and $N(z)$ profiles were gridded to bins of the vertical axis

4.2 The Equations of Propagation of Internal Waves

The phenomenon of wave transformation caused by bottom variation is explained theoretically using nonlinear equations for shallow water waves. Internal waves can propagate along the interface of two homogeneous layers or in a continuously stratified fluid. We considered a rotated basin of variable depth filled with stratified water at the latitude of 44°N. The linearized equations of motion under the Boussinesq approximation are as follows

$$\rho\left(\frac{\partial u}{\partial t} - fv\right) = -\frac{\partial P}{\partial x}, \quad \rho\left(\frac{\partial v}{\partial t} + fu\right) = -\frac{\partial P}{\partial y}, \quad \rho\frac{\partial w}{\partial t} + \rho' g = -\frac{\partial P}{\partial z}, \qquad (1)$$
$$\frac{\partial u}{\partial x} + \frac{\partial v}{\partial y} + \frac{\partial w}{\partial z} = 0,$$

where where $\rho'(z)$ is the background density perturbation; P is the pressure; u, v, and w are the velocity components along the standard orthogonal coordinates x, y, and z, respectively;—is offshore distance coordinate; $f = 1.01 \cdot 10^{-4}$ s^{-1}—is the Coriolis parameter. For internal waves propagating along the bottom, the governing system (1) can be reduced to one equation for the vertical velocity component $w = W(z)exp[i(kx + \omega t)]$, where k is the wavenumber along the horizontal plane, ω—is the wave frequency, and $W(z)$ satisfies:

$$\frac{\partial^2 W}{\partial^2 z} + k^2\left(\frac{N^2(z) - f^2}{\omega^2 - f^2} - 1\right) = 0. \qquad (2)$$

with the following boundary conditions

$$W(0) = W(H) = 0, \qquad (3)$$

where H is the water depth. Using the dispersion relation $\omega^2 = C^2k^2 + f^2$, (2) can be rewritten as $\frac{\partial^2 W}{\partial^2 z} + \left(\frac{N^2(z) - f^2}{C^2}\right) W = k^2 W$. Approximating the derivatives with the finite differences discretized over Δz yields a standard form of eigenvalue problem. The internal waves dispersion characteristics and the normal mode vertical profiles were estimated by solving numerically the following boundary value problem [6]:

$$\frac{\partial^2 W}{\partial^2 z} + \left(\frac{N^2(z) - \omega^2}{C_{in}^2}\right) W = 0, \quad \text{where} \quad C_{in}^2 = \frac{\omega^2 - f^2}{k_n^2} \qquad (4)$$

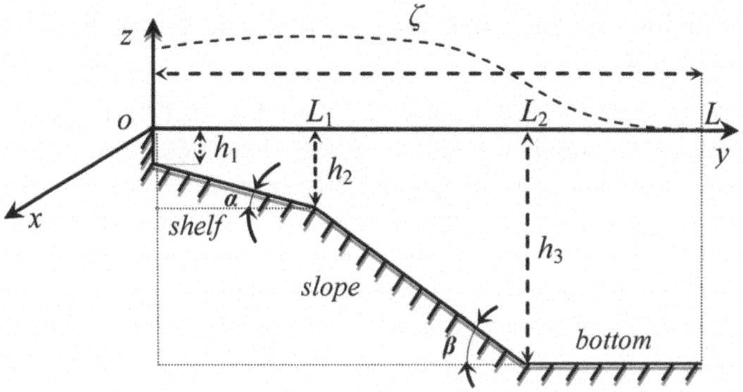

Fig. 3. Model geometry profile.

with the boundary conditions (3). Here, C_i refers to the internal wave phase speed in a nonrotating ocean, and $n = 1, 2, \ldots$ is the vertical mode number.

In our study we used the topography of the Black Sea from the *GEBCO* database (http://www.gebco.net). It has a spatial resolution for the longitude with 0.008334°E step, latitude with 0.008342°N step. The topography of the continental shelf and slope is approximated as a two–slope depth profile adjacent to a deep ocean of a constant depth. The parameters of the transects in the considered region depend only on one spatial variable (Fig. 3).

The depth profile is characterized by the width of the shelf, slope, and deep ocean (L_1 and L_2) as well as by the depth of the coastal wall, shelf break, and deep ocean (h_1, h_2 and h_3). Table 1 contains the parameters of some bottom shape profiles; rows for each profile indicate the depth values at the shore (h_1), slope (h_2), and offshore (h_3) and the width values of the shelf (L_1), and continental slope (L_2). It also shows the inclination angle for the shelf (α) and slope (β).

Table 1. Profiles geometry.

Profiles	Shelf			Slope			Bottom
	h_1, m	L_1, km	α, deg	h_2, m	L_2, km	β, deg	h_3, m
1	2	75,2	32	158	16,0	45	1685
10	11	46,2	38	189	27,7	44	1331
12	74	33,4	34	158	17,6	45	1831
18	22	25,0	39	146	14,9	45	1419

Shelf of the Crimean coast is quite wide (profiles 1–4) reaching 67.3–79 km in width. A sharp transition to the continental slope (where the maximum $\alpha = 45$) is observed at profile 5. Profiles 13–18 are the most special, where the shelf is very narrow (L_1 is 23.3–28.2 km), the bottom inclinations are small, and $h(y)$ profile is close to the quasi–linear one.

4.3 Observations of Internal Waves Near the Black Sea Coast of the Crimea

The available Sentinel-2 and Landsat-8 data for 2016–2017 were analyzed. Images of satisfactory quality (cloudiness less than 30%) were selected. Internal waves appeared only at 6 images for 2016 (out of 32) and at 25 images for 2017 (out of 53). Thus the period from May to September, 2017 was selected for the analysis. The following operations were performed with the selected images:

(A) Preparation of composite images with enhanced contrast.

(B) Localization of internal waves appearance, separation from ship-generated waves.

(C) Obtaining georeferenced perimeter of 6 anchor points for each wave (3 frontal and 3 rear).

(D) Saving the waves parameters (front width, wavelength, number of waves in packet, direction, date and time).

After that, the results were visualized with QGIS 2.18 and statistically analyzed with MS Excel.

5 Results and Discussion

5.1 Analysis of Buoyancy Frequency Profiles

The water stratification profile (constructed as initial conditions for numerical experiments) was characterized by a relatively sharp interface at the depth of 50 m and a less pronounced main pycnocline located between 500 and 1200 m deep (see Fig. 4). The vertical distribution of buoyancy frequency in spring and summer is different (left vs central panel). In the summer, the $N(z)$ curves differ little in different parts of the coast

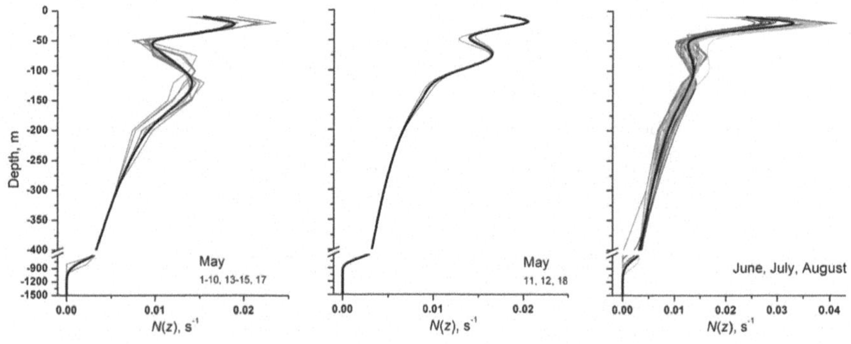

Fig. 4. Vertical profiles of buoyancy frequency $N(z)$. In the left and central panels, lines corre-spond to the transects (1–10, 13–15, 17) and (11, 12, 18), respectively, during the spring. In the right panel lines correspond to the transects 1–18 during the summer. Black lines show the averaged $N(z)$.

5.2 Theoretical Estimate of the Fundamental Mode Parameters

For the interpretation of observations, we calculate wave properties (Table 2) and eigenvectors for the vertical velocity component $W(z)$ of baroclinic Poincar waves in the ocean of depth H described by equation (4).

Figure 5 shows the results of calculation of volume waves maps for three offshore profiles. Calculations were made for seasonally different buoyancy frequency profiles—late spring and summer. Spatial distribution of the vertical velocity of the first mode waves was analyzed using the maps in Fig. 5, left panel. Significant vertical movement of water was found in the deep sea in spring, and in the area of the sharp continental slope in summer, which was caused by a more powerful flow of the Rim current jet at the edge of the shelf. At the shelf with a sharper slope ($\alpha = 45$), the maximum vertical velocity for both zones was found in the deep sea. In profile 10 one can see that in summer internal waves appear above the edge of the shelf and propagate in both directions from the disturbances sources.

Table 2. Parameters of the three main baroclinic modes in the central part of Kalamita Bay for the spring profile of $N(z)$.

Mode	Wavelength, m	Phase speed, ms^{-1}	Period, h
Deep part, maximum depth 1500 m			
1	2416	0.93	7.21
2	950	0.35	5.86
3	459	0.27	4.51
Shelf part, maximum depth 300 m			
1	770	0.58	0.37
2	290	0.25	0.32
3	210	0.18	0.32

The highest vertical velocities for the second mode wave (Fig. 5, central panel) take place in the surface layer. The zero–amplitude curve, running at a depth of 150–300 m, lies almost horizontally in spring, and bends to the bottom in summer (profiles 12, 17). Maps of the third mode vertical wave velocity distribution (Fig. 5, right panel) confirm the influence of bottom topography on their structure. The position of the first nodal line of the third mode is much closer to the sea surface (50–200 m) in compari-son with the zero amplitude curve of the second mode and largely repeats its shape.

The second nodal line is located at a depth of 400–800 m and in the summer it usually repeats the configuration of the steep slope (profiles 10, 17) or sometime differs from the shape of the first nodal line (profile 12). The highest values of the velocity of the third mode waves occur in the surface (100–200 m) and deep (800–1200 m) layers of the sea. For the deep part of the slope the phase velocity

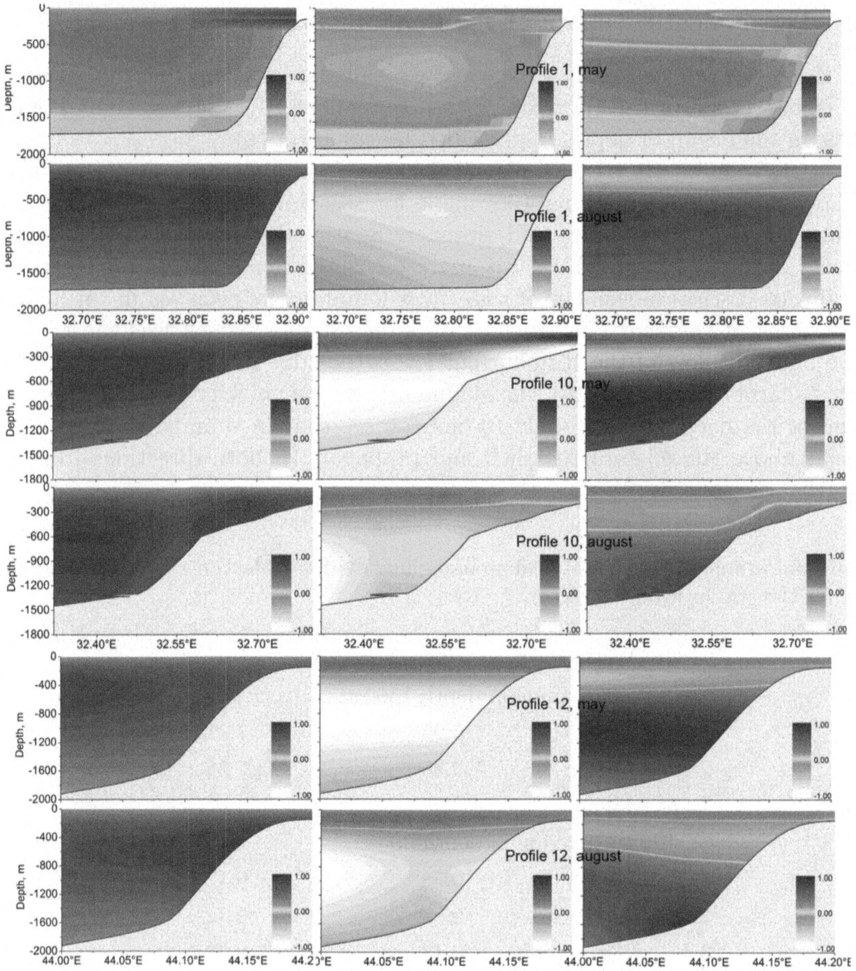

Fig. 5. Volume waves maps for the vertical velocity field $W(z)$ along the profiles 1, 10 and 12. Results of numerical experiment for a spring and summer buoyancy frequency of $N(z)$: 1 mode (left), 2 mode (center) and 3 mode (right).

of the internal waves of the first mode at the given profiles varies from 2.6 to 5 ms^{-1} (C) depending on the season, their length from 9 to 180 km (λ); for the waves of the second mode—1.1–2.3 ms^{-1} (C), 4–42 km (λ); for the waves of the third mode—0.7–1.4 ms^{-1} (C), 1–27 km (λ).

5.3 Comparison of Model Calculations with Satellite Observations

Qualitative analysis of the maps of waves distribution and direction showed that the areas of the most recent waves appearance were located to the North–West of the Heracles Peninsula (where the quasi–steady Sevastopol anticyclone is known to be formed) and to the North–East of it, in the shallow Kalamita bay. The waves detected over the wide shelf oriented mainly shoreward and those found at the slope—along the shelf.

129 cases of internal waves propagation were identified from May to September, 2017. The distribution over the months is not equal. In June, 50 waves were observed, which is the highest number, while in May and September only 12 and 6 respectively. From July to August the number of waves decreased from 42 to 19. Wavelengths varied from 0.05 to 1.14 km with a mean value of 0.4 km and a median value very close to it: 0.375 km. Thus most waves were 0.15 km to 0.6 km long. The longest waves, around 0.8 km, were more often observed between Sevastopol and Yevpatoria along the 50 m isobaths.

6 Conclusion

The main spatial and temporal characteristics of internal waves at the shelf of the Crimean coast were determined on the basis of satellite sensing data and the results of analytical calculations. The assumed reason for generation of intense internal waves through the interaction of the Rim current jet with the shelf edge was confirmed by the results of numerical calculations of the vertical distribution of the first three modes velocity. The mathematical modeling allowed us to obtain theoretical estimates of the parameters of internal waves depending on the season and shape of the longshore relief. Comparing the wavelengths of the observed and calculated waves, we can conclude that during 2017 the waves of 2 and 3 modes were observed from the satellite in the area of study the most frequently. They propagate up to 1200 m deep and typically manifest themselves the most intensively at the distance 20–30 km from the coast, depending on the shelf width and vertical water density profile.

Acknowledgments. The reported study was funded by RFBR and Government of Sevastopol according to the research project No 18–45–920036. The authors thank Professor Aleksander E. Yankovsky (University of South Carolina) for supporting this research and consultations in creating the applications for processing information on the intra–wave oscillations.

References

1. Ivanov, V.A., Konyaev, K.V., Serebryanyy, A.N.: Groups of intensive internal waves on the sea shelf. Izvestiya AN SSSR Fizika atmosfery i okeana **17**(12), 1302–1309 (1981)
2. Klimov, D.M., Karev, V.I., Kovalenko, Y.F., Ustinov, K.B.: Mechanical-mathematical and experimental modeling of well stability in anisotropic media. Mech. Solids **48**, 357–363 (2013)
3. Ivanov, V.A., Serebryanyy, A.N.: Short-period internal waves in the coastal zone of a non-tidal sea. Izvestiya AN SSSR Atmos. Ocean. Phys. **21**(6), 648–656 (1985)
4. Colosi, J.A., Beardsley, R.C., Lynch, J.F.: Observations of nonlinear internal waves on the outer New England continental shelf during summer Shelfbreak Primer study. Geophys. Res. **106**(C5), 9587–9601 (2001)

5. Liu, A.K., Chang, Y.S., Hsu, M.K., Liang, N.K.: Evolution of nonlinear internal waves in the East and South China Seas. Geophys. Res. **103**(C4), 7995–8008 (1998)
6. Yankovsky, A.E., Zhang, T.: Scattering of a semidiurnal barotropic kelvin wave into in-ternal waves over wide continental shelves dynamics. Phys. Ocean. **47**, 2745–2762 (2017)
7. Polonsky, A.B., Shokurova, I.G., Belokopytov, V.N.: Decadal temperature and salinity fields at standard levels in the Black Sea. Phys. Ocean. **6**, 27–41 (2013)
8. Algorithms for computation of fundamental properties of seawater. UNESCO technical papers in marine science, vol. 44, p. 53. UNESCO (1983)

Stress-Dependent Permeability of Reservoir Rock and Its Influence on Well Flow Rate: Experiment and Simulation

Larisa Nazarova[1(✉)], Leonid Nazarov[1], and Nikita Golikov[2]

[1] Chinakal Institute of Mining, Siberian Branch RAS, 630091 Novosibirsk, Russia
lanazarova@ngs.ru

[2] Trofimuk Institute of Petroleum Geology and Geophysics, Siberian Branch RAS, 630090 Novosibirsk, Russia

Abstract. Thermobaric tests are performed on Bazhenov oil-bearing shale containing above 10% of kerogen in view to determine its rheological properties. The test specimen under a statistical axial load is heated stepwise at temperatures $T_n = 60$, 100 and 150 °C with measurements of height $H(t)$. The temperature is risen if relative velocity of H gets less than 0.05. Deformation of the specimen is described by Kelvin–Voigt model, within which the inverse problem is stated and solved to determine Young module E_n and effective viscosity η_n of rocks by $H(t)$. The results of solution are approximated by two-parameter exponential functions, $E = E(T)$ and $\eta = \eta(T)$ relationships are established.

The lab-scale test bench is designed and manufactured to study the relationship of granular geomaterials permeability versus stresses. A polyurethane measurement cell of parallelepiped shape is filled with sized sand; controllable vertical stress σ is applied to different sections of the top edge. The constant gas pressure p is created on one of vertical edges, while gas flow rate $Q(p, \sigma)$ is recorded on the opposite edge. The mathematical model of experimental is developed. Hypothesizing that permeability depends on effective stress according to the exponential law with coefficient β in index, the analytical solution is found to the problem on stationary filtration in a cell under a non-uniform stress state. The researchers propose the process for quantitative estimation of β based on minimization of the relative discrepancy functional between Q and the theoretical value of gas flow rate. It is appeared that β can be determined without reference to gas viscosity and an initial permeability of geomaterial packing. According to the present test results β magnitude used to reduce slightly with stress growth.

The poroelastic model describing evolution of geomechanical and hydrodynamic fields in a near-well zone was employed for comparative analysis of metering characteristics of a well in terms of the routine deformation and poroperm parameters of production oil-bearing beds. It is demonstrated that a disregard of the established empirical relationships of permeability versus effective stresses and temperature can result in an appreciable upward bias of oil production prediction.

V. I. Karev et al. (Eds.): *Physical and Mathematical Modeling of Earth and Environment Processes (2018)*, SPEES, pp. 93–106, 2019.
https://doi.org/10.1007/978-3-030-11533-3_10

Keywords: Porous and granular media · Rock · Reservoir · Filtration · Lab experiment · Stress-dependent permeability · Rheology · Temperature · Inverse problem

1 Introduction

Substantiation of the oil pool development conception, estimation of underground reservoir production characteristics, GIS data processing are items of the far from complete list of problems which solution rests on the reliable information on poroperm properties of rocks, and regularities of their variability in the course of deposit exploitation [1–3]. Cross-disciplinary evolution models of different-nature physical fields are applied to study deformation and mass transfer in the well environment [4–11]. Intercoupling of these fields is described by empirical relationships, established, as a rule, under laboratory conditions. The relations of rheological and poroperm properties of reservoir rock versus stresses [12–14] and temperature [15] can serve study cases of such dependences being occasionally of the dominant factor for flow characteristics of a well. The present paper deals with establishment of empirical relationships of permeability vs. stresses and temperature based on the laboratory data for different-type geomedia, as well as the effect of such relationships on flow characteristics of a well is assessed in the frame of the geomechanical-hydrodynamic model [16] of fracture-porous medium.

2 Experimental Investigation of Bazhenov Shale Rheological Properties at Different Temperatures

The world hydrocarbon reserves are mostly represented with problematic resource such as low-gravity oil and bitumen [17, 18]; oil deposit of Bazhenov formation being surely of the said type. Such oil fields are conventionally developed by the thermal oil-bed treatment with both positive and negative effects: fluid viscosity reduction is of positive value, while decrease in permeability owing to fall of deformation parameters of rocks with temperature rise is a negative after-effect [19, 20]. Bazhenov rock (bazhenite) series contain up to 10% of organic matter functioning as a binder, thereto, their mechanical characteristics tend to alter wildly under thermal treatment.

The research objective is to establish empirical relationships of bazhenite rheological properties versus temperature. Under investigation was a collection of five cylindrical specimens of radius $R = 14$ mm and height $H_0 = 35.8$ mm, sampled from Bazhenov formation (Salym petroleum deposit, West Siberia). The tests were executed at the original test bench (Fig. 1a), which design is presented in Fig. 1b. A specimen placed between two aluminium punches was charged to a tube-type furnace. The system was subjected to vertical compression by a load of $M = 23$ kg in mass, followed by a stepwise heating. The specimen temperature was controlled by a heat sensor (red line in Fig. 2), variations in height versus time $H(t)$ were recorded by a dial-test indicator (absolute precision being 1 μm). Given a relative variation in H for 12 h becoming less than 5%, the temperature was increased up to a magnitude of the follow-on level for a few minutes, therewith the specimen used to elongate because of the thermal expansion of the rock.

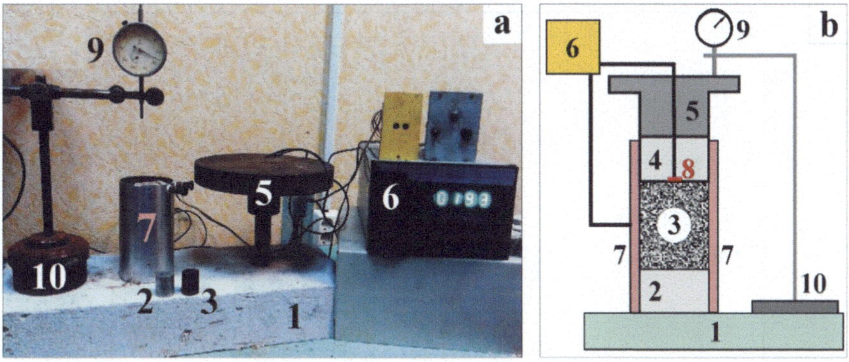

Fig. 1. Test bench (a) and experimental design (b): 1–heat-sealing base; 2, 4–bottom and top punches; 3–specimen; 5–load; 6–temperature-control module; 7–tube-type furnace; 8–heat sensor; 9–dial-test indicator; 10–stand.

Test data – variations in a relative height of the specimen $D(t) = H(t)/H_0$ – are shown by green circles in Fig. 2.

Let us consider axial strain

$$\varepsilon(t) = 1 - H(t)/H(t_n)$$

at successive time intervals $[t_n, t_{n+1})$ $(n = 1, 2, 3)$, where t_n – time moment when constant temperature T_n is reached in the test specimen (Fig. 2). In each of these intervals ε tends to a constant magnitude, so the deformation process can be described by Kelvin-Voigt viscoelastic model [21, 22]

$$E_n\varepsilon + \eta_n\dot{\varepsilon} = S, \quad \varepsilon(t_n) = 0, \tag{1}$$

where E_n and η_n are Young modulus and viscosity of geomaterial, respectively; $S = Mg/(\pi R^2)$ is constant stress applied to the specimen, g is gravity acceleration. Solution of (1) is

$$\varepsilon(t) = \varepsilon_n^r[1 - \exp(-E_n(t - t_n)/_n], \tag{2}$$

$\varepsilon_n^r = S/E_n$ is residual strain.

Let state the coefficient inverse problem: to find E_n and η_n from thermobaric test data (Fig. 2). Introduce the objective function

$$\Phi_n(E_n, \eta_n) = \frac{\sqrt{I_n \sum_i [\varepsilon(t_i, E_n, \eta_n) - \varepsilon_i]^2}}{\sum_i \varepsilon_i},$$

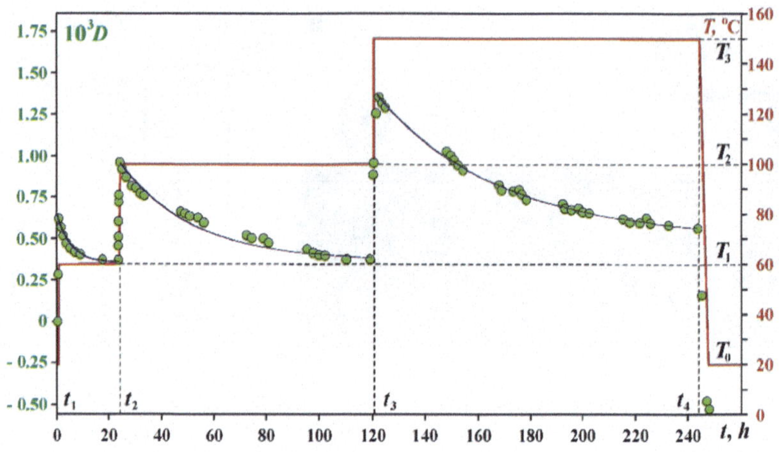

Fig. 2. Temperature graph and experimental data (Specimen 4).

where $\varepsilon(t_i, E_n, \eta_n)$ is a value of function (2) at some values of Young module and viscosity; input data $\varepsilon_i = 1 - H(t_i)/H(t_n)$ are calculated by a measured specimen height at moment t_i; summing is performed relative to i, for which $t_i \in [t_n, t_{n+1})$ (I_n – their amount, Table 1, line 2). A priory estimation of Young modulus $E_n^0 = SH_0/(H_0 - H_{min})$ (H_{min} – the minimum specimen height at the study time interval) is reported in Table 1 (line 3).

Table 1. Rheological properties of Specimen 4 at different temperatures.

	N	1	2	3
1	T_n, °C	60	100	150
2	I_n	11	36	52
3	E_n^0, MPa	211.5	66.0	50.8
4	E_n^*, MPa	220.6	59.8	47.1
5	η_n^*, TPa · c	1.48	7.38	7.95

According to numerical experiments, the objective functions are unimodal: Fig. 3 demonstrates level lines of Φ_n. The minimum point (E_n^*, η_n^*) of Φ_n (Table 1, lines 4 and 5), bringing a solution to the inverse problem was found by a modified conjugate-gradient method [23].

Approximation of the data in Table 1 (line 4) gives the relationship of Young modulus versus temperature

$$E(T) = E_0 \exp(-\alpha T/T_0), \tag{3}$$

where $T_0 = 20$ °C, $E_0 = 647.9$ MPa and $\alpha = 0.384$, the relative error being 19%.

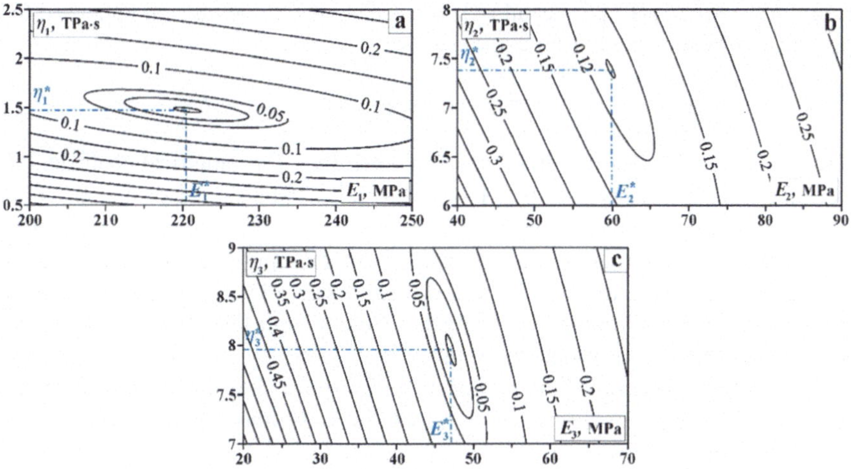

Fig. 3. Level line of objective functions: Φ_1 (a); Φ_2 (b); Φ_3 (c).

It is known [24, 25], that permeability k and porosity m of rocks are related through an empirical law

$$k \cong m^{3/2} \tag{4}$$

and m depends on mean normal stress σ [26]

$$m(\sigma) = m_0 \exp(-\sigma/K), \tag{5}$$

where K – bulk modulus, positive values of σ correspond to compression. Given $3K = E/(1-2\nu)$ (ν – Poisson ratio), it is possible to find the relationship of permeability vs. stresses and temperature from (3), (4), and (5)

$$k(T) = k(T_0)\exp\left[\frac{9(1-2\nu)\sigma}{2E_0}(\exp(\alpha) - \exp(\alpha T/T_0))\right], \tag{6}$$

value $k(T_0)$ can be determined by standard procedures [27, 28].

3 Experimental Determination of Stress–Permeability Relationship in Granular Geomaterials

The gas-hydrate reservoirs are usually granular [29] or weakly consolidated [30] media, which poroperm properties depend on grain-size composition, porosity, packing and other characteristics [31–33]. The stress state of well environment, packing of a granular matter, and, as a consequence, poroperm properties suffer changes as soon as a producing horizon is exposed and gas-hydrates production is launched. A good deal of laboratory [34–38] and field [12, 39, 40] data is stored on the relationship of rock (mainly oil-bearing rock) permeability k versus stresses. Experimental data used to be approximated by exponential function [13, 14]

$$k(\sigma_e) = k_0 \exp(-\beta\sigma_e), \quad (7)$$

where $\sigma_e = p - \sigma$ is effective stress, p is fluid pressure, β is empirical coefficient. For rocks of 15–20% porosity $\beta \cong 0.001 - 0.01\,\text{MPa}^{-1}$. Publications containing similar test data on granular media are much scarce in number [41, 42]. However consideration of permeability–stress relationship of type (7) in multiphysical models, describing deformation and mass transfer in a near well zone [43], appreciably influences hydrodynamic and electromagnetic fields and therethrough GIS data inversion results when estimating potential productivity [44].

In the present paragraph, the approach to determine the relationship of granular material permeability versus stresses is substantiated theoretically and evaluated on a physical model.

3.1 Experimental Design and Test Procedure

A measurement cell is a polyurethane sleeve of parallelepiped shape of $l = 105$ mm in length, $h = 30$ mm in width and height with end closed with metal flanges with connectors to a measurement system (Fig. 4). The cell is filled with calibrated sand of 160–250 μm in size, 1650 kg/m^3 of bulk density and 0.165 kg of total weight. The sand was subjected to the maximum consolidation in the wetted state and follow-on drying in the sleeve. The vertical stress S_V was individually applied by means of a loading device to every of three b_1, b_2, b_3 sections of the top edge under control of stress gauges S_1, S_2 and S_3 (Fig. 5). The side edge $x = l$ was under the constant air pressure $p = p_i$, on the opposite edge $p = p_0$ (p_0 – atmosphere pressure); the flow rate was measured in the stationary filtration mode.

Flow rate Q_0 was recorded at $p_i = 2.0, 2.2, \ldots, 3.0$ bar for the initial packing. Stress $S_V = 9$ bar was applied at section b_1; flow rates $Q_{11}, \ldots, Q_{16}$ were measured at the same values of p_i, next section b_2 was loaded and finally the load was applied to section b_3. The relative test values of $\bar{Q}_{in} = Q_{ni}/Q_0$ are summarized in Table 2.

The cell was shaked before every next series of measurements. The initial filtration characteristics of the test packing were controlled by measuring of flow rate $Q_0(p_i)$. Then the loading of sections in the top edge of the physical model was repeated in the above sequence with flow rate measurements: the data on $S_V = 10.5$ and 12 bar are cited in Tables 3 and 4, respectively.

3.2 Model of Experiment

The stationary filtration in the cell was described by a 1D model, including continuity equation

$$div\vec{V} = 0 \quad (8)$$

and Darcy law

Fig. 4. Test bench: 1–cell; 2–flanges; 3–punches; 4–loading device; 5–flowmeter.

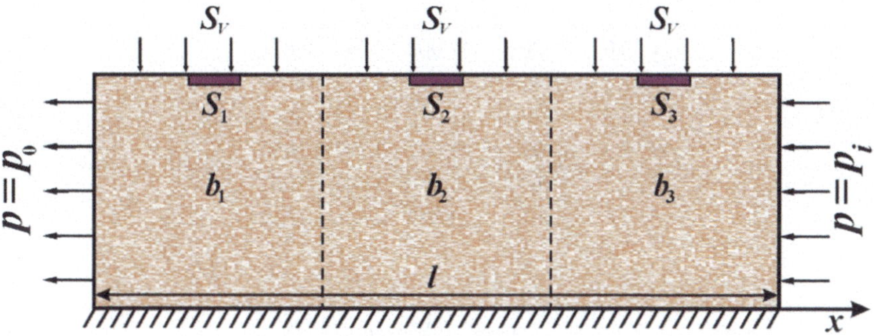

Fig. 5. Experimental design and boundary conditions of the model.

Table 2. Relative flow rate under $S_V = 9$ bar

i	p_i, bar	Q_0, cc/s	Section of loading		
			b_1	b_1, b_2	b_1, b_2, b_3
			$n = 1$	$n = 2$	$n = 3$
			$\bar{Q}_{in}$		
1	2.0	123	1.031	0.894	0.715
2	2.2	154	1.089	0.913	0.717
3	2.4	208	1.074	0.907	0.736
4	2.6	224	1.149	0.925	0.758
5	2.8	266	1.123	0.934	0.823
6	3.0	307	1.174	0.982	0.835

Table 3. Relative flow rate under $S_V = 10.5$ bar

i	p_i, bar	Q_0, cc/s	Section of loading		
			b_1	b_1, b_2	b_1, b_2, b_3
			$n = 1$	$n = 2$	$n = 3$
			$\bar{Q}_{in}$		
1	2.0	121	1.046	0.794	0.649
2	2.2	150	1.056	0.826	0.685
3	2.4	205	1.044	0.834	0.688
4	2.6	223	1.077	0.854	0.718
5	2.8	265	1.099	0.882	0.724
6	3.0	305	1.122	0.907	0.746

Table 4. Relative flow rate under $S_V = 12$ bar

i	p_i, bar	Q_0, cc/s	Section of loading		
			b_1	b_1, b_2	b_1, b_2, b_3
			$n = 1$	$n = 2$	$n = 3$
			$\bar{Q}_{in}$		
1	2.0	120	1.002	0.743	0.627
2	2.2	147	1.011	0.774	0.639
3	2.4	204	1.018	0.784	0.641
4	2.6	222	1.022	0.792	0.657
5	2.8	264	1.057	0.797	0.661
6	3.0	305	1.073	0.866	0.692

$$\vec{V} = -k\nabla p/\mu, \tag{9}$$

where $\vec{V} = (w, 0, 0)$, w is velocity in x direction (Fig. 5), μ is gas viscosity, k is permeability, depending on effective stress under (7), p is pressure. System (7)–(9) is reduced to equation

$$\frac{\partial}{\partial x}\left(\frac{k_0}{\mu} e^{\beta(\sigma - p)} \frac{\partial p}{\partial x}\right) = 0 \tag{10}$$

with boundary conditions (Fig. 5)

$$p(0) = p_0,\ p(l) = p_i. \tag{11}$$

Solution of system (10), (11) is

$$p(x) = \frac{1}{\beta} ln\left[e^{\beta p_0} + \left(e^{\beta p_i} - e^{\beta p_0}\right) G(x, \beta)/G(l, \beta)\right], \quad (12)$$

$G(x, \beta) = \int_0^x e^{\beta\sigma(\xi)} d\xi$, $\sigma(\xi)$ is distribution of mean normal stress within interval $[0, l]$. In the case under consideration σ is a piecewise-constant function, in particular, if a load is applied at section b_1, then $\sigma(\xi) = S_V Z(l/3 - \xi)/3$ (Z being Heaviside step function). It should be note that at $\beta \to 0$ from (12) it follows

$$p(x) = p_0 + (p_i - p_0)x/l.$$

Using (12), it is possible to determine a flow rate at $x = 0$

$$Q(p_i, \beta) = Q_0 F(p_i, \beta),$$
$$Q_0 = \frac{k_0 h^2}{\mu} \frac{p_i - p_0}{l}, \quad F(p_i, \beta) = \frac{\exp(\beta p_i) - \exp(\beta p_0)}{\beta(p_i - p_0)} \frac{l}{G(l, \beta)}$$

thereto Q possesses the following properties

$$\lim_{p_i \to p_0} Q(p_i, \beta) = 0 \quad \lim_{\beta \to 0} Q(p_i, \beta) = Q_0.$$

Level lines $F(p_i, \beta)$ at $S_V = 6$ bar and different loading schemes n are shown in Fig. 6 in dimensionless coordinates $(p_i/p_0, \beta p_0)$. It is explicit that pressure p_i at which $F > 1$ is observed at any β values; it should be considered as a consequence of (7): provided $p > \sigma$, then permeability exceeds initial value k_0.

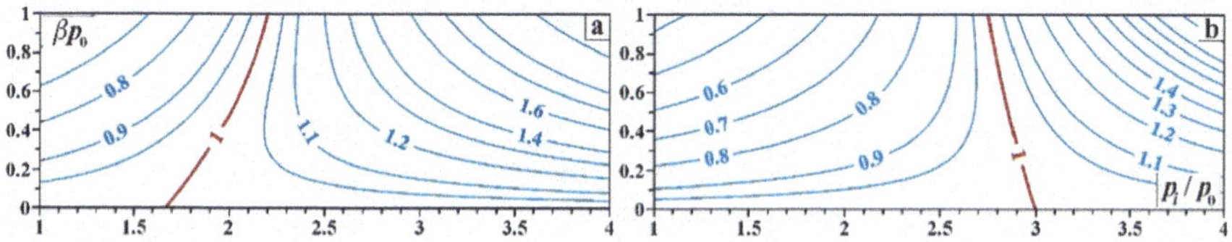

Fig. 6. Level lines of F: $n = 2$ (a); $n = 3$ (b)

3.3 Interpretation of the Experimental Data

Let assess empirical parameter β in terms of test results on relative flow rate $\bar{Q}_{in}$ under different loading schemes n and pressure p_i (Tables 2, 3, and 4). To do this, let consider the objective function

$$\Psi(\beta) = \frac{\sqrt{18 \sum_{n=1}^{3} \sum_{i=1}^{6} [F(p_i, \beta) - \bar{Q}_{in}]^2}}{\sum_{n=1}^{3} \sum_{i=1}^{6} \bar{Q}_{in}}$$

represented a relative discrepancy between theoretical and test values of flow rate. The minimum points of Ψ are required β, cited in Table 5 for different stress values S_V. The tendency turned out is though insignificant, but reduction in parameter β in (7) with growth of stresses in a granular medium alternatively to the case with rocks.

Table 5. Parameter β in (7) from the experimental data

S_V, bar	β_*, 1/bar	Δ, %
9.0	0.215	4.5
10.5	0.198	6.9
12.0	0.189	5.6

$\Delta = \Psi(\beta_*)$ – relative error

Figure 7 demonstrates the new-obtained relationships $F(p_i, \beta)$ at $\beta = \beta_*$, digits nearby the lines indicate the model-loading schemes; circles are test data.

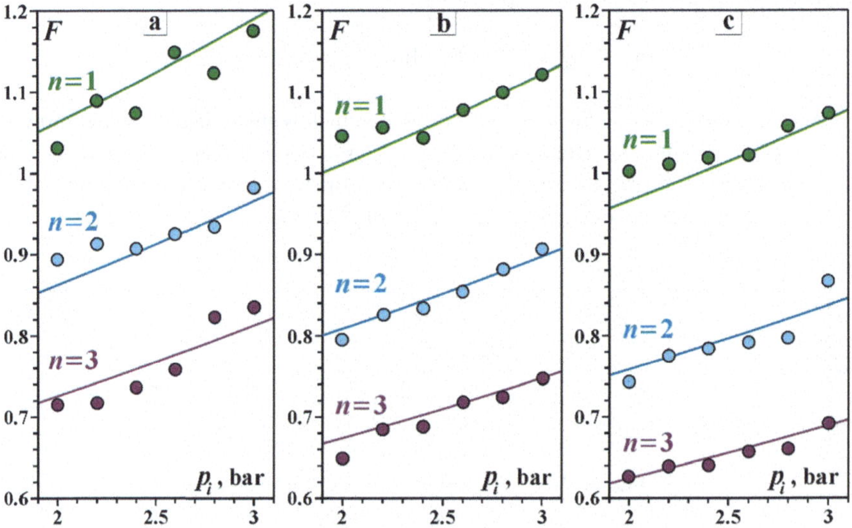

Fig. 7. Functions $F(p_i, \beta)$: $S_V = 9$ bar (a); $S_V = 10.5$ bar (b); $S_V = 12$ bar (c)

It is essential to lay emphasis that the new-proposed method enables to determine parameter β in empirical relationship (7) disregarding a viscosity of a fluid used in the tests and initial permeability of a packing practiced for a granular geomaterial.

4 Analysis of Well Output Characteristics Under Stress-Dependent Permeability and Thermal Treatment of Oil-Bearing Formation

A rise of temperature in a production bed leads to reduction in viscosity μ [19, 44, 45] and growth of compressibility C [46] of oil, and a fall of the matrix bulk module K (given that kerogen content being high) along with accompanying decline in permeability. Let analyze the effect of the above factors on metering characteristics of a well in terms of an axis-symmetric model of a fracture-porous rock mass [47]. The effective compressibility of the oil-filled formation can be evaluated as

$$C_e = m_0 C + (1 - m_0)/K, \tag{13}$$

where K is calculated by (6). In Table 6 the laboratory test data on Bazhenov highly-viscous oil at different temperatures are reported in cursive [45, 46].

Table 6. Viscosity and compressibility of oil at different temperatures.

T, °C	20	40	50	60	80	100	120	150
μ, Pa · s	3.11	***2.47***	***2.06***	***1.74***	***1.44***	1.07	0.82	0.55
$10^9 C$, Pa^{-1}	0.91	***1.28***	1.45	***1.67***	***2.09***	***2.64***	3.29	4.76

The present data are approximated with exponential functions:

$$\mu(T) = \mu_0 \exp(-\delta_1 T/T_0) \tag{14}$$

($\mu_0 = 4.064\,\text{Pa} \cdot \text{s}$, $\delta_1 = 0.266$, relative error $\Delta = 3.4\%$) and

$$C(T) = C_0 \exp(\delta_2 T/T_0) \tag{15}$$

($C_0 = 0.797 \cdot 10^9 \text{Pa}^{-1}$, $\delta_2 = 0.238$, $\Delta = 1.5\%$), which are used to calculate the rest values of μ and C in Table 6.

Let apply the thermal treatment to an oil-bearing bed within an external reservoir boundary (for example, by steam injection [48]); constant temperature T_s is gained. Introduce empirical relationships (13), (14), and (15) (Case I) and next additionally (6) (Case II) into the poroelastic model of borehole environment [16, 47] and simulate an annual oil production volume V.

The computation is based on routine deformation and poroperm properties of a productive oil bed [3, 49], occurring at 2 km depth. Figure 8a demonstrates time variation of V at different temperatures T_s (in relation to the maximum value V_{max} at $T_s = 150°\text{C}$) in terms of the model (Case I). Figure 8b shows the same function, but computed with regard to variations in permeability under (3) (Case II). It is explicit that the selected numerical parameters of the geomechanical-hydrodynamic model can really give the double overestimated production rate prediction in the case of

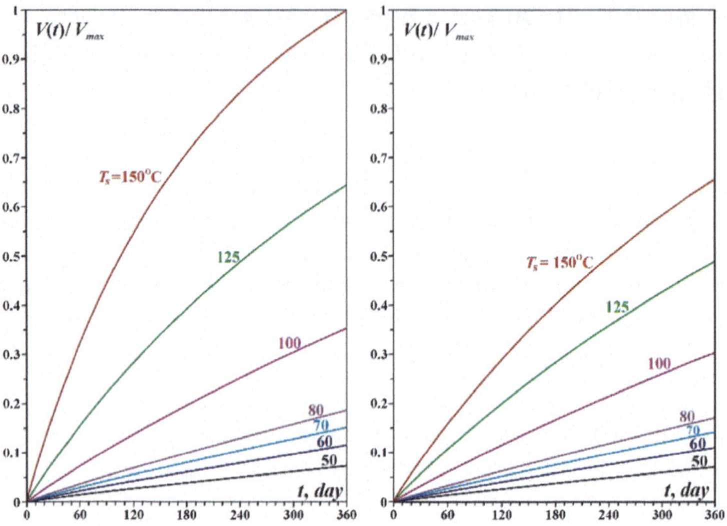

Fig. 8. Annual oil production volume V: Case I (a); Case II (b)

disregarding the thermal effect on elastic characteristics of an oil-bearing bed, viz., the resultant permeability decline as a consequence.

5 Conclusion

The processes enabling to establish a quantitative relation between filtration properties of oil-bearing beds consisting of granular and high-organic rocks versus stresses and temperature are theoretically justified and tested. The processes rest on solving the coefficient inverse problems using test data obtained at original laboratory test facilities. The comparative analysis of metering well characteristics revealed that a disregard of the empirical permeability–effective stress relationships might lead to appreciably overestimated oil production prediction.

Acknowledgement. The investigations was partially supported by the Russian Foundation for Basic Research: Project No. 16-05-00573 (paragraphs 2&4) and Project No. 18-05-00830 (paragraph 3).

References

1. Speight, J.G.: An Introduction to Petroleum Technology, Economics, and Politics. Wiley, New York (2011)
2. Dake, L.P.: The Practice of Reservoir Engineering. Elsevier, The Netherlands (2001)
3. van Golf-Racht, T.D.: Fundamentals of Fractured Reservoir Engineering. Elsevier, Amsterdam (1982)

4. Rutqvist, J., Wu, Y.-S., Tsang, C.-F., et al.: A modeling approach for analysis of coupled multiphase fluid flow, heat transfer, and deformation in fractured porous rock. Int. J. Rock Mech. Min. Sci. **39**, 429–442 (2002)
5. Coussy, O.: Mechanics and Physics of Porous Solids. Wiley, New York (2010)
6. Nie, R.-S., Meng, Y.-F., Jia, Y.-L., et al.: Dual porosity and dual permeability modeling of horizontal well in naturally fractured reservoir. Transp. Porous Media **92**(1), 213–235 (2012)
7. Zhou, X., Ghassemi, A.: Finite element analysis of coupled chemo-poro-thermo-mechanical effects around a wellbore in swelling shale. Int. J. Rock Mech. Min. Sci. **46**(4), 769–778 (2009)
8. Zhuang, X., Huang, R., Liang, C., Rabczuk, T.: A coupled thermo-hydro-mechanical model of jointed hard rock for compressed air energy storage. Mathematical Problems in Engineering, ID 179169 (2014)
9. Wu, K., Olson, J.E.: Simultaneous multifracture treatments: fully coupled fluid flow and fracture mechanics for horizontal wells. SPE J. **20**(02), 337–346 (2015)
10. Wu, Y.-S.: Multiphase fluid flow in porous and fractured reservoirs. Elsevier, Amsterdam (2016)
11. Yeltsov, I.N., Nazarov, L.A., Nazarova, L.A., et al.: Logging interpretation taking into account hydrodynamical and geomechanical processes in an invaded zone. Dokl. Earth Sci. **445**(2), 1021–1024 (2012)
12. Daigle, H., Rasromani, E., Gray, K.E.: Near-wellbore permeability alteration in depleted, anisotropic reservoirs. J. Petrol. Sci. Eng. **157**, 302–311 (2017)
13. Zoback, M.D., Nur, A.: Permeability and effective stress. Bull. Am. Assoc. Petrol. Geol. **59**, 154–158 (1975)
14. Ma, J.: Review of permeability evolution model for fractured porous media. J. Rock Mech. Geotech. Eng. **7**(3), 351–357 (2015)
15. Nazarova, L.A., Nazarov, L.A., Golikov, N.A.: Assessment of rheological properties of bazhenov formation by thermobaric test data. J. Min. Sci. **53**, 434–440 (2017)
16. Yeltsov, I.N., Nesterova, G.V., Sobolev, A.Y., Epov, M.I., Nazarova, L.A., Nazarov, L.A.: Geomechanics and fluid flow effects on electric well logs: multiphysics modeling. Russ. Geol. Geophys. **55**(5–6), 775–783 (2014)
17. BP Statistical Review of World Energy (2018)
18. Zapivalov, N.P.: Geological and technological peculiarities in development of difficult mineral resources. Oil Ind. **6**, 57–59 (2005)
19. Butler, R.M.: Thermal Recovery of Oil and Bitumen. Prentice-Hall, Englewood Cliffs (1991)
20. Bennion, D.B., Ma, T., Thomas, F.B. et al.: Laboratory procedures for optimising the recovery from high temperature thermal heavy oil and bitumen recovery operations. In: Petroleum Society's 8th Canadian International Petroleum Conference, paper 2007–206, Canada (2007)
21. Cristescu, N.: Elastic-viscoplastic constitutive equation for rock. Int. J. Min. Sci. Geomech. Abs. **24**(5), 271–282 (1987)
22. Wu, L., Zuo, Q., Lu, Z.: Study on the constitutive model of visco-elasticity-plasticity considering the rheology of rock mass. Adv. Mater. Res. **639–640**, 567–572 (2013)
23. Nazarov, L.A., Nazarova, L.A., Karchevsky, A.L., Panov, A.V.: Estimation of stresses and deformation properties of rock masses which is based on the solution of an inverse problem from the measurement data of the free surface displacement. J. Appl. Ind. Math. **7**, 234–240 (2013)
24. Dullien, F.A.L.: Porous Media: Fluid Transport and Pore Structure. Academic Press, New York (1979)
25. Carman, P.C.: Flow of Gases Through Porous Media. Butterworths Scientific Publications, London (1956)
26. Shchelkachev, V.N.: Selected Works in Two Volumes, vol. 1. Nedra, Moscow (1990)
27. GOST N. 26450.2-85. Rocks. Method for determination of absolute gas permeability coefficient by stationary and non-stationary filtration (1985)

28. Vafai, K.: Handbook of Porous Media, 2nd edn. Taylor & Francis Group, USA (2005)
29. Istomin, V.A., Yakushev, V.S.: Gas Hydrates in Natural Conditions. Nedra, Moscow (1992)
30. Isaeva, O.A., Bruzhes, L.N., Vavilin, V.A., et al.: Lithology, mineralogy and pore space of the ryabchik-type AV1 horizon in the Urevskoe oilfield. In: Sedimentary Basins, Sedimentation and Post-Sedimentation Processes in Geological History. In: Proceedings of VII All-Russian Lithological Conference, vol. 1, pp. 388–392. INGG SO RAN, Novosibirsk (2013)
31. Urumovic, K., Urumovic, Sr.K.: The effective porosity and grain size relations in permeability functions. Hydrol. Earth Syst. Sci. Discuss. **11**, 6675–6714 (2014)
32. Knackstedt, M.A., Duplessis, J.P.: Simple permeability model for natural granular media. Geophys. Res. Lett. **23**(13), 1609–1612 (1996)
33. Kuhn, M.R., Sun, W., Wang, Q.: Stress-induced anisotropy in granular materials: fabric, stiffness, and permeability. Acta Geotech. **10**(4), 399–419 (2015)
34. Holt, R.M.: Permeability reduction induced by a nonhydrostatic stress field. SPE Formation Eval. **5**, 444–448 (1990)
35. Heller, R., Vermylen, J., Zoback, M.: Experimental investigation of matrix permeability of gas shales. Bull. Am. Assoc. Petrol. Geol. **98**(5), 975–995 (2014)
36. Roi, R., Paredes, X., Holtzman, R.: Reactive transport under stress: Permeability evolution in deformable porous media. Earth Planet. Sci. Lett. **493**, 198–207 (2018)
37. Pan, Z., Connell, L.D.: Modelling permeability for coal reservoirs: a review of analytical models and testing data. Int. J. Coal Geol. **92**, 1–44 (2012)
38. Geng, Y., Tang, D., Xu, H., et al.: Experimental study on permeability stress sensitivity of reconstituted granular coal with different lithotypes. Fuel **202**, 12–22 (2017)
39. Rhett, D.W., Teufel, L.W.: Effect of reservoir stress path on compressibility and permeability of sandstones. In: SPE Annual Technical Conference and Exhibition. SPE Paper No. 24756, Washington DC (1992)
40. Randall, M.S., Conway, M., Salter, G., Miller, S.: Pressure-dependent permeability in shale reservoirs implications for estimated ultimate recovery. In: AAPG Search and Discovery Article N 9012.2011, AAPG Hedberg Conference, Austin, Texas (2010)
41. Chapuis, R.P., Gill, D.E., Baass, K.: Laboratory permeability tests on sand: influence of the compaction method on anisotropy. Can. Geotech. J. **26**(4), 614–622 (1989)
42. Wong, R.C.K.: Strain-induced anisotropy in fabric and hydraulic parameters of oil sand in triaxial compression. Can. Geotech. J. **40**(3), 489–500 (2003)
43. Nazarova, L.A., Nazarov, L.A., Epov, M.I., Eltsov, I.N.: Evolution of geomechanical and electro-hydrodynamic fields in deep well drilling in rocks. J. Min. Sci. **49**, 704–714 (2013)
44. Farouq Ali, S.M., Jones, J.A., Meldau, R.F.: Practical heavy oil recovery. Mont Resour Inc., Bakersfield (1997)
45. New Chemist's and Engineer's Handbook, V.1, Basic properties of inorganic, organic, and elementary-organic compounds. Mir i Semiya, Moscow (2002)
46. Bilinkin, G.P., Kuvandikov, I.Sh.: Dependence of volume coefficient, coefficient of compressibility, density of deep–subsided bed oils from pressure, temperature, composition and gas content. Geology of Oil and Gas, 9 (1992)
47. Nazarova, L.A., Nazarov, L.A.: Evolution of stresses and permeability of fractured-and-porous rock mass around a production well. J. Min. Sci. **52**, 424–431 (2016)
48. Mozaffari, S., Nikookar, M., Ehsani, M.R., et al.: Numerical modeling of steam injection in heavy oil reservoirs. Fuel **112**, 185–192 (2013)
49. Nazarova, L.A., Nazarov, L.A.: Evolution of hydrodynamical and stress fields in near-well zone in fractured porous media. In: Proceedings of the 6th Biot Conference on Poromechanics 6, Paris, pp. 287–294 (2017)

Development of Model of Sea Surface Elevations Distributions Created by Wind Waves and Swell

A. S. Zapevalov(✉)

Marine Hydrophysical Institute of RAS, Sevastopol, Russian Federation
sevzepter@mail.ru

Abstract. A new approach to solving the fundamental problem, which is the construction of the probability density function of the parameters of sea surface waves, is solved. Sea waves belong to the class of quasi-Gaussian processes. Probability density functions of such processes are described by models based on truncated Gram-Charlier series, which allow us to describe the probability density of a random variable only in a limited range. To overcome this limitation, a combined model is used in which either the Gram-Charlier distribution or the Gaussian distribution correspond to different ranges. The transition from one distribution to another is described by a function that acts as a filter. A new criterion for determining the correctness of the combined model based on the evaluation of the cumulants obtained with its help is introduced. It is also shown that the presence or absence of negative values in the truncated Gram-Charles distribution should be taken into account when choosing filter parameters.

Keywords: Sea surface · Waves · Probability density functions · Combined model

1 Introduction

A wide range of problems related to remote sensing of the ocean from space, the interaction of surface and acoustic waves, the impact of waves on ships and others, requires a detailed description structures and variability of sea surface [1–3].

A model that represents the sea surface as a linear superposition of sinusoidal waves dominated in oceanology for a long time. According to the central limit theorem, the Gaussian distribution corresponds to this model. The development of nonlinear models began after the discovery of wave-wave interactions [4]. It was shown that the skewness of the elevation distribution depends on the average slope of the sea surface [5].

The nonlinearity of sea waves is small, so deviations from the Gaussian distribution are also small [6]. The distributions of the elevations and slopes of the sea surface belong to the class of quasi-Gaussian distributions. Quasi-Gaussian distributions are usually described by truncated Gram-Charlier series. Typically, models that include the first five members of the Gram-Charlier series are used [7]. The restriction is caused by the fact that in the experiments the statistical moments are determined only up to the fourth order.

V. I. Karev et al. (Eds.): *Physical and Mathematical Modeling of Earth and Environment Processes (2018)*, SPEES, pp. 107–113, 2019.
https://doi.org/10.1007/978-3-030-11533-3_11

The disadvantage of the models constructed using truncated Gram-Charlier series is the appearance of negative values of the probability density function [7, 8]. To exclude negative values in [9], it was suggested to use a combined model. In the range where the values of the random variable are small, the probability density function is described by a truncated Gram-Charlier distribution. Outside this area, it approaches the Gauss distribution.

2 Truncated Gram-Charlier Distribution

The Gram-Charlier series are based on the expansion in a series of derivatives of the function

$$PN(x) = \frac{1}{\sqrt{2\pi}} \exp\left(-\frac{1}{2}x^2\right). \tag{1}$$

Derived function $PN(x)$ are defined by the expression

$$\frac{d^n}{dx^n} PN(x) = (-1)^n H_n(x) \cdot PN(x), \tag{2}$$

where $H_n(x)$ are Hermite polynomials that have the property of orthogonality. The probability density of a quasi-Gaussian distribution of a random variable x constructed on the basis of Gram-Charlier series can be represented as

$$P_{G-C}(x) = \sum_{n=0}^{\infty} a_n H_n(x) \frac{1}{\sqrt{2\pi}} \exp\left(-\frac{1}{2}x^2\right). \tag{3}$$

where a_n are the coefficients of the series The coefficients a_n are expressed in terms of statistical moments of a random variable or its cumulants. Usually, when modeling the probability density function of sea surface elevations Edgeworth's A-type form of the Gram-Charlier series is used. The probability density function of sea surface elevations η is described by expression

$$P_{G-C}(\eta) = \frac{\exp\left(-\frac{\eta^2}{2D}\right)}{\sqrt{2\pi D}} \left[1 + \frac{A}{6} H_3\left(\frac{\eta}{\sqrt{D}}\right) + \frac{E}{24} H_4\left(\frac{\eta}{\sqrt{D}}\right)\right], \tag{4}$$

where D is variance, A is skewness, E is kurtosis, $P_G(\eta) = \exp\left(-\frac{\eta^2}{2D}\right) / \sqrt{2\pi D}$ is Gaussian distribution. It is assumed that the mean value η is zero.

The values of skewness and kurtosis in sea lie mainly within

$$0 \le A \le 0.4, \quad -0.4 \le E \le 0.4. \tag{5}$$

Measurements conducted in laboratory and sea have shown that cumulants A and E are not correlated with each other.

We introduce a normalized random variable $\tilde{\eta} = \eta/\sqrt{D}$, and will analyze the changes $P_{G-C}(\tilde{\eta})$ in the range $|\tilde{\eta}| > 2$. The results of this analysis are shown in Fig. 1. Curves corresponding to different skewness values are plotted in steps of 0.1.

The problem of the appearance of negative values also takes place when modeling the probability density function of sea surface slopes ξ. Approximations constructed on the basis of the Gram-Charlier series correctly describe the slope probability density function only in the range [10]

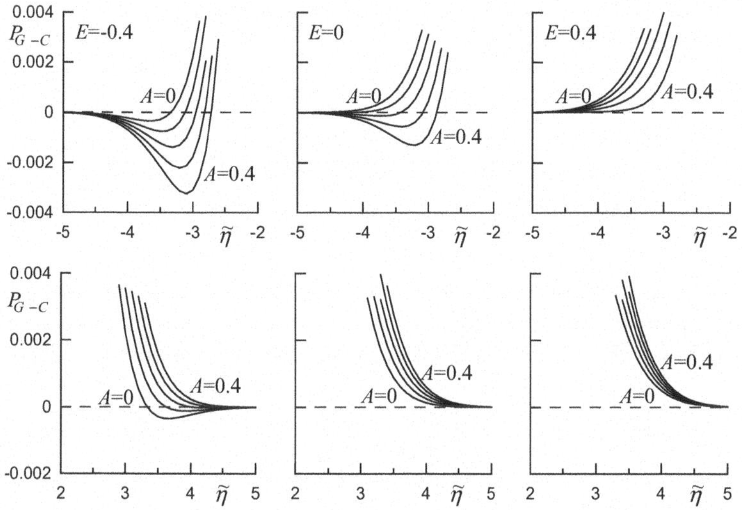

Fig. 1. Fragments of the probability density function (4) in the range $|\tilde{\eta}| > 2$.

$$|\xi| < 2.5\sigma_{\xi}, \tag{6}$$

where σ_{ξ} is the root mean square. A combined model is used to avoid negative values [9]. For surface elevations, the combined model can be introduced in the form

$$P_C(\eta) = \frac{\exp\left(-\frac{\tilde{\eta}^2}{2}\right)}{\sqrt{2\pi D}} \left\{1 + \left[\frac{A}{6} H_3(\tilde{\eta}) + \frac{E}{24} H_4(\tilde{\eta})\right] F(\tilde{\eta})\right\}, \tag{7}$$

where the function $F(\tilde{\eta})$ acts as a filter. The combined model must be nonnegative and satisfy the "smoothness" condition, i.e. the derivative $dP_C(\tilde{\eta})/d\tilde{\eta}$ must not change in an abruptly.

In [9] the symmetric filter in the form of

$$F(\tilde{\eta}) = \exp[-(|\tilde{\eta}|/d)^n] \tag{8}$$

was used. The parameter d defines the range where the functions are $P_{G-C}(\eta)$ and $P_C(\eta)$ close. Outside of this range, the function $P_C(\eta)$ tends to $P_G(\eta)$. The parameter n determines how fast the functions $P_C(\eta)$ and $P_G(\eta)$ approach.

3 Cumulants

In the sea it is practically impossible to obtain an estimate of the probability density function of sea surface elevations in the range $|\tilde{\eta}| > 2.5 - 3$. Therefore, you need to create a criterion for selecting a filter $F(\tilde{\eta})$. Let's consider two types of filters: symmetrical $F_S(\tilde{\eta})$ and asymmetrical $F_A(\tilde{\eta})$

$$F_S(\tilde{\eta}) = \exp[-(|\tilde{\eta}|/d_S)^{n_S}], \tag{9}$$

$$F_A(\tilde{\eta}) = \begin{cases} \exp[-(|\tilde{\eta}|/d_1)^{n_1}] & npu\ \tilde{\xi} \leq 0 \\ \exp[-(|\tilde{\eta}|/d_2)^{n_2}] & npu\ \tilde{\xi} > 0. \end{cases} \tag{10}$$

As a new additional criterion, we will use the proximity of cumulants calculated from the combined model, given the values of skewness A and E kurtosis, i.e. we will compare these parameters with the cumulants calculated by the combined model

$$\lambda_3 = \int_{-b}^{b} \tilde{\eta}^3 P_C(\tilde{\eta})\,d\tilde{\eta}, \tag{11}$$

$$\lambda_4 = \int_{-b}^{b} \tilde{\eta}^4 P_C(\tilde{\eta})\,d\tilde{\eta} - 3. \tag{12}$$

The lower and upper limits of integration in (11) can be taken equal respectively to −5 and 5. The following will show that increasing the limits of integration in comparison with the indicated values does not influence on the values λ_3 and λ_4.

When the values of asymmetry and excess of sea surface elevations given by conditions (5), three situations are possible: negative values in model (4) appear both at $\tilde{\eta} < 0$ and with $\tilde{\eta} > 0$; negative values appear only when $\tilde{\eta} < 0$; the model (4) has no negative values. In the first situation, the appearance of negative values occurs at close values $|\tilde{\eta}|$ (see Fig. 1). Therefore, it is advisable to use a symmetric filter $F_S(\tilde{\eta})$. Unfortunately, in this situation it is not possible to achieve the proximity of the parameters A and λ_3, as well as the proximity of the parameters E and λ_4.

In the second situation when the condition $P_{G-C}(\tilde{\eta}) < 0$ is satisfied only at $\tilde{\eta} < 0$, it is necessary to use an asymmetric filter $F_A(\tilde{\eta})$. The shape of the symmetric and asymmetric filters is shown in Fig. 2. For the symmetric filter, the following parameters are selected $d_S = 3.5$ and $n_S = 2.5$; for the asymmetric filter – $d_1 = d_S$, $d_2 = 5$, $n_1 = n_S$, $n_2 = 1.9$.

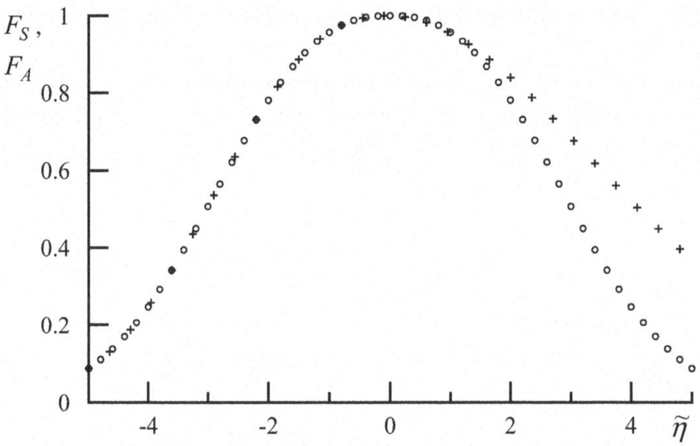

Fig. 2. Symmetric F_S (O) and asymmetric F_A (+).

The filters shown in Fig. 2 were constructed for the situation when $A = 0.4$ and $E = 0.2$. In the range $\tilde{\eta} < 0$ where the approximation (4) has negative values of the filters $F_S(\tilde{\eta})$ and $F_A(\tilde{\eta})$ coincide. The cumulants obtained for the combined model with a symmetric filter are $\lambda_3 = 0.19$, $\lambda_4 = 0.05$. The cumulants obtained for the combined model with a asymmetric filter are $\lambda_3 = 0.24$, $\lambda_4 = 0.19$. The use of the asymmetric filter allowed to significantly bring together the fourth-order cumulant estimates and slightly improve the third-order cumulant estimates. Fragments of probability density functions calculated for the range $2.5 < |\tilde{\eta}| < 4.5$ are shown in Fig. 3.

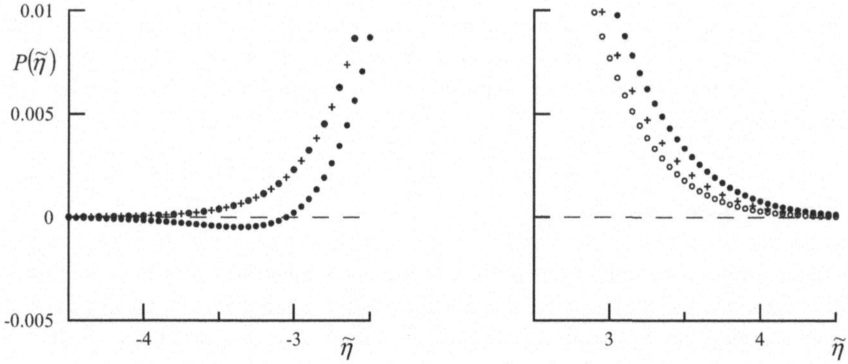

Fig. 3. Fragments of probability density functions $P(\tilde{\eta})$. Combined model with symmetrical filter – O; combined model with asymmetric filter – +; model (4) – ●

In the third situation when $P_{G-C}(\tilde{\eta}) > 0$ it is advisable to use the symmetrical filter $F_S(\tilde{\eta})$ again. The filter parameters are selected in such a way that it is close to unity in the entire region in which the combined model is constructed. Let us consider how in this case the integration limits in (11) and (12) affect the calculated values of cumulants.

The model dependencies $\lambda_3 = \lambda_3(b)$ and $\lambda_4 = \lambda_4(b)$ are shown in Fig. 4. At $b = 5$ the skewness recovery error is less than one percent. At the same value of b, the kurtosis recovery error is less than two percent.

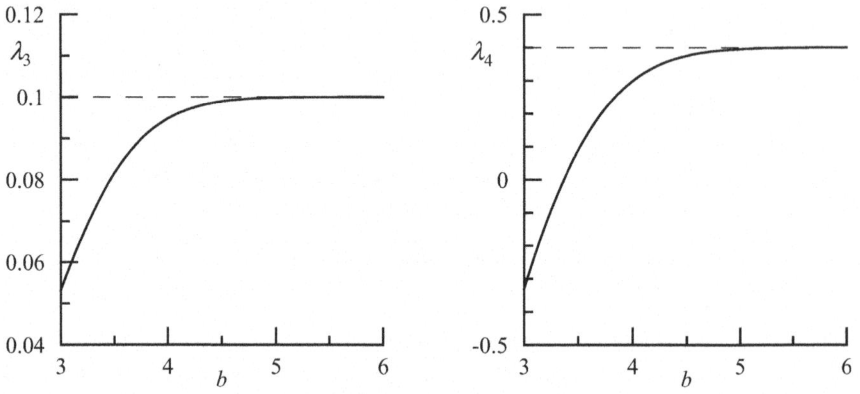

Fig. 4. Dependences of cumulants λ_3 and λ_4 on the limits of integration b.

One of the main applications of the combined model is ocean remote sensing [2]. In [11] an attempt was made to retracing the skewness of the elevation of the surface from altimetric measurements from spacecraft. This procedure is carried out within the framework of the Brown model [12]. In the framework of this model, the altimeter signal is described by the convolution of three functions, two of which are determined by the characteristics of the altimeter and the spacecraft, the third function is determined by the distribution of the elevations of the sea surface. The analysis carried out here shows the limitations of the possibility of restoring the highest statistical moments of sea surface elevations.

4 Conclusion

The main problem modeling the probability density function using truncated Gram-Charlier series is the limitation of the range in which changes in the random variable are correctly described. To solve this problem, a combined model is used in which the Gram-Charlier distribution and the Gaussian distribution correspond to different ranges. The transition from one distribution to another is carried out by means of a filter. The filter should ensure the smoothness and non-negativity of the combined model. An additional criterion for determining the filter parameters based on the assessment of the sea surface elevation cumulants obtained using the combined model is introduced. It is shown that when selecting the filter parameters it is necessary to take into account the presence or absence of negative values in the distribution based on the truncated Gram-Charlier series.

In conclusion, it should be added that although the approach to the construction of the probability density function considered here was developed for the field of sea surface waves, it can also be implemented for other weakly nonlinear processes.

Acknowledgments. This work was carried out in the context of the State project No. 0827-2018-0003.

References

1. Hayne, G.S.: Radar altimeter mean return waveforms from near-normal-incidence ocean surface scattering. IEEE Trans. Antennas Propagat AP **28**(5), 687–692 (1980)
2. Pokazeev, K.V., Zapevalov, A.S., Pustovoytenko, V.V.: The simulation of a radar altimeter return waveform. Mosc. Univ. Phys. Bull. **68**(5), 420–425 (2013)
3. Zapevalov, A.S., Pokazeev, K.V.: Modeling the spectrum of infrasonic hydroacoustic radiation generated by the sea surface under storm conditions. Acoust. Phis. **62**(5), 554–558 (2016)
4. Longuet-Higgins, M.S.: The effect of non-linearities on statistical distribution in the theory of sea waves. J. Fluid Mech. **17**(3), 459–480 (1963)
5. Phillips, O.M.: On the dynamics of unsteady gravity waves of finite amplitude. Part 2. J. Fluid Mech. **11**(1), 143–155 (1961)
6. Babanin, A.V., Polnikov, V.G.: On the non-Gaussian nature of wind waves. Phys. Oceanogr. **6**(3), 241–245 (1995)
7. Zapevalov, A.S., Bol'shakov, A.N., Smolov, V.E.: Simulating of the probability density of sea surface elevations using the Gram-Charlier series. Oceanology **51**(3), 407–414 (2011)
8. Tatarskii, V.I., Tatarskii, V.V.: Statistical non-Gaussian model of sea surface with anisotropic spectrum for wave scattering theory. Progr. Electromagn. Res. **25**, 259–313 (1999)
9. Zapevalov, A.S., Pustovoitenko, V.V.: Modeling of the probability distribution function of sea surface slopes in problems of radio wave scattering. Radiophys. Quantum Electron. **53** (2), 100–110 (2010)
10. Cox, C., Munk, W.: Measurements of the roughness of the sea surface from photographs of the sun glitter. J. Optical. Soc. America. **44**(11), 838–850 (1954)
11. Gómez-Enri, J., Gommenginger, C.P., Srokosz, M.A., Challenor, P.G.: Measuring global ocean wave skewness by retracking RA-2 Envisat waveforms. J. Atmos. Oceanic Technol. **24**, 1102–1116 (2007)
12. Brown, G.S.: The average impulse response of a rough surface and its applications. IEEE Trans. Antennas Propagat AP **25**, 67–74 (1977)

Analytical Estimation of Pollutants Transported by Wind Currents in the Shallow Sea

S. M. Khartiev, A. R. Ioshpa(✉), and I. I. Sknarina

Southern Federal University, 105/42 Bolshaya Sadovaya Str., Rostov-on-Don 344006, Russia
aioshpa@yandex.ru

Abstract. The transportation of non-conservative pollutants by wind currents has been studied in the paper on the basis of the equations of hydrodynamics and turbulent diffusion. The transfer has been considered purely by drift currents in open areas of the sea. In the framework of the theory of small-scale and medium-scale turbulent processes, analytical estimates have been obtained for the size of the region of concentrations of pollutants not exceeding the maximum acceptable concentrations for open areas of the shallow sea and in a coastal zone, taking into account wind surges. The maximum transportation distances of non-conservative pollutants have been calculated for open areas of the Azov Sea using data from coastal hydrometeorological Taganrog, Berdyansk, and Kerch stations. The results obtained allow one to predict and assess the level of danger and risks associated with the spread of pollutants in the marine environment.

Keywords: Wind surges · A coastal zone · Pollutant transportation · Thermodynamic equations · The Sea of Azov

1 Introduction

The implementation of the projects connected with anthropogenic impact on water bodies should be preceded by the creation of a mathematical model followed by a series of numerical experiments that allow to assess all aspects of this impact both in the future and in various emergency situations [1, 2]. Modern methods of mathematical modeling, based on satellite information and a number of applied programs [3, 4], make it possible to analyze the situation data associated with pollutants which spread in the marine environment [5, 6]. However, the application of numerical methods requires the operative and sufficiently detailed parameters of mathematical models; that is not always possible to obtain even with the help of modern monitoring systems [6, 7]. So, during the fuel oil spill in the Kerch Strait on November 11, 2007, the first satellite images were available only five days later after the disaster. Regarding this, the possibility of obtaining information analytically from the hydrodynamic and turbulent diffusion equations formulas is studied that allow us to estimate (at first approximation) the size of the concentration areas of pollutants exceeding the ACL for a wide range of changes of hydro physical parameters.

V. I. Karev et al. (Eds.): *Physical and Mathematical Modeling of Earth and Environment Processes (2018)*, SPEES, pp. 114–121, 2019.
https://doi.org/10.1007/978-3-030-11533-3_12

2 Method of Calculation

Let us first consider the transfer of pollutants by purely drift currents in open areas of the sea. We choose the origin of the rectangular coordinate system XYZ on the sea surface, the Z axis is directed vertically down, the X axis is in the direction of the wind force. In the case of shallow water ($H \ll h$), the solution of the hydrodynamic equations is [8, 9]:

$$u = \tau H(1 - z/H)/(\rho A) \tag{1}$$

where: u - current velocity; H is the depth; h is the friction depth ($h = \sqrt{(f/(2A))}$); ρ is the density of water, τ is the wind stress tangent, f is the Coriolis parameter; A is the coefficient of vertical turbulent viscosity.

According to [8, 10], the following is true for the shallow sea

$$A = \gamma VH/(4\rho k).\ \tau = \gamma V^2. \tag{2}$$

Here: V is the wind speed; $\gamma = 0.00325\ kg/m^3$, $k = 0.015$.

Substituting expression (2) into (1), we obtain:

$$u = 4kV(1 - z/H), u_{max} = u(0) = u_0 = 4kV \tag{3}$$

Let us estimate the maximum distance of transported pollutants in the direction of the wind force R_{ACL}, at which the concentration of the pollutants S does not exceed the value of S_{ACL} (admissible concentration limit). We place the source of pollutants with the concentration S_0 at the origin of the coordinate system; the X-axis is oriented in the direction of the maximum value of the wind speed V. The boundary-value task, describing the transportation of pollutants by wind currents, suppose as follows (3):

$$\partial S/\partial t + u_0 \partial S/\partial x = D\,\partial^2 S/\partial x^2 - mS \tag{4}$$

$$S(0,0) = S_0;\ S \to 0\ \text{if}\ t \to \infty; S \to 0\ \text{if}\ x \to \infty \tag{5}$$

where S0 = const; D и m – coefficients of turbulent diffusion and destruction (decay) accordingly,

$$D > 0, m > 0, u_0 > 0; 0 \leq x \leq u_0 t \tag{6}$$

Let us analyze the stationary solution S(x) of the problem (4) and (5). We write it as:

$$S(x) = S_0 \exp(-\lambda x). \tag{7}$$

$$\text{here}\ \lambda = (\sqrt{1+\varepsilon} - 1)u_0/(2D), \varepsilon = 4mD/u_0^2 \tag{8}$$

It comes from (7) that the concentration decreases when the distance from the pollution source increases. Let us define the distance RACL, at which the concentration reaches its maximum permissible value SACL, that is: S(RACL) = SACL. Substituting x = RACL into the expression (7), we obtain SACL, = S0 exp(-λ RACL). From that we find: RACL = λ-1ln (S0/SACL), where S0 > SACL. Since at v > 1 m/s in real conditions of the shallow sea for small-scale and medium-scale turbulent processes ε ≪1 [8], from (8) exactly to ε^2 values we have:

$$\lambda = m/u_0, \quad \mathrm{S} = \mathrm{S}_0 e^{-\frac{m}{u_0}x} \mathrm{R}_{\mathrm{ACL}} = \frac{u_0}{m} \ln \frac{S_0}{S_{ACL}} \tag{9}$$

Thus, it can be claimed that in the region x > R_{ACL}, where R_{ACL} is determined by the formula (9), the concentration of pollutants will correspond to the sanitary norms.

Now we study the nonstationary solution of problem (4)–(6). Imagine it as:

$$\mathrm{S} = \mathrm{S}_0 \exp(-\mathrm{m}x/\mathrm{u}_0)(1 + \mathrm{m}\theta)/\exp(m\theta) \tag{10}$$

here $\theta = t - x/u_0$. According to (6), $\theta \geq 0$.

Expression (10) is an exact solution of the boundary value problem (4)–(6). We rewrite it as follows:

$$\mathrm{S} = \mathrm{S}_0 \mathrm{F}(\theta) \exp(-\mathrm{m}x/\mathrm{u}_0), \mathrm{F} = (1\theta)/\exp(\mathrm{m}\theta) \tag{11}$$

Since $0 \leq \mathrm{F}(\theta) \leq 1$, then the estimates (9) obtained in the previous case are valid. Taking into account (3), we write (9) it as follows:

$$\mathrm{R}_{\mathrm{ACL}} = (4\mathrm{k}/\mathrm{m}) \ln(\mathrm{S}_0/\mathrm{S}_{\mathrm{ACL}}) \tag{12}$$

It is seen from (12) that RACL is directly proportional to V (the wind speed) and inversely proportional to m (the destruction coefficient).

We consider now the case when S_0 is a function of time t and satisfies the following conditions:

$$\mathrm{S}_0(0) = 0, \ \mathrm{S}_0(\mathrm{t}) \to 0 \,\mathrm{if}\, \mathrm{t} \to \infty; \quad \max \mathrm{S}_0(\mathrm{t}) = \mathrm{C} \tag{13}$$

We formulate the initial and boundary conditions for Eq. (4), when relations (6) are satisfied, in the following:

$$\mathrm{S}(\mathrm{x}, 0) = 0, \ \mathrm{S}(0, \mathrm{t}) = \mathrm{S}_0(\mathrm{t}), \ \mathrm{S} \to 0 \,\mathrm{if}\, \mathrm{x} \to \infty, \mathrm{S} \to 0 \,\mathrm{if}\, \mathrm{t} \to \infty \tag{14}$$

Then the exact solution of the boundary-value problem (4), (6), (14) for D = 0 can be represented as follows:

$$\mathrm{S} = \mathrm{S}_0(\theta) \exp(-\mathrm{m}x/\mathrm{u}_0), \quad \theta = \mathrm{t} - \mathrm{x}/\mathrm{u}_0 \tag{15}$$

Since S0(θ) is bounded from above (see (13)), then, by analogy with the previous cases, for an arbitrary form of the function S0(t) satisfying conditions (13), the following equality holds:

$$R_{ACL} = 4k\ (V/m)\ln(\max S_0(t)/S_{ACL}) \tag{16}$$

If $D \neq 0$, we use the Laplace transform to solve the boundary value problem (4), (6), (14). Thus we have:

$$\bar{S}(x, v) = \bar{S}_0(v)\exp[\lambda(v)x], \tag{17}$$

$$\text{where } \lambda = (1 - \sqrt{1 + 4(v+m)D/u_0^2})u_0/(2D),\ (\mathrm{Re}\,\lambda < 0) \tag{18}$$

$$\bar{S}(x, v) = \int_0^\infty e^{-vt} S(x,t)dt, \bar{S}_0(v) = \int_0^\infty e^{-vt} S_0(t)dt.$$

Applying the inversion formula [11] to (17), we obtain:

$$S(x, t) = \frac{1}{2\pi i}\int_{v-i\infty}^{v+i\infty} e^{vt}\bar{S}_0(v)e^{\lambda(v)x}dt \tag{19}$$

Further investigation of the integral (19) is carried out for a specific form of the function S0(t) satisfying the conditions (13) (see Fig. 1):

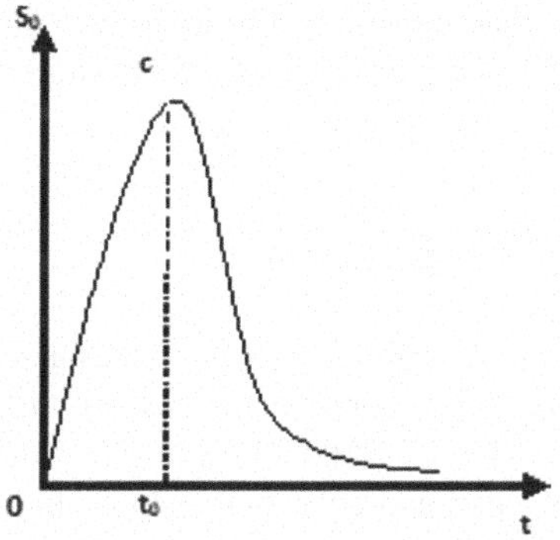

Fig. 1. $S_0(t)$, describing the entry of pollutants into the sea

$$S_0 = C_0\alpha/\left(1 - e^{-\beta t}\right)e^{-\alpha t}, \quad \alpha > 0, \beta > 0, \alpha + \beta < m \tag{20}$$

$$\bar{S}_0 = \alpha C_0/[(v+\alpha)(v+\alpha+\beta)] \tag{21}$$

Taking into account (21) and using the theory of residues [11] to calculate the integral (19), we find:

$$S = C_0 \frac{\alpha}{\beta}\left[e^{-\alpha t} e^{\lambda(-\alpha)x} - e^{-(\alpha+\beta)t} e^{\lambda(-\alpha-\beta)x}\right] \tag{22}$$

As it has been noted above, if v > 1 m/s, for real conditions of small-scale and medium-scale turbulent processes in the shallow sea, the inequality ε§1 is valid (see (8) and (20)). Then, according to (18), with exact quantities of order ε^2, we have:

$$\lambda(\alpha) = -(m-\alpha)/u_0; \quad \lambda(\alpha+\beta) = -(m-\alpha-\beta)/u_0 \tag{23}$$

Substituting (23) into (22) and taking into account (20), we obtain the Eq. (15), where the function $S_0(0)$ is given by formula (20). Then Eq. (16) will be valid, where, according to (20), we determine:

$$max\, S_0(t) = C_0[(\alpha/(\alpha+\beta))]^{(\alpha+\beta)/\beta} \tag{24}$$

We can note that, in contrast to the previous case (D = 0), the solution found is not exact and will be valid only for specific functions of the form (20).

Thus, R_{ACL} obtained for open sea areas in stationary (12) and nonstationary (16) cases show that for small conditions if V > 1 m/s the small-scale and medium-scale turbulent processes of surface transport and decay of pollutants dominate over diffuse processes. R_{ACL}, within the approximations adopted, does not depend on the depth of the sea and is completely determined by the hydrometeorological conditions (wind speed and direction), as well as the destructive features and conditions of the arrival of non-conservative pollutants into the sea.

3 Analysis of Numerical Experiments

Let us now study the transfer of pollutants in the coastal zone of the sea, taking into account surging phenomena. Let us consider the case of wind surges caused by a wind field moving along the shore. Let us set the OX axis along the shoreline in the direction of wind speed, the axis OY - perpendicular to the shore. In the framework of the linear theory of long waves averaged over the depth of the hydrodynamic equation, we represent it as follows [12, 13]:

$$\partial u/\partial t - fv = \tau/\rho H, \partial v/\partial t + fu = -g\partial\xi/\partial y \tag{25}$$

$$(\partial u/\partial x + \partial v/\partial y) + \partial\xi/\partial t = 0 \tag{26}$$

The solution of the system (25) and (26) corresponding to the stationary transfer v (y) directed perpendicular to the shore (v(0) = 0), being independent of x is the following [12]:

$$v = -\tau[1 - exp(-y/r)]/(\rho Hf) \quad (27)$$

$$u = \tau te(-y/r)/(\rho H), \xi = \tau t\, exp(-y/r)/(\rho c) \quad (28)$$

$$\text{here } c = \sqrt{gH}, \mathrm{r} = \mathrm{c/f} - \text{Rossby radius}$$

Equations (27) and (28) show that the existence of a shore boundary leads to surges, and since the velocity is perpendicular to the shore and constant, water at the shore will continuously accumulate and the level ξ(y, t) will increase proportionally to time.

We estimate the transfer of pollutants in the coastal zone, using the formula (27). We place the source of contamination with the initial concentration S0 at a distance *l* from the shore at the origin, the X axis is oriented in the direction of the Ekman flow, perpendicular to the shoreline (parallel to the OY axis in the hydrodynamic equations (25) and (26)). Then, according to (2) and (27), the maximum value for the velocity of the Ekman flow *v* can be written as follows:

$$u_0 = max|v| = \gamma V^2[1 - exp(-l/r)]/(\rho Hf).$$

In this case, taking into account (9), we obtain:

$$R_{ACL} = [\gamma/(m\rho Hf)][1 - exp(-l/r)]ln(S_0/S_{ACL})V^2 \quad (29)$$

Comparing the expressions (12) and (29), it should be noted that for the open sea regions there is a linear dependence of the R_{ACL} on the wind speed V, and in case if there is a shoreline and storm surges, it is quadratic. Equation (29) also shows that in the conditions of storm surges, R_{ACL} essentially depends on the geographic features of the given area of the coastal zone (depth H, remoteness of the pollution source from shore l, Coriolis parameter f).

Figure 2 shows the results of R_{ACL} calculations performed for open areas of the Sea of Azov according to the meteorological stations Taganrog, Berdyansk, Kerch [14] if при $\frac{S_0}{S_{ACL}} = e^n$, n = 2, m = $25 \cdot 10^{-5} \cdot c^{-1}$, ZV BPK_5.

Table 1 shows the values of R_{ACL} calculated for the case of surges, when the source of pollutants BPK_5 is 700 m from the shore (n = 2, l = 700 m).

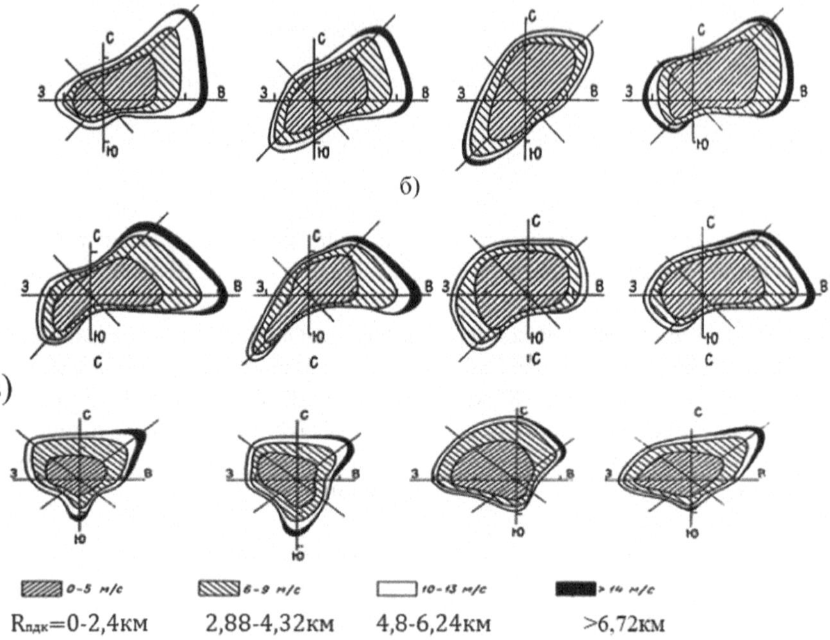

Fig. 2. R_{ACL} calculated for the Sea of Azov with given distributions of the wind speed (1 line of the scale) in the areas of stations (a) Taganrog, (b) Berdyansk, (c) Kerch.

Table 1. The results of R_{ACL} calculations for characteristic ranges of maximum speed values of the north-east wind in the areas of the Berdyansk and Taganrog stations (l = 700 m)

Months 2016 г	Berdyansk		Taganrog	
	V(м/s)	R_{ACL}(M)	V(м/s)	R_{ACL}(M)
January	13–14	189–220	14–16	403–526
April	14–15	220–287	13–14	347–403
July	9–10	91–112	10–13	206–347
September	13–14	189–220	14–17	403–594

4 Conclusion

Thus, the results obtained allow us to forecast and assess the level of hazard and risks associated with the pollutants that spread in the marine environment. For example, as it can be seen from the table $L = 1 - R_{ACL}$, which determines the size of the sanitary zone ($S < S_{ACL}$), can be negative even if V = 19 m/s.

Acknowledgments. The work has been carried out within the framework of RFBR grant 18-05-80082 called "Laws of formation of dangerous coastal processes in the Sea of Azov and their socio-economic consequences".

References

1. Matishov, G.G., et al.: Modern Hazardous Exogenous Processes on the Azov Sea Coast. Southern Federal University Press, Rostov-on-Don (2015)
2. Shustova, V.L., Khartiev, S.M., Baselyuk, A.A., et al.: Prediction of water resources conditions in the region. (On the example of the Lower Don River). High school of Southern Caucasus Scientific Center. Sciences, № 4, pp. 106–109 (1998)
3. Blumberg, A.F., Mellor, G.L.: A description of three dimensional coastal ocean circulation model. In: Heaps, N. (ed.) Three-Dimensional Coastal Ocean Models, vol. 4, pp. 1–16. American Geophysical Union, Washington, D.C. (1987)
4. Wannawong, W., Humphries, U.W., Wongwises, P., Vongvisessomjai, S.: Mathematical modeling of storm surge in three dimensional primitive equations. Int. Comput. Math. Sci. **5**, 44–53 (2011)
5. Khartiev, S.M.., Surkov, F.A., Zaporozhets, V.Yu.: Model study of spatial distributions of impurities in sea water from deep-water sources of various designs. High School of Southern Caucasus Scientific Center. Issue: Sciences, № 4, pp. 50–55 (1991)
6. Matishov, G.G., Shulga, T.Ya., Khartiev, S.M., Ioshpa, A.R.: Studies of particulate matter distribution by aqua modis data and simulation results. Earth Sci. Rep. **481**(Part 1), 967–971 (2018)
7. NASA Goddard Space Flight Center, Ocean Ecology Laboratory, Ocean Biology Processing Group. Moderate-resolution Imaging Spectroradiometer (MODIS) Aqua Ocean Color Data; 2014 Reprocessing. NASA OB.DAAC, Greenbelt, MD, USA. https://doi.org/10.5067/aqua/MODIS_OC.2014.0. Accessed 03 Nov 2017
8. Mamaev, O.I.: Physical Oceanography: Selected Works. RSRRS Press, Moscow (2000)
9. Khartiev, S.M., Ioshpa, A.R.: Fundamentals of Ocean Hydrology. Southern Federal University Press, Rostov-on-Don (2014)
10. Felzenbaum A.I. Theoretical bases and methods of estimation of steady sea currents, Moscow (1960)
11. Sveshnikov, A.G., Tikhonov, A.N.: The theory of functions of a complex variable. In: Higher Mathematics and Mathematical Physics Course, 2nd edn., vol. 4 (1967)
12. Ivanov, V.A., Pokazeev, K.V., Shreider, A.A.: Fundamentals of Oceanology. Lan, Saint-Petersburg (2008)
13. Cherkesov, L.V., Ivanov, V.A., Khartiev, S.M.: Introduction to Hydrodynamics and Wave Theory. Hydrometeo Press, Saint-Petersburg (1992)

Far Fields of Surface Gravity Waves Under Unsteady Generation Regimes

Vitaly Bulatov(✉) and Yury Vladimirov

Ishlinsky Institute for Problems in Mechanics RAS, 119526 Moscow, Russia
internalwave@mail.ru

Abstract. The far fields of surface wave perturbations excited by an oscillating localized source moving in a heavy liquid of infinite depth are studied. It is shown that the excited fields are a sum of two wedge-like ship waves located insider the corresponding wave wedges. Each of the excited two waves is a complicated wave system of transverse and longitudinal perturbations. The properties of the dispersion curves are studied and the phase pictures describing the structure of wave surface perturbations are calculated. The characteristics of the excited wave fields are studied depending on the basic parameters of the wave generation such as the velocity of motion of the perturbation source and the frequency of its oscillations. Uniform asymptotic solutions are constructed in terms of the Airy function and its derivative, which permits describing the far fields of surface perturbations both outside and inside the corresponding wave wedges.

Keywords: Gravity surface waves · Far fields · Uniform asymptotics

1 Introduction

The surface wave motions in the marine environment can either originate due to natural causes (wind waves, flow past underwater obstacles, bottom relief variations, density and flow fields) or be generated by the flow past natural obstacles (platforms, underwater pipelines, complex hydraulic facilities). The general system of hydrodynamic equations describing the surface perturbations is a rather complicated mathematical problem from the standpoint of proving the existence and uniqueness theorems for solutions in the corresponding function classes and from the computational standpoint [1–6]. In the framework of the linear theory, the surface wave perturbations are analytically studied by integral representation methods and various asymptotic methods [7–11]. The main results of solving the problems of generation of surface wave perturbations are represented in most general integral form, and to obtain the integral solutions, it is thus necessary to develop asymptotic methods for their investigation which admit a qualitative analysis and rapid estimations of the obtained solutions. Moreover, to analyze the data of the sea surface remote sensing, it is required to know the causes of various surface phenomena. To obtain a detailed description of a wide class of physical phenomena related to the dynamics of surface perturbations in inhomogeneous and unsteady natural environments, it is necessary to have sufficiently developed mathematical models. The fact that the structure of the heavy sea surface is

V. I. Karev et al. (Eds.): *Physical and Mathematical Modeling of Earth and Environment Processes (2018)*, SPEES, pp. 122–131, 2019.
https://doi.org/10.1007/978-3-030-11533-3_13

three-dimensional is also significant, and there are currently no possibilities for large-scale computational experimental modeling of three-dimensional ocean flows at large times with a sufficient accuracy. But in several cases, the initial qualitative concept of the considered class of wave phenomena can be obtained by using simpler asymptotic models and analytic methods for studying them. In this connection, it is necessary to mention the classical hydrodynamic problems of constructing asymptotic solutions which describe the evolution of surface perturbations excited by sources of various nature in heavy homogeneous liquids. The model solutions permit further obtaining representations of surface wave fields with regard to variability and unsteadiness of real natural environments. So several results of asymptotic analysis of linear problems describing different regions of generation and propagation of surface perturbations also underlie the currently actively developing nonlinear theory of generation of ocean waves of extremely large amplitude, the so-called rogue waves [3]. The contemporary state of the art in the study of linear and nonlinear surface perturbations can be found in [5]. In [12, 13], the problem of constructing uniform asymptotics of far fields of surface perturbations due to an oscillating source moving with a bounded velocity was considered. The goal in the present paper is to construct uniform asymptotics of far fields of surface perturbations excited by the fast motion of a localized oscillating source of perturbations in a heavy homogeneous liquid of infinite depth.

2 Problem Formulation and Integral Forms of Solutions

We consider the steady-state pattern of wave perturbations on the surface of an ideal heavy liquid of infinite depth when the perturbation source moves with velocity V in the positive direction of the axis x. The waves are generated by a moving oscillating point source of perturbations located at the depth H (the axis z is directed upwards from the unperturbed liquid) whose capacity varies by the law $q = \exp(i\omega t)\exp(\varepsilon t)$ $(-\infty < t < \infty)$. Further, we seek the limit as $\varepsilon \to 0$ in the obtained solution. The perturbation of the potential $\Phi(x, y, z, t)$ with respect to the homogeneous flow moving with velocity V ($\nabla\Phi = (u, v, w)$, where u, v, w are components of the vector of perturbations of the velocity $(V, 0, 0)$) is described by the following equation with an appropriate linearized boundary condition on the surface of the liquid [1, 5, 12, 13]

$$\begin{gathered}\delta\Phi(x,y,z,t) = \exp(i\omega t)\exp(\varepsilon t)\delta(x)\delta(y)\delta(z+H), \quad z<0\\ \left(\frac{\partial}{\partial t} + V\frac{\partial}{\partial x}\right)^2 \Phi + g\frac{\partial\Phi}{\partial z} = 0, \quad z=0\end{gathered} \tag{1}$$

Here Δ is the three-dimensional Laplace operator, and $\delta(x)$ - is the Dirac delta function. We seek the solution of problem (1) in the for $\Phi(x, y, z, t) = \exp(i\omega t)\exp(\varepsilon t)\varphi(x, y, z)$, where the function $\varphi(x, y, z)$ is determined from the problem

$$\Delta\varphi(x, y, z) = \delta(x)\delta(y)\delta(z+H), \quad z<0,$$
$$\left(i\omega + \varepsilon + V\frac{\partial}{\partial x}\right)^2 \varphi + g\frac{\partial\varphi}{\partial z} = 0, \; z = 0.$$

The Fourier transform of the potential $\varphi(x, y, z)$,

$$\Omega(\mu, \nu, z) = \int_{-\infty}^{\infty} \exp(i\mu x)dx \int_{-\infty}^{\infty} \exp(i\nu y)\varphi(x, y, z)\, dy$$

is determined from the boundary-value problem

$$\frac{\partial^2\Omega(\mu, \nu, z)}{\partial z^2} - k^2\Omega(\mu, \nu, z) = \delta(z+H), \qquad z<0,$$
$$(i\omega + \varepsilon - i\mu V)^2\Omega(\mu, \nu, z) + g\frac{\partial\Omega(\mu, \nu, z)}{\partial z} = 0, \quad z = 0,$$
$$\Omega(\mu, \nu, z) \to 0, \quad z \to -\infty, \; k^2 = \mu^2 + \nu^2,$$

whose solution in the domain $-H<z<0$ has the form

$$\Omega(\mu, \nu, z) = -\frac{(\omega - \mu V)^2 sh(kz) + gk\; ch(kz)}{k\exp(kH)((\varepsilon + i(\omega - \mu V))^2 + gk)}.$$

The free surface elevation $\eta(x, y, t)$ is related to the potential $\Phi(x, y, z, t)$ by the condition [1, 5, 12, 13]

$$\eta(x, y, t) = -\frac{1}{g}\left(\frac{\partial}{\partial t} + V\frac{\partial}{\partial x}\right)\Phi(x, y, z, t) =$$
$$= \frac{-\exp(i\omega t + \varepsilon t)}{g}(i(\omega - i\varepsilon)\varphi(z, y, z) + V\frac{\partial\varphi(x, y, z)}{\partial x}), z = 0$$

Then the Fourier transform $\Lambda(\mu, \nu, t)$ of the function $\eta(x, y, t)$ becomes

$$\Lambda(\mu, \nu, t) = \frac{i(\omega - \mu V)\exp(i\omega t)\exp(-kH)}{(\varepsilon + i(\omega - \mu V))^2 + gk}$$

In the obtained expression, the parameter ε is preserved only in the denominator, which is necessary to determine the displacement of the pole of the integrand with respect to the real axis (into the upper or lower half-plane). Then we can obtain the inverse Fourier transform

$$\eta(x, y, t) = \frac{i\exp(i\omega t)}{4\pi^2}\int_{-\infty}^{\infty} \exp(-i\nu y)d\nu \int_{-\infty}^{\infty} \frac{(\omega - \mu V)\exp(-kH - i\mu x)d\mu}{(\varepsilon + i(\omega - \mu V))^2 + gk} \tag{2}$$

The zeros of the denominator in the integrand in (2) determine the dispersion relation: $B(\omega, \mu, \nu) \equiv (\omega - \mu V)^2 - g\sqrt{\mu^2 + \nu^2} = 0$. The set of values of the parameter $V\omega/g > 0$ is divided by two characteristic values $1/4$ and $\sqrt{6}/9$ into three intervals. For $V\omega/g < 1/4$, the dispersion curve consists of three branches, one closed and two non-closed; this case was considered in [12]. In this case, the wave picture is a sum of two wedge-like (longitudinal) waves with half-opening angle of the wave wedge less than $\pi/2$ and the annular-like (transverse) waves around the source. For $V\omega/g > \sqrt{6}/9$, the dispersion curve consist of two unclosed branches without extrema. In this case, the wave picture is a sum of two wedge-like ship waves with the half-opening angle of the wave wedge less than $\pi/2$. If $1/4 < V\omega/g < \sqrt{6}/9$, then the dispersion curve consists of two unclosed curves one of which has two local extrema. One branch of the dispersion curve corresponds to the usual wedge-like waves with the half-opening angle of the wave wedge less than $\pi/2$, and the second branch corresponds to the ship waves with the half-opening angle of the wave wedge greater than $\pi/2$ (the wave front is directed upstream away from the source). This system of hybrid waves simultaneously has the features of both the annular (transverse) and wedge-like (longitudinal) waves [13]. Further, we consider the case $V\omega/g > \sqrt{6}/9$. Then the integrand in the inner integral in (2) has two real poles $\mu_{1,2}$. Figure 1 illustrates the results of calculations of the corresponding dispersion curves $\mu_{1,2}(\nu)$. Here and below, the following parameters of calculations were used: $H = 5$ m, $\omega = 0.5$ s^{-1}, and $V = 30$ m/s.

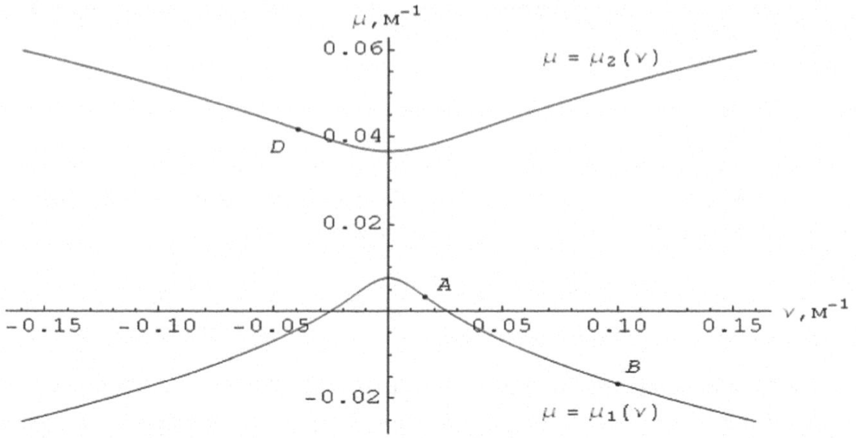

Fig. 1. Dispersion curves $\mu_1(\nu)$ and $\mu_2(\nu)$, A and D are deflection points, and B is a root of the equation $\mu_1'(\nu) = \mu/\nu$.

3 Construction of Solution Asymptotics

To calculate the inner integral in (2), it is necessary to determine the displacement of the poles $\Delta\mu$ for $\varepsilon > 0$. For this, we equate the determinant of the integrand in the inner integral in (2) with zero for $\varepsilon = 0$: $B(\omega, \mu, \nu) = 0$, and for $\varepsilon > 0$: $B(\omega - i\varepsilon, \mu + \Delta\mu, \nu) = 0$. Then we obtain: $\Delta\mu = i\varepsilon \frac{\partial B}{\partial \omega} / (\frac{\partial B}{\partial \mu})$. Under the assumption that $\omega = \omega(\mu, \nu)$, we have $\frac{\partial B}{\partial \omega} / (\frac{\partial B}{\partial \mu}) = -1/(\frac{\partial \omega}{\partial \mu})$ and $\delta\mu = -i\varepsilon/(\frac{\partial \omega}{\partial \mu})$. With regard to $\frac{\partial \omega}{\partial \mu} > 0$, we see that both poles move into the lower half-plane. Then, for $x < 0$, the contour of integration over the variable μ in (2) becomes closed in the upper half-plane, and the poles $\mu_{1,2}(\nu)$ do not make contributions to the wave field. For $x > 0$, the contour of integration over μ in (2) becomes closed in the lower half-plane and, taking into account the contributions of the poles, we obtain

$$
\begin{aligned}
&\eta(x, y, t) = I_1(x, y, t) + I_2(x, y, t), \\
&I_m(x, y, t) = \int_{-\infty}^{\infty} F(\mu_m(\nu), \nu) \exp(-ixS_m(\nu, \alpha)) d\nu, \\
&F(\mu_m(\nu), \nu) = \frac{\exp(i\omega t)}{2\pi} \frac{(\omega - \mu V) \exp(-kH)}{2V(\omega - \mu V) + \mu g/k}, \\
&S_m(\nu, \alpha) = \mu_m(\nu) + \nu tg\alpha, tg\alpha = y/x, m = 1, 2.
\end{aligned} \tag{3}
$$

For large value of $x > 0$, the asymptotic behavior of the integrals in the sum (3) is completely determined by the stationary points of the phase function $S_m(\nu, \alpha)$ which are determined from the equation $\mu'_m(\nu) = -tg\alpha$. First, we consider the first term in (3). The function $\mu'_1(\nu)$ has one maximum on the interval of integration over the variable ν associated with the corresponding value of the argument α, which is further denoted by A_1. The value A_1 determines the boundaries of the wave wedge (Kelvin wedge) which are described by the equation $y = \pm x \, tg \, A_1$. For $0 < \alpha < A_1$, the phase function $S_1(\nu, \alpha)$ has two stationary points on the real axis ν: $0 < \nu_2(\alpha) < \nu_1(\alpha)$. For $A_1 < \alpha < \pi/2$, there are two complex conjugate stationary points $\nu_1(\alpha), \nu_2(\alpha)$, and for definiteness, we assume that $\mathrm{Im}\, \nu_1(\alpha) > 0$.

We introduce the notation: $\Phi_1 = -\mu_1(\nu) - \nu y + \omega t$. Then from the phase stationarity condition $\mu'_1(\nu) = -\mathrm{tg}\alpha$, we can obtain the parametric equations of the family of constant phase lines $\Phi_1 = C$ ($C = \mathrm{const}$) for different values of C

$$
x(\nu) = \frac{\omega t - C}{\mu_1(\nu) - \nu\mu'_1(\nu)}, \quad y(\nu) = -\frac{\mu'_1(\nu)(\omega t - C)}{\mu_1(\nu) - \nu\mu'_1(\nu)}
$$

Figure 2 shows the lines of equal phase for $t = 0$, $C = 2\pi n$, $n = -5, -4, \ldots, 5$. The right-hand branch ($\nu > 0$) of the dispersion curve $\mu_1(\nu)$ corresponds to with the upper part of the picture ($y > 0$). Since the phase portrait of the waves is symmetric with respect to the axis x, we further consider only this domain. Point A in Fig. 1 is the deflection point of the curve $\mu_1(\nu)$, i.e., a root of the equation $\mu''_1(\nu) = 0$. Therefore, the value A is associated with the wave wedge boundary (dashed line in Fig. 2) inside which the traveling waves described by the integral I_1 are propagating. Point B in Fig. 1 is a root of the equation $\mu'_1(\nu) = \mu/\nu$ and is associated with dashed line 4 in Fig. 2. The part of the dispersion curve from zero to point A (Fig. 1) is associated with transverse crests of the waves (solid lines 2 in Fig. 2). The part of the dispersion curve from point A to point B in Fig. 1 is associated with longitudinal crests (solid lines 1a in Fig. 2) located between dashed lines 3 and 4 in domain I. The part of the dispersion curve to the right of point B (Fig. 1) is associated with longitudinal crests (solid lines 1b in Fig. 2) located between the dashed line 4 and the axis x (domain II in Fig. 2). On the crests of longitudinal waves in domain I and on the crests of the transverse waves, the phases Φ_1 take the values $2\pi l$ $(l = -1, -2, .., -5)$. The phases of the crests of longitudinal waves in domain II in Fig. 2 are equal to $2\pi k$ $(k = 1, 2, .., 5)$. On dashed line 4, the wave phase Φ_1 is zero for $t = 0$. The longitudinal waves in domain I and the transverse waves propagate from the origin to infinity. The longitudinal waves in domain II propagate in the direction of dashed line 4. We present the basic characteristics of the wave field for the following parameters of calculations: the wave length along the horizontal axis is $\lambda_1 = 2\pi/\mu_1(0) = 828.8$ m, the half-opening angle of the wave wedge is $A_1 = 20.3°$, and the wave front is given by the equation $y = x \text{ tg } A_1$.

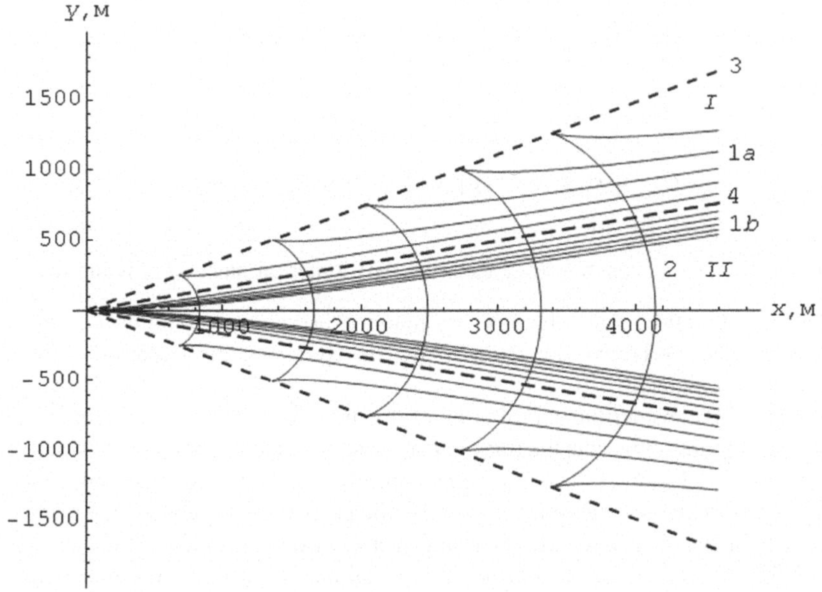

Fig. 2. Lines of equal phase for the integral I_1: lines 1a correspond to longitudinal waves in domain I, lines 1b correspond to longitudinal wave in domain II, lines 2 correspond to transverse waves, line 3 indicates the wave front, and line 4 separates domains I and II.

Inside the wave wedge, the field can be calculated by the method of stationary phase, then the contribution is made by both of the stationary $\nu_1(\alpha)$, $\nu_2(\alpha)$, and the field is exponentially small outside the wave wedge. The asymptotics of the integral $I_1(x, y, t)$ for large $x > 0$ calculated by the method of stationary phase has the form [14, 15]

$$I_1(x,y,t) \approx T_1 + T_2,$$
$$T_i = \left(\frac{2\pi}{x\left|(\mu_1'' \nu_i(\alpha))\right|}\right)^{1/2} F(\mu_1(\nu_i(\alpha)), \nu_i(\alpha))\, E_i,$$
$$E_i = \exp(-ix(\mu_1(\nu_i(\alpha)) + \nu_i(\alpha)tg\alpha)) - i\pi sign(\mu_1''(\nu_i(\alpha)))/4), i = 1,2.$$

The asymptotics calculated by the method of stationary phase is not uniform, because the stationary points merge on the wave front: $\nu_1(A_1) = \nu_2(A_1)$ and $\mu_1''(\nu_1(A_1)) = \mu_1''(\nu_2(A_1)) = 0$. Therefore, the asymptotics calculated by the method of stationary phase cannot be applied near the boundary of the Kelvin wave wedge. The uniform asymptotics of $I_1(x,y,t)$ for $x > 0$ applicable at far a distance from the wave front and near it has the form [14, 15]

$$\begin{aligned} I_1(x,y,t) \approx \frac{2\pi \exp(i(r\lambda(\alpha) + \omega t))}{x^{1/3}} (0.5(G(\sqrt{\sigma(\alpha)}) + G(-\sqrt{\sigma(\alpha)}))Ai(x^{2/3}\sigma(\alpha)) - \\ -i\frac{(G(\sqrt{\sigma(\alpha)}) - G(-\sqrt{\sigma(\alpha)})}{2x^{1/3}\sqrt{\sigma(\alpha)}} Ai'(x^{2/3}\sigma(\alpha))), \\ \lambda(\alpha) = (S_1(\nu_1(\alpha),\alpha) + S_1(\nu_2(\alpha),\alpha))/2, \\ \sigma(\alpha) = (3(S_1(\nu_2(\alpha),\alpha) - S_1(\nu_1(\alpha),\alpha))/4)^{2/3}, \\ G(\sqrt{\sigma(\alpha)}) = F(\mu_1(\nu_1(\alpha)), \nu_1(\alpha))\sqrt{\frac{2\sqrt{\sigma(\alpha)}}{\mu''_1(\nu_1(\alpha))}}, \\ G(-\sqrt{\sigma(\alpha)}) = F(\mu_1(\nu_2(\alpha)), \nu_2(\alpha))\sqrt{\frac{2\sqrt{\sigma(\alpha)}}{\mu''_1(\nu_2(\alpha))}}, \end{aligned} \tag{4}$$

where $Ai(\tau) = \frac{1}{2\pi}\int_{-\infty}^{\infty} \cos(\tau t - t^3/3)dt$ is the Airy function and $Ai'(\tau)$ is the derivative of the Airy function. The obtained asymptotics becomes non-uniform if the Airy function and its derivative are replaced by the corresponding expansions for large values of the argument.

Further, we consider the contribution of the dispersion curve $\mu_2(\nu)$ to the resultant wave field. Figure 3 presents the lines of equal phase of the integral $I_2(x,y,t)$ for $t = 0$, $C = 2\pi n$, $n = -1, -2, \ldots, -5$. The upper half of Fig. 3 ($y > 0$) corresponds to the left branch of the dispersion curve $\mu_2(\nu)$ ($\nu < 0$). The part of the dispersion curve from zero to point D in Fig. 1 is associated with transverse waves (solid lines 2 in Fig. 3). The part of the dispersion curve from point D to infinity is associated with longitudinal waves (solid lines 1 in Fig. 3). The deflection point is associated with the wave front $y = x$ tg A_2 (dashed line 3 in Fig. 3), where the half-opening angle of the wave wedge is $A_2 = 10.1°$. All lines of equal phase go from the origin to infinity. The length of the transverse wave along the horizontal axis x is $\lambda_2 = 2\pi/\mu_2(0) = 171.5$ m. The uniform asymptotics of the integral $I_2(x,y,t)$ for large $x > 0$ are estimated similarly to (4). We note that the waves described by the integral I_2 significantly (approximately by a factor of three) exceed in amplitudes the waves determined by the integral I_1.

The above numerical calculations show that an increase in the velocity V of motion of the source (for a fixed frequency ω of the source oscillations) leads to a decrease in the half-opening angles of both of the wave wedges. In this case, the distance between the neighboring wave crests increases; in particular, there is an increase in the lengths of transverse waves λ_1 and λ_2 along the axis x. Figures 4 and 5 presents the results of

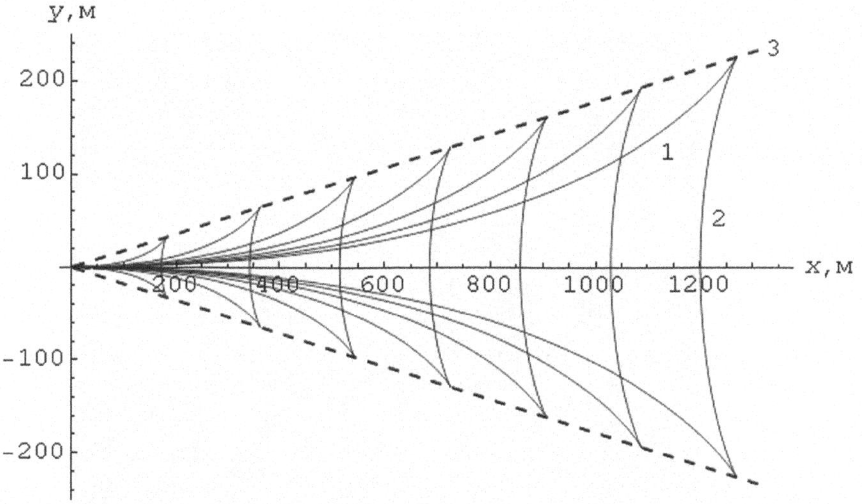

Fig. 3. Lines of equal phase for the integral I_2: lines 1 are longitudinal waves, lines 2 are transverse waves, and line 3 is the wave front.

calculations of uniform asymptotics of the integrals $I_1(x, y, t)$ and $I_2(x, y, t)$ for $t = 0$. The sum of these terms describes the total field of the free surface elevation at a far distance from the moving oscillating source of perturbations.

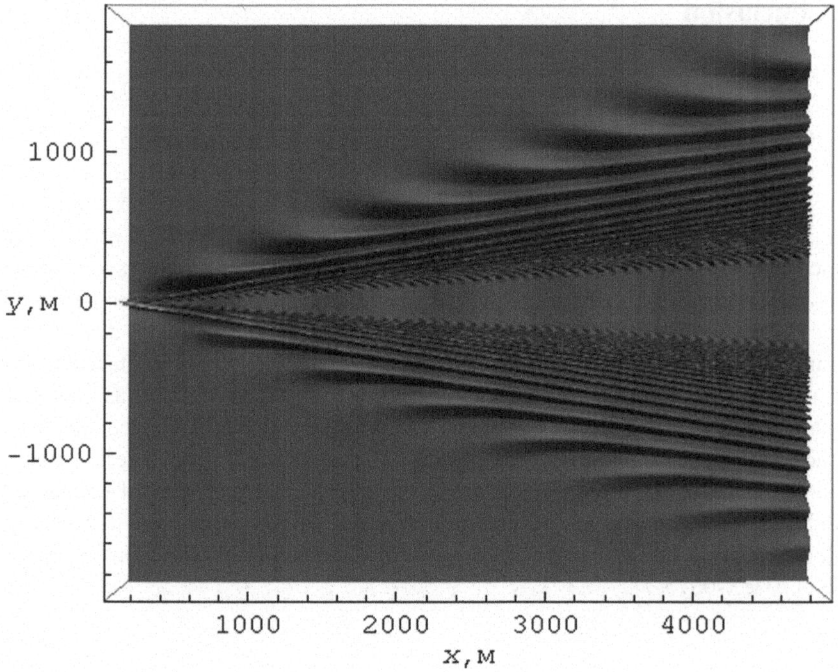

Fig. 4. Uniform asymptotics of the integral I_1 at a far distance from the moving source.

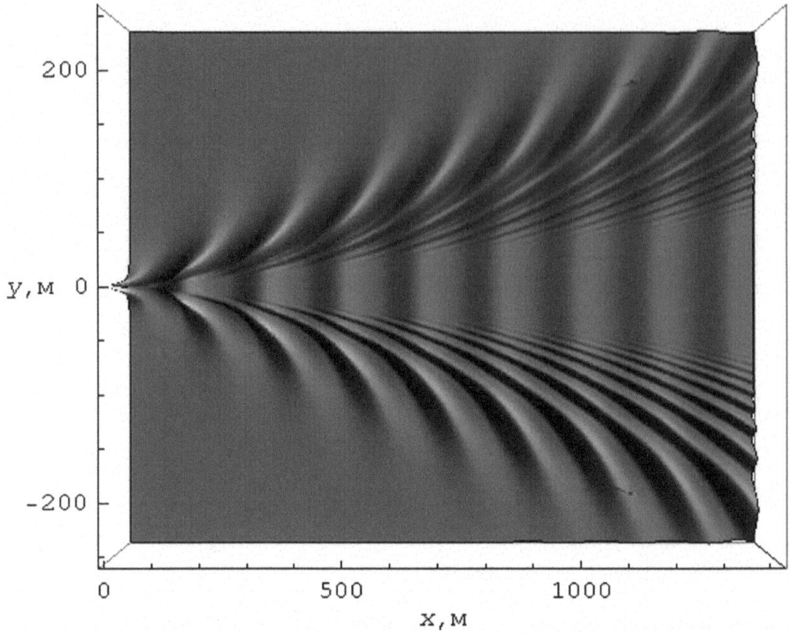

Fig. 5. Uniform asymptotics of the integral I_2 at a far distance from the moving source.

4 Conclusion

We have shown that the far fields of surface perturbations due to a rapidly moving localized oscillating source in a heavy liquid of infinite depth are sums of two wedge-like (ship) waves each of which is contained inside the corresponding Kelvin wave wedge. It is shown that the amplitude of one wave is several times greater than the amplitude of the other wave. Each of the excited two waves is a complex wave system of transverse and longitudinal wave perturbations. The characteristics of the excited surface perturbations were studied depending on the basic parameters of the wave generation such as the velocity of motion of the source of perturbations and the frequency of its oscillations. The constructed asymptotic solutions permit describing the far-range fields of surface perturbations excited by a moving localized unsteady source both outside and inside the corresponding wave wedges. The obtained asymptotics of far-range fields of surface wave perturbations allow one efficiently to calculate the basic characteristics of wave fields and, in addition, qualitatively to analyze the obtained solutions, which is important for obtaining the well-posed statements of mathematical models of wave dynamics of surface perturbations of real natural environment. The work was carried out within the framework of the state assignment (project AAAA-A17-117021310375-7).

References

1. Lighthill, J.: Waves in Fluids. Cambridge University Press, Cambridge (1978)
2. Pedlosky, J.: Waves in the Ocean and Atmosphere: Introduction to Wave Dynamics. Springer, Heidelberg (2010)
3. Kharif, C., Pelinovsky, E., Slynyaev, A.: Rogue Waves in the Ocean. Springer, Berlin (2009)
4. Bulatov, V., Vladimirov, Yu.: Waves Dynamics of Stratified Mediums. Nauka Publisher, Moscow (2012)
5. Mei, C., Stiassnie, M., Yue, D.: Theory and Applications of Ocean Surface Waves. Advanced Series of Ocean Engineering, vol. 42. World Scientific Publishing, London (2017)
6. Morozov, E.: Oceanic Internal Tides. Observations, Analysis and Modeling. Springer, Berlin (2018)
7. Brown, M.: The Maslov integral representation of slowly varying dispersive wavetrains in inhomogeneous media. Wave Motion **32**, 247–266 (2000)
8. Chen, X., Wu, G.: On singular and highly oscillatory properties of the Green function for ship motions. J. Fluid Mech. **445**, 77–91 (2001)
9. Zakharov, V., Dyachenko, A., Vasilyev, O.: New method for numerical simulations of a nonstationary potential flow of incompressible fluid with a free surface. Eur. J. Mech. B Fluids **21**, 283–291 (2002)
10. Lannes, D.: Well-posedness of the water waves equations. J. Am. Math. Soc. **18**, 605–654 (2005)
11. Dobrokhotov, S., Grushin, V., Sergeev, S., Tirozzi, B.: Asymptotic theory of linear water waves in a domain with non-uniform bottom with rapidly oscillating sections. Russ. J. Math. Phys. **23**, 455–475 (2016)
12. Bulatov, V., Vladimirov, Yu., Vladimirov, I.: Far fields of the surface disturbances produced by a pulsating source in an infinite-depth fluid. Fluid Dyn. **52**(5), 617–622 (2017)
13. Bulatov, V., Vladimirov, Yu.: Hybrid surface waves from a harmonic perturbation source. Atmos. Oceanic Phys. **54**(2), 196–200 (2018)
14. Borovikov, V.: Uniform stationary phase method. In: IEE Electromagnetic Waves Series, vol. 40, London (1994)
15. Haberman, R.: Applied Partial Differential Equations. Pearson Educations, Upper Saddle River (2013)

Simulation of Coupled Flow and Damage in Porous Medium

Alexey Meretin[1](✉) and E. B. Savenkov[2]

[1] Moscow Institute of Physics and Technology, Dolgoprudny, Moscow Region, Russia
meretin.as@cet-mipt.ru
[2] Keldysh Institute of Applied Mathematics, Russian Academy of Sciences, Moscow, Russia
e.savenkov@googlemail.com

Abstract. In this paper, we propose mathematical model and a computational algorithm for simulation of thermoporoelastic medium with damage. The model considers a two-phase medium consisting of a porous skeleton and a mobile fluid. The medium state is described by the system of mass, momentum and energy conservation laws. To take into account the energy consumption for material damage, the additional term in energy conservation law is introduced. Constitutive relation are obtained by using the Coleman-Noll procedure, which ensures the fulfillment of the thermodynamic consistency principle. Damage of the medium is modeled within the framework of the continuum damage theory. The computational algorithm in fully three-dimensional formulation is based on the finite element method. Mesh is built using tetrahedral Taylor-Hood elements. We present some verification tests and the results of the synthetic test calculation demonstrating effects arising from thermal formation treatment.

Keywords: Thermoporoelasticity · Continuum damage mechanics · Thermodynamic consistency principle · Finite element method

1 Introduction

Nowadays, one of the most recent problems in oil engineering is estimation of efficiency of various enhanced oil recovery (EOR) techniques. With regard to hard-to-recover oil formations, one of the most advanced enhanced oil recovery method is thermal treatment, which consist in supplying heat to the reservoir by heat-transfer fluid injection or by conductive heat exchange.

Simulation of the reservoir processes induced by these EOR techniques can only be performed using coupled approaches which account for multiphysics nature of the reservoir processes,—including but not limited to fluid flow with temperature depended properties, chemical reactions, geomechanical processes. In addition, under the high pressures and temperatures in the reservoir it can be

V. I. Karev et al. (Eds.): *Physical and Mathematical Modeling of Earth and Environment Processes (2018)*, SPEES, pp. 132–141, 2019.
https://doi.org/10.1007/978-3-030-11533-3_14

observed significant deformations affecting the structure of the pore space, which leads to the formation of microcracks and reservoir damage. These processes have a significant impact on the rock's elastic properties and flow in the reservoir.

Currently, one of the most common approaches, implemented in various commercial software accurately resolve different temperature dependent flow effects, is the so-called "compositional" modeling. These models are widely used in applied simulation but they do not account for deformation and damage. Consideration of geomechanical effects, if happen, usually is based on simplified "split" models which are not themodynamically consistent [1,2]. In turn, thermodynamically based models for coupled flow, geomechanics and damage are exists but rarely used in industrial strength settings [3].

The present paper proposes mathematical model for description of the thermoporoelastic medium taking into account damage processes. The model is thermodynamically consistent, i.e. it satisfies second law of thermodynamics. Also, in the work we presented computational algorithm based on the finite element method.

2 Mathematical Model

2.1 Conservation Laws

Consider the elementary volume Ω which contains two continuums—skeleton and fluid. At the time moment $t = t_0$ the positions of the material volumes of the skeleton $\Omega_s(t_0)$ and fluid $\Omega_f(t_0)$ are the same, and the phase velocities are $\boldsymbol{v}_s$ and $\boldsymbol{v}_f$ respectively. Let us write the basic conservation laws for both continuums at the time moment $t = t_0$.

The fluid mass conservation law has the form:

$$\frac{\partial m_f}{\partial t} + \mathrm{div}(\rho_f \boldsymbol{w}) = 0, \tag{1}$$

where ρ_f is true fluid density, i.e. fluid mass related to the volume occupied by it, $\boldsymbol{w}$—flux vector equaled to $\boldsymbol{w} = m_f/\rho_f\,(\boldsymbol{v}_f - \boldsymbol{v}_s)$.

Let $\boldsymbol{f}$ be external force, $\boldsymbol{b}_\alpha^{\mathrm{int}}$—interaction forces density of phase α ($\alpha = s$ for skeleton, $\alpha = f$ for fluid) with other phases, $\boldsymbol{\sigma}_\alpha$—partial stress tensor. In this case, the momentum conservation law has the form of [4]:

$$\mathrm{div}\,(\boldsymbol{\sigma}_\alpha) + m_\alpha \left(\boldsymbol{f} - \frac{d\boldsymbol{v}_\alpha}{dt}\right) + \boldsymbol{b}_\alpha^{\mathrm{int}} = 0. \tag{2}$$

Denote total stress tensor by $\boldsymbol{\sigma} = \boldsymbol{\sigma}_f + \boldsymbol{\sigma}_s$. Then, given that $\boldsymbol{b}_f^{\mathrm{int}} + \boldsymbol{b}_s^{\mathrm{int}} = 0$, after summing up the Eq. (2) for skeleton and fluid we get:

$$\mathrm{div}\,(\boldsymbol{\sigma}) + m_f \left(\boldsymbol{f} - \frac{d\boldsymbol{v}_f}{dt}\right) + m_s \left(\boldsymbol{f} - \frac{d\boldsymbol{v}_s}{dt}\right) = 0. \tag{3}$$

Consider a thermoporoelastic medium in which, as a result of deformations, can occur a fractured zone—diffuse damage, which is the micro-fracture zones formation in the solid phase. It is significant that the characteristic dimension of the cracks is substantially smaller than the dimension of the medium representative volume. For this reason, the damage is described by a quantity defined in space (having a scalar form in the simplest case).

In accordance with the classical concepts of the crack propagation mechanics, it is necessary to supply a certain amount of energy to form a unit surface area of a crack. For this reason, it is natural to introduce into consideration the part of the total energy of the system associated with the damage of the medium. Then for domain volume $\Omega(t) = \Omega_s(t) = \Omega_f(t)$ the equation describing the energy conservation law and taking into account the damage of the material has the form of:

$$\frac{d\,(K+U)}{dt} = P + Q + \Pi, \tag{4}$$

with

$$K = \int\limits_{\Omega(t)} \left(\frac{1}{2} m_s \boldsymbol{v}_s^2 + \frac{1}{2} m_f \boldsymbol{v}_f^2\right) d\Omega, \quad U = \int\limits_{\Omega(t)} \left(m_s e_s + m_f e_f\right) d\Omega,$$

$$P = \int\limits_{\Omega(t)} \left(m_s \boldsymbol{v}_s^2 \boldsymbol{f} + m_f \boldsymbol{v}_f^2 \boldsymbol{f}\right) d\Omega_\alpha + \int\limits_{\partial\Omega} \left(\boldsymbol{t}_s \boldsymbol{v}_s + \boldsymbol{t}_f \boldsymbol{v}_f\right) d\omega_\alpha,$$

$$Q = -\int\limits_{\partial\Omega} \boldsymbol{q}\boldsymbol{n}\, d\omega, \quad \Pi = -\int\limits_{\Omega(t)} \boldsymbol{Y} : \frac{d\boldsymbol{D}}{dt} d\Omega.$$

Here K—kinetic energy of the system, U—internal energy, P—external forces power, Q—heat flux, Π—energy consumption for material damage per unit of time, e_α – phase specific internal phase energy, $\boldsymbol{q}$—heat flux density vector, $\boldsymbol{Y}$—energy dissipation rate associated with material damage, $\boldsymbol{D}$—damage variable tensor.

Passing from the skeleton partial stress tensor to the total stress tensor, we get the following expression:

$$\begin{aligned} &\frac{\partial}{\partial t}\left[m_s\left(e_s + \frac{\boldsymbol{v}_s^2}{2}\right) + m_f\left(e_f + \frac{\boldsymbol{v}_f^2}{2}\right)\right] + \\ &\quad + \operatorname{div}\left[m_s\left(e_s + \frac{\boldsymbol{v}_s^2}{2}\right)\boldsymbol{v}_s\right] + \operatorname{div}\left[m_f\left(e_f + \frac{\boldsymbol{v}_f^2}{2}\right)\boldsymbol{v}_f\right] = \\ &\quad = \operatorname{div}\left[\boldsymbol{\sigma}_f\left(\boldsymbol{v}_f - \boldsymbol{v}_s\right) + \boldsymbol{\sigma}\boldsymbol{v}_s\right] + \left(m_s\boldsymbol{v}_s + m_f\boldsymbol{v}_f\right)\boldsymbol{f} - \boldsymbol{Y} : \frac{d\boldsymbol{D}}{dt} - \operatorname{div}(\boldsymbol{q}). \end{aligned} \tag{5}$$

2.2 Derivation of Constitutive Relations

We write the second law of thermodynamics without external sources in the form:

$$\frac{\partial S}{\partial t} + \operatorname{div}\left(m_f s_f \boldsymbol{v}_f\right) + \operatorname{div}\left(m_s s_s \boldsymbol{v}_s\right) \geq -\operatorname{div}\left(\frac{\boldsymbol{q}}{T}\right), \tag{6}$$

where $S = m_s s_s + m_f s_f$ is total entropy,

$$m_s \frac{ds_s}{dt} + m_f \frac{ds_f}{dt} + \operatorname{div}\left(\frac{\boldsymbol{q}}{T}\right) \geq 0. \tag{7}$$

Define the Helmholtz free energy for the phase α as $\psi_\alpha = e_\alpha - Ts_\alpha$. Then, applying the laws of momentum and energy conservation, entropy inequality can be represented as the sum of skeleton δ_s, fluid δ_f and thermal dissipation δ_t:

$$\delta_s + \delta_f + \delta_t \geq 0, \tag{8}$$

where

$$\begin{aligned}\delta_s = &-\left(m_s \frac{d\psi_s}{dt} + m_s s_s \frac{dT}{dt} + m_f \frac{p}{\rho_f^2}\frac{d\rho_f}{dt}\right) + \\ &+ \left[\boldsymbol{\sigma}_f : \operatorname{grad}(\boldsymbol{v}_f - \boldsymbol{v}_s) + \boldsymbol{\sigma} : \operatorname{grad}(\boldsymbol{v}_s)\right] - \boldsymbol{Y} : \frac{d\boldsymbol{D}}{dt}\end{aligned}, \tag{9}$$

$$\delta_f = (\boldsymbol{v}_f - \boldsymbol{v}_s)\left[m_f\left(\boldsymbol{f} - \frac{\partial \boldsymbol{v}_f}{\partial t}\right) + \operatorname{div}(\boldsymbol{\sigma}_f)\right], \tag{10}$$

$$\delta_t = -\frac{\boldsymbol{q}}{T}\operatorname{grad}(T). \tag{11}$$

Suppose that the stress tensor of a fluid is spherical, fluid flow follows the Darcy's law, and the heat flux is determined from the Fourier law. In this case, it can be shown that $\delta_f + \delta_t \geqslant 0$. Then, according to the Coleman-Noll procedure [6], to satisfy the entropy inequality (8) it is enough that:

$$\begin{aligned}\delta_s = &-\left(m_s \frac{d\psi_s}{dt} + m_s s_s \frac{dT}{dt} + m_f \frac{p}{\rho_f^2}\frac{d\rho_f}{dt}\right) + \\ &+ \left[\boldsymbol{\sigma}_f : \operatorname{grad}(\boldsymbol{v}_f - \boldsymbol{v}_s) + \boldsymbol{\sigma} : \operatorname{grad}(\boldsymbol{v}_s)\right] - \boldsymbol{Y} : \frac{d\boldsymbol{D}}{dt} \geqslant 0.\end{aligned} \tag{12}$$

In accordance with the principle of equipresence, we obtain an expression for skeleton dissipation:

$$\begin{aligned}\delta_s = &\left(\boldsymbol{\sigma} - m_s \frac{\partial \psi_s}{\partial \boldsymbol{\varepsilon}}\right)\frac{d\boldsymbol{\varepsilon}}{dt} + \left(p\frac{d(m_f/\rho_f)}{dt} - m_s \frac{\partial \psi_s}{\partial p}\frac{dp}{dt}\right) \\ &+ \left(s_s - \frac{\partial \psi_s}{\partial T}\right)\frac{dT}{dt} - \left(\boldsymbol{Y} + m_s \frac{\partial \psi_s}{\partial \boldsymbol{D}}\right)\frac{d\boldsymbol{D}}{dt} \geqslant 0,\end{aligned}$$

where $\boldsymbol{\varepsilon}$—infinitesimal strain tensor, $\boldsymbol{\varepsilon} = (\nabla \otimes \boldsymbol{\xi} + (\nabla \otimes \boldsymbol{\xi})^2)/2$, $\boldsymbol{\xi}$—displacement vector.

We introduce the Gibbs energy g_s such that $m_s g_s = m_s \psi_s - p m_f/\rho_f$. Then:

$$\begin{aligned}\delta_s = &\left(\boldsymbol{\sigma} - m_s \frac{\partial g_s}{\partial \boldsymbol{\varepsilon}}\right) d\boldsymbol{\varepsilon} - \left(\frac{m_f}{\rho_f} + m_s \frac{\partial g_s}{\partial p}\right) dp \\ &+ \left(s_s - \frac{\partial g_s}{\partial T}\right) dT - \left(\boldsymbol{Y} + m_s \frac{\partial g_s}{\partial \boldsymbol{D}}\right) \boldsymbol{D} \geqslant 0.\end{aligned} \tag{13}$$

The implementation of the thermodynamic consistency principle requires the validity of entropy inequality (6) for any history of states, as well as the independence of the constitutive relations from the choice of reference configuration. For this it is suffices to fulfill the inequality (13), setting:

$$\boldsymbol{\sigma} = m_s \frac{\partial g_s}{\partial \boldsymbol{\varepsilon}}; \quad m_f = -\rho_f m_s \frac{\partial g_s}{\partial p}; \quad s_s = \frac{\partial g_s}{\partial T}; \quad \boldsymbol{Y} = -m_s \frac{\partial g_s}{\partial \boldsymbol{D}}, \tag{14}$$

which provides $\delta_s = 0$.

In this case, after linearization the constitutive relations for the skeleton and fluid are of the form:

$$\begin{aligned}
\Delta\boldsymbol{\sigma} &= \boldsymbol{C} : \Delta\boldsymbol{\varepsilon} - \boldsymbol{b}\Delta p - \boldsymbol{C} : \boldsymbol{\alpha}_T \Delta T - \boldsymbol{\gamma} : \Delta\boldsymbol{D}, \\
\Delta m_f &= \rho_f \boldsymbol{b} : \Delta\boldsymbol{\varepsilon} + \rho_f \frac{\Delta p}{M} - \alpha_m \rho_f \Delta T + \rho_f \boldsymbol{\omega} : \Delta\boldsymbol{D}, \\
\Delta S_s &= \boldsymbol{C} : \boldsymbol{\alpha}\Delta\boldsymbol{\varepsilon} - \alpha_\phi \Delta p + \frac{C_{ps}}{T^0}\Delta T + \frac{1}{m_s}\boldsymbol{\theta} : \Delta\boldsymbol{D}, \\
\Delta\boldsymbol{Y} &= \boldsymbol{\gamma} : \Delta\boldsymbol{\varepsilon} + \boldsymbol{\omega}\Delta p + \boldsymbol{\theta}\Delta T + \boldsymbol{\eta} : \Delta\boldsymbol{D}, \\
\frac{\rho_f}{\rho_{f0}} &= 1 + \frac{\Delta p}{K_f} - \alpha_f \Delta T, \\
\Delta S_f &= -\alpha_f m_f^0 \frac{\Delta p}{\rho_f^0} + C_{pf}\frac{\Delta T}{T^0},
\end{aligned} \tag{15}$$

where $\boldsymbol{\gamma}, \boldsymbol{\omega}, \boldsymbol{\theta}, \boldsymbol{\eta}$—tensors of damage coefficients, also the following notations are used $\Delta f = f - f^0$ for all f:

$$\frac{1}{M} = \frac{1}{N} + \frac{m_f}{\rho_f}\frac{1}{K_f}, \qquad \alpha_m = \alpha_\varphi + \frac{m_f}{\rho_f}\alpha_f, \quad S_\alpha = m_\alpha s_\alpha.$$

Here M, N are Biot modulus, K_f stands for fluid bulk modulus, $\alpha_{f,\varphi}$ temperature expansion coefficients.

To proceed further we introduce the following assumptions: the influence of inertial forces is negligible; the skeleton velocity is negligible compared to fluid velocity; the kinetic energy of the fluid is neglected in comparison with the value of the internal energy; the influence of external forces is neglected; damage variable is a scalar function of the strain tensor and affects only the elastic coefficients tensor by the formula $C(D) = C(1 - D)$.

With regard to these assumptions, the system of Eqs. (1), (3), (5), (15), has the form:

$$\begin{aligned}
&\frac{\partial m_f}{\partial t} + \operatorname{div}(\rho_f \boldsymbol{w}) = 0, \quad \operatorname{div}\boldsymbol{\sigma} = 0, \\
&\frac{\partial\left(m_s e_s + m_f e_f\right)}{\partial t} + \operatorname{div}\left(\rho_f e_f \boldsymbol{w}\right) = \operatorname{div}\left(-p\boldsymbol{w}\right) - \operatorname{div}(\boldsymbol{q}), \\
&\boldsymbol{w} = \frac{\boldsymbol{k}}{\mu}\left[-\operatorname{grad}(p)\right], \quad \boldsymbol{q} = -\boldsymbol{\kappa}\operatorname{grad}(T).
\end{aligned} \tag{16}$$

To close this system of equations, the constitutive relations (15) and damage propagation law [7] are used:

$$D = \begin{cases} 0, & \tilde{\varepsilon} < \tilde{\varepsilon}_c \\ \dfrac{D_{\text{off}}}{\tilde{\varepsilon}_{\text{off}} - \tilde{\varepsilon}_c}\tilde{\varepsilon} - D_{\text{off}}\dfrac{\varepsilon_c}{\tilde{\varepsilon}_{\text{off}} - \tilde{\varepsilon}_c}, & \tilde{\varepsilon}_c \leq \tilde{\varepsilon} \leq \tilde{\varepsilon}_{off}, \\ D_{\text{lim}} - (D_{\text{lim}} - D_{\text{off}})\dfrac{\tilde{\varepsilon}_{\text{off}}}{\tilde{\varepsilon}} & \tilde{\varepsilon} > \tilde{\varepsilon}_{off}, \end{cases} \tag{17}$$

where $\tilde{\varepsilon}$ is given by:

$$\tilde{\varepsilon} = \sqrt{\sum_{i=1}^{3} \langle \varepsilon_i \rangle^2}, \quad \langle \varepsilon_i \rangle = \frac{\varepsilon_i + |\varepsilon_i|}{2},$$

with ε_i being principal strains.

The system (16) is solved by the finite element method. Implicit scheme was used for temporal discretization of the system. To solve the non-linear system arose at each time step, the Newton method was utilized. To increase the stability of solution, the mass matrices were lumped [8].

Space discretization of the Eq. (16) was performed on tetrahedral mesh with quadratic basis functions for displacements and linear for pressure and temperature (Taylor-Hood elements [9]). A feature of this type of finite elements is the satisfaction of the so-called inf-sup conditions (Ladyzhenskaya–Babuska–Brezzi condition). The fulfillment of these conditions provides stability of the solution for the poroelasticity problem. The numerical integration was performed using second-order Gauss quadrature formulas.

3 Numerical Results

3.1 Mandel's Problem

Mandel [10] presented an analytical solution for the three-dimensional consolidation of poroelastic material in which there is a non-monotonic pore pressure evolution (Mandel-Cryer effect). In the original formulation, a sample of poroelastic material with a length of $2a$ and a height of $2b$ saturated with fluid is considered. The sample is fixed between two rigid impermeable plates, and a distributed load with a force of $2F$ is applied to the upper face.

In this test, the linear dimensions of the sample were $1 \times 1 \times 1$ m. The vertical stress applied on the top face is 1 kPa. The initial pressure and displacements are vanishing. The pore pressure distribution along the x axis and displacement projections along the x and z axes were calculated. The obtained values were compared with a known analytical solution. The results are presented in Figs. 1 and 2.

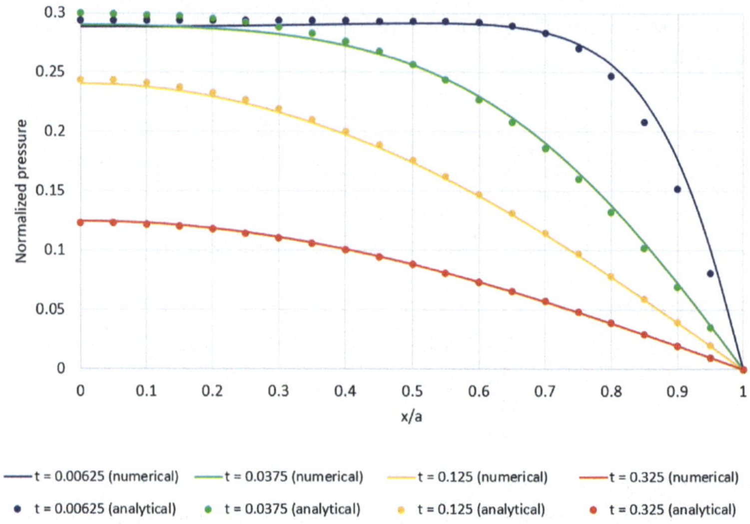

Fig. 1. Normalized pore pressure distribution along x axis.

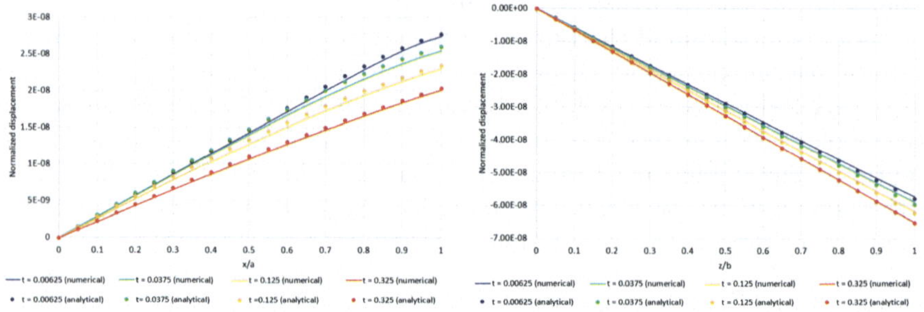

Fig. 2. Normalized projection of x (left) and z (right) displacement components.

3.2 One-Dimensional Non-isothermal Expansion

The problem of one-dimensional sample thermal deformation is considered. Sample dimensions are $a \times a \times h$. The base of the sample is fixed on the surface, the lateral borders allow only vertical displacement, the upper boundary is free. It is assumed that the sample is thermally isolated and has a temperature T_0 at the initial time. The upper boundary has a constant temperature equal to $T_0 + \theta_0$.

The problem has an analytical solution obtained using the Laplace transform [11]. Testing was performed for various time points from 10 to 200 h. The results are presented in Fig. 3.

3.3 Damage Evolution Near Injection Well

Simulation the damage evolution in a small region near the well at the initial times was carried out on a model of size $1 \times 1 \times 1$ m. The initial reservoir pressure

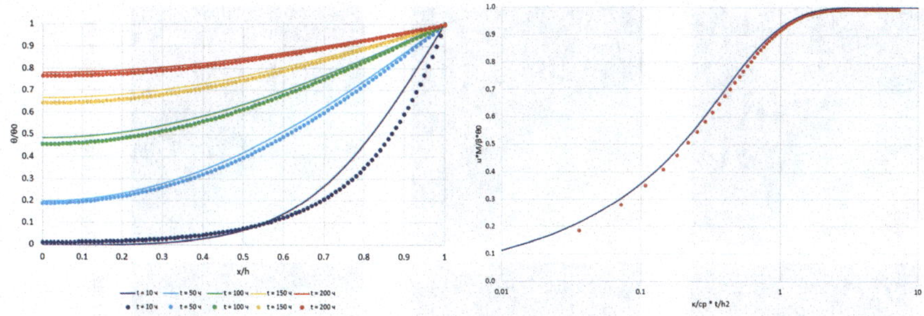

Fig. 3. Temperature distribution over the sample (left) and vertical displacement at the upper boundary of the sample (right).

was assumed to be 200 bar, the initial temperature 100°C, and the initial stress in the reservoir in the x and y directions was 300 and 330 bar respectively. In the corner point of the domain, there is an injection well with a cylindrical shape with a radius of 0.1 m, operating with a constant injection rate of 0.5 m^3/day.

To simulate the damage evolution, the following parameters were used: D_{lim} D_{off} assumed to be equal 1, ε_c and ε_{off} are equal to 0.0002 and 0.015 respectively. Other model parameters are listed in the Table 1. Dynamic parameters distribution for the moments of 2 h, 1 and 10 days after the start of the well are presented in the Fig. 4.

Table 1. Input parameters for damage evolution test

Parameter	Value
Young's modulus, E	20 GPa
Poisson's ratio, ν	0.3
Biot modulus, N	10 GPa
Fluid bulk modulus, K_f	3.3 GPa
Biot coefficient, b	0.79
Permeability, k	$1 \cdot 10^{-16}\,\text{m}^2$
Porosity, φ	0.1
Viscosity, μ	1 mP · s
Skeleton density, ρ_s	2100 kg/m^3
Fluid density, ρ_s	1000 kg/m^3
Skeleton volume temperature expansion coefficient, α_s	$1 \cdot 10^{-6}$ 1/K
Fluid volume temperature expansion coefficient, α_f	$1 \cdot 10^{-4}$ 1/K
Skeleton Specific Heat, c_{ps}	1000 J/(kg · K)
Fluid Specific Heat, C_{pf}	4200 J/(kg · K)
Effective thermal conductivity, κ	2 W/(m · K)

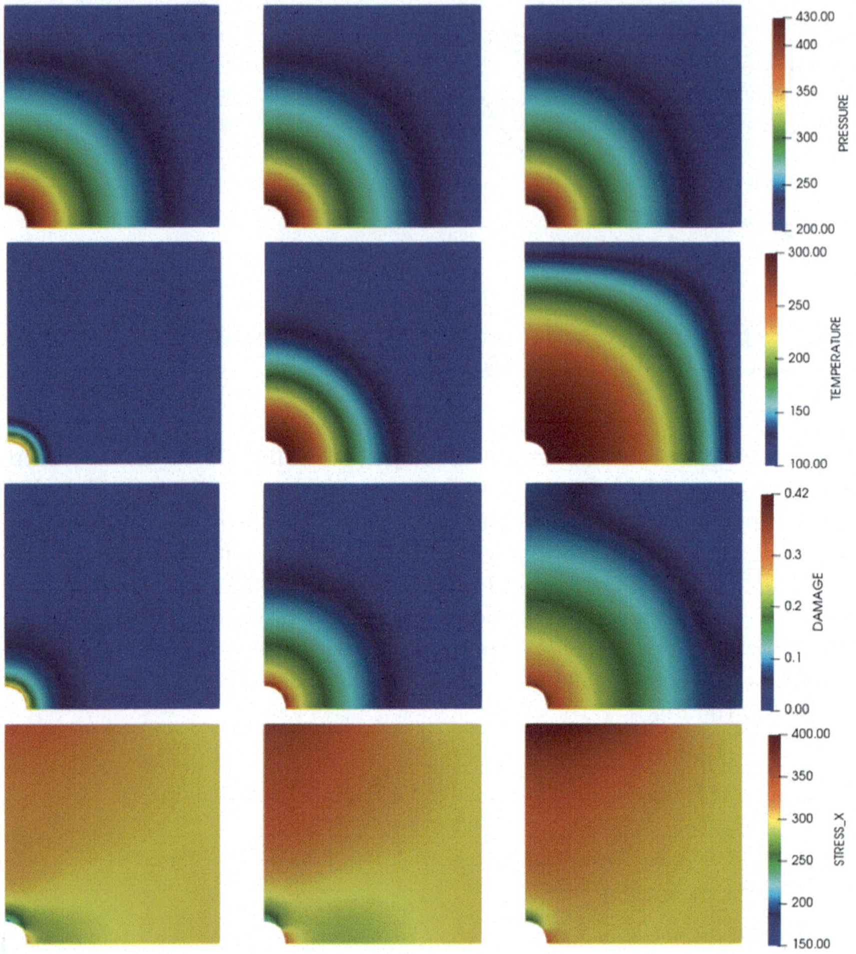

Fig. 4. Distribution of pressure p (first row), temperature T (second row), damage D (third row) and horizontal component of stress tensor σ_{xx} (fourth row) after 2 h (left), 1 day (center) 10 days (right).

4 Conlusion

We presented a thermodynamically consistent mathematical model and a computational algorithm for calculating damage in a thermoporoplastic medium. The damage of the medium is modeled within the framework of the continuum damage theory. The model takes into account the effects of skeleton deformation, fluid flow, and non-isothermal effects. The system of equations consists of the basic conservation laws (mass, momentum, energy) and the kinetic equation for the damage variable. For the closure of a system of equations, the constitutive relations obtained by Coleman-Noll procedure are used.

This paper also presents a numerical algorithm based on the finite element method in a fully three-dimensional formulation. The validity of the algorithm is confirmed in a number of test examples. Also an example of synthetic model test is presented for estimating the propagation of damage near injection well, which inject fluid at high temperature. According to the calculations in the near-wellbore zone, a significant damage of the rock is observed.

References

1. Murakami, S., Ohno, N.: A continuum theory of creep and creep damage. In: Ponter, A.R.S., Hayhurst, D.R. (eds.) Creep in Structures, pp. 422–444. Springer, Heidelberg (1981). https://doi.org/10.1007/978-3-642-81598-0_28
2. Lemaitre, J. (ed.): Handbook of Materials Behavior Models, Three-Volume Set: Nonlinear Models and Properties. Elsevier, Amsterdam (2001). https://doi.org/10.1007/978-3-7091-2806-0
3. Kondaurov, V.I., Fortov, V.E.: Osnovy Termomekhaniki Kondensirovannoi Sredy (Fundamentals of the Thermomechanics of a Condensed Medium). Izd. MFTI, Moscow (2002). [In Russian]
4. Kondaurov, V.I.: Mechanics and termodynamics of saturated porous medium. M.: MFTI (2007). [In Russian]
5. Bazarov, I.P., Gevorkyan, E.V., Nikolaev, P.N.: Nonequilibrium Thermodynamics and Physical Kinetics. Izd. Mosk. Univ, Moscow (1989). [In Russian]
6. Coleman, B.D., Noll, W.: The thermodynamics of elastic materials with heat conduction and viscosity. Arch. Ration. Mech. Anal. **13**(1), 167–178 (1963). https://doi.org/10.1007/978-3-642-65817-4_9
7. Pogacnik, J., OSullivan, M., OSullivan, J.: A Damage Mechanics Approach to Modeling Permeability Enhancement in Thermo-Hydro-Mechanical Simulations. In: Proceedings, pp. 24-26 (2014)
8. Neuman, S.P.: Saturated-unsaturated seepage by finite elements. In J. HYDRAUL. DIV., PROC., ASCE (1973)
9. Taylor, C., Hood, P.: A numerical solution of the Navier-Stokes equations using the finite element technique. Comput. Fluids **1**(1), 73–100 (1973). https://doi.org/10.1016/0045-7930(73)90027-3
10. Mandel, J.: Consolidation des sols (tude mathmatique). Geotechnique **3**(7), 287–299 (1953)
11. Carter, J.P., Booker, J.R.: Finite element analysis of coupled thermoelasticity. Comput. Struct. **31**(1), 73–80 (1989). https://doi.org/10.1016/0045-7949(89)90169-7

Modeling of a Sun Glint on a Sea Surface for Remote Sensing Purposes

Nick Eugen Lebedev(✉) and A. S. Zapevalov

Marine Hydrophysical Institute RAS, Sevastopol 299029, Russian Federation
{nick_leb, sevzepter}@mail.ru

Abstract. According to calculations of the upward radiation at the TOA level made on the base of the Cox-Munk model of sea surface slopes distribution and Sobolev model of atmospheric scattering, presented are the sun glint quantitative characteristics such as length, maximal width and maximal contrast for satellite scanner measurements in the visible and near-IR ranges of 0.4–1.6 μm for sea wind velocities of 2–14 m/s at a 10 m height at zonal and meridional directions and aerosol optical depth of 0.03, 0.21, corresponding to the cases of pure atmosphere and with presence of marine aerosol above moderate level, for a sun-synchronous orbit at an altitude of 866 km. For other satellite heights the maximal contrast and length are almost unchanged, and for the maximal width corresponding corrections are given. Presented results may be used for the purposes of planning of satellite measurements of sea and atmosphere parameters sensitive to the sun glint manifestations and for analyzing of satellite images.

Keywords: Satellite optical sounding · Sun glint · Scanner · Sea surface

1 Introduction

Satellite measurements of upward radiation are used to solve a wide range of tasks. Among them are finding of the sea color, of the content of dissolved organics and suspensions, of the aerosol optical thickness of the atmosphere, of the content of water vapor in the atmosphere. Herewith, a part of the satellite image in some degree is covered by a sun glint, which is a serious obstacle in solving these tasks [1, 2]. On the contrary, for a number of others, among them are determining of the sea waves parameters or the sea wind velocity [3], detection of internal waves and oil slicks [2, 4], the sun glint is an informative factor.

Therefore, for the purposes of both planning of satellite measurements and analyzing of satellite images, it is desirable to have a general representation of the glint parameters, among which it is worth to select it's width, length and brightness contrast. The first two ones characterize the area occupied by the glint (where the solution of one group of tasks is difficult, and of the other is the only possible), the last characterizes the glint intensity and, thus, it is directly related to the residual error levels in solving of the first-type tasks in the glint zone after the glint affect correction [1, 2]. Although the sun glint manifestations are known from the very beginning of satellite observations, its parameters similar to mentioned above have not yet been analyzed.

V. I. Karev et al. (Eds.): *Physical and Mathematical Modeling of Earth and Environment Processes (2018)*, SPEES, pp. 142–149, 2019.
https://doi.org/10.1007/978-3-030-11533-3_15

Manifestation of the sun glint depends on the surface roughness (determined mainly by the sea wind), the geometry of the observation, and the transparency of the atmosphere which significantly depends on the optical properties of atmospheric aerosol.

The purpose of this work is to present a quantitative assessment of the sun glint parameters for scanning satellite measurements in the visible and near-IR ranges of 0.4–1.6 μm for sea wind speeds from 2 to 14 m/s of zonal and meridional directions and optical densities of aerosol scattering of 0.03 and 0.21. The range of wavelengths is based on those involved in the relevant measurements [5], wind speeds - on the capabilities of existing models of sea surface slopes distribution [6], optical densities of aerosol scattering - from their characteristic values over the oceans (which are correspond to a very clean atmosphere and with the apparent presence of a marine type aerosol [7, 8]). The calculations were performed for an orbit altitude of 866 km.

Preliminary results obtained for the simple model of isotropic sea surface slopes distribution are presented in [9].

2 Mathematical Formulation of the Task

Brightness of the upward radiation detected by the satellite, J, is the sum of several terms [2], which can be reduced to three ones. The first term, I, describes the reflection by the sea surface towards the satellite of downward direct solar irradiation passed through the atmosphere; this term itself is responsible for the sun glint formation [2, 3]. The second term, A, is caused by the processes of light scattering in the atmosphere with participation of multidirectional reflection from the sea surface [2]. The third term, W, is associated with the upward scattering of part of solar radiation that has penetrated into the seawater column.

In the near infrared the W value is negligibly small, in visible it depends on the amount of organics and suspensions. Estimations made on data of upward radiation measurements near Hawaiian islands [10], according to which W is at the level of 0,4 (λ = 0,75 mm) - 13% (λ = 0,4 mm) from A magnitude show that at most situations, its inclusion leads to a decrease of the glint parameters values by units of percent. In this paper, the influence of the W addendum is not taken into account; that is, it is assumed that $J = I + A$.

The I value was calculated according to [3], A value - by the semi-empirical Sobolev formula [11], which showed agreement at the 10% level with the field measurements [12]. The total sea surface albedo was calculated in a form which takes into account its optical properties and roughness [13].

3 Glint Parameters Definitions

Let's determine that on each scan of a satellite image the glint boundary falls on those sea surface points, where the glint brightness I is ten times less than the non-glint component A:

$$I = 0.1A. \tag{1}$$

Such a way of determining the glint boundary is not the only possible one, but it is most convenient in the context of task of dividing the entire satellite image into areas exposed to glint affect and free from it.

Let's call a scan passing through a sub-sun point (directly or by its continuation) as a zero scan. Glint section through the zero scan is wider and brighter than through other scans.

With the disposal of the satellite and, respectively, the sub-satellite point with current scan crossing this point, from the zero scan, the value of max(I) decreases, whereas slowly changing through the scan A value slightly increases. With this, the glint boundary narrows and ultimately tightens to a single point which we'll call as polar point. There are two such points - northern and southern; so, there is no glint above the northern and below the southern polar points.

We'll define the glint length in the northern/southern direction, Φ_N/Φ_S, as the angular distance between the zero scan and the northern/southern polar point. The Φ_N, Φ_S values are equal in cases where the sea surface slopes distribution is symmetric about the zonal direction, that is, for westerly or easterly winds.

Next, we'll define the maximal width, D, as the angular distance between the glint border points ξ_1, ξ_2 on the zero scan. Last, we'll define the maximal glint contrast, R, as the ratio on the zero scan of M/m values: $R \equiv M/m$, where $M = \max(J)$, $m = \text{mean}(J(\xi_1), J(\xi_2))$ (see Fig. 1).

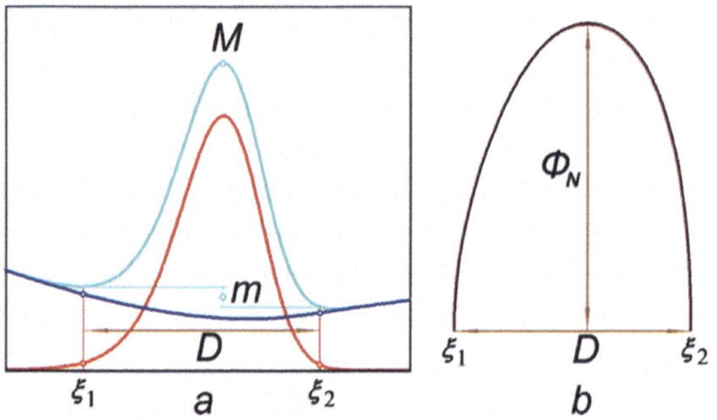

Fig. 1. To the calculation of the glint parameters: a - points ξ_1, ξ_2 of the glint border on the zero scan (abscissa) and parameters D, $R \equiv M/m$. The I value is shown by red line; A - blue; $J = I + A$ - cyan; b - the glint length in the northern direction, Φ_N. The northern part of the glint boundary is shown here.

The glint limits ξ_1, ξ_2 on arbitrary scan do not coincide generally with the inflection points of the J curve on this scan.

4 Calculations and Results

Calculations of the D, F, and R parameters were carried out in the visible and near-infrared range $\lambda = 0.4–1.6$ μm for typical aerosol-scattering optical thickness values in sea conditions $\tau_a = 0.03$ and 0.21 (for very pure atmosphere and in presence of marine type aerosol with above average value) in the wind speed range $V = 2–14$ m/s at a standard 10 m height (related with wind speed at 12.5 m being used in Cox-Munk model [3] accordingly to [14] as $V_{12.5} = 1.026V_{10}{}^{-}0.016$) under a number of conditions and assumptions:

- the atmosphere is cloudless and does not contain absorbing components (as clouds mask sea surface in visual and IR ranges);
- the optical fields of the atmosphere and wind are uniform;
- the sub-sun point is located at the equator;
- the satellite orbit is directed along the zero meridian;
- seawater is free from pollutants (as pollutants substantially affect the results through changing of the sea surface optical and elastical properties).

Among the general features of the dependence of the A, I quantities on the τ_a, λ, V task input parameters, it should be noted an increase of A and decrease of I with τ_a increasing (due to the growth of scattering by aerosol of a downward sunlight irradiance); a decrease of A with λ increasing (due to a decrease of the molecular scattering level); practical independence of A from V (due to the weakly expressed connection of A with albedo, and albedo with V); a decrease and broadening of I with V increasing (due to a more uniform redistribution of the sun irradiation reflected by the sea surface into the upward solid angle). These features lead to the fact that with V increasing and/or τ_a decreasing, there is an increase of D and a decrease of R. Also with V increasing there is an increase in the asymmetry of the glint boundaries due to increase in the asymmetry of the sea surface slopes probability density along the wind direction. These features are traced in Figs. 2 and 3.

Figure 2 shows the contours of the glint boundaries for $\lambda = 0.4$, 0.8 and 1.6 μm, $\tau_a = 0.03$ and 0.21, $V = 2$, 8 and 14 m/s of the southern and eastern directions.

Figure 3 shows the dependences of the D, Φ and R values on the meridional and zonal winds at $\tau_a = 0.03$, 0.21 for $\lambda = 0.4$, 0.5, 0.6, 0.8, 1, 1, 2, 1.4, 1.6 μm.

5 Discussion

The boundaries of glint having length Φ less than 45° at near-meridional winds are close to an ellipse. With Φ increasing, the upper and lower boundaries of the glint become flattened, which is associated with the features of the Mercator projection. In reality, the satellite crosses the equator not exactly at noon local time, so the sub-sun point is placed out of track. This leads, on the one hand, to an additional distortion of the ellipticity of the glint boundary and, on the other, to enlarging of the D and Φ parameters. For example, at $\tau_a = 0.03$, $\lambda = 0.8$ μm, $V = 14$ m/s and sub-sun point shifted relative to the satellite track by 24°, the D value increases by 18%, and Φ by 4%. The calculations presented here, as well as satellite observations accomplished

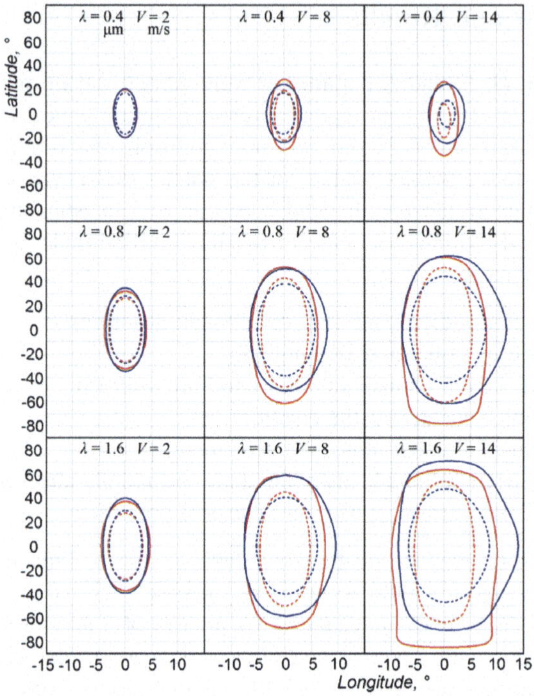

Fig. 2. The contours of the glint boundaries for λ = 0.4, 0.8 and 1.6 μm (from top to bottom), τ_a = 0.03 (solid lines) and 0.21 (dashed lines), V = 2, 8 and 14 m/s (from left to right) of the southern (red) and eastern (blue) wind directions.

with sea wind and aerosol data, show that the most significant factor influencing the shape of the glint is the heterogeneity of the wind module field.

If to separate the glint zone from the non-glint one more rigidly (by lowering a coefficient in the expression (1)), then the glint boundaries will expand, and the D, Φ values will increase. For instance, if in (1) the parameter 0.1 is replaced by 0.05, then D will increase by 5–20% (and more for $\lambda < 0.5$ μm), and Φ will increase by 1–9% (and more for $\lambda < 0.5$ μm). The R value as a rule increases up to 3–5% (for $\lambda < 0.5$ μm it always increases, by tens of percent and the more, with growth of τ_a and/or V).

With V increasing, as noted above, the glint component I expands and decreases. At that, if the A value is smaller than I, then the expansion of the latter leads to the glint expansion. Another case could be for smaller λ (therefore, for larger value of molecular scattering): here, as V increases, the A value increases with respect to expanding and decreasing I so expressively that instead of widening, the glint may be narrowed. Such situation occurs here for λ = 0.4 μm at τ_a = 0.21 and $V > 5$ m/s (see Fig. 3).

The Φ value is proportional to D, with the proportionality coefficient is of 3.5–5.05, which being decreased while V and/or τ_a increase. The R value decreases with: V increasing (due to I decrease), τ_a increasing/λ decreasing (due to A increase because of increasing of aerosol/molecular scattering). The D, Φ, and R parameters sets for wavelengths within the range of 1–1.6 μm at τ_a = 0.03 are close due to a much lower

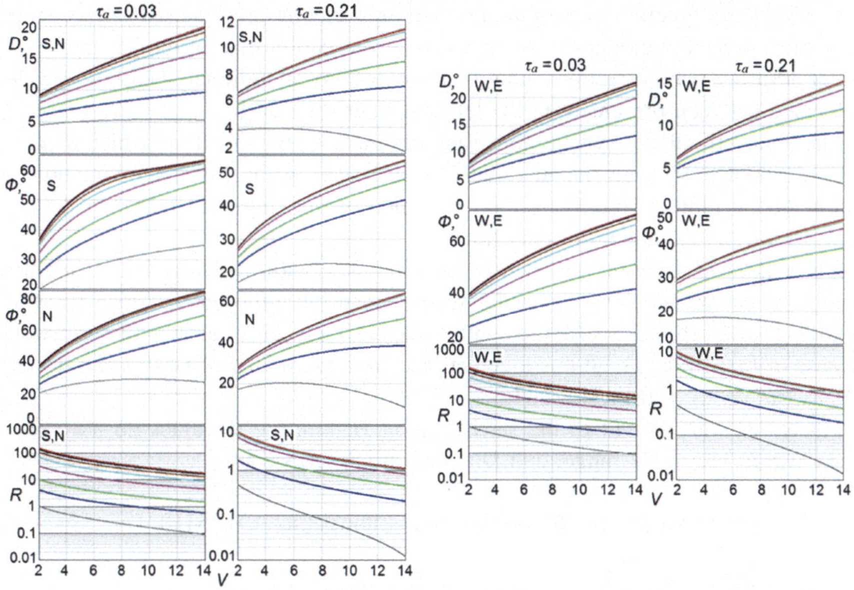

Fig. 3. *D*, *Φ* and *R* values depending on wind values of the meridional direction (two left columns of graphs), zonal direction (two right columns) with $\tau_a = 0.03$ and 0.21 for $\lambda = 0.4$ (grey lines); 0.5 (blue); 0.6 (green); 0.8 (magenta); 1 (cyan); 1.2 (brown); 1.4 (black); 1.6 μm (red), bottom to top on each graph.

level of molecular scattering optical depth for $\lambda > 0.8$ μm than minimal τ_a value of 0.03; at $\tau_a = 0.21$, they are almost coincide (see Fig. 3).

Let's consider some changes in the obtained results when deviating from the conditions and assumptions within which the calculations were carried out.

Atmospheric gases and water vapor of the real atmosphere in the 0.4–1.6 μm range have some strong absorption bands (for example, 0.9–1, 1.1–1.2 μm), where the true results will differ significantly from those obtained. But the channels of satellite instruments are outside these bands, - where absorption is of continual character and is much less. This absorption leads to a decrease of the calculated *D*, *Φ*, *R* values. If it corresponds to a decrease in the vertical transmittance of the atmosphere by 10%, then the *D* value will decrease by 1 - 3%, *Φ* - by 0.2 - 1.5%, *R* - by 7.5 - 9.5%.

For arbitrary latitude of the sub-sun point φ_s, the true values of *Φ* are equal to the calculated ones plus φ_s.

For $V < 2$ m/s, the calculations of *D*, *Φ*, *R* values were not done because of the lower accuracy of the sea surface slopes distribution model due to the greater influence of such uncontrolled factors as surface currents and the appearance of internal waves. For this case, estimates of *D*, *Φ*, *R* can be obtained by extrapolating of their graphical values on Fig. 3.

In Fig. 2 it is seen that the glint boundary is broadened to the windward side. This is due to the fact that the sea surface slopes probability density decreases more steeply in the windward direction than in the leeward and, accordingly, the sea surface slopes, on average, are more elongated in the windward direction than in the leeward.

For arbitrary wind directions other than the zonal or meridional, the D, Φ, R values are located between presented on the left and right sides of Fig. 3.

The polar-orbital sun-synchronous satellites move in the direction, which makes an angle of about 8–9° with the meridian. In this case, in the first approximation, the contour of the glint is rotated by the magnitude of this angle, maintaining the same values of all the parameters of the glint.

Differences in the glint parameters for different λ with unchanged V, τ_a are associated with the influence of molecular scattering in the atmosphere. Since its optical thickness is proportional to λ^{-4}, then with λ decreasing it causes growth of A values and decrease of I (see Fig. 1); all this together leads to a narrowing of D, Φ and a decrease in R, as can be seen in Figs. 2 and 3.

For an arbitrary height of the satellite orbit H, km, the Φ values do not change, R vary within –3 - +1% range, and D values are described by

$$D(H,\mathrm{km}) = D(866\,\mathrm{km})\left[1 + (H - 866)(8.82 \pm 0.55)10^{-4}\right].$$

From here, for example, for Aqua and Terra satellites:

$$D(705\,\mathrm{km}) = D(866\,\mathrm{km})[0.86 \pm 0.009].$$

6 Conclusion

The results of numerical modeling of the sun glint parameters in the visible and near-IR range, such as length, maximal width and maximal brightness contrast are presented. The dependences of these parameters on the aerosol optical depth in cases of very pure atmosphere and in presence of marine type aerosol with above average value, the wavelength of electromagnetic radiation from 0.4 to 1.6 μm, and the wind speed and direction from 2 to 14 m/s, zonal and meridional, are built. The obtained results can be used in planning of satellite measurements of the sea and the atmosphere parameters that are sensitive to the manifestation of the sun glint, in analysis and processing of satellite images.

The work was done within the framework of the state task on the topic № 0827-2018-0002.

References

1. Mobley, C.D., Werdell, J., Franz, B., Bailey, S.: Atmospheric correction for satellite ocean color radiometry, 85 p. (2016). https://oceancolor.gsfc.nasa.gov/docs/technical/NASA-TM-2016-217551.pdf
2. Kay, S., Hedley, J.D., Lavender, S.: Sun glint correction of high and low spatial resolution images of aquatic scenes: a review of methods for visible and near-infrared wavelengths. Remote Sens. **1**, 697–730 (2009). https://doi.org/10.3390/rs1040697

3. Cox, C., Munk, W.: Measurement of the roughness of the sea surface from photographs of the sun's glitter. J. Opt. Soc. Am. **44**(11), 838–850 (1954)
4. Lu, Y., Zhou, Y., Liu, Y., Mao, Z., Qian, W., Wang, M., Zhang, M., Xu, J., Sun, S., Du, P.: Using remote sensing to detect the polarized sunglint reflected from oil slicks beyond the critical angle. J. Geophys. Res. Oceans **122**(8), 6342–6354 (2017). https://doi.org/10.1002/2017jc012793
5. Li, J., Menzel, W.P., Yang, Z., Frey, R.A., Ackerman, S.A.: High-spatial-resolution surface and cloud-type classification from MODIS multi-spectral band measurements. J. Appl. Meteorol. **42**, 204–226 (2003)
6. Lebedev, N.E., Zapevalov, A.S. Comparison of empirical sea-surface slopes probability densities for the purposes of satellite sounding. In: Physical and mathematical modeling of earth and environment processes. PMMEEP 2017. Springer Geology. Springer, Cham. (2017). https://doi.org/10.1007/978-3-319-77788-7_21
7. Kiliyanpilakkil, V.P., Meskhidze, N.: Deriving the effect of wind speed on clean marine aerosol optical properties using the A-Train satellites. Atmos. Chem. Phys. **11**, 11401–11413 (2011). https://doi.org/10.5194/acp-11-11401
8. Myhre, G., Stordal, F., Johnsrud, M., Diner, D.J., Geogdzhayev, I.V., Haywood, J.M., Holben, B.N., HolzerPopp, T., Ignatov, A., Kahn, R.A., Kaufman, Y.J., Loeb, N., Martonchik, J.V., Mishchenko, M.I., Nalli, N.R., Remer, L.A., Schroedter-Homscheidt, M., Tanré, D., Torres, O., Wang, M.: Intercomparison of satellite retrieved aerosol optical depth over ocean during the period September 1997 to December 2000. Atmos. Chem. Phys. **5**, 1697–1719 (2005). https://doi.org/10.5194/acp-5-1697-2005
9. Lebedev, N.E., Zapevalov, A.S.: Characteristic geometric features of the sun glint on sea surface at scanning satellite measurements, pp. 179–182 (2018). http://symp.iao.ru/files/symp/aoo/24/C.pdf. (in Russian)
10. http://d33.infospace.ru/d33_conf/tarusa2018/03.pdf. (in Russian)
11. Sobolev, V.V.: Light Scattering in Planetary Atmospheres, 256 p. Pergamon, New York (1975)
12. Zibordi, G., Voss, K.J.: Geometrical and spectral distribution of sky radiance - comparison between simulations and field measurements. Remote Sens. Environ. **27**, 343–358 (1989). https://doi.org/10.1016/0034-4257(89)90094-1
13. Jin, Z., Qiao, Y., Wang, Y., Fang, Y., Yi, W.: A new parameterization of spectral and broadband ocean surface albedo. Opt. Express **19**(27), 26429–26443 (2011). https://doi.org/10.1364/OE.19.026429
14. Smith, S.D.: Coefficients for sea surface wind stress, heat flux, and wind profiles as a function of wind speed and temperature. J. Geophys. Res. **93**(C12), 15467–15472 (1988). https://doi.org/10.1029/JC093iC12p15467

Geomechanical Approach to Improving the Efficiency of the Operation of Underground Gas Storages

V. I. Karev(✉)

Ishlinsky Institute for Problems in Mechanics of Russian Academy of Sciences, Ave. Vernadskogo 101, Bldg 1, Moscow 119526, Russia
wikarev@ipmnet.ru

Abstract. The paper describes the scientific problems associated with the operation of underground gas storages (UGS), existing methods for modeling mechanical and filtration processes in UGS, the ways to improve these models: accounting the mutual influence of stresses and filtration, the effect of cyclical pressure change in a wide range at gas injection and extraction on strength of rocks. The results of an experimental determination of the dependence of the permeability of rocks on the stresses acting in them, of a theoretical study of the effect of filtration flow on the stress-strain state in the vicinity of the wells and the size of the plastic zone, an experimental study of the effect of cyclic loading on the strength of rocks are presented.

Keywords: Underground gas storage · Rock · Borehole · Test apparatus · Stress-strain state · Seepage · Strength · Cyclic loading

1 Introduction

The efficiency of operation of underground gas storages is determined by the productivity of production wells in the gas extraction cycle and their injectivity in the gas injection cycle, as well as the duration of the time without borehole overhaul mainly related to the fracture intensity of the bottomhole formation zone leading to sand flow into the well, mudding and abrasive depreciation of underground equipment. The fracture of the bottomhole zone is promoted by gradients of high pressure on the borehole wall during gas filtration, breakthrough of reservoir water at unjustified pressure drawdown, change in direction of filtration flows during gas extraction and injection, unregulated sharp increase in well production (for example, during gas-dynamic well studies).

To improve the efficiency of UGS and reduce the number of complications at the operation of wells, it is necessary to have reliable data on the geomechanical and filtration properties of rocks and their dependence on changes in the stress-strain state in the reservoir during operation. The presence of such knowledge is of great importance, given the wide range of changes in reservoir pressure in formations during the process of cyclic regimes "extraction-injection". Particular attention should be paid to the study of changes in the filtration characteristics of the host rocks to assess the risk of

V. I. Karev et al. (Eds.): *Physical and Mathematical Modeling of Earth and Environment Processes (2018)*, SPEES, pp. 150–158, 2019.
https://doi.org/10.1007/978-3-030-11533-3_16

deterioration of the insulating properties of overlying layer during the operation of the UGS. For this purpose, studies of the stress-strain state in the near-wellbore zone and technogenic changes in its properties should be performed. The inclusion of the obtained results in the geomechanical and hydrodynamic models of UGS will allow a more complete and detailed study of the deformation and dynamic processes during the cyclic operation of underground gas storage facilities and to ensure their more efficient operation.

2 Contextualization and Goals of the Study

Underground gas storage is a complex of engineering structures in reservoirs of geological formations, as well as in workings-tanks created in rock salt deposits, and intended for injection, storage and subsequent gas extraction.

According to the International Gas Union, the global gas storage capacity is more than 400 billion m^3 (operating capacity, i.e., the volume of gas injected and pumped out, unlike the buffer gas, which is constantly in the reservoir to maintain pressure) [1].

The underground gas storage system performs the following functions:

- regulation of seasonal irregularity of gas consumption;
- storage of gas reserves in case of abnormally cold winters;
- regulation of uneven export gas supplies;
- ensuring gas supply in case of emergency situations in the gas supply system;
- creation of long-term gas reserves in case of force majeure in the production or transportation of gas.

There are three types of UGS.

1. In depleted gas and oil fields. This is the best object for creating UGS. The field has been fully explored: the geometric dimensions and shape of the gas-bearing area, the tightness of the overlying strata are known. It is possible to use the wells and equipment of old fields.
2. In aquifers. Highly permeable porous reservoirs with reliable impermeable overlying strata can be used as UGS.
3. In salt caverns. Both gas and liquid hydrocarbons can be stored in salt caverns. The salt cavern is a perfectly sealed reservoir, because the salt dome is impermeable to gas. UGSs in salt caverns provide the highest daily gas supply.

Underground gas storage facilities are an integral part of the Unified Gas Supply System of Russia and are located in the main areas of gas consumption. There are 27 underground gas storage facilities located in Russia, of which 18 were built in depleted fields, 8 in aquifers, 1 in rock salt deposits.

For example, the Kasimovskoye UGS, the world's largest underground gas storage facility created in an water bearing bed, provides 30–35% of the daily need for Moscow and the Moscow Region, and also largely addresses the issues of gas supply in the Central Region of Russia [2]. The active capacity of the Kasimovskoye UGS is 12 billion cubic meters of gas, the maximum daily withdrawal is 100 million cubic meters

Fig. 1. The Kasimovskoye UGS

of gas. The photo shows how the Kasimovskoye UGS looks from a bird's eye view, Fig. 1.

The vast majority of UGSs are created in depleted beds or in water bearing beds. This paper is devoted to scientific problems associated with the operation of these types of UGSs.

New gas storages are under construction, modernization and improvement of operating technologies for existing ones is required. The efficiency of the UGS operation, reduction of the number of complications in the operation of wells should be based on the use of adequate geomechanical and hydrodynamic models describing the processes in the reservoir taking into account the large range of pressure changes in the production wells in the process of "extraction-injection" cyclic modes. The models should include reliable data on the geomechanical and filtration properties of rocks and their dependence on changes in the stress-strain state in the reservoirs.

In order to improve such models and provide them with parameters, the Institute for Problems of Mechanics of the Russian Academy of Sciences began studies of reservoir systems, deformation processes, and technogenic changes in the properties of the near-wellbore zone for the conditions of underground gas storages.

3 Existing Models of UGS and Ways to Improve Them

Today, an isotropic poroelastic body model is used for mathematical modeling of mechanical processes in reservoirs of underground gas storage (Hooks – Terzaghi – Bio model) [3]

$$\sigma_{ij} = \lambda \cdot e \cdot \delta_{ij} + G \cdot \varepsilon_{ij} + \beta_s \cdot K \cdot p \cdot \delta_{ij} \tag{1}$$

here σ_{ij} are effective stresses; λ, G are Lame coefficients; K is bulk modulus; ε_{ij} is strain tensor ($e = \varepsilon_{ii}$); β_s is soil grain compressibility factor; δ_{ij} is Kronecker symbol, p is gas pressure. When the stresses egress on the surface of the flow, determined by the equation (in the transition to plastic deformation)

$$f(\sigma_{ij}, \kappa) = 0, \tag{2}$$

along with elastic deformations ε_{ij}^{e}, calculated by (1), there are plastic deformations ε_{ij}^{p}:

$$\varepsilon_{ij} = \varepsilon_{ij}^{e} + \varepsilon_{ij}^{p} \tag{3}$$

At $f(\sigma_{ij}, \kappa) < 0$ the rock is in an elastic state (here κ is a hardening parameter). The defining relations in the plasticity zone are given as

$$v^{p} = -\lambda \partial g / \partial \sigma \tag{4}$$

here v^p is strain rate tensor; $g(\sigma)$ is flow potential.

If g = f, then (4) is the so-called associated flow law. Coulomb-Mohr ($f = f_{C\text{–}M}$) and Drucker-Prager ($f = f_{D\text{–}P}$) criteria are commonly used to determine the flow surface:

$$f_{C-M} = C + k \cdot \sigma_n \tag{5}$$

$$f_{D-P} = \alpha \cdot I_1 + \sqrt{J_2} - K \tag{6}$$

here σ_n is normal effective stress at the oriented surface with unit normal vector $\boldsymbol{n}$, I_1 и J_2 are the first invariant of the stress tensor and the second invariant of the stress tensor deviator. Constants k, C, α, K are expressed by using the cohesion (S_C) and the angle of internal friction (ϕ).

The hydrodynamic model includes the material balance equations for phases and Darcy's law for flows:

$$\partial(mS_\alpha \rho_\alpha)/\partial t + \mathrm{div}(\rho_\alpha q_\alpha) = Q_\alpha, \quad \alpha = g, l, w, \tag{7}$$

$$q_\alpha = -K \cdot k_\alpha / \mu_\alpha (\mathrm{grad}\, P_\alpha + \rho_\alpha g) \tag{8}$$

here g, l, w are phase designation indices (in general case, hydrocarbon gas, liquid hydrocarbon and water); K is absolute permeability of the porous medium; k_α, μ_α are relative permeability and viscosities of phases; Q_α is mass flow rate of extraction/injection phase (at the source/sink); q_α – filtration rate (in reservoir).

For the numerical solution of the adjoint problem (1)–(8), two approaches are proposed. The most commonly used approach is based on a separate calculation. Geomechanical fields are calculated at fixed hydrodynamic fields, and hydrodynamic fields are calculated at fixed geomechanical fields. The second, more rigorous methodically approach is to simultaneously calculate the associated geomechanical and hydrodynamic fields (it is associated with great computational difficulties).

What are the ways to improve the existing geomechanical models of UGS. First of all, it is necessary to take into account the mutual influence of stresses acting in the reservoir and filtration flows. It is known that stress strain state has a significant effect on the permeability of rocks, and this relationship is a property of the rock and can only be determined experimentally or on the basis of well logging or gas-hydrodynamic well tests. The filtration flow in turn makes a significant contribution to the stresses acting on the soil skeleton in result of action of the mass forces due to the fluid pressure gradient. As well as an essential feature of the UGS operation is the cyclic mode of pressure change in the wells, which ensures constantly alternating extraction and injection of gas into the reservoir. Cyclic impact has a significant effect on the strength of reservoir rocks and can lead to enhanced sand formation rate, destruction of equipment and wells. The reaction to cyclic loading is also a property of the rock and cannot be determined by calculation, only experimentally.

4 Experimental Apparatus

Experimental investigations are carried out by using the unique Triaxial Independent Load Test System (TILTS) created at the Institute of Problems of Mechanics of the Russian Academy of Sciences [4]. The apparatus allows studying the strain, strength and filtration properties of rocks by testing of a cubic specimens of rock with an edge of 40 or 50 mm, Fig. 2.

Fig. 2. Triaxial Independent Load Test System (TILTS)

An original kinematic scheme of the TILTS loading unit allows the pressure plates to approach in three directions without creating obstacles to each other, and to load the specimen independently along each of the three axes. This makes it possible to recreate in the course of the experiments any stress states that arise in rock massifs during conducting technological operations and to study the deformation processes and filtration properties of rocks.

TILTS belongs to the class of electro-hydraulic testing machines, it is equipped with an automated control system with feedback. TILTS allows to control the loading of the specimen both by force and by displacement of pressure plates. Control of loading by the magnitude of the displacement provides the possibility to plastic deform of the sample, not bringing it to fracture. The maximum load that the machine provides is 500 kN.

5 The Effect of Stress on Rock Permeability

Numerous experimental studies and field observations show that stresses have a significant effect on the permeability of rocks. At IPMech RAS, much attention was paid to the study of this dependence. The rocks of various lithological composition, from various regions, occurring in various geological conditions were studied. The stress-strain state that arises in the vicinity of the well with an increase in pressure drawdown was simulated on rock samples using TILTS. It was revealed how various rocks react to changes in the stresses acting in them, and a classification of reservoir rocks by this feature [5] was created. The widest class of rocks, which also includes highly porous, highly permeable sandstones usually forming productive USG reservoir, are characterized by the following properties. At the initial stage, when simulating a decrease in pressure in a well and an increase in tangential stresses acting in a rock, its permeability gradually decreases. However, when tangential stresses reach the certain level, the rock begins to deform inelastically, to creep, which ultimately leads to its cracking and a sharp irreversible increase in permeability.

As an example, the results of tests of this type of rock are shown at Figs. 3 and 4. This is sandstone from the reservoir of the Cheremukhovskoye field in Tatarstan, from the depth 982 m. The loading program that simulates a decrease in pressure in the bottom hole and a curve of permeability changes during the course of the experiment are presented at Fig. 3. Strain curves of the sample are at Fig. 4.

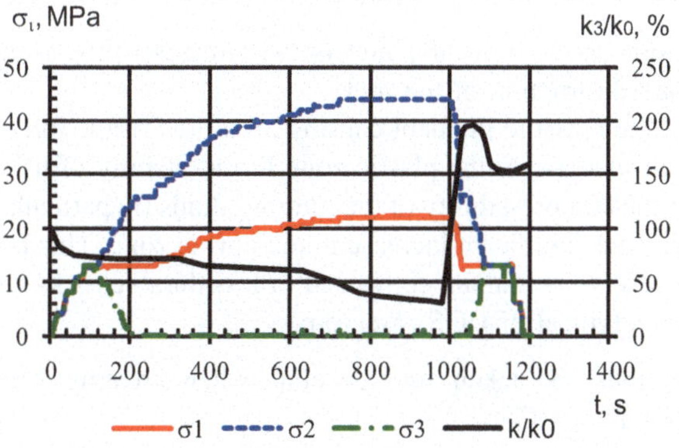

Fig. 3. The loading program and the change in permeability of the sample

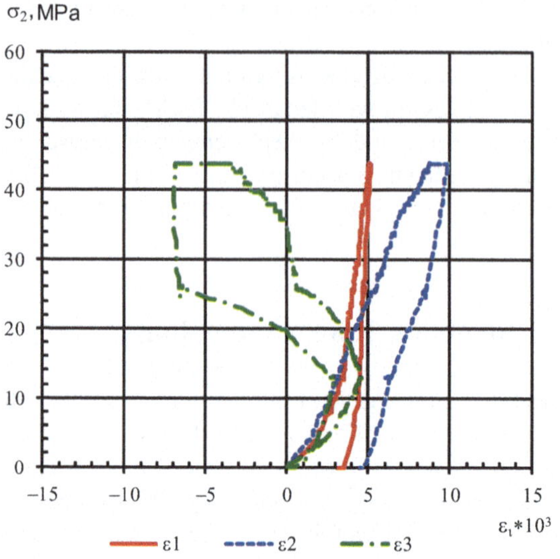

Fig. 4. Strain curves of the sample

Using such experiments, dependences of the permeability of rocks on the stress-strain state for a specific reservoir of the UGS can be obtained for inclusion in the geomechanical model of this object.

6 The Effect of the Filtration Flow on the Stress-Strain State

At the Geomechanics Laboratory of IPMech RAS, the issue of the effect of filtration flow on the stress-strain state of rock in the vicinity of the well [6] was theoretically investigated. Analytical solutions were obtained to determine the stresses and the size of the damaged zone (plastic zone) in the vicinity of the well, taking into account the compressibility of the fluid for stationary and non-stationary filtration regimes. The analysis of the obtained solutions allowed us to draw the following conclusions.

1. The filtration process significantly affects the stresses acting in the reservoir, the deformation and destruction of the rock.
2. An increase in the pressure gradient causing an inflow of the filtering fluid into the well leads to an increase in the plastic zone in the vicinity of the well.
3. An increase in the compressibility of the filtering fluids (in particularly in the case of gas) leads to a more significant increase in the plastic zone. This factor is especially significant for reservoirs that are composed of less durable rocks with low adhesion moduli, which is typical of UGS reservoirs.

These factors must be taken into account when creating geomechanical and hydrodynamic models of UGS.

7 Modeling the Effect of Cyclic Impact on the Strength Properties of the Rock

A simulation of cyclic pressure change in the well was carried out at rocks of the terrigenous reservoir of the Siberian field in the Perm region of CJSC LUKOIL-PERM. Figures 5 and 6 presents the results of tests of this type sample. The sample was taken from the depth 2341 m. The initial permeability of the sample was 11 mD. Figure 5 shows the loading program that simulates the cyclic pressure change in the well (3 cycles), and the change in permeability during the experiment. Figure 6 shows the strain curves of the sample.

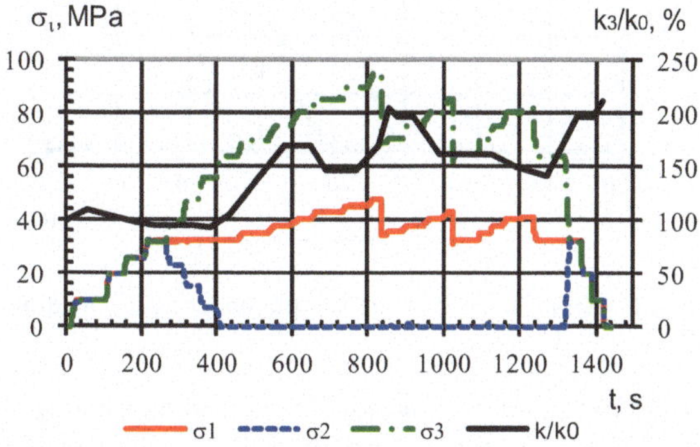

Fig. 5. The loading program and the change in the permeability of the sample under cyclic loading

The deformation curves show that the rock strength decreases at each next loading cycle—the sample passes into the plastic deformation stage during each subsequent cycle at a smaller value of pressure drawdown in the well, although the difference in limiting values decreases. The obtained data allow us to determine the dependence of the strength properties of the rock on cyclic loading.

8 Results

To improve the UGS performance, it is necessary to develop the geomechanical and hydrodynamic models, on the basis of which the optimal operating conditions are calculated.

1. On the one hand, the injection and pumping out of gas from the wells must be carried out with maximum productivity, on the other hand, it is necessary to control the sand formation and prevent the destruction of the equipment and well.
2. The model should take into account the mutual influence of stresses acting in the reservoir and filtration flows. A method for determining these dependencies is proposed.

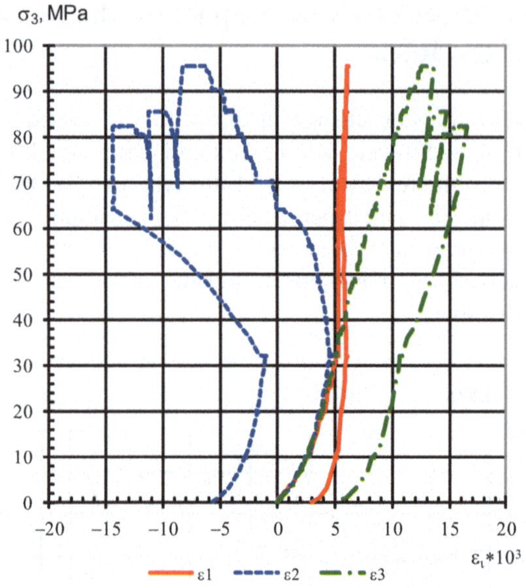

Fig. 6. Strain curves of the sample under cyclic loading

3. The reaction of the rock to the cyclic loading is also an important parameter of the model, which must be determined experimentally for the specific field conditions.

Acknowledgements. The work was done under financial support of the Russian Scientific Foundation, grant No. 16-11-10325.

References

1. http://www.unece.org/fileadmin/DAM/energy/se/pdfs/wpgas/pub/Report_UGS_Study_www.pdf
2. http://www.gazprom.com/press/news/2007/november/article64030/
3. Pyatakhin, M.V.: Geomechanical problems in the operation of wells, 266 p. VNIIGAS, Moscow (2011). (in Russian)
4. Karev, V.I., Kovalenko, Yu.F.: Well stimulation on the basis of preliminary triaxial tests of reservoir rock. In: Rock Mechanics for Resources, Energy and Environment. Proceedings of EUROCK 2013. The ISRM International Symposium, Wroclaw, 23–26 September 2013, pp. 935–940. CRC Press/Balkema, Leiden (2013)
5. Klimov, D.M., Karev, V.I., Kovalenko, Y.F.: Experimental study of the influence of a triaxial stress state with unequal components on rock permeability. Mech. Solids **50**(6), 633–640 (2015)
6. Zhuravlev, A.B., Karev, V.I., Kovalenko, Y.F., Ustinov, K.B.: The effect of seepage on the stress-strain state of rock near a borehole. J. Appl. Math. Mech. **78**(1), 56–64 (2014)

Experimental Study of Substance Transfer in Vortex and Wave Flows in Multicomponent Media

T. O. Chaplina[1,2(✉)]

[1] Faculty of Physics, Lomonosov Moscow State University, Moscow, Russia
chaplina_to@inbox.ru
[2] A.Yu. Ishlinsky Institute for Problems in Mechanics RAS, Moscow, Russia

Abstract. The fine structure of the phase boundary surfaces between the oil body and water and also between the liquid (water or immiscible hydrocarbons) and air in a compound vortex is studied experimentally, including the beginning of the emulsions formation. The experiments showed that a uniform layer of hydrocarbons on the water free surface is pulled together into the oil body located on the flow axis. Some portions of the oil remain on the free surface in the form of spiral arms, either continuous or decomposed into separate drops. At high angular speeds of rotation the oil droplets can be found all over the liquid – air contact surface. The contact surface geometry depends on the parameters of the vortex flow, determined by angular velocity of inductor rotation and quantitative composition of medium. The dynamics of the solid-state transfer marker - plastic or ice in the compound vortex of two fluid water-oil analyzed. It is shown that the motion of the marker, inscribed on the surface of the composite vortex complex and includes a tangential displacement, radial displacement and spinning about its own axis. Character movement depends on the experimental conditions, the type of liquid and form a marker. Radial velocity of motion increases to handle liquids with oil slick and at the same time, the angular velocity of its rotation around the vertical axis of the flow is markedly reduced.

Keywords: Vortex · Wave · Flow · Marker · Visualization

1 Introduction

Submerged jets and shear flows are often found in nature and various technical applications, such as mixing, burning, generation of sound disturbances, etc. Vortex structures are quite common, we are faced with such currents as when using different technical devices, and in natural systems - in the atmosphere and the ocean. Scales of such currents fluctuate in the widest range and can reach huge sizes. On Earth, the largest vortices are the synoptic - atmospheric cyclones of medium latitudes, the characteristic transverse dimensions of which are 1500–2000 km; vortices in the open ocean, reaching a size of 100–200 km across. Also, vortices are formed at the ends of the wings of aircraft and when the flow is separated from the sharp edges of aircraft and ships (Fig. 1).

V. I. Karev et al. (Eds.): *Physical and Mathematical Modeling of Earth and Environment Processes (2018)*, SPEES, pp. 159–173, 2019.
https://doi.org/10.1007/978-3-030-11533-3_17

a *b* *c*

Fig. 1. Vortex structures of different scales: (*a*) spiral galaxy M83 [1]; (*b*) cyclone over the Atlantic Ocean [2]; (*c*) vortices from the wings of aircraft [3].

The study of the processes of matter transport in vortex currents is one of the traditional problems of fluid mechanics, the practical value of its results for the problems of ecology and the development of industrial technologies is steadily growing. In recent years, one of the main products of surface contamination of water bodies, both small and the entire World Ocean, are fragments of plastics in the form of raw materials, products and packaging.

Garbage in various forms enters the aquatic environment, where it is transported, degraded and accumulated at certain levels of the ecosystem.

In connection with the growth of the volume and nomenclature of pollutants entering the environment, it is of interest to study the transfer of markers in various types of vortex flows developing both in pure liquid and in a multilayer formed by mixing and immiscible components. Of special interest is the study of the transfer of markers in a layer of a liquid covered with an oil or oil film formed by the flow of substances from natural and anthropogenic sources [4]. The ecological effect of the film depends on its thickness and coating density, the presence of "clean windows" has a positive effect on gas exchange in polluted waters.

Observations of marine systems have made it possible to distinguish a number of general features of the propagation of impurities. Solid-state remnants are destroyed by the action of waves, wind and solar radiation, scattered, submerged to the bottom or thrown ashore, partially transported to ocean circulation centers, where they can assemble into compact "garbage islands" [5].

To clarify the dynamics of the processes of solid impurities transport in the ocean and to register the main characteristics of currents, vortices and waves of various nature, freely drifting drifters on the surface and buoys of neutral buoyancy in the marine environment are actively used in the ocean (only the "Argo" project is currently monitoring the movement more than 3800 buoys [6]. Interpretation of the buoyed data is based on the assumption of complete entrainment of the suspended body by currents. However, the adequacy of the assumed assumption needs to be refined taking into account the regularities of the real behavior of bodies in the vortex flows of a homogeneous, multilayered or continuously stratified fluid. Experimental studies of the dynamics of free bodies of neutral buoyancy were carried out in [7], in this paper the focus is on the transfer of markers in a two-layer fluid.

The increased concentration of oil coming from natural and industrial sources under natural conditions was observed in the form of separate, narrow bands, paradoxically oriented in intense wave fields [8]. The length of such bands, causing serious pollution of the coast, can reach many tens of kilometers [4, 9].

Observations in marine conditions provide important factual material. However, due to the complexity and variety of concurrent natural processes, the data obtained can not be unambiguously interpreted.

In many cases, the amount of data is insufficient to identify the underlying mechanisms based on different flow models and to determine their parameters. In addition to observations of the transfer of markers under natural conditions (in the atmosphere and in the hydrosphere), laboratory modeling of the transport process, including wave and eddy currents, began to develop in recent years.

The transport of matter in wave channels of simple geometry with a wave-product set at one of the ends was studied in [10]. Studies of the movement of markers are often carried out in circular cylindrical containers, a rotational disk in the top cover [11] or the bottom of the working volume creates a vortex flow [12].

As shown earlier experiments, a liquid marker, in particular oil, can be located on the surface of such a vortex in the form of spiral bands and forms a compact oil body in the vicinity of the rotation axis [11].

In this case, the shape of the free surface and the interface of the components in multilayer fluids depends on many factors: the type of liquid, container dimensions, angular velocity of rotation and the position of the inductor [13, 14]. In a two-layer liquid with an activator rotating at moderate angular velocities in the layer of oil lying above, a number of characteristic forms of an oil body are identified: a hill, a pointed peak, a smoothed peak (Fujiyama), a bell [10]. The axisymmetric cavern at high angular velocities of the disk installed in the thickness of the liquid acquires an angular shape reflecting the change in the symmetry of the velocity field in the thickness of the liquid [15, 16].

An equally complex picture is recorded during the transfer of a solid-state marker, which in flows with a shift not only rotates around the center of the vortex, but also revolves around its own axis [5]. Such forced movements of the body perturb vortex currents and additionally affect the transport of liquid markers.

In this paper, the fine structure of the oil-water-liquid interface and liquid (water or immiscible hydrocarbons) - air in a composite vortex, including the regime of the formation of emulsions, has been experimentally studied. As a research object, massively used liquids (sunflower and aviation oils, petroleum, diesel fuel, as well as their mixtures in different proportions) were chosen. Also in the article dynamics of the transfer of a solid marker - plastic or ice in a composite vortex in a two-layer liquid water-oil/petroleum was investigated.

2 Experimental Setup and Flow Parameters

Our experiments were performed on the Rotational Vortex Flow setup (Fig. 2*a*). The open cylindrical container *1* with diameter 29:4 cm is filled with water. The water level is set 40 cm. To decrease optical distortions of the recorded flow pattern, the cylindrical container is put into the rectangular tank *2* with plane walls, the size of the tank is 63.6 × 44.6 × 70.0 cm. The activator of vortex motion, the flat disk *3* (2 mm thickness), is located in the center of the tank bottom. To flatten the bottom of the cylindrical container on the level of the upper edge of the disk the false bottom *4* is mounted. The disk is activated by an electric motor with the frequency from 200 to 1200 rpm, with a signal converter block.

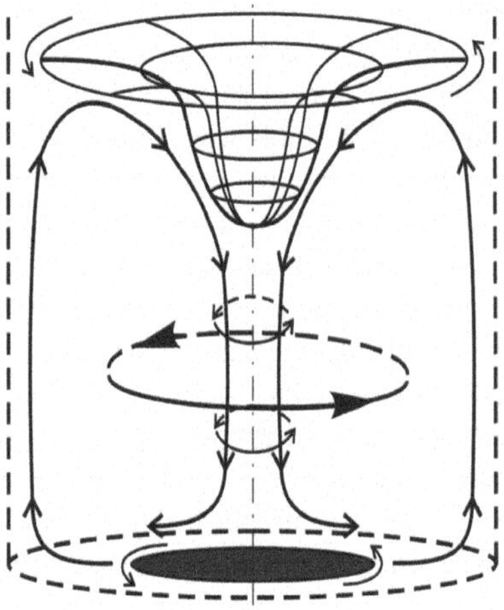

Fig. 2. The experimental setup (*a*) and the scheme of a compound-vortex (*b*) induced by the rotating disk in the cylindrical container.

The photo- and video-recordings of the flow pattern are performed simultaneously from the top 5 and side 6 views. The control of the experiment and the data recording is committed to PC 7. The water volume is illuminated by a white-light source with a dispersive sheet 8. The container is equipped with a hydraulic system. In all experiments, the illumination and location of the cameras are chosen to provide visibility of all details of the liquid free surface.

The complex flow in the container, the schematic pattern of which is shown in Fig. 1b, contains both the vortex and wave components. The uniformly rotating disk swirls the liquid about the vertical axis due to no-slip condition and rejects it towards the vertical container sidewall. The accelerated liquid uprises along the container sidewall and is then displaced towards the centre along the free surface, and descends near the rotation axis, forming a flow, which compensates the constant transport along the disk surface (Fig. 2*b*).

Free surface made it possible to apply a wide range of markers and to control the conditions for their introduction. As ice and solid-state markers, plastic, with a diameter of 1 cm or 2 cm and a height of 0.5 cm were used, immiscible liquids - aviation fuel and oil in a volume of 30 ml. In experiments with the addition of oil and aviation oil, an immiscible impurity was added to the surface of the resting liquid, then the inductor was set in motion and markers were placed in the steady vortex flow. Each new experiment began after the attenuation of all visible movements in the basin.

The composite vortex flow is characterized by a set of dimensional and dimensionless parameters, the values of which are given in Tables 1, 2, where ρ_w, ρ_o, ρ_a – density of water, oil and air, respectively, $R = 7.5\,\text{cm}$, Ω – rotational speed of the inductor drive, g – acceleration of gravity, ν_w - kinematic viscosity of water, coefficients of surface tension on contact surfaces - water-air σ_w^a, oil-air σ_o^a, water-oil σ_w^o, $L_\Omega = g/\Omega^2$ - the characteristic length scale (under the conditions of these experiments $0.02 < L_\Omega < 10$), $\delta_\Omega^\nu = \sqrt{\nu_w/\Omega}$ – scale of the Ekman layer on the surface of a rotating disk (under the conditions of these experiments is sufficiently small and is within $0.02\ \text{cm} < \delta_\Omega^\nu < 0.1\ \text{cm}$).

Table 1. Basic flow parameters.

Parameter	Minimum value	Maximum value	Characteristic value
Reynolds number $\mathrm{Re}_\Omega^w = (R_d^2\Omega)/\nu_w$	$5.6 \cdot 10^3$	$1.4 \cdot 10^5$	$7.3 \cdot 10^5$
Froude number $\mathrm{Fr} = (R_d^2\Omega^2)/gH$	10^{-3}	0.9	0.4415
Bond numbers $\mathrm{Bo}_w^a = gH(\rho_w - \rho_a)/\sigma_w^a$	–	–	0.055
$\mathrm{Bo}_o^a = gH(\rho_o - \rho_a)/\sigma_o^a$	–	–	0.109
$\mathrm{Bo}_w^o = gH(\rho_w - \rho_o)/\sigma_w^o$	–	–	0.257
Atwood numbers $\mathrm{At}_w^a = (\rho_w - \rho_a)/(\rho_w + \rho_a)$	–	–	0.998
$\mathrm{At}_o^a = (\rho_o - \rho_a)/(\rho_o + \rho_a)$	–	–	0.997
$\mathrm{At}_w^o = (\rho_w - \rho_o)/(\rho_w + \rho_o)$	–	–	0.099

The values of Reynolds numbers are typical for vortex and nonstationary-vortex flows. At given values of Froude numbers, the wave deformations of the free surface play an appreciable, and at the smallest values Fr - decisive role. Moderate Bond numbers indicate the possibility of small-scale effects in flow patterns.

Table 2. Characteristic scales of the problem.

Parameter	Value
$\xi_H = R_d/H$	0.38
$\xi_0 = R_d/R_c$	1.0
$\xi_b = 2R_k/h_k$	0.8
$L_\Omega = g/\Omega^2$	$3{,}9 \times 10^3$
$\delta_\Omega = \sqrt{\nu/\Omega}$	3×10^{-4}

3 Experimental Study of the Transport of Immiscible Liquid Admixture in a Vortex Flow

Comparison of the size of the oil body, depending on the amount and physical properties of the immiscible impurity added to the stream, shows that increasing the viscosity of the marking additive leads to a decrease in the vertical size of the oil body. Also, the vertical dimension of the area occupied by the immiscible liquid in the flow is affected by the surface tension coefficient. With its growth, the depth of retraction of the impurity increases (Fig. 3).

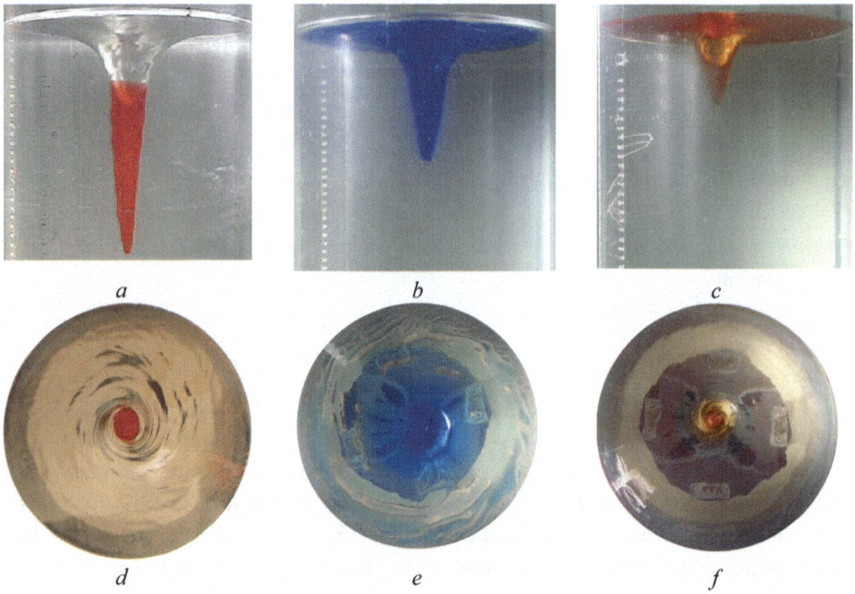

Fig. 3. Flow patterns in a composite vortex with addition immiscible liquid admixture ($V_k = 150$ ml $H_w = 40$ cm, $R_d = 7.5$ cm): (*a* and d) sunflower oil, $\Omega_d = 770$ rev/min; (*b* and *e*) – mixture of equal parts of sunflower oil and diesel fuel, $\Omega_d = 750$ rev/min; (*c* and *f*) – diesel fuel, $\Omega_d = 820$ rev/min.

Dependences of the vertical dimensions of the surface cavity and oil body depend on the rotational speed of the activator are interpolated by linear functions $h_t = A\,\Omega + \mathrm{B}$, the coefficients included in them are given in Table 3.

Table 3. Coefficients of linear interpolations of the dimensions of the oil body.

	A	B
Sunflower oil	2.33 ± 0.13	−11.66 ± 1.30
Sunflower oil and diesel fuel	2.09 ± 0.27	−8.44 ± 0.67
Petroleum	1.88 ± 0.16	−8.01 ± 0.55
Diesel fuel	1.66 ± 0.21	−7.74 ± 1.48

There is a clear tendency to reduce the vertical dimensions of the oil body with a decrease in the density and surface tension coefficient of the impurity added to the flow, which is confirmed by the approximations obtained.

Saving forms the central part of the oil body when changing the angular speed of disc rotation allows to characterize the relative change in the ratio of one of its dimensions $\xi_m = 2R_m/h_m$ as a function of Reynolds number (Re) [9].

An increase in the rotation speed of the disk leads to an increase in the vertical component of the flow velocity at the center of the container and, correspondingly, to the shear stress at the boundary between the water and oil body. In the range of parameters Re from 500 to 12000, the dependence is represented by two linear functions $\xi_m = a\ \mathrm{Re}\ + b$: for small Reynolds numbers (less than 2,500), the angular coefficient is large $a = (-10.3 \pm 2.4) \times 10^{-4}, b = 2.2 \pm 0.4$ and rapid change in the ratio of the height of the oil body to its diameter near the free surface is observed, at large Reynolds numbers the line becomes flatter $a = (4.0 \pm 2.0) \times 10^{-5}$, $b = 0.7 \pm 0.1$, the change in the ratio of the diameter and height of the oil body is insignificant (Fig. 4).

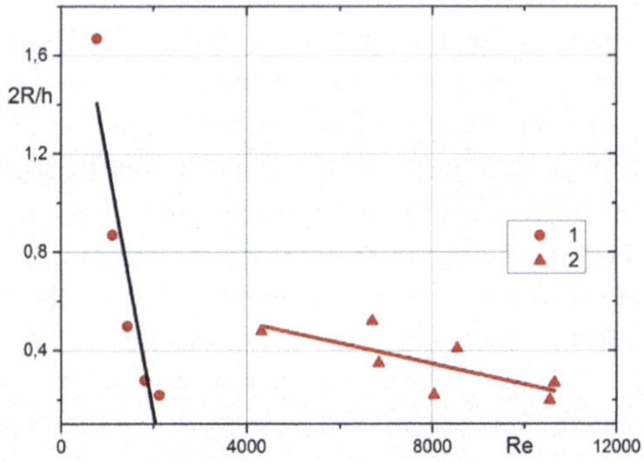

Fig. 4. Dependence of the oil body acuity coefficient ξ_b on the flow parameters: *1* – Sunflower oil; *2* – Petroleum.

The stability of the interface between the water and the oil body and the shape of the contact area of the liquids at the bottom of the cavity depend on the type of marker. At identical angular velocities of rotation of the disk, the water-oil boundary remains unchanged (Fig. 5*a*), and drops start to separate from the oil body in the water-aviation oil system (Fig. 5*b*). The rotating oil body is preserved when the rotational speed of the disk is increased (Fig. 5*c*), while in the water-aviation oil system an invert emulsion-cells with water, contoured by an angular oil shell-begins to form on the flow axis.

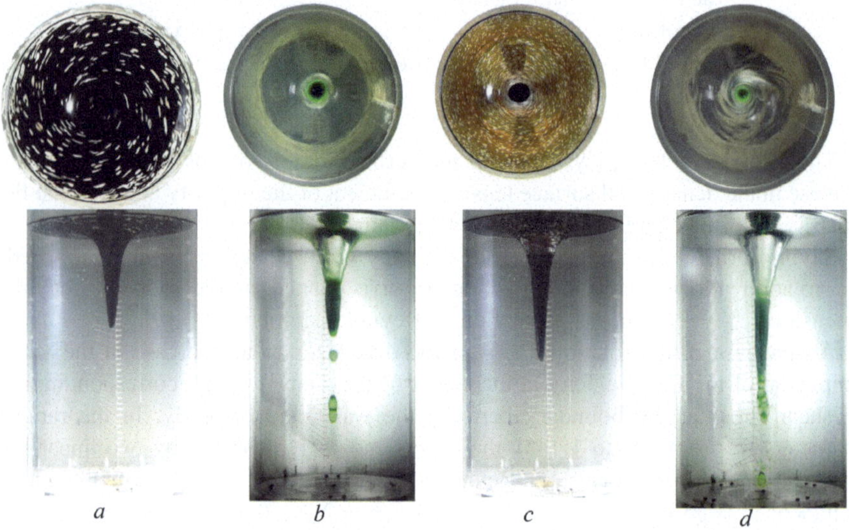

Fig. 5. Different types of immiscible marker in a vortex flow under different experimental conditions ($V_k = 150\,\text{ml}$, $H_w = 40\,\text{cm}$): (*a*) petroleum, $\Omega_d = 770$ rev/min; (*b*) aviation oil, $\Omega_d = 820$ rev/min; (*c*) petroleum, $\Omega_d = 1020$ rev/min; (*d*) aviation oil, $\Omega_d = 1050$ rev/min.

The general picture of the flow in the region of the contact of the oil body with the air cavern at various angular velocities of disk rotation is shown in Fig. 6. When the oil layer covers the entire free surface, the walls of the cavity are smooth, the transition of the oil from the inner to the outer surface of the cavity occurs smoothly, without singularities (Fig. 6*a*). With an increase in the disk rotation frequency, spiral waves are observed on the surface of the cavern, and in the region of the marker transition from the outside of the cavern to other types of markers. Separating from the lower edge of the oil body, spherical drops combine into groups and partially merge, forming pear-shaped bodies. With a further increase in the speed of rotation, the step in the region of the oil transition line from the outer to the inner side of the surface of the liquid becomes more pronounced, and instead of the drops of the direct emulsion, invert cells are formed at the lower edge of the body (an enlarged picture of the flow pattern is shown in the sidebar on the right in Fig. 5*c*). The dimensions of the emulsion cells range from 0.8 to 3.5 cm.

Fig. 6. Change in the shape of the oil body (aviation oil) when the rotational speed of the activator $(V_k = 150\,\text{ml}, H_w = 40\,\text{cm}, R_d = 7.5\,\text{cm})$: (*a–c*) $\Omega_d = 670, 820, 1050$ rev/min.

4 Motion of Solid Markers in the Compound Vortex

Our experiments (consisting of three separate experiments with the same boundary conditions for each marker) were carried out for ice and solid-state markers, plastic, with a diameter of 1 cm or 2 cm and a height of 0.5 cm were used. One of the objectives of this work was to determine whether the shape of the active scalar will affect its displacement in the fluid flow. The cylindrical container was filled with liquid to a predetermined level (40 cm). Then, the activator was turned on. After the decay of transients, the markers were placed on the surface of the liquid at a distance equal to half the radius of the container. At the same time, the video recording was begun. The captured video was divided into still images during processing. Each still image shows the position of the center marker, its orientation in space and trajectory.

The shapes of the marker trajectory depend on its initial position. There is a separatrix, which separates the region of different marker motion [5]. From one region, a marker will move to the vortex center, and from the other one, a marker will move to the side walls of the container.

The marker in the container rotates around the center of the vortex and around its own axis, forces and moments act on it (Fig. 7). Friction on the water surface twists the marker around its own axis (due to the difference in speed). Also under the surface there is a current directed toward the center of the free surface, which also creates friction.

The motion of the marker is described in several coordinate systems (Fig. 8). As a base coordinate system - Cartesian XOY is chosen whose center is on the axis of flow and the axes are orthogonal to the walls of the basin, and cylindrical (r, ψ).

The radial position of the marker is given by the radius vector r. With the center of the marker is a mobile Cartesian coordinate system $X_a0_0Y_a$, which allows to determine the rotation angle of the marker around the proper axis φ. The rotation is investigated using the applied risk located along the axis 0_0X_a. The marker radially moved to the

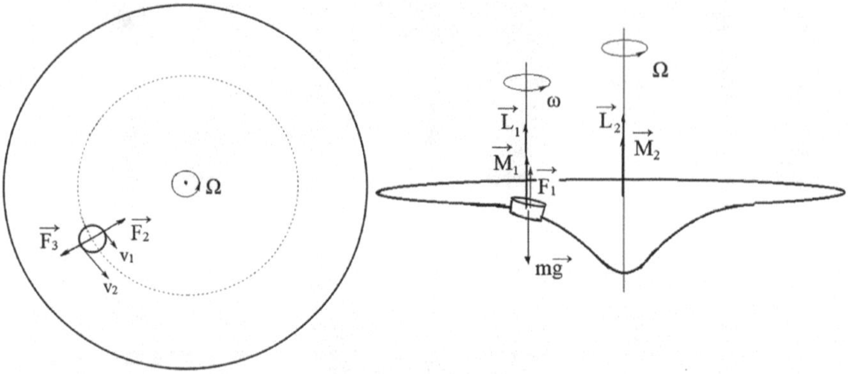

Fig. 7. Marker motion: (*a*) - top view: (*b*) - side view. F_1 – the force of Archimedes, F_2 – the centripetal force, F_3 – the centrifugal force, M_1 – the moment of force relative to the center of the marker, L_1 – the angular momentum relative to the center of the marker, M_2 – the moment of force relative to the vortex center, L_2 – the angular momentum relative to the vortex center, mg – the gravity, v_1 и v_2 – the velocities at the edge of the marker.

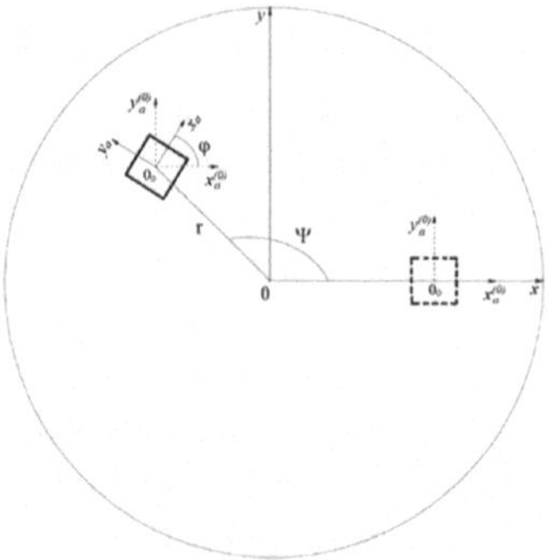

Fig. 8. Coordinate system.

center of the vortex flow and rotated around its own axis in all the experiments carried out. Differences in the character of motion of the marker in the composite vortex in pure water and on the surface of a two-layer water-oil liquid (oil volume V_k = 50 ml, thickness of unperturbed layer Δ = 1.5 mm) are illustrated by samples from the videograms shown in Fig. 9.

In pure water, the cylindrical marker moves along with the fluid flow around the central vertical axis, moves slowly toward the center and simultaneously rotates around its own axis. The tangential component of the velocity of the center of the marker in pure water at $\Omega = 9.2\,\mathrm{s}^{-1}$ at a distance of r = 8.1 cm from the center is $v_\varphi = 2.26\,\mathrm{cm/s}$ and changes slowly as it moves to the center of the flow. During the time of observation between the ones shown in Fig. 9, a, b frames (40 s) the marker makes 11 revolutions and shifts to the center by 4.3 cm and 48 turns around its own axis.

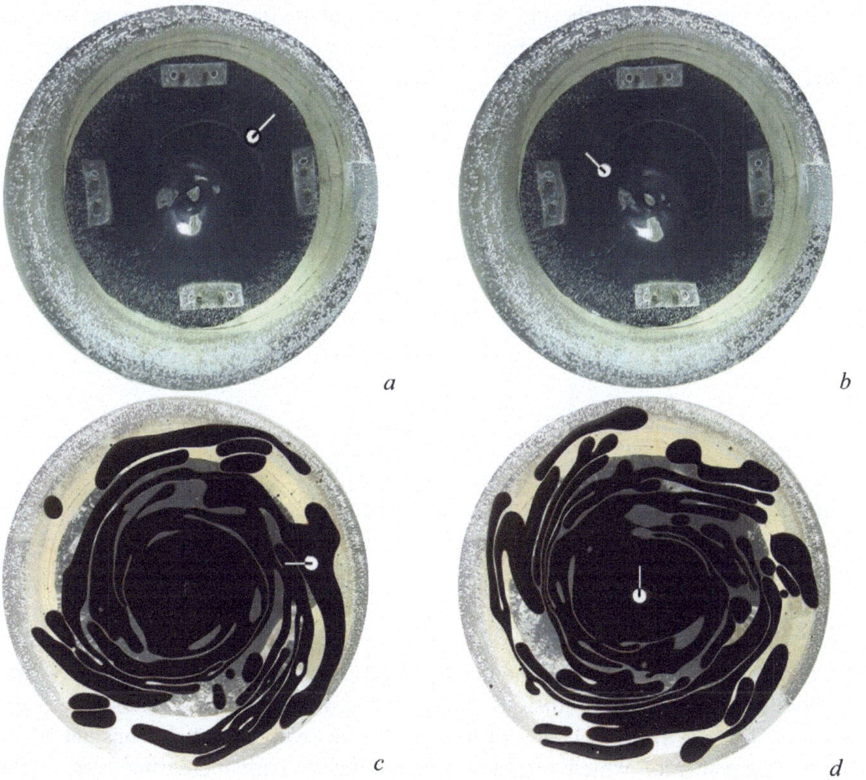

a b

c d

Fig. 9. Transferring a circular marker in a composite vortex ($H_w = 40\,\text{cm}, R_d = 7.5\,\text{cm}$, $\Omega_d = 9.2\,\text{s}^1$: (*a*, *b*) – pure water ($t = 10, 50\,\text{s}$), $\text{Re}^w_\Omega = 51562.5$, $\text{Fr} = 0.119$; (c, d) – two-layer liquid (water and petroleum $V_k = 50\,\text{ml}$) ($t = 10, 50\,\text{s}$) (for clarity, the risks are continued by white lines in the figures).

Under the same conditions $(H_w = 40\,\text{cm}, R_d = 7.5\,\text{cm}, \Omega_d = 9.2\,\text{s}^1)$ two-layer liquid (water and petroleum the cylindrical marker within t = 40 s has time to make 8 revolutions around the center ($\psi = 16\pi$), 17 turns around its own axis ($\varphi = 34\pi$) and falls into the center of rotation at the bottom of the surface cavity, passing the distance $\Delta r = 8.1\,\text{cm}$ (in a homogeneous fluid at the same angular velocity of rotation of the disc, the marker is displaced to a shorter distance $\Delta r = 4.3\,\text{cm}$).

In order to obtain the motion characteristics of the markers, the processing of video films was carried out, during which the temporal variability of the following parameters was determined: the distance between the marker and flow centers $r(t)$, the angular displacement of the marker's center in the laboratory coordinate system $\psi(t)$, the rotation angle of the marker around its own center $\varphi(t)$, which allows determining the functional connection between the angles φ and ψ.

Each experiment was repeated at least three times, the graphs show average values for the whole set of experiments. Temporal variability of the radial position of the markers in a homogeneous (curve 1) and two-layer (curve 2) liquid is shown in Fig. 10.

In a homogeneous liquid, the circular marker oscillates near a certain position and slowly moves to the edge of the basin, while in the liquid with the oil film it moves quickly enough to the center.

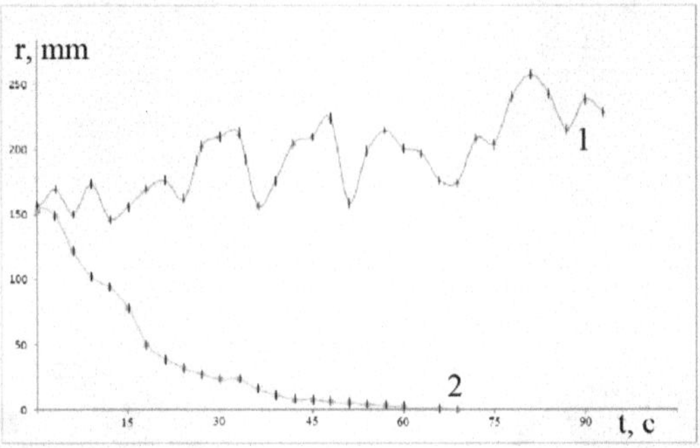

Fig. 10. Radial position of the circular marker in the composite vortex ($H_w = 40\,\text{cm}$, $R_d = 7.5\,\text{cm}$, $\Omega_d = 10.8\,\text{s}^{-1}$, $\text{Re}_{\Omega}^{w} = 60927.5$, $\text{Fr} = 0.165$): *1* – pure water, *2* – two-layer liquid (water and petroleum $V_k = 50\,\text{ml}$).

The nature of the rotation of the marker depends on its shape and angular velocity of rotation of the activator disk Ω_d. The round marker after a short interval of involvement (t = 5 s) in the center of the flow rotates with an almost constant angular velocity, the value of which increases monotonically with increasing disk rotation speed: $\dot{\varphi}_1^c = d\varphi_1^c/dt = 66.6$; $\varphi_2 = d\varphi_2^c/dt = 99.3$; $\dot{\varphi}_3 = d\varphi_3^c/dt = 134.8$.

The measured values of the rotation angle φ along the entire marker track are approximated by functions $\varphi_c = 0.07\,t^{2.9}$ (for $\Omega_d = 3.3\,\text{c}^{-1}$), $\varphi_c = 55.82\,t^{1.4}$ (for $\Omega_d = 9.2\,\text{c}^{-1}$), $\varphi_c = 185.42\,t^{1.34}$ (for Ω_d =16.7 c^{-1} (Fig. 11).

The square marker also rotates with an almost constant angular velocity at a small angular velocity of rotation of the disk $\dot{\varphi}_1^s = d\varphi_1^s/dt = 30.1(\Omega_d = 3.3\,\text{s}^{-1})$ and with a large velocity $\dot{\varphi}_3 = d\varphi_3^s/dt = 128.2(\Omega_d = 16.7\,\text{s}^{-1})$. At intermediate values of the angular velocity of the activator disc $\Omega_d = 9.2\,\text{s}^{-1}$, the square marker is not verified unequally (curve 2, Fig. 11*b*), which reflects the complex nature of the interaction of the dragged square marker with the main current.

Perhaps the irregular nature of the rotation is due to the interaction of the marker with inertial and spiral waves that arise arbitrarily on the surface of a rotating fluid [14], the most pronounced for a given flow regime.

The trajectories of a square marker are approximated by functions $\varphi_s = 40.8\,t^{1.4}$ (for $\Omega_d = 3.3\,\text{s}^{-1}$), $\Omega_s = 0.77\,t^{2.68}$ (for $\Omega_d = 9.2\,\text{s}^{-1}$), $\varphi_s = 1.4\,t^{2.4}$ (for $\Omega_d = 16.7\,\text{s}^{-1}$).

The features of the motion of various markers in a liquid with an oil film are illustrated by photographs of flow patterns shown in Fig. 12. The experiments showed that the very distribution of oil over the surface of the composite vortex is not homogeneous, oil can spread into a film and collect in an oil body in the vicinity of the

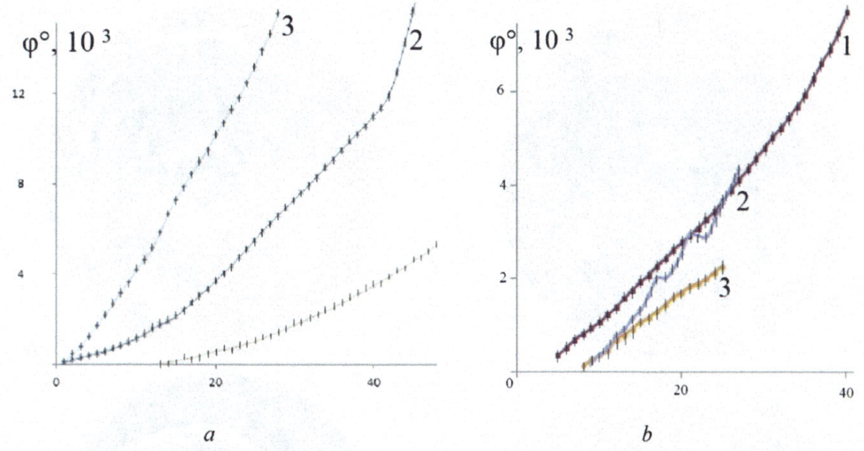

Fig. 11. Dependence of the angle of rotation around its own axis on time ($H_w = 40\,\text{cm}, R_d = 7.5\,\text{cm}$): *1* - $\Omega_d = 3.3\,\text{s}^{-1}$ ($\text{Re}^w_\Omega = 18750$ Fr = 0.016), - $\Omega_d = 9.2\,\text{s}^{-1}$ ($\text{Re}^n_Q = 51750$, Fr = 0.119), 3 – $\Omega_d = 16.7\,\text{s}^{-1}$ ($\text{Re}^w_\Omega = 93750$, Fr = 0.392): (*a*) – round marker, (*b*) - square marker.

rotation axis [17]. Petroleum covered the surface of the vortex almost completely in experiments with moderate angular rotational speeds of the activator disk.

With an increase in the angular velocity of rotation of the disk, the thickness and nature of the distribution of the oil film along the surface of the water decreases. The inserted marker leaves a trace that is practically free of oil (Fig. 12). At the same time, the own rotation of the marker relative to its own center remains.

The geometry of the marker trace depends on all the parameters of the problem (the angular velocity of the inductor rotation, the depth of water, the amount of oil - the thickness of the initial layer), and also the shape of the marker.

A round marker rotates around the center, revolves around its own axis, and forms characteristic structures including compact regions without an oil film, the shape of which depends on the duration of the motion. The square, like the round marker, also rotates and leaving a trace of variable thickness behind it (Fig. 12*c*).

Qualitative features of the trace are preserved even in the case of the motion of the ice marker to the stage of the onset of melting (Fig. 12*d*). In this case, the contours of the track are more even than for the solid-state marker, in addition, in this case the marker is also rotated around its own axis.

5 Conclusion

The experiments showed that a uniform layer of hydrocarbons on the water free surface is pulled together into the oil body located on the flow axis. Some portions of the oil remain on the free surface in the form of spiral arms, either continuous or decomposed into separate drops. At high angular speeds of rotation the oil droplets can be found all

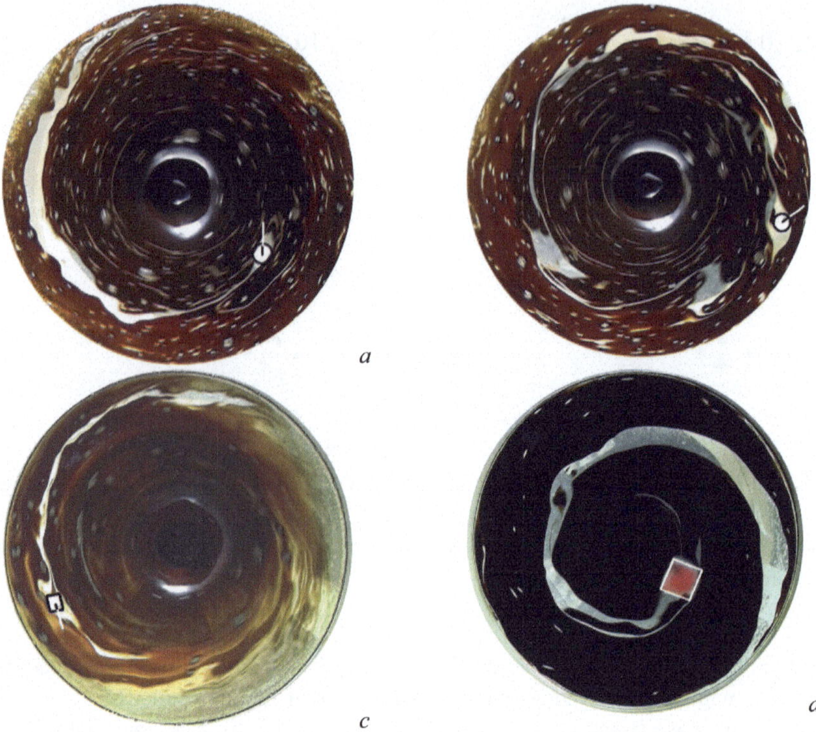

Fig. 12. Rupture of oil film by a free drift marker ($H_w = 40\,\text{cm}, R_d = 7.5\,\text{cm}, V_k = 50\,\text{ml}$): (*a*, *b*) – round marker ($\Omega_d = 13.7\,\text{s}^{-1}$, $t = 6, 10\,\text{s}$), $\text{Re}_\Omega^w = 77062.5$, $\text{Fr} = 0.261$; (*c*) – square marker ($\Omega_d = 3.3\,\text{s}^{-1}$, $t = 3\,\text{s}$, $\Omega = 18750, \text{Fr} = 0.016$); (*d*) – ice ($\Omega_d = 10.8\,\text{s}^{-1}$, $t = 4$ s, $\text{Re}_\Omega^w = 60927.5, Fr = 0.165$).

over the liquid – air contact surface. The contact surface geometry depends on the parameters of the vortex flow, determined by angular velocity of inductor rotation and quantitative composition of medium.

The conditions for the loss of continuity of the interface depend on the type of impurity. At average angular velocities of rotation, a direct emulsion is formed, both types are observed with increasing speed, and at high rates - an invert emulsion.

The experiments showed that the movement of the marker introduced on the surface of the composite vortex is complex and includes tangential displacement (rotation relative to the center of the flow), radial displacement and rotation relative to its own axis. The nature of the movement depends on the conditions of the experiments (the depth of the liquid and the rotational speed of the activator disk), the type of liquid (homogeneous or two-layered with oil film), and the shape of the marker.

The inserted marker, both solid and melting ice, slides relative to the upper layer of the liquid and leaves a trace of a complex shape behind it.

The angular velocities of rotation and rotation around the flow center are related by a functional dependence determined by the flow parameters and the properties of the marker (volume, mass, size and shape). In this case, circular markers develop over time a higher speed of rotation around the center of the vortex, as well as their own.

Acknowledgment. This work is supported by Russian Foundation for Basic Research (Grant 18-01-00116).

References

1. http://www.dailygalaxy.com/my_weblog/2014/01/page4
2. http://finheaven.com/threads/40279
3. http://a2zhdwallpapers.com
4. Montero, P., et al.: Oil spill monitoring and forecasting on the Prestige-Nassau accident. In: Arctic and Marine Oil Spill Program Technical Seminar, vol. 2, pp. 1013–1030. Environment/Canada (2003)
5. Budnikov, A.A., Chashechkin, Yu.D: Marker transfer in a settled composite vortex. Mosc. Univ. Phys. Bull. **69**(3), 270–274 (2012)
6. http://www.argo.net/
7. Chashechkin, Yu.D., Prihodko Yu.V.: Structure of the currents formed with free vibrations of bodies on the horizons of neutral buoyancy in a continuously stratified fluid. Dokl. Phys. **406**(5), 622–625 (2006). (in Russian)
8. Redondo, J.M., Platonov, A.K.: Self-similar distribution of oil spills in European coastal waters. Environ. Res. Lett. **4**, 140–148 (2009)
9. Hu, C., Pichel, W.G., Muller-Karger, F.E.: Detection of natural oil slicks in the NW Gulf of Mexico using MODIS imagery. Geophys. Res. Lett. **36**, L01604 (2009)
10. Fujimoto, S., Takeda, Y.: Topology changes of the interface between two immiscible liquid layers by a rotating liquids. Phys. Rev. E **80**, 015304 (2009). (in Russian)
11. Chaplina, T.O., Stepanova, E.V., Chashechkin, Yu.D.: Specific features of admixture transport in a steady-state vortex flow. Vestnik MSU **3**(4), 73–79 (2012)
12. Stepanova, E.V., Chaplina, T.O., Chashechkin, Yu.D.: Transport of oil in a compound vortex. Fluid Dyn. **46**(2), 214–224 (2011). (in Russian)
13. Fujimoto, S., Murai, Y., Tasaka, Y., Takeda, Y.: Visualization of transient interfacial waves induced by spin-up of two immiscible fluid layers. J. Vis. **13**(1), 17–23 (2010)
14. Chaplina, T.O., Chashechkin, Yu.D., Stepanova E.V.: Deformation of a compact oil patch at a compound vortex surface. Dokl. Phys. **55**(5), 238–241 (2010)
15. Jansson, T., Haspang, M., Jensen, K., Hersen, P., Bohr, T.: Polygons on a rotating fluid suface. Phys. Rev. Lett. **96**, 174502 (2006)
16. Hirsa, A.H., Lopez, J.M., Miraghaie, R.: Symmetry breaking to a rotating wave in a lid driven cylinder with a free surface: experimental observation. Phys. Fluids **14**(6), 29–32 (2002)
17. Chaplina, T.O., Chashechkin, Yu.D, Stepanova, E.V.: Two forms of contact surface breakup for immiscible liquids in the compound vortex with a cavern. Monit. Sci. Technol. 1, 77–83 (2016)

Convective Motions in Water: Linear and Nonlinear Models, Criteria of Convection Onset

Anatoly Kistovich(✉)

All-Russian Scientific Research Institute of Physico-Technical and Radio-Technical Measurements, Moscow, Russia
kavmendeleevo@mail.ru

Abstract. The problem of convective motions adequate description on the base of linear or nonlinear mathematic models is considered. The criterion of convection onset is formulated in the form $|\mathbf{J}_{conv}|/|\mathbf{J}_{cond}| > 1$ where the $|\mathbf{J}_{conv}|$ and $|\mathbf{J}_{cond}|$ are the convective and conductive fluxes of heat correspondingly. As a result of governing equations solution for initial time moments of the problem the characteristic scales are chosen. It is shown that for Rayleigh-Bénard convection the Rayleigh number $Ra = \alpha g|\nabla T|d^4/\chi\nu$ should be attributed not to the onset, but to the intensity of convection and to the rate of its development after the onset.

Keywords: Convection · Heat fluxes · Onset criterion

1 Introduction

During more than one century after the basic experimental work by Bénard [1] and the prime theoretical investigation by the Lord Rayleigh [2], the problem of the heat convection occupies one of the fundamental place in hydrodynamic studies. The results of linear approach to the announced problem were summarized in [3]. The further progress in convection phenomena analysis was developed by means of weak nonlinear models for which the different mathematical approaches were applied [4–6].

The results of experiments [7] and correct form of mathematical model equations *sufficiently* depend on the substance in which the convection is developed. The reason is the dependence of thermodynamic characteristics of liquid on the temperature particularly in the regimes with great overheating [8]. Hence, it is necessary at once to declare the substance for which the problem is considered.

The main goal of the presented work is to investigate the problem of convection in the water without admixtures.

V. I. Karev et al. (Eds.): *Physical and Mathematical Modeling of Earth and Environment Processes (2018)*, SPEES, pp. 174–188, 2019.
https://doi.org/10.1007/978-3-030-11533-3_18

2 The Governing Equations

The classic system of governing equations and boundary conditions of the problem under consideration has the form

$$\begin{gathered}\rho\left(\partial \mathrm{v}_i/\partial t+(\mathbf{v}\cdot\nabla)\,\mathrm{v}_i\right)=-\partial p/\partial x_i+\partial\sigma'_{ik}/\partial x_k-\rho g\delta_{iz}\\ \rho T\left(\partial s/\partial t+\mathbf{v}\cdot\nabla s\right)=\sigma'_{ik}\partial \mathrm{v}_i/\partial x_k+\nabla\cdot(\kappa\nabla T)\\ \partial\rho/\partial t+\nabla\cdot(\rho\mathbf{v})=0,\qquad \rho=\rho\,(p,s)\end{gathered}\tag{1}$$

Here $\sigma'_{ik}=\eta\left(\partial \mathrm{v}_i/\partial x_k+\partial \mathrm{v}_k/\partial x_i-2\delta_{ik}\nabla\cdot\mathbf{v}/3\right)+\zeta\delta_{ik}\nabla\cdot\mathbf{v}$ is the tensor of viscose tension, η and ζ are the first and second dynamic viscosity correspondingly, ρ is the liquid density, p is the pressure, T is the temperature, s is the entropy per unit volume, g is gravitational acceleration, κ is heat conductivity coefficient.

The system (1) consists of Navier-Stokes, heat transport, continuity and state equations. In the general case, it is regarded that the density, kinetic coefficients and coefficient of thermal expansion depend on the temperature.

From the state equation it follows

$$d\rho=\left(\frac{\partial\rho}{\partial p}\right)_s dp+\left(\frac{\partial\rho}{\partial s}\right)_p ds=\frac{1}{c^2}dp+\left(\frac{\partial\rho}{\partial T}\right)_p\left(\frac{\partial T}{\partial s}\right)_p ds=\frac{1}{c^2}dp-\frac{\rho T}{c_p}\alpha ds$$

where c is adiabatic speed of sound, c_p is heat capacity under constant pressure, $\alpha=-\left(\partial\rho/\partial T\right)_p/\rho$ is the coefficient of thermal expansion which also depend on the temperature.

Because in convective regimes the generation of sound may be neglected then the continuity equation reforms into equation

$$\begin{gathered}\frac{d\rho}{dt}+\rho\nabla\cdot\mathbf{v}=\rho\nabla\cdot\mathbf{v}-\frac{\rho\alpha T}{c_p}\frac{ds}{dt}=0\Rightarrow\\ \Rightarrow\nabla\cdot\mathbf{v}=\frac{\alpha T}{c_p}\frac{ds}{dt}=\frac{\alpha}{\rho c_p}\left(\sigma'_{ik}\partial \mathrm{v}_i/\partial x_k+\nabla\cdot(\kappa\nabla T)\right).\end{gathered}\tag{2}$$

On the base of thermodynamic relation $ds=c_p dT/T$ and Eq. (2) it is possible to rewrite the system (1) in the form

$$\begin{gathered}\rho\left(\partial \mathrm{v}_i/\partial t+(\mathbf{v}\cdot\nabla)\,\mathrm{v}_i\right)=-\partial p/\partial x_i+\partial\sigma'_{ik}/\partial x_k-\rho g\delta_{iz}\\ \nabla\cdot\mathbf{v}=\alpha\left(\partial T/\partial t+\mathbf{v}\cdot\nabla T\right)=\frac{\alpha}{\rho c_p}\left(\sigma'_{ik}\partial \mathrm{v}_i/\partial x_k+\nabla\cdot(\kappa\nabla T)\right).\end{gathered}\tag{3}$$

The boundary conditions on the rigid surfaces consist of the condition of zero velocity field and either fixing value of temperature

$$T|_\Sigma=T_\Sigma\tag{4}$$

or fixing thermal flux

$$\chi \partial T/\partial n + \gamma_T T|_\Sigma = q_T \tag{5}$$

where $\chi = \kappa/\rho c_p$ is the thermal diffusion coefficient, γ_T is the coefficient of heat exchange of boundary surface and water, q_T is heat flux and n is the unit normal to the boundary. In the presented investigation the boundary conditions

$$T|_{z=0} = T_b \qquad T|_{z=d} = T_c \qquad T_b > T_c \tag{6}$$

will be used, where T_b and T_c are the temperatures of bottom and cover respectively.

3 The Onset of Convection

As it was proclaimed in the **Introduction** the reason for convection onset is nonuniform temperature distribution along the bottom or/and cover. Let in real situation the temperature of bottom may be presented as the sum of mean temperature T_b and some disturbances $\widetilde{T}(x, y, t)$ with zero mean value

$$T = T_b + \widetilde{T}(x, y, t), \qquad \langle \widetilde{T}(x, y, t) \rangle = \lim_{S\to\infty} \frac{\int_S \widetilde{T}(x, y, t)\, dx\, dy}{S} = 0 \tag{7}$$

where S is a square of heater.

The density of water without sub mixtures at atmospheric pressure is defined by the relation

$$\rho(T) = \rho_0 \exp\left(T\left(\alpha_0 - \frac{\alpha_1}{1+\alpha_2}\left(\frac{T}{T_r}\right)^{\alpha_2}\right)\right) \tag{8}$$

where T is the water temperature in centigrade degree, $T_r = 273.15\,K$ is reference temperature (ice melting temperature in absolute thermodynamic scale), and the constants are $\rho_0 = 999.87$ kg/m^3, $\alpha_0 = 1.83 \cdot 10^{-4}\,K^{-1}$, $\alpha_1 = 1.6 \cdot 10^{-3}\,K^{-1}$, $\alpha_2 = 0.54$. The relative error of the above relation is not greater than 10^{-3} [9] for whole diapason of liquid water phase.

Substitution of (7) into (8) permits to describe the distribution of water density near the heater by the relation

$$\begin{aligned} \rho = \rho_0 \exp\left(\alpha(1+\widetilde{T}/T_b)\left(1 - \beta\left(1 + \frac{\widetilde{T}}{T_b}\right)^{\alpha_2}\right)\right) \approx \\ \approx R_0(1 + \gamma x + \delta x^2), \quad x = \frac{\widetilde{T}}{T_b} \end{aligned} \tag{9}$$

where $R_0 = \rho_0 \exp(\alpha(1-\beta))$ is the density of water on the heater surface in the case of absense of temperature disturbances $\widetilde{T}$, $\alpha = \alpha_0 T_b$, $\beta = \dfrac{\alpha_1}{\alpha_0(1+\alpha_2)}\left(\dfrac{T_b}{T_r}\right)^{\alpha_2}$, $\gamma = \alpha(1-(1+\alpha_2)\beta)$, $\delta = \dfrac{\alpha}{2}(\alpha - (1+\alpha_2)(\alpha_2+2\alpha)\beta + \alpha(\alpha+\alpha_2)^2\beta^2)$. The relative error of the (9) is not greater than 10^{-2}.

The goal of further actions is to calculate the dispersion and root mean square deviation of density variations from ideal value R_0 near heater surface. Accordingly to (9) the difference of density is defined by relation

$$\Delta\rho = \rho - R_0 \approx R_0(\gamma x + \delta x^2) \tag{10}$$

then with regard to $\langle x\rangle = 0$ (because $\langle \widetilde{T}\rangle = 0$) the relation is valid

$$\overline{\Delta\rho} = R_0\delta\overline{x^2} \quad \Rightarrow \quad \overline{\Delta\rho}^2 = R_0^2\delta^2 a\overline{x^2}^2 \tag{11}$$

From the other hand

$$(\Delta\rho)^2 \approx R_0^2(\gamma x + \delta x^2)^2 = R_0^2(\gamma^2 x^2 + 2\gamma\delta x^3 + \delta^2 x^4) \approx R_0^2\gamma^2 x^2 \tag{12}$$

where the last approximate relation is provided by the smallness of x and δ/γ for real experimental regimes.

On the base of (11, 12) the dispersion of density variations is calculated

$$\sigma_\rho^2 = \overline{(\Delta\rho)^2} - (\overline{\Delta\rho})^2 = R_0^2\gamma^2\overline{x^2} - R_0^2\delta^2\overline{x^2}^2 \approx R_0^2\gamma^2\overline{x^2} \tag{13}$$

also as density rms deviation

$$\sigma_\rho = R_0|\gamma|\sigma_T \tag{14}$$

where $\sigma_T = \sqrt{\overline{x^2}}$ is rms deviation of heater surface temperature.

Because the buoyancy force is defined as product of gravitational acceleration and water density deviation from its mean value then rms deviation of Archimedes force equals to the value

$$\sigma_A = g\sigma_\rho = gR_0|\gamma|\sigma_T. \tag{15}$$

In Boussinesq approximation the inertial acceleration of liquid element at initial time moments is defined by the ratio of buoyancy force to mean water density, i.e. by the value $a_{in} = g|\gamma|\sigma_T$. When the rms deviation of temperature is fixed then the more the temperature of the heater, the more the rms of buoyancy force (and inertial acceleration). It means that the local perturbations of conductive mechanism of heat transfer due to convective flux are rising under heater temperature increasing even through immutable quality of the heater temperature stabilization.

Let's consider some small spherical liquid element of radius R near the bottom heater (analogous procedure can be done also for cover colder). Accordingly to probability theory its more probable overheating temperature is $T_b + \sigma_T$. The mean temperature on the same horizontal level is T_b then the buoyance force will be act on this element. Under action of this force the liquid element will float up and as the development of motion the viscose force will arise. Let $\bar{T}(t)$ is the mean temperature of element, so its density is defined by the relation

(8) $\bar{\rho}(t) = \rho(\bar{T}(t))$. The initial temperature distribution of ambient liquid is described by the formula

$$T_{st}(z) = T_b - |\nabla T| z, \quad |\nabla T| = \frac{T_b - T_c}{d} \tag{16}$$

where d is the distance between bottom and cover.

The density of stratified liquid is also defined by (8) so $\rho_{st}(z) = \rho(T_{st}(z))$. The vertical coordinate of floating up element will be signed by the symbol $z(t)$. During its vertical motion the considered element will be surrounded by ambient stratified liquid which temperature and density are equal to $T_{st}(z(t))$ and $\rho_{st}(t) = \rho(T_{st}(z(t)))$ correspondingly. As a result both the mean temperature and the mean density of the element will be changed due to thermal conductivity with surrounding liquid. Thus the system of equations describing the motion of considered element is the following

$$\begin{gathered}
\bar{\rho}(t) V \ddot{z} = \big(\bar{\rho}(t) - \rho_{st}(t)\big) V g - \pi K_f \bar{\rho}(t) R \nu \dot{z}, \quad \dot{z}|_{t=0} = z|_{t=0} = 0 \\
\dot{\bar{T}}(t) = -\tilde{\chi}\big(\bar{T}(t) - T_{st}(t)\big), \quad \bar{T}(t)|_{t=0} = T_b + \sigma_T
\end{gathered} \tag{17}$$

where $V = 4\pi R^3/3$ is the volume of the element, K_f is the coefficient of profile drag of liquid sphere in the same liquid (experimental data show that $K_f \in [1..3]$), L_χ is some characteristic scale of diffusion for spherical element and $\tilde{\chi} = \chi / L_\chi^2$.

The solution of the second initial problem of (17) has the form

$$\begin{aligned}
\bar{T}(t) &= \left(T_b + \sigma_T + \tilde{\chi} \int_0^t T_{st}(\tau) e^{\tilde{\chi}\tau}\, d\tau \right) e^{-\tilde{\chi} t} = \\
&= T_{st} + \left(\sigma_T + |\nabla T| \int_0^t \dot{z}(\tau) e^{\tilde{\chi}\tau}\, d\tau \right) e^{-\tilde{\chi} t}
\end{aligned} \tag{18}$$

which was produced with regard to (16).

The substitution of the result received into the first initial problem of (17) permits to get in the crude approximation the initial problem for coordinate $z(t)$ of liquid element

$$\ddot{z} = \alpha_0 g \left(\sigma_T + |\nabla T| \int_0^t \dot{z}(\tau) e^{\tilde{\chi}\tau}\, d\tau \right) e^{-\tilde{\chi} t} - \tilde{\nu} \dot{z}, \quad \dot{z}|_{t=0} = z|_{t=0} = 0 \tag{19}$$

where $\tilde{\nu} = 3K_f \nu / 4R^2$.

The solution of (19) produces the required estimation for vertical coordinate and velocity of the floating liquid element on the initial stage of motion

$$\begin{gathered}
z(t) = \frac{\alpha_0 g \sigma_T}{k^2 - \Omega^2} \left(1 - e^{-kt} \big(\cosh(\Omega t) + \frac{k}{\Omega} \sinh(\Omega t) \big) \right) \\
\dot{z}(t) = \frac{\alpha_0 g \sigma_T}{\Omega} e^{-kt} \sinh(\Omega t)
\end{gathered} \tag{20}$$

where

$$k = \frac{\tilde{\nu} + \tilde{\chi}}{2}, \quad \Omega = \sqrt{\left(\frac{\tilde{\nu} - \tilde{\chi}}{2}\right)^2 + \alpha_0 g |\nabla T|}\,. \tag{21}$$

As it follows from (20, 21) the character of liquid element floating depends on relation between parameters k and Ω. If $\Omega > k$ the height of floating is restricted only by cover so $z_{max} = d$. The dependencies of $z(t)$ for some ratios $\Omega/k > 1$ are shown on Fig. 1.

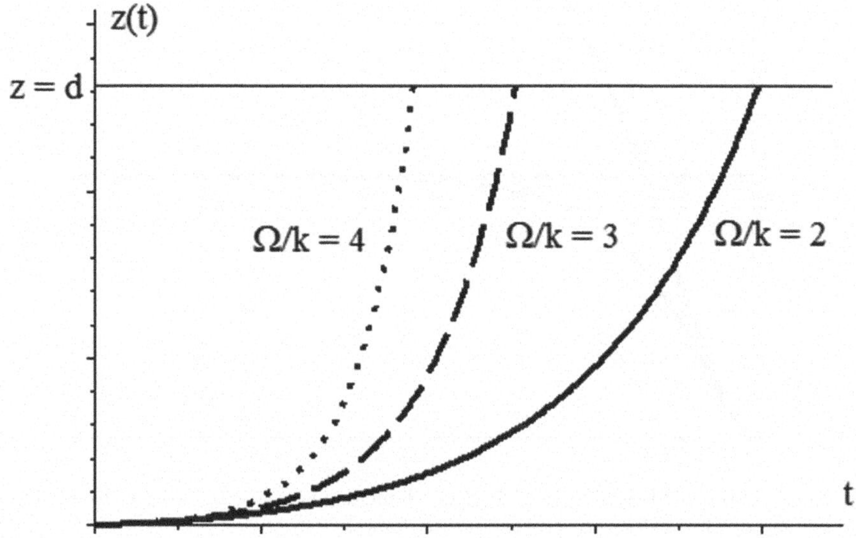

Fig. 1. The graphs of height $z(t)$ of floating element when $\Omega > k$ (solid, dash and dot lines). The thin horizontal line is the level of cover $z = d$.

The greater is the ratio Ω/k the greater is the value of standard Rayleigh number Ra and the quickly the floating liquid element will reach the cover.

In the opposite case when $k > \Omega$ the height of floating has the maximum defined by the condition

$$\dot{z}(t) = 0 \quad \Rightarrow t \to \infty$$

and described by the relation

$$z_{max} = \lim_{t\to\infty} z(t) = \frac{\alpha_0 g \sigma_T}{k^2 - \Omega^2}. \tag{22}$$

The greater is the ratio k/Ω the small is the height of the liquid element rising which always should be less than the liquid layer thickness d. The dependences of $z(t)$ for some ratios $k/\Omega > 1$ are shown on Fig. 2.

On the base of (21) the condition $\Omega > k$ may be rewrite in equivalent form, namely

$$\widetilde{Ra} = \frac{\alpha_0 g |\nabla T|}{\tilde{\chi}\tilde{\nu}} > 1 \tag{23}$$

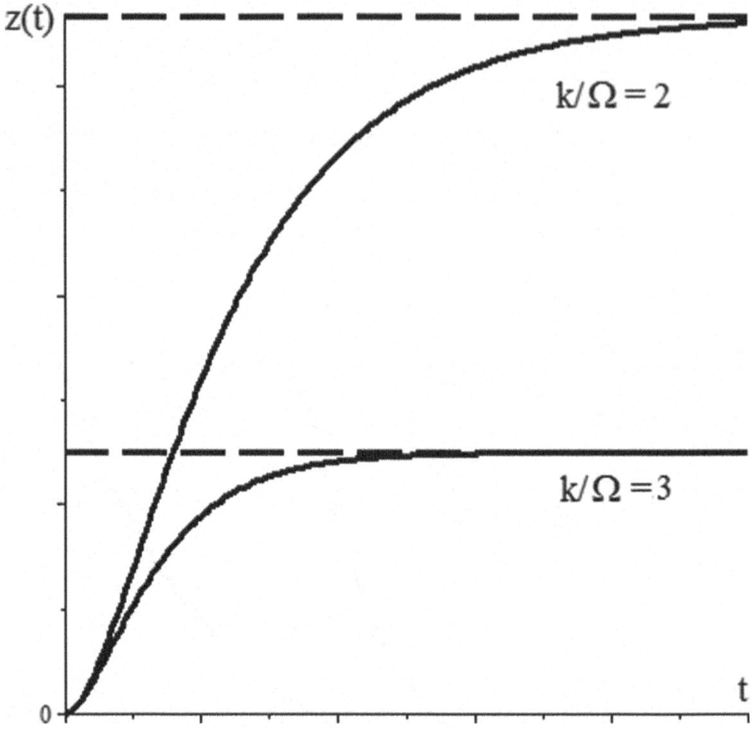

Fig. 2. The graphs of height $z(t)$ of floating element when $k > \Omega$ (solid lines). Dashed lines are asymptotic values for corresponding curves.

It is evident that received modified Rayleigh number $\widetilde{Ra}$ is proportional to the standard defined $Ra = \alpha_0 g|\nabla T|d^4/\chi\nu$ but don't coincide with it and has other physical sense which is usually attributed to the Rayleigh number.

By means of consequent differentiation the Eq. (19) may be transformed into equation

$$\ddot{z} + (\tilde{\nu} + \tilde{\chi})\dot{z} + \left(\tilde{\nu}\tilde{\chi} - \alpha_0 g|\nabla T|\right)z = \text{const}. \tag{24}$$

The characteristic equation correspondent to (24) has the form and solutions

$$\lambda^2 + (\tilde{\nu} + \tilde{\chi})\lambda + \tilde{\nu}\tilde{\chi} - \alpha_0 g|\nabla T| = 0$$
$$\lambda_{\pm} = -\frac{\tilde{\nu} + \tilde{\chi}}{2} \pm \sqrt{\left(\frac{\tilde{\nu} - \tilde{\chi}}{2}\right)^2 + \alpha_0 g|\nabla T|} \tag{25}$$

The instability of solutions of (19) has place when $\lambda_+ > 0$ that corresponds to $\alpha_0 g|\nabla T|/\tilde{\nu}\tilde{\chi} > 1$. Thus the critical value of the new Rayleigh number is defined by the relation

$$\widetilde{Ra}_c = 1. \tag{26}$$

It necessary to emphasize here that Rayleigh number $\widetilde{Ra}$ defines the velocity of growth of convective transport and its intensity but not the onset of convection. The onset of convection is defined by the value of rms deviation of the heater temperature σ_T. If $\sigma_T = 0$ then accordingly to physical sense no convective motion may be observed in water. When $\sigma_T \neq 0$ then convection always has place and the criterion of convection onset is only *subjective* conception as it was proclaimed in **Introduction**.

Prior to formulate the criterion of convective onset it is necessary to give some subjective definition of convective motion presence. Let the definition is the following: *The convective motion in the observed space point is proclaimed as existing if the absolute value of convective heat flux is not less than conductive one.*

For the problem under consideration it means that the ratio should be valid

$$(T_b - T_c)\overline{\dot{z}(t)} \geq \chi |\nabla T| \quad \equiv \quad \overline{\dot{z}(t)} \geq \frac{\chi}{d}. \tag{27}$$

In (27) the difference $T_b - T_c$ has place because from the cover the negative heat flux due to the same reasons as a positive flux from the bottom is presented. The over line in (27) is the symbol of averaging through the time of floating.

In accordance to results presented on Fig. 2 there are two possible cases. The first one is the case $k > \Omega$ when the liquid element doesn' t reach the cover and its maximum height is defined by relation (22). But the time of rising equals to infinity so the average velocity is equal to zero. But the such approach is incorrect for the case under consideration. The fact of the matter is that at the first time moments the liquid element is accelerated upward. Through some time period t_* its acceleration change its sign and the element is breaking. But during the acceleration period the possibility of (27) performance may have place. The time t_* of acceleration period follows from condition

$$\ddot{z}(t_*) = 0$$

from which it follows

$$t_* = \operatorname{arctanh}(k/\Omega)/\Omega, \quad z(t_*) = \frac{\alpha_0 g \sigma_T}{k^2 - \Omega^2}\left(1 - \frac{k \exp(-kt_*)}{\sqrt{k^2 - \Omega^2}}\right) \tag{28}$$

so the condition (27) obtains the form

$$\overline{\dot{z}(t)} = \frac{z(t_*)}{t_*} \geq \frac{\chi}{d}. \tag{29}$$

In the second case $k < \Omega$ the liquid element float up to the cover so $z_{max} = d$ and the time t^* of element rising is defined as solution of the equation

$$d = \frac{\alpha_0 g \sigma_T}{k^2 - \Omega^2}\left(1 - e^{-kt^*}\left(\cosh(\Omega t^*) + \frac{k}{\Omega}\sinh(\Omega t^*)\right)\right) \tag{30}$$

and the condition (27) obtains the form

$$t_\chi / t^* \geq 1 \tag{31}$$

where $t_\chi = d^2/\chi$ is the characteristic diffusive time of the liquid layer.

The inverted value to $\tilde{\chi}$, earlier introduced in (17), is the estimation of heat loss time by the liquid element. Accordingly to well-known solution [10] for dynamic sphere cooling in the ambient environment the time of heat losing is proportional to square of sphere radius so one can write

$$L_\chi \sim R \Rightarrow \tilde{\chi} = K_\chi \frac{\chi}{R^2} \tag{32}$$

where K_χ is some constant.

So the value of the $\widetilde{Ra}$ may be presented in the form

$$\widetilde{Ra} = K_{Ra} \frac{\alpha_0 g |\nabla T| R^4}{\chi \nu}, \quad K_{Ra} = \frac{4}{3K_f K_\chi} \tag{33}$$

It follows from (33) that the smaller is floating liquid element the smaller is the Rayleigh number $\widetilde{Ra}$ and the smaller is the intensity of convective motion in liquid. The existence of the fluid motion when standard Rayleigh number $Ra = 0$ (homogeneously temperature distribution) is shown on Fig. 3. For common imaging all curves are normalized on its own maximum.

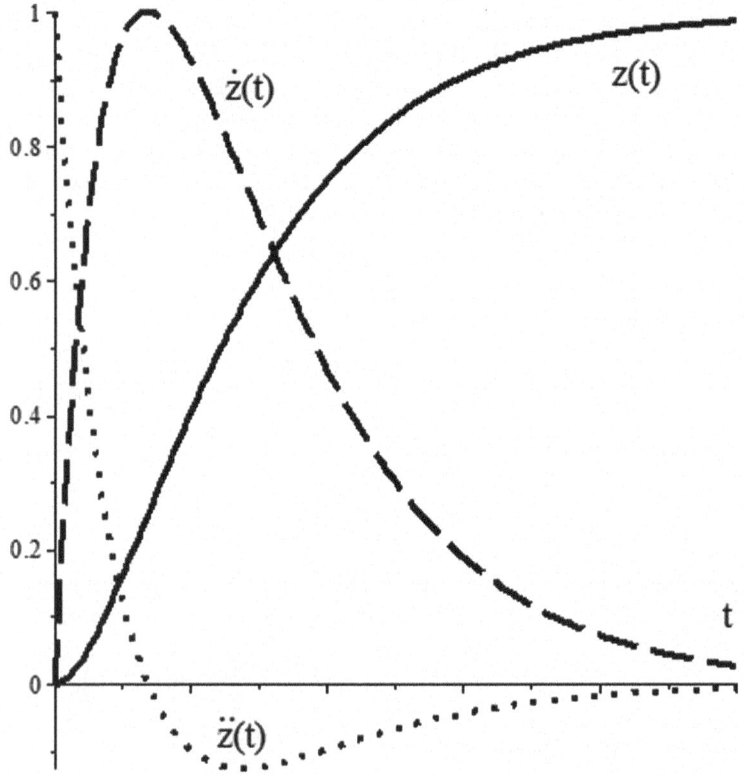

Fig. 3. The acceleration $\ddot{z}(t)$ (dot), velocity $\dot{z}(t)$ (dash) and height $z(t)$ (solid) of the floating liquid element when $Ra = 0$.

The explicit analytical forms of (29, 31) are some tedious to be presented here, but for concrete situations it is easy to check out numerically the announced conditions. If the condition (29) is valid then the relations (20 − 22, 28) describe the characteristics of vertically limited convection. For the case of (31) the characteristics of fully developed convection are described by (20 − 22, 30).

4 The Transformation of Governing System

For the further investigations it is useful to transform the system (3) to convenient form by means of some preliminary actions. Because the generation of heat due to viscose effects is sufficiently ineffective the term $\sigma'_{ik}\partial \mathrm{v}_i/\partial x_k$ in the second equation of (3) may be neglected. The velocity field is divided on two parts

$$\mathbf{v} = \bar{\mathbf{v}} + \tilde{\mathbf{v}} \quad \nabla \cdot \bar{\mathbf{v}} = 0, \qquad \nabla \cdot \tilde{\mathbf{v}} = \frac{\alpha}{\rho c_p} \nabla \cdot (\kappa \nabla T)\,. \tag{34}$$

Because the convective transport of temperature and advective transport of momentum as well as viscose force are stipulated basically by solenoidal part $\bar{\mathbf{v}}$ of velocity then the approximation

$$\bar{\sigma}'_{ik} \approx \eta\,(\partial \bar{\mathrm{v}}_i/\partial x_k + \partial \bar{\mathrm{v}}_k/\partial x_i) \tag{35}$$

is used and the system (3) transforms into the new approximate system

$$\begin{gathered} \rho\,(\partial \bar{\mathrm{v}}_i/\partial t + (\bar{\mathbf{v}} \cdot \nabla)\,\bar{\mathrm{v}}_i) = -\partial p/\partial x_i + \partial \bar{\sigma}'_{ik}/\partial x_k - \rho g \delta_{iz} \\ \partial T/\partial t + \bar{\mathbf{v}} \cdot \nabla T = \nabla \cdot (\chi \nabla T),\ \nabla \cdot \bar{\mathbf{v}} = 0, \quad \nabla \cdot \tilde{\mathbf{v}} = \alpha \nabla \cdot (\chi \nabla T). \end{gathered} \tag{36}$$

For the reason of very slowly varying of the term ρc_p for water in the presence of temperature variations its value was introduced under action of ∇ operator.

The standard approach to the analysis of nonlinear hydrodynamic equations is to describe the phenomenon in dimensionless form. For this reason, some scales are usually introduced for differential (space and time) and field (velocity, pressure, temperature etc.) variables. The correct definition of these scales is very complex problem on the solution of which the final results are strongly dependent. For example, the frequently used scale for velocity field (namely χ/d in [3–5] etc.) is in incompatible contradiction with experimental results which show the value of velocity in dozen times greater. As a result of the announced scale application the convective and advective terms of equations are considered as negligible or very small in comparison with linear ones and the solutions received (for velocity and temperature fields) are far from the observed in experiments.

The complexity of the problem of the suitable scale choice is conditioned by the fact that the numerous experiments show the different partial and time scales for different convection regimes and furthermore on the different stages of the certain regime. The standard used scales ([3–5] etc.) at once transform the nonlinear governing system of equations into the form of linear or weakly nonlinear mathematical model which is not adequate to the problem under consideration. This fact points to the necessity of *solving of spatial and time scales problem* for correct construction of dimensionless variant of equations. And this problem is arisen for certain regime of convection.

4.1 The Initial Stage of Convection

To investigate the characteristic features of the initial stage of convection it is necessary to transform the system (36) into dimensionless form. The time and space scales differ for the cases $\Omega > k$ and $k > \Omega$. For this reason the results will be presented separately.

In the case $\Omega > k$ the characteristic time scale follows from (20) in the form

$$t' = (\Omega - k)t \tag{37}$$

where t' is the dimensionless time.

When the time moments are so small that the floating height of liquid element is less than d then accordingly to (20) the action of differentials have the forms

$$\frac{\partial}{\partial z} = \frac{\Omega(\Omega - k)}{\alpha_0 g \sigma_T} e^{-t'} \frac{\partial}{\partial t'} \quad \frac{\partial}{\partial t} = (\Omega - k)\frac{\partial}{\partial t'}. \tag{38}$$

The scale of the vertical velocity component follows from (20)

$$\mathrm{v_z} = \frac{\alpha_0 g \sigma_T}{\Omega} e^{t'} \mathrm{v}' \tag{39}$$

where v' is the dimensionless velocity.

The temperature of water is presented in the form

$$T = T_{st}(z) + \left(\Delta T + \sigma_T\right)\theta \tag{40}$$

where θ is the dimensionless temperature perturbation.

The above introduced dimensionless field values have the order $\mathrm{v}', \theta \sim O(1)$.

On the base of (37–40) the heat transport equation of the system (36) transforms into equation

$$\theta'_{t'} + \lambda \mathrm{v}\left(1 + \lambda^{-1}\theta'_{t'}\right) = \lambda^{-2}\left(\chi\theta''_{t't'} + \chi'_T {\theta'_{t'}}^2\right) - \chi\lambda^{-1}\theta'_{t'} \tag{41}$$

where $\lambda = \alpha_0 g \sigma_T \exp(t')/\Omega(\Omega - k)$.

From (41) follows that the greater term of the ratio of convective and conductive heat fluxes is defined by relation

$$\frac{|\mathbf{J}_{conv}|}{|\mathbf{J}_{cond}|} \approx \chi\lambda^2 \mathrm{v}. \tag{42}$$

For the water and $\sigma_T > 10^{-2}\,K$ this ratio is greater then 1 at the initial moment $t = 0$ and grows with a time.

When the time momets are sufficiently great so the floating height of liquid element equals d then the spatial vertical scale is defined by relation

$$z = dz'. \tag{43}$$

For this situation the time, velocity and temperature scales remain the same and the heat transport equation of the system (36) transforms into equation

$$\theta'_{t'} + \frac{\lambda \mathrm{v}}{d}(1 + \theta'_{z'}) \approx \frac{1}{d^2(\Omega - k)}\left(\chi'_T \Delta T (1 + \theta'_{z'})^2 + \chi \theta''_{z'z'}\right) \tag{44}$$

from which follows

$$\frac{|\mathbf{J}_{conv}|}{|\mathbf{J}_{cond}|} \approx \frac{\lambda \mathrm{v} d(\Omega - k)}{\chi} \tag{45}$$

which for water is greater than 1 and tends to growth with a time.

In the case $k > \Omega$ the characteristic vertical scale follows from (20) in the form

$$z = \frac{\alpha_0 g \sigma_T}{k^2 - \Omega^2} z'. \tag{46}$$

Let the time scale is described by the relation

$$t = m_t t' \tag{47}$$

where m_t is undefined positive time scale.

So velocity and temperature scales remain the same then on the base (46, 47) the heat transport equation of the system (36) transforms into equation

$$\begin{aligned} m_t \theta'_{t'} + \frac{\alpha_0 g \sigma_T}{d\Omega} \mathrm{v}\Big(1 + \frac{d(k^2 - \Omega^2)}{\alpha_0 g \sigma_T} \theta'_{z'}\Big) = \\ = \left(\frac{k^2 - \Omega^2}{\alpha_0 g \sigma_T}\right)^2 \left(\chi'_T \Delta T (1 + \theta'_{z'})^2 + \chi \theta''_{z'z'}\right) \end{aligned} \tag{48}$$

From (48) it follows the result

$$\frac{|\mathbf{J}_{conv}|}{|\mathbf{J}_{cond}|} \approx \frac{(\alpha_0 g \sigma_T)^2}{d\chi \Omega(\Omega - k)} \mathrm{v} \exp((\Omega - k) m_t t') \tag{49}$$

which value for water is less than 1 in the starting moments and decrease with a time.

Thus in the case $\Omega > k$ the heat transport equation is strongly nonlinear and no perturbation methods can be applied to analyze it. Such methods may be valid only for the opposite case $k > \Omega$.

4.2 The Quasi-stationary Regime of Convection

Because in the regime of stationary full-developed convection the velocity of convective transport sufficiently exceeds the diffusive velocity then the system (36) in regard of (35) may be rewrite in the form

$$\begin{aligned} \rho \bar{\mathrm{v}} \cdot \nabla \bar{\mathrm{v}}_i = -\partial p / \partial x_i + \eta \Delta \bar{\mathrm{v}}_i + \partial \eta / \partial x_k \left(\partial \bar{\mathrm{v}}_i / \partial x_k + \partial \bar{\mathrm{v}}_k / \partial x_i\right) - \rho g \delta_{iz} \\ \bar{\mathbf{v}} \cdot \nabla T \approx 0,\ \nabla \cdot \bar{\mathbf{v}} = 0. \end{aligned} \tag{50}$$

The second equation of (50) means that in the regime of stationary convection the fields of velocity and temperature gradient are orthogonal to each other. Naturally this orthogonality has an approximate character for the reason of the previously approximations done. From this equation the representation of velocity field it follows

$$\bar{\mathbf{v}} = \mathbf{A} \times \nabla T \tag{51}$$

which transforms the announced equation into identity.

The substitution of (51) into the third equation of (50) leads to the consequence of relations

$$\nabla \cdot \bar{\mathbf{v}} = \nabla \cdot (\mathbf{A} \times \nabla T) = \nabla T \cdot \nabla \times \mathbf{A} - \mathbf{A} \cdot \nabla \times (\nabla T) = \nabla T \cdot \nabla \times \mathbf{A} = 0 \tag{52}$$

The first way to satisfy to the last equality in consequence (52) is to proclaim $\mathbf{A} = \nabla \Phi$ where Φ is some arbitrary scalar function. In this case the velocity field is described by relation

$$\bar{\mathbf{v}} = \nabla \Phi \times \nabla T \tag{53}$$

The second way is to use the representation for vector field $\mathbf{A}$ in the form

$$\mathbf{A} = A\mathbf{e}_x + B\mathbf{e}_y + C\mathbf{e}_z \tag{54}$$

where A, B and C are the arbitrary functions.

As a result the velocity field has the form

$$\bar{\mathbf{v}} = (BT'_z - CT'_y)\mathbf{e}_x + (CT'_x - AT'_z)\mathbf{e}_y + (AT'_y - BT'_x)\mathbf{e}_z \tag{55}$$

and the condition (52) of solenoidal field type transforms into relation

$$(C'_y - B'_z)T'_x + (A'_z - C'_x)T'_y + (B'_x - A'_y)T'_z = 0. \tag{56}$$

In the case of $2D$-motion (the roll convection) the conditions $T'_y = \bar{\mathrm{v}}_y = 0$ are valid from (55) it follows

$$\bar{\mathrm{v}}_x = BT'_z, \quad \bar{\mathrm{v}}_z = -BT'_x, \quad CT'_x - AT'_z = 0. \tag{57}$$

In the first case when $\mathbf{A} = \nabla \Phi$ it follows from (54) that , $A = \Phi'_x$, $B = \Phi'_y$, $C = \Phi'_z$ and the last relation of (57) transforms into equation

$$\Phi'_z T'_x - \Phi'_x T'_z = 0 \tag{58}$$

and because $T'_y = 0$ then the first integral of (58) is the temperature T so $\Phi = \Phi(T)$.

In this case $B = \Phi'_y = \Phi'_T T'_y = 0$ (in accordance to $2D$ problem) and in the result $\bar{\mathrm{v}}_x = \bar{\mathrm{v}}_z = 0$. It means that the first way can't be realized.

In the second case for $T'_y = \bar{v}_y = 0$ from (56) it follows

$$(C'_y - B'_z)T'_x + (B'_x - A'_y)T'_z = 0, \quad CT'_x - AT'_z = 0 \tag{59}$$

and then the result is valid: the introduction of stream function $\Psi = \Psi(T)$ is valid so

$$\bar{\mathrm{v}}_x = \Psi'_z = \Psi'_T T'_z, \quad \bar{\mathrm{v}}_z = -\Psi'_x = -\Psi'_T T'_x. \tag{60}$$

Thus for $2D$-motion the presentation (11) provides the realization of the second and third relations of the system (50). The exclusion of pressure from the first equation of (50) leads to the *unique* equation in which (in regard to (60)) the *two* unknown values are presented namely the temperature field T and some stream function $\Psi(T)$. For the definite substance (water in the problem under consideration) the temperature dependencies $\rho(T)$ and $\eta(T)$ are known. The necessary derivatives of these values are calculated by means of relations

$$\frac{\partial \rho}{\partial x_i} = \rho'_T \frac{\partial T}{\partial x_i}, \quad \frac{\partial \eta}{\partial x_i} = \eta'_T \frac{\partial T}{\partial x_i}.$$

The explicit form of the announced *unique* equation does not presented for the reason of its tedious size.

In $3D$-motion for the regimes of the developed cell convection the description of velocity field into the cell it is convenient by means of toroidal-poloidal potential because the axis z is singled out so

$$\bar{\mathrm{v}}_x = \Phi'_y + \Psi''_{xz}, \quad \bar{\mathrm{v}}_y = -\Phi'_x + \Psi''_{yz}, \quad \bar{\mathrm{v}}_z = -\Psi''_{xx} - \Psi''_{yy} = -\Delta_\perp \Psi. \tag{61}$$

So the experimental results point to absence of axial rotation in the individual cell then the toroidal potential may be neglected and (61) are transformed into

$$\bar{\mathrm{v}}_x = \Psi''_{xz}, \quad \bar{\mathrm{v}}_y = \Psi''_{yz}, \quad \bar{\mathrm{v}}_z = -\Delta_\perp \Psi. \tag{62}$$

The representations (61, 62) transform the condition $\nabla \cdot \bar{\mathbf{v}}$ into identity and substitution of (62) into (51) reforms the equation of heat transport into equation

$$T'_y \Delta \Psi'_x - T'_x \Delta \Psi'_y = 0. \tag{63}$$

The experimental observations also show that in a good approximation the convective structures by Rayleigh-Bénard are characterized by zero helicity i.e. the condition

$$\bar{\mathbf{v}} \cdot \nabla \times \bar{\mathbf{v}} = 0 \tag{64}$$

is valid.

In the case of $2D$ roll convection this condition is valid automatically don't supply additional relations. But in the regime of cell convection

$$\nabla \times \bar{\mathbf{v}} = -\Delta \Psi'_y \mathbf{e}_x + \Delta \Psi'_x \mathbf{e}_y \tag{65}$$

so with regard the condition (64) the additional kinematic condition for velocity field is formed

$$\Delta \Psi'_x \Psi''_{yz} - \Delta \Psi'_y \Psi''_{xz} = 0. \tag{66}$$

From (63) and (65) also follows $\nabla \times \bar{\mathbf{v}} \cdot \nabla T = 0$. It means that in any space point the vectors $\bar{\mathbf{v}}$, $\nabla \times \bar{\mathbf{v}}$ and ∇T form the orthogonal triplet. Hence, the representation for temperature field is valid

$$\nabla T = f(x,y,z)\bar{\mathbf{v}} \times \nabla \times \bar{\mathbf{v}} \Rightarrow T'_x = f\Delta_\perp \Psi \Delta_\perp \Psi'_x,$$
$$T'_y = f\Delta_\perp \Psi \Delta_\perp \Psi'_y, \quad T'_z = f\big(\Psi''_{yz}\Delta_\perp \Psi'_y + (\Psi''_{xz}\Delta_\perp \Psi'_x\big). \tag{67}$$

The relations (66, 67) may be very helpful for the solving of nonlinear equations which arise after pressure elimination from Navier-Stokes equation.

5 Conclusions

Presented here the new approach to formation of convective flows in liquids (particularly in water) shows that the standard Rayleigh number doesn't characterize the condition for convection start but defines the intensity and velocity of growth of convective motion. The proposed new criterion based on new form of the Rayleigh number has the explicit physical sense.

The one more general conclusion follows from the investigations done that the linear or slightly nonlinear mathematical models may be applied when the modified Rayleigh number $\widetilde{Ra}$ is less than unity. In the opposite case the convective motion is described by strongly nonlinear equations.

References

1. Bénard, H.: Les tourbillons cellulaires une nappe liquide transportant de la chaleur par convection en régime permanent. Ann. Chim. Phys. s7. **23**, 62–144 (1901)
2. Rayleigh, L.: On convection currents in a horizontal layer of fluid when the higher temperature is on the underside. Phil. Mag. **32**(6), 529–546 (1916)
3. Chandrasekhar, S.: Hydrodynamic and Hydromagnetic Stability. Clarendon, Oxford (1961)
4. Malkus, W.V.R., Veronis, G.: Finite amplitude cellular convection. J. Fluid Mech. **4**, 225–260 (1958)
5. Serrin, J.: On the stability of viscose fluid motions. Arch. Rat. Mech. Anal. **3**, 1–119 (1959)
6. Joseph, D.D.: On the stability of the Boussinesq equations. Arsh. Rat. Math. Anal. **20**, 59–87 (1965)
7. Berg, J.C., Boudart, M., Acrivos, A.: Natural convection in pools of evaporating liquids. J. Fluid Mech. **24**(Pt. 4), 721–735 (1966)
8. White, D.B.: The planforms and onset of convection with temperature-dependent viscosity. J. Fluid Mech. **191**, 247–286 (1988)
9. Kistovich, A.V.: The model of stable water stratification with regard to dependence of thermodynamic parameters on temperature and salinity. Almanac Mod. Metrol. **4**, 22–38 (2015). (In Russian)
10. Koshljakov, N.S., Gliner, E.B., Smirnov, M.M. The equations in partial differentials of mathematical physics. "High School", Moscow (1970). (In Russian)

Stage Pole Distribution for Meso-Cenozoic Lithospheric Plate Rotation

A. A. Schreider(✉) and A. E. Sazhneva

Shirshov Institute of Oceanology, Russian Academy of Sciences, Moscow, Russia
aschr@ocean.ru

Abstract. The main features of stage pole distribution are investigated. Three areas of localization of stage poles related to the Mesozoic-Cenozoic plate kinematics are established. These areas coincide with the areas of upwelling flows in the lower mantle. Using statistical centers of the cluster, the major continental fractures and sea-floor transform faults were outlined. The influence of mantle upwelling flows and zones of weakness on plate kinematic is proved.

Keywords: Stage pole distribution · Mesozoic-Cenozoic plate kinematics · Upwelling flows in the lower mantle · Euler poles

1 Introduction

The concept of plate tectonics postulates that the Earth's lithosphere is divided into rigid plates which are moving with respect to one another. The most striking evidence for relative plate motion consist of the ocean linear magnetic anomalies showing the orientation of spreading centers and transform faults which stretch along the direction of plate motion. The relative motion of any pair of plates may be described by finite and stage Euler poles [1]. Finite poles depict the resulting plate displacements over any long time period and are widely used for plate-tectonic reconstructions. In opposite to finite poles, stage poles describe the real relative plate rotations for limited periods of time which are characterized by a constant direction of plate motion.

Distribution of the linear magnetic anomalies and transform faults [2–5] has revealed that over the broad areas of oceanic floor general spreading directions have been stable for long periods of geological time. It supposes spatial restriction of stage pole distribution because stage pole positions determine plate-kinematic parameters. Stage poles should abruptly change their positions during periods of great reorganizations in the spreading system and then regain stability other regions after restabilization of the system. So, an investigation of regularities of stage pole distribution for various pairs of plates and for various geological time intervals seems to be very significant since it would reveal global characteristics of plate kinematics.

V. I. Karev et al. (Eds.): *Physical and Mathematical Modeling of Earth and Environment Processes (2018)*, SPEES, pp. 189–195, 2019.
https://doi.org/10.1007/978-3-030-11533-3_19

2 Approach and Methods

Poles of relative motion are the most reliable for the pairs of plates separated by a divergent boundary and '° only such pairs of plates were taken into account. So, we considered the distribution of stage poles related with spreading system. Stage poles were calculated using published finite poles based on linear magnetic anomalies [6–21]. The shorter the time interval involved, the better stage poles describe real plate motion, so they were calculated over the shortest possible time interval, but no more than 25 Ma years. At first, stage poles were obtained with respect to Africa and then rotated with respect to the absolute frame. Data on used finite poles and calculated paleopositions of stage poles are presented in [4, 5, 22–26]. All computations are provide with taken into account the geochronological scale [27].

3 Discussion

Analysis of stage pole spatial distribution on the Earth surface shows their concentration in three restricted areas (Fig. 1). The meansquare statistical centers of these areas are presented in Table. The first area embraces the Norwegian-Greenland Region of the North Atlantic. More than 44% of considered stage poles are concentrated in this area. The second area coincides with the Central Atlantic and includes about 21% of points. The third area corresponds to the African-Arabian Region and includes about 16% of points (Table 1; Fig. 2).

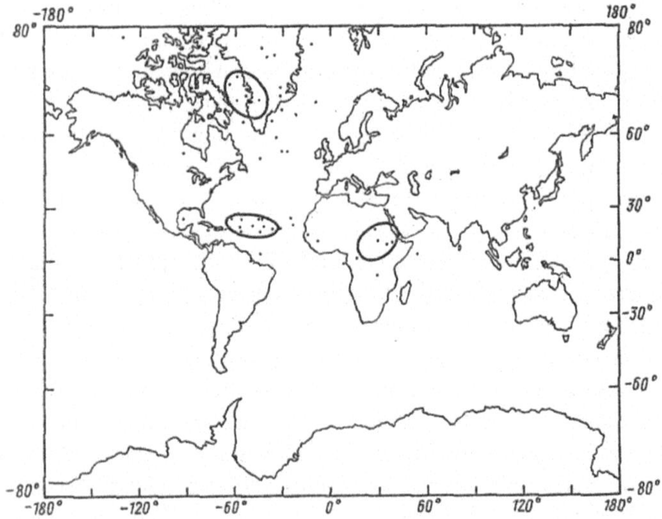

Fig. 1. Distribution of stage poles on the Earth surface. Ellipses show the areas of 95% probability for mean square statistical centers (according [4] with modification)

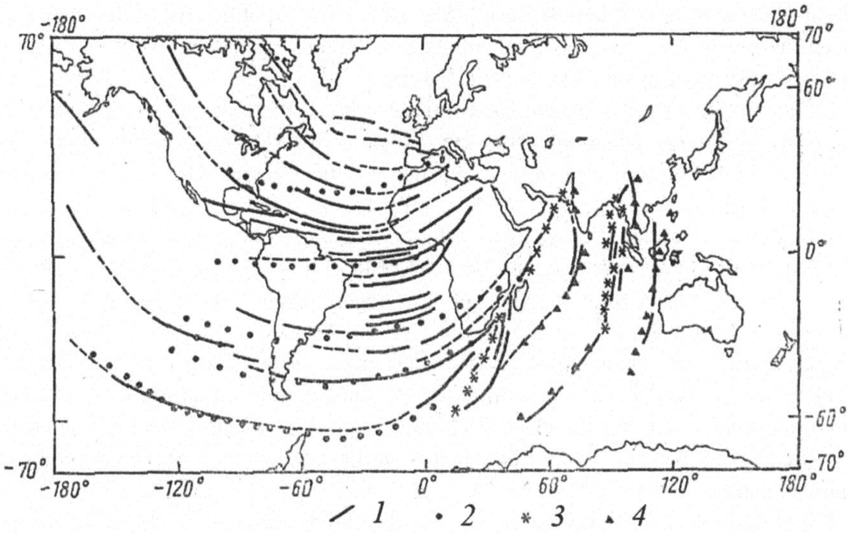

Fig. 2. Comparison of the strikes of continental fracture zones and ocean transform faults/1/ with the circle arcs around the centers of the first/2/, second/3/ and third/4/ areas of stage pole distribution (according [4] with modification)

Table 1. Rotation parameters

Number of claster	The coordinates of statistical center		Meansquaie error		Amount of points
	Latitude deg	Longitude deg	Latitude deg	Longitude deg	
1	71.14	−56.57	1319	28.64	37
2	18.43	−46.66	11.43	24.78	19
3	12.71	25.19	13.46	15.71	15

The stage poles for the Pacific and the Cocos plates and also for the Eurasian and the North American plates for the periods from 6 until 21 Ma years belong to none of the areas mentioned above. Such disagreement may be caused by the uncertainties of finite pole determinations.

The first and third distinguished areas of stage pole concentration are attached to the supposed areas of mantle upwelling flows [4] and the centers of the former are nearby the centers of the latter.

The first and third distinguished areas of stage pole concentration are attached to the supposed areas of mantle upwelling flows [4] and the centers of the former are nearby the centers of the latter. These areas are within so-called 'hot fields', which are characterized by intensive intraplate volcanism, high heat flow, positive geoid anomalies and uplift of the asthenosphere surface [28]. The Central Atlantic area of the stage poles concentration is not related to any upwelling area. The time space of the poles existence there is shorter than in the first and third areas.

The stage poles of plate rotation are mostly situated there during the periods of the plate movement reorganizations. The stage poles for the slowly diverging plates, such as African and Antarctic plates, are attached to the Central Atlantic area as well. Comparison of the time intervals of stage pole activity shows that the earliest among them, being active about 150 Ma, belong to the first area. The most ancient stage poles within the second area were active about 120 Ma and within the third area—85 Ma. So, the relative plate rotation, first of all, began around the stage poles of the Norwegian-Greenland area, then—of the Central Atlantic area, and at last—of the African-Arabian area. This supposes that stage poles of considered areas occurred in the plate kinematics through the interval of 30 Ma years. According to [29], the Norwegian-Greenland and the Central Atlantic centers of mantle flows are older than the African-Arabian one and or originated at the stage of initial breakup of the Gondwanaland.

The stage pole distribution suggests that the upper convection system is strongly related to the isometric lower mantle plums. The lower mantle material possibly spills at the asthenosphere level as linear flows, marking the potential spreading zones. Any branch of this system may transfer into active spreading center through formation of the secondary raising flows, which tear the lithosphere and move apart the lithospheric plates. Longitudinal flows below mid-ocean ridges were described by [30, 31] and others. Such phenomenon was supposed to characterize the ridge segments adjacent to hot spots. The obtained results evidence that the longitudinal flows must be typical for the whole mid-ocean ridge system. With increasing distance from hot spots, flows below mid-ocean ridges become weaker, being partially dampened by the great transform faults. Stage poles localization suggests that the longitudinal flow system is mainly controlled (or influenced) by two main lower mantle plums.

The analysis of general strike of transform faults [2, 4, 19, 32] may be used as an independent test of a revealed pattern of stage pole distribution. The greatest transform faults should coincide with circle arcs around statistical centers of the obtained point clusters. In fact, transform faults of the Atlantic equatorial zone, Atlantis, Vema, Saint Helen, Agulhas with their continental continuations, as well as Eltanin, Chilean and Nazca in the Pacific are traced sufficiently well from the center of the Norwegian-Greenland area. General strikes of the Owen fracture zone, its south-western continuation, Mozambique and Prince Edward fracture zones, northern and southern segments of the fracture zones along the Ninetyeast Ridge can be countered from the center of the Central Atlantic area. The Chagos and the Investigator transform faults, the fracture zones in the Pert and the Gascogne Basins, as well as the Saint Paul and the Amsterdam transform faults can be outlined from the center of the African-Arabian area. Not only transform faults, but the Galapagos Rift can also be traced from the center of the Norwegian-Greenland area. This infers that this young spreading center follows the ancient weak zone in the lithosphere. The fact, that fractures surrounding the Ninetyeast Ridge can not be outlined by a single line, may be evidence for the complex nature of this zone that is supported by the existence of the Osborn Knoll near the central part of the Ridge.

It should be noted that only some segments of transform faults and structural lineaments can be described from the center of the second cluster. The exceptions are the Prince Edvard and Mozambique fracture zones related to the slowly spreading at West-Indian Ridge.

4 Conclusions

The obtained results favor the close connection of lithosphere plate kinematics and internal dynamics of the Earth. The coincidence of distinguished areas of stage pole concentration with the areas of intensive volcanic activity, high heat flow, positive geoid anomalies and uplift of the asthenosphere surface indicates that relative plate motion is caused by upwelling flows in the lower mantle. The epicentral zones of such flows should be characterized by isotropy convection of magma under the lithosphere and, consequently, by the absence of horizontal plate displacements. Thus, the stage poles, are localized in such epicentral zones where there is no relative plate motion.

We are, then, observing the superficial manifestation of a complicated mantle convection system. It may be supposed that this system mainly consists of isometric lower mantle plums, asthenosphere flows below mid-ocean ridges and the secondary raising flows along the active spreading centers, which move the lithospheric plates apart. Since the areas of stage pole distribution were in. place for many tens of millions years, the stable "railways" of lithospheric plate movement should exist. Therefore, the greatest transoceanic and transcontinental transform faults should stretch along such stable "railways". Besides great transform faults, the great nontransform fracture zones are also traced from the centers of the areas of stage pole distribution. Fractures of the Great Lakes and the Labrador Trough in the North America, the Brahma and the Kangting fractures in the South-East Asia are the examples of such fracture zones. Their coincidence with the circles around the centers of distinguished clusters may be caused by their connection with mantle zones of weakness and horizontal displacements along them within the continents. Mantle zones of weakness may determine the main directions of lithospheric plate motion and result in localization of stage poles. The investigation of the hierarchy of mantle upwelling flows and zones of weakness in their influence on the plate kinematics is of great interest.

Acknowledgements. The work was done in the framework of the State assignment, project 0149-2019-0005. While methodological issues of combining conjugate isobaths was supported by the Russian Foundation for Basic Research Project № 17-05-00075. Authors are very appreciating to N. Yu. Bocharova for helping in joint computations of stage pole positions in the absolute frame according to plate movement with respect to hot spots according investigations [4, 22, 33–35].

References

1. Le Pichon, K., Francheteau, J., Bonnin, G.: Plate Tectonics, 287 p. Mir, Moscow (1977). [in Russian]
2. Cande, S., Labreque, J., Larson, R., et al.: Magnetic Lineations of the World Ocean Basins. AAPG Map Series, 1 sheet, Tulsa, OK (1989)
3. Weissel, J., Hayes, D., Herron, E.: Plate tectonic synthesis: the displacement between Australia, New Zealand and Antarctica since the Late Cretaceous. Mar. Geol. **25**, 231–277 (1977)
4. Bocharova, NYu., Schreider, A.A.: Peculiarities of lithospheric plate stage pole distribution. Bul. MOIP. Geol. **6**, 20–28 (1992). [in Russian]
5. Schreider, A.A.: Geomagnetic investigations of the Indian ocean, 319 p. Nauka (2001). [in Russian]
6. McKenzie, D., Sclater, J.: Evolution of the Indian ocean since the late cretaceous. Geophys. J. Roy. Soc. London **25**, 437–528 (1971)
7. Norton, J., Sclater, J.: A model of the evolution of the Indian Ocean and breakup of Gondwanaland. J. Geophys. Res. **84**, 6803–6829 (1979)
8. Sclater, J., Fisher, R., Patriat, P., et al.: Eocene recent development of the Southwest Indian Ridge a consequence of the evolution of the Indian Ocean triple junctions. Geophys. J. Roy. Astron. Soc. **64**, 527–604 (1981)
9. Fisher, R., Sclater, J.: Tectonic evolution of the Southwest Indian Ocean since the Mid-Cretaceous: plate motion and stability of the pole of Antarctica/Africa for last 80 Myr. Geoph. J. Roy. Astr. Soc. **73**, 553–576 (1983)
10. Martin, A., Hartnady, C.: Plate tectonic development of the Southwest Indian Ocean: a revised reconstruction of the East Antarctica and Africa. J. Geophys. Res. **91**, 4767–4786 (1986)
11. Lawver, L., Scotese, C.: A revised reconstruction of Gondwana. AGU Geoph. Mon. **40**, 17–23 (1987)
12. Patriat, P.: Reconctitution de revolution du systeme de dorsales de l'Ocean Indien par le metodes de la cinematique des Plaques. Terres Aust.Antract. Fr., Paris, 308 p. (1987)
13. Thompson, R., Prentice, N.J.: An alternative method of calculating finite plate rotation. Phys. Earth Planet. Inter. **48**(1–2), 79–83 (1987)
14. Bocharova, N.Yu., Kononov, M.V.: Numerical methods of determination of stage and finite poles of plate rotation. VINITI, N4128-B88 (1988). [in Russian]
15. Zonenshain, I.P., Lomize, M.G., Riabuhin, A.G.: A practical guide for geotectonic. MSU, 95 p. (1990). [in Russian]
16. Patriat, P., Segoufin, J.: Reconstructions of the Central Indian Ocean. Tectonophysics **155**, 211–234 (1988)
17. Powell, C., Roots, S., Veevers, J.: Prebreakup continental extention in East Gondwanaland and the early opening of the Eastern Indian Ocean. Tectonophysics **155**, 261–283 (1988)
18. Royer, J., Patriat, P., Bergh, H., et al.: Evolution of the Southwest Indian Ridge from the Late Cretaceous (anomaly 34) to Middle Eocene (anomaly 20). Tectonophysics **155**, 235–260 (1988)
19. Minster, J., Jordan, T.: Present day plate motion. J. Geoph. Res. **83**, 5331–5354 (1978)
20. Scotese, C., Gahagan, L., Larson, R.: Plate tectonic reconstructions of the Larson et al. Map of the age of the ocean basins. Tectonophysics. **155**, 27–48 (1985)
21. Schreider, A.A., Zonenshain, L.P., Bocharova, N.Yu., et al: Paleogeodynamic reconstructions of the Indian Ocean for the last 70 mln. years. Geological history of the USSR territory and plate tectonics, Moscow, Nauka, pp. 147–160 (1989)

22. Schreider, A.A.: Formation of the Black sea deep basin: Scientific word, 216 p. (2012). [in Russian]
23. Schreider, A.A.: Marvin spoor separation from Lomonosov ridge in the Arctic ocean. Oceanology **54**, 530–537 (2014). [in Russian]
24. Schreider, A.A.: Peleogeodynamics of the Bransfield strait. Oceanology **55**, 840–849 (2015). [in Russian]
25. Glebovsky, V.Y., Kaminsky, V.D.: Formation of the Eurasia Basin in the Arctic Ocean as inferred from geohistorical analysis of the anomalous magnetic field. Geotectonics. **40**, 263–281 (2006). [in Russian]
26. www.earthbyte.org/Resources/Agegrid/Muller_etal_Agegrid_Rotations_GCubed08.pdf
27. Gradstein, F., Ogg, J., Schmitz, M., et al: The Geologic Timescale, 1139 p. Elsevier, Amsterdam (2012)
28. Zonenshain, L.P., Kuzmin, M.I., Bocharova, NYu.: Hot field tectonics. Tectonophysics **199**, 165–192 (1991)
29. Sorokhtin, O.G.: The Earth global evolution, Moscow, Nauka, 182 p. (1974)
30. Vogt, P.R., Johnson, G.L.: Mantle plumes and transform fractures. EOS Trans. AGU **54**, 239 (1973)
31. Vogt, P.R., Johnson, G.L.: Transform faults and longitudinal flow below the midoceanic ridge. J. Geoph. Res. **80**, 1399–1428 (1975)
32. Neev, D., Hall, J.: A global system of spiraling geosutures. J. Geoph. Res. **82**, 10689–10708 (1982)
33. Morgan, W.: Hotspot tracks and opening of the Atlantic and Indian Oceans. In: Emiliani, C. (ed.) The Sea, vol. 7, pp. 443–487. Wiley Interscience, New York (1981)
34. Dunkan, R.A.: Age progressive volcaniszm in the New England Seamounts and the opening of the Central Atlantic Ocean. J. Geoph. Res. **89**(B12), 9980–9990 (1984)
35. Engebretson, D.C., Cox, A., Gordon, P.G.: Relative motions between oceanic and continental plates in the Pacific Basin. Geol. Soc. Amer. Bull. paper 206, 59 p. (1985)

A Model Approach to Comparing and Selecting the Optimal Equations of State

G. G. Petrik(✉)

Institute for Geothermal Research DSC RAS, Makhachkala, Russia
galina_petrik@mail.ru

Abstract. In connection with the unresolved, all three components of the problem of low-parametric equations of state retain their relevance: obtaining new ones, improving known ones and choosing optimal equations. New analytical and calculated results obtained in the framework of the model of interacting point centers are presented. A procedure for the choice of equations of state in a one-parameter family has been proposed and tested by calculations of the critical isotherms of argon and perfluorocyclobutane. The comparison for the two substances showed that the equation of state selected by the new method turns out to be better than the "simplest and best" proposed by Martin on the basis of volume translation. The mean absolute deviations, in percent, were for Ar (reduced densities from 0.01 to 1.8): 3.3 (our equation of state), 5.25 (Martin), 7.3 (Redlich-Kwong). For C_4F_8 (reduced densities from 0.06 to 1.5): 1.33% is our equation, 2.76—Peng-Robinson, 3.92-Martin, 7.46—Redlich-Kwong. There are no adjustable parameters in the point center model.

A method for comparing equations and determining equations-analogues in different families based on new characteristic data, which reflects the ratio of configurational contributions to pressure determined by the forces of attraction and repulsion, is proposed.

The molecular level information confirms the predicted intervals of the values of the new parameter, which is another physical rationale for the model being developed.

Keywords: Low-parametric equations of state · Molecular model of interacting point centers · Critical isotherms · Equations-analogues

1 Introduction

1.1 The Method of Unified Equations of State

This year marks the 145th anniversary of the van der Waals equation of state (EOS), which became the ancestor of a powerful area of research in thermal physics and related fields of basic science and practical applications. An example is the work related to the systems, the importance of which at the present time cannot be overestimated—to natural hydrocarbons [1]. The main method of mathematical modeling of phase equilibrium of such complex systems from 70s of the last century was the method of unified EOS [2, 3]. A significant role in its development was played by the EOS Redlich-Kwong [4], which revived active interest in the van der Waals EOS. The result

V. I. Karev et al. (Eds.): *Physical and Mathematical Modeling of Earth and Environment Processes (2018)*, SPEES, pp. 196–212, 2019.
https://doi.org/10.1007/978-3-030-11533-3_20

was the emergence of a large number of modifications, which are now referred to as a vdw-type EOS — low-parametric (from two to five parameters) cubic with respect to volume equations.

1.2 Low Parameter EOS

Specifying the number of parameters makes sense, because at the present stage, equations with the number of parameters up to 10–20 are related to similar EOS. And despite the fact that it is an order of magnitude smaller than the number of parameters that characterize modern multi-constant EOS—up to 100–150, these are fitting parameters that do not make sense. At the same time, only the found physical meaning of all the parameters of the EOS and their connection with the parameters of the molecular model underlying the EOS obtained will be able to provide a way out of empiricism into the area of fundamental concepts. This, in our opinion, is especially important and promising work with low-parameter EOS. Although, on the other hand, the significance of cubic EOS is preserved taking into account the specific requirements not only for accuracy, but also for the efficiency of repeatedly performed engineering calculations.

1.3 Three Components of the Problem. Comparison and Selection of EOS in Three Sets of EOS

And because of this—the creation of new, improved known ones and the choice of the optimal EOS among them—still represent the three actual components of the big problem of low-parameter equations. The goal of our work is to present new results obtained in the framework of the molecular-thermodynamic model being developed and related mainly to the problem of comparing and choosing the optimal EOS.

Over the past century and a half, hundreds of independent simple EOS of various shapes have been proposed [1–5]. Obviously, this made the problem of choosing the optimal EOS even sharper and, therefore, comparing them. At this stage, the quality of the EOS is estimated by the results of the calculation of the properties. However, if several EOS give almost the same results, then how to compare the quality of the equations themselves? Is it possible to determine which one is the best? How to do it? For us, it is obvious that the answers must be sought at the molecular level, since in the EOS as a model of the thermodynamic level, the properties of molecules that manifest themselves in their interaction should be reflected. Probably, the comparison would be greatly facilitated if the EOS could be reduced to a single form, and the parameters would have a similar meaning. It is from these positions we analyze here the connections and the possibility of transitions between the three sets of EOS. Two sets are formed by the EOS, which are based on the simplest molecular models—point centers and hard spheres. The third is the family of the EOS by Martin [6], a mathematical model that is not associated with any physical models. At the same time, the assertion that this is the simplest and best of the well-known two-element EOS requires a substantiation at the level of physics.

2 Equations of State van der Waals (vdw)-Type

2.1 The Most Famous EOS, Which Became the Basis of the First Set

Recall the form of the most famous thermal cubic EOS, which became the basis of the considered set. EOS van der Waals (1) (based on the model of hard spheres)

$$P = \frac{RT}{(V-b)} - \frac{a}{V^2}, \tag{1}$$

EOS Redlich—Kwong (2):

$$P = \frac{RT}{(V-b)} - \frac{a}{T^{1/2}V(V+b)}, \tag{2}$$

The empirical modifications of these famous equations (in which both the second and the first contribution can be changed—separately or simultaneously) have formed a set of independent EOS vdw-type, two-membered cubic with respect to the volume of the equation (see reviews [1–3]). In our opinion, a special relationship to the van der Waals model is determined by the fact that it includes not only the EOS, but also the molecular model, which is considered to be the simplest and most physically visual.

The structure of the vdw-type EOS is determined by two contributions associated with repulsion and attraction of rigid spheres

$$P = P(rep) + P(attr), \tag{3}$$

2.2 Problems of EOS vdw-Type

The general form of a vdw-type EOS is absent or not revealed. Of particular interest to us in this study is a large group of EOS, the first contribution of which retains the same shape as in (1) and/or (2). It is quite remarkable that already at the very beginning—when trying to determine its meaning—it is required to make a mutually exclusive choice. It is now generally accepted that the first contribution takes into account the repulsion of hard spheres. At the same time, it is a long-known in the literature EOS with covolume—which takes into account the volume of non-interacting spheres. The second contributions of EOS (3) differ in shape, which creates a new problem. How to compare EOS-modifications that differ in form? As a rule, the classification (and comparison) goes according to the number of parameters—2.3… Of these, the authors of the equations think so—the first two have the same meaning as by van der Waals EOS: b is related to the size of the molecules (according to the same formula van der Waals, $b = 4v_{mol}N_{Av} = 2/3\pi\sigma^3 N_{Av}$), a—“cohesive”—is responsible for the attraction of objects. With such a tight binding already for the third parameter “there is no sense” and this creates a problem of the same name — a series of questions for which no answers have been received. This list opens the question—what does the third parameter of the model lead to in general? As a result, cognitive problems remain unresolved, and the equations, as mentioned above, are compared according to the results of calculations, from which it is impossible to draw a unambiguous conclusion.

2.3 The Main Disadvantage of a vdw-Type EOS and the Possibility of Its Elimination in the New Model

Recall the basic idea of a systematic approach to the problems identified. The properties of molecules are manifested in their interaction, including the ratio of attractive and repulsive forces, which should manifest itself at the thermodynamic level of modeling in the structure, shape, and parameters of the EOS. At the same time, it is precisely the lack of communication with the micro level that is considered at this stage to be the main disadvantage of a vdw-type EOS. In our works, we show [7–9] that this relationship can be revealed after a definite reformulation of the vdw-type EOS and their inclusion in a new physically sound model.

3 EOS Martin Received by Volume Translation

3.1 "All Inclusive" EOS by Martin

In the same 70s, when there was an information boom in the field of simple EOS caused by the appearance of the Redlich-Kwong EOS, Martin [6] proposed a general cubic equation, which he called "all inclusive"

$$P = \frac{RT}{V} - \frac{\alpha}{(V+\beta)(V+\gamma)} + \frac{\delta}{V(V+\beta)(V+\gamma)}$$

α, β, γ, δ are parameters that can be constant values or depend on temperature. In the case of a limitation by a two-member EOS ($\delta = 0$), it was shown that any of the known vdw-type EOSs represent a particular form of this "all-inclusive" EOS. Martin's approach was based on his own idea of volume translation [10]. (Recognized by a number of researchers, formal and diverting from the right direction when constructing EOS [1], it became for many [11, 12] the basis for improving the calculated capabilities of simple EOS. It is noteworthy that the number of works of this kind clearly exceeds today the number of papers in which new EOS are proposed).

3.2 «The Simplest and Best» EOS, Translation on Martin

The result of Martin's many years of research was the conclusion that the simplest and best EOS is an equation in the form proposed by Clausius [13] exactly 100 years before his work:

$$P = \frac{RT}{(V-b)} - \frac{a}{(V+c)^2}, \tag{4}$$

General view of the reduced (relative to the critical parameters P_C, V_C, T_C) two-term three-parameter—*A, B, C*—Martin's EOS after the translation

$$P_r = \frac{T_r}{z_C V_r - B} - \frac{A(T_r)f(T_r)}{(z_C V_r + C)^2}, \tag{5}$$

$z_c = P_C V/RT_C$ is the experimental value of the critical compressibility factor (CCF) of a substance (Martin substantiates such a choice—as opposed to the conventional approach, when it is a parameter/characteristic of EOS); the function determining the dependence on the temperature of the second contribution, $f\ (T_r) = T_r^n$; B is the numerical value of the translation t; $C = 1/8 - B$. When considering a specific EOS, when the parameter $A = 27/64$, the EOS differ in the translation values. (In view of what was said above about translation, it seems to us that the problem of the physical substantiation of this equation deserves attention).

In order to compare the EOS (5) of Martin with others, among which were the Redlich—Kwong, Peng—Robinson, virial, Lee—Edmister, Usdin—MacAulife, the author calculated the *PVT*-properties of a number of substances, including argon, isopentane, ethylene, ammonia and others. It was concluded that the EOS with translation in all cases gives a forecast, if not better, then in no way worse than the known equations considered.

3.3 On the Connection of the Three-Set EOS with Molecular Level and EOS-Analogs

Before proceeding to the next part, we summarize. Investigating the EOS (5), and making a conclusion about its optimality and simplicity, Martin did not affect his connection with molecular concepts. With regard to the vdw-type EOS, when referring to the molecular model underlying the new equation, the authors refer to the van der Waals model, which is considered the most simple and physically visual. However, the simplest realistic molecular-level model is the interacting point centers (IPC). It was on its basis that we obtained physically grounded cubic equations of state [14, 15]. Our task is to identify the links of the EOS of all three sets and show the possibilities of mutual transitions and interchangeability. Solving such a problem can save considerable funds. For example, in order to confirm the adequacy of the proposed EOS, as a rule, calculations of properties are performed on it and the most well-known EOS. Moreover, the greater the number of comparisons, the better. It takes time, finances, and human intelligence. At the same time, if we are able to identify similar in quality EOS-analogs, the comparison circle will sharply narrow, which should give significant savings.

4 EOS Based on a Molecular Model of Interacting Point Centers

4.1 EOS and Its Control Parameter

On the basis of the IPC molecular model (the molecule is modeled by a material point, the repulsion of the PC is rigid, attraction is optimized) the three-parameter EOS (in standard notation for one mole) was obtained [16]—the first of a cluster of three-member equations:

$$P = \frac{RT}{V} + \frac{RTb}{V(V-b)} - \frac{a}{V(V+c)}, \tag{6}$$

The first contribution is the EOS of non-interacting point centers, the other two are configurational, determined by the acting forces, repulsion and attraction of the center. Consequently, the structure of the EOS IPC:

$$P = P(PC, no\text{int}) + \Delta P(rep) + \Delta P(attr), \tag{7}$$

All three parameters *b, c, a* of the Eq. (6) make sense. Two of them are equal to the changes in the volume available for a center caused by the action of attraction and hard repulsion, respectively: $c = -\Delta V_f(attr)$, $b = \Delta V_f(rep)$, the third parameter *a* is associated with *c* and is the result of replacing the *RTc* simplex in the numerator, if there was a complete similarity of the two configuration contributions. The same meaning of the two parameters of the EOS allowed us to introduce a factor comparing the manifestations of the forces of attraction and repulsion [16]

$$\chi = c/b$$

For the case when the parameters *c, b* are constant and, therefore, $\chi = const$ expressions are obtained for all parameters of the above EOS [17] in the form of explicit functions from χ (CCF is a characteristic of the EOS $Z_C = \beta * \alpha$):

$$\beta = \frac{1}{\chi}\left(\sqrt[3]{(1+\chi)} - 1\right), \; \sigma = \left(\sqrt[3]{(1+\chi)} - 1\right), \; Z_C = \frac{\chi}{\sqrt[3]{(\chi+1)}(\chi-1)+2\chi+1}, \tag{8}$$

As a result, the three-parameter EOS of the IPC turns into a one-parameter family of the EOS:

$$P_R = \frac{1}{Z_c(\chi)}\left[\frac{\tau}{V_R} + \frac{\tau\beta(\chi)}{V_R(V_R - \beta(\chi))} - \frac{\alpha(\chi)}{V_R(V_R + \chi\beta(\chi))}\right], \tag{9}$$

This allows us to distinguish and compare the EOS by the value of one parameter χ, the controlling parameter of the thermodynamic level, in which the force ratio is directly manifested.

4.2 General EOS Within the IPC Model

In the case when the condition of constancy was removed from the parameter c, the expression for the parameter in question was generalized as

$$c = b(k_1 + k_2 \frac{b}{V}) \quad \text{or} \quad c = b(k_1 + k_2 b\rho), \tag{10}$$

Here k_1, k_2 are certain numbers. In the IPC model under consideration, one is at least clear: these numbers must be related to the manifestation of intermolecular interaction forces—by virtue of determining the parameters c and b, through manifestations of the forces of attraction and repulsion in relation to the volume available to the interacting centers.

Now the parameter χ should be supplied with an index marking its dependence on volume

$$\chi_v = c/b = \kappa_1 + \kappa_2 b\rho,$$

for the critical point we get: $\chi_C = \kappa_1 + \kappa_2 b/V_C = \kappa_1 + \kappa_2 \beta$.

As a result, we obtain a "structured" general EOS IPC [18]:

$$P = \frac{RT}{V} + \frac{RTb}{V(V-b)} - \frac{a}{V(V + b(k_1 + k_2 b/V))}, \tag{11}$$

We draw attention to the fact that, in the identified form of EOS, the "physics" of the model is emphasized. The study of EOS (11) by conventional methods gives a general cubic equation for the parameter β as a function of two defining numbers

$$\beta^3(k_2 - k_1(k_2 + k_1)) - 3\beta^2(k_1 + k_2) - 3\beta + 1 = 0, \tag{12}$$

Formulas for calculating other parameters of the EOS can also be found in [18].

4.3 On the Analytical Capabilities of the IPC Model

Until recently, it was the analytical capabilities of the IPC model that were of particular interest to us—and we looked for answers to a number of questions asked for a long time, realizing that without solving these problems, we cannot count on moving forward. Many of them managed to answer. Due to lack of space, we cannot allow a detailed description of the search for these answers. We present only a few questions and references to works where the answers were given in the framework of the IPC model. Moreover, according to the method of logical abduction, this fully proves the adequacy of the new approach and the new model. On the meaning of the first contribution. The problem of the third parameter [19]. The correct value of the parameter b in the Redlich – Kwong EOS—why is its value exactly two times smaller than that calculated by the van der Waals formula [20]? Why successful EOS are not determined by a set of physical meaning of the values—CCF Z_C, b, B_C, but by a set of meaningless numbers k_1, k_2 [21]?

It was—if we refer to the obtained formulas (8) of the model—the solution of the direct problem—the search for the values of the parameter χ, its connection with the molecular level and possible calculations of all 4 parameters based on it.

5 Methods of Choice of EOS and Its Approbation

5.1 Methods of Choosing a EOS in the IPC Family

However, quite recently—it was helped by the connection to the analysis of the Martin EOSs—the same set of (8) 4 formulas became the basis of the methodology for choosing a EOS in the IPC family, which, as shown by calculations and comparisons, can claim to be optimal [22]. The sequence is as follows: using the known (reference) value of the CCF of the substance in question, applying formula (8), we find the value of the parameter χ. Based on the χ found, we calculate three parameters of the EOS and this consistent set of values highlights, determines the specific EOS of the IPC family, which will be used to calculate the properties.

5.2 Testing Methods

The described approach was implemented by us recently. Let us consider two examples of testing the method.

Example 1: Calculation of the Critical Argon Isotherm. From the known value of CCF $Z_C = 0.291$, using formula (8), we found that the parameter $\chi = 3.3$. For the remaining parameters, we obtained: $\beta = 0.18973$, $\sigma = 0.62615$, $\alpha = 1.53374$. Given the EOS IPC for the critical isotherm of Ar is

$$P_R = \frac{1}{0.291 V_R}\left[1 + \frac{0.18973}{V_R - 0.18973} - \frac{1.53374}{V_R + 0.62615}\right], \tag{13}$$

The given densities varied from 0.01 to 0.625. Comparison of the calculation results for this IPC EOS (13), for EOS (21) according to Martin's proposed for argon as the best (see Sect. 8.3) and EOS (2) Redlich-Kwong showed the following: the average deviations as a percentage of the EOS, selected by Martin as reference in [6], respectively, amounted to 1.67, 4.3 and 5.78%. The allocated EOS IPC turned out to be "better than the best".

Example 2: Calculation of the Critical Isotherm of Perfluorocyclobutane. C_4F_8, CCF $Z_C = 0.278$. According to the formula (8) found the corresponding value of the parameter $\chi = 4.5$. Selected in the IPC family EOS has the form

$$P_R = \frac{1}{0.278 * V_R}\left[1 + \frac{0.17}{V_R - 0.17} - \frac{1.63626}{V_R + 0.765}\right], \tag{14}$$

In Table 1 shows the results of calculations of the critical isotherm by EOS (14). The average deviation for the EOS of the IPC (14) was 1.33%. For EOS of Martin—3.92; Redlich-Kwong—7.46; Peng-Robinson—2.76%. And again, the EOS IPC turns out to be "better than the best" EOS Martin and better than the Peng-Robinson EOS, which is recognized as the most successful in modeling hydrocarbons. At the same time, in our model, unlike the standard ones, there are no fitting parameters at all. So this is a EOS, rather, can be described as giving a reasonable forecast.

Table 1. Calculation of the critical isotherm C_4F_8 by 4 EOS

V_R	P_R	$1/Z_CV_R$	P(rep)	P(attr)	P(res)	Δ, %	M-n	R-K	P-R
1	*2*	*3*	*4*	*5*	*6*	*7*	*8*	*9*	*10*
15.82	0.205	0.2274	0.01086	0.09866	0.2074	1.18	0.7	1.0	0.14
8.5	0.3545	0.4232	0.02041	0.17660	0.35709	0.72	0.88	1.4	0.04
4.13	0.6390	0.8710	0.04293	0.33427	0.61722	3.4	0.49	0.15	1.9
2.069	0.9166	1.7386	0.08952	0.57737	0.89042	2.8	0.07	0.76	1.1
1.552	0.9799	2.3179	0.1232	0.70681	0.96515	1.5	0.32	0.71	0.27
1.164	1.0007	3.0944	0.17103	0.84826	0.99878	0.18	0.33	0.14	0.03
0.887	1.0007	4.4574	0.26688	1.04088	1.00736	0.02	2.5	5.0	1.3
0.776	1.0083	4.6355	0.28053	1.06182	1.01384	0.55	10.	18.	6.4
0.690	1.0757	5.2336	0.32711	1.12481	1.05876	1.56	20.	40.	14.
Average absolute deviation, %						1.33	3.92	7.46	2.76

6 Transitions Between Different Families of EOS

As has been shown [23], many vdw-type EOSs can be incorporated into the framework of the IPC model. This is especially simple for EOSs, the first contribution of which has the form RT/(V − b). By returning to it the "legitimate" meaning of the equation with a covolume for noninteracting spheres, we are able to go into the model of a point center with a hard repulsion, which corresponds to the following transformation EOSs

$$\frac{RT}{V-b} = \frac{RT}{V} + \frac{RTb}{V(V-b)}$$

The attractive contributions are reduced to the form corresponding to the EOS IPC (11). Information about the parameter χ will be found by the type of EOS. As a result, we obtain a new EOS, in which configuration contributions are clearly highlighted. And the set of independent EOS turns into equations of one IPC family, the difference between them is determined by the difference in the ratio of manifestations of intermolecular forces, which is provided by different values of parameters χ. As a result, a physical basis appears for a direct comparison of the quality of EOSs.

Formally, the same transformations are possible for EOS of Martin. However, we will look at them in detail in another section.

As for the IPC EOS family itself, the EOSs structure itself pushes to introduce another parameter—comparing two configuration contributions to the EOS IPC. We first applied it in [24], where we considered problems of the single compressibility line (Zeno, or Bachinsky lines).

7 New Comparison Parameter and Its Capabilities (Value)

7.1 Comparison of Configuration Contributions

As a characteristic of the EOS, we introduce a value that compares the configurational contributions to the pressure

$$\chi_P = \Delta P(npum)/\Delta P(omm), \tag{15}$$

In the following Table 2, we present the details of the calculations performed for the three EOSs — the values of the configuration contributions are given separately (columns 2, 3—EOS (16), t = 0.06, columns 5, 6—EOS IPC, k_1 = 3, columns 8, 9—EOS (21), t = 0.082, in form IPC-three terms). For the argon critical isotherm, we write Martin's EOS (t = 0.06) in the form of EOS IPC

Table 2. Calculation of the critical argon isotherm for three EOS. Introduction of a new similarity parameter

V_R	$P_R(rep)$	$P_R(attr)$	χ_P	$P_R(rep)$	$P_R(attr)$	χ_P	$P_R(rep)$	$P_R(attr)$	χ_P
1	2	*3*	4	5	*6*	7	*8*	9	*10*
100	0.00206	0.01442	7	0.00217	0.015676	7.25	0.00282	0.01444	5.1
20	0.01041	0.07025	6.8	0.01092	0.076540	7.0	0.01427	0.07133	5
5	0.04297	0.01655	6.2	0.04519	0.281512	6.2	0.05957	0.2729	4.6
2.5	0.08982	0.48839	5.4	0.09466	0.508398	5.3	0.1269	0.51666	4.0
1.25	0.19736	0.83439	4.2	0.20913	0.851557	4.07	0.29078	0.92703	3.2
10/9	0.22765	0.90412	3.97	0.2416	0.920601	3.8	0.33946	1.01584	3.0
1	0.25951	0.96225	3.7	0.27583	0.984456	3.57	0.39196	1.09999	2.8
10/11	0.29307	1.0272	3.5	0.3120	1.04369	3.34	0.44876	1.17937	2.6
10/12	0.32847	1.0815	3.3	0.3503	1.09870	3.14	0.51079	1.25451	2.5
10/14	0.40540	1.1774	2.9	0.4340	1.19816	2.76	0.65079	1.39283	2.1
2/3	0.44731	1.2196	2.73	0.4799	1.24314	2.59	0.73127	1.45646	2.0
10/16	0.4922	1.2589	2.56	0.5289	1.28535	2.43	0.82102	1.51718	1.8
10/18	0.59017	1.3275	2.24	0.6371	1.36247	2.14	1.02937	1.6280	1.6
1/2	0.70093	1.3849	1.97	0.7618	1.43116	1.88	1.28931	1.72688	1.34

$$P_R = \frac{1}{Z_c V_R}\left[1 + \frac{0.06}{(Z_C V_R - 0.06)} - \frac{27/64}{Z_C V_R (1 + 0.065/Z_C V_R)^2}\right], \tag{16}$$

The new parameter χ_p was calculated by the obtained results (see Table 2, columns 4, 7, 10).

Let's compare the intervals of the obtained values for the first two EOSs: they turned out to be very close: 7—3.7—1.9 and 7.25—3.57—1.88. As for the third EOSs, (which, as was shown by the calculation, practically coincides with the Redlich-Kwong EOSs), the intervals of the parameter values are (5.1—2.8—1.34) for the EOSs Martin (21), t = 0.082) and very close to it for the EOS of Redlih-Kwong (4.89—2.67—1.56). The number in the middle of each set is the value of the parameter at the critical point. It turns out that comparing this parameter with several identical density values allows us to determine the degree of closeness of the EOS.

The same ratio of contributions can be estimated differently by calculating it from the general expression of this characteristic, obtained from the type of EOS (9) IPC:

$$\chi_P = \frac{\alpha}{\beta} \frac{\varphi - \beta}{\varphi + \chi_V \beta}, \tag{17}$$

Ultimately, the value of the new characteristic at the critical point is also a function of the control parameter of the IPC model

$$\chi_P = \frac{\alpha}{\beta} \frac{1 - \beta}{1 + \chi_V \beta}$$

Comparison of the data for different EOSs and their coincidence—especially parameters χ_p—shows which EOSs of the two families—IPC and Martin—can be considered analogs. As for the newly introduced comparative parameter, special attention is required on its value at the critical point. Comparing the values of this most important factor, in which the ratio of the forces of attraction and repulsion of objects of the molecular level is manifested, a preliminary conclusion can be made about which EOSs are likely to turn out to be analogs. The solution of such a problem of choice, obviously, will exclude from consideration the deliberately "distant" from the similarity the EOS of different families, for which they also carry out calculations for the purpose of comparing the EOS.

7.2 Molecular Level of the Model. Force Contributions

In this regard, by comparing the individual contributions to the pressure, we give our old result obtained in [25] when simulating at the molecular level. Introducing an inflection point into an explicit description of the intermolecular potential curve (IMPC) [26], we obtained and used the ability to describe a force curve for which this point simultaneously captures the maximum force and estimated the ratio ζ of two

Table 3. The prediction of the ratio of the IM forces at the molecular level

n	11	12	13	17	18	20	22	27	34	48
ζ	1.714	1.857	2	2.57	2.71	3	3.28	4.0	5	7

contributions—from the side of attraction and repulsion for different values of the index n of the model potential Mie-Lennard-Jones (m = 6), considered the most likely for different molecules. These data are presented in tabular form (Table 3).

Comparison of data for "force" contributions at two levels shows their coincidence, which, firstly, is of obvious undoubted interest from the point of view of the molecular-thermodynamic model being built, and, secondly, serves as yet another proof of the IPC model's physical validity.

8 «A Simple» EOS by Martin as a Special Case of the Family of EOSs IPC

8.1 On the Relationship Between the Parameters of the EOS Martin and the Common EOS IPC

Since, on the one hand, many vdw-type EOSs can be included in the IPC model, and at the same time, according to Martin, are special cases of the "all-inclusive" EOSs, it is of interest to find out whether Martin EOS and EOS IPC family can be linked. Referring to the translation two-element EOS. Let's translate the EOSs to the form of a three-member EOSs of the IPC (as in the case of a vdw-type EOSs, when from non-interacting hard spheres pass to a center with a hard repulsion)

$$P = \frac{RT}{V} + \frac{RTb}{V(V-b)} - \frac{a(T)}{(V+d)^2}, \tag{18}$$

Suppose that the parameters of the EOS (18) are related by the relation $\lambda = d/b$. Then the denominator of the EOSs (18) and the equation itself, respectively, take the form:

$$(V+d)^2 = V(V + b(2\lambda + \lambda^2 b/V)),$$
$$P = \frac{RT}{V} + \frac{RTb}{V(V-b)} - \frac{a(T)}{V(V+b(2\lambda+\lambda^2 b/V))}, \tag{19}$$

Comparison of the received EOSs (19) and the general—"structured" IPC (11) allows estimating the value of the control parameter χ, according to the IPC model associated with the manifestation of the forces of attraction and repulsion of objects

$$c/b = \chi = 2\lambda + \lambda^2 b/V$$
$$\chi_C = k_1 + k_2\beta,\ \beta = b/V_C$$

It follows that the parameters of the two models—IPC and Martin—are bound by the conditions:

$$k_1 = 2\lambda, k_2 = \lambda^2 \quad \text{or} \quad k_1^2 - 4k_2 = 0, \tag{20}$$

Here k_i are the generating numbers that form all the parameters of the selected EOS IPC. Thus, the Martin EOSs are included in the IPC EOS family, and each of them is characterized by a certain value of the parameter χ, on which the EOSs is ultimately compared. However, it makes sense to involve for comparison not any EOS IPC, but only those determined by the conditions found.

8.2 Determination of Translation by Volume from an Agreed Set of Parameters Allocated to the EOS IPC

Let us turn to our work [18], where the general cubic equation was studied and the results of parameter calculations were presented in the form of a table. According to the obtained relation, the values of the parameters of the EOS are determined by pairs of generating numbers, which were chosen in the intervals: $k_1(0-5), k_2(-4-5)$, giving

Table 4. Control parameters of the interacting point centers and Martin models

λ	k_1	k_2	χ	Z_C^{at}	$t = B$	Z_C (эксп)
1	2	*3*	*4*	*5*	*6*	*7*
0	0	0	0	0.375	0.125 (=1/8)	
½	1	0.25	1.063	0.3333	0.08	0.291 (аргон)
1	2	1	2.2	0.312	0.062	0.270 (изопентан)
1.5	3	2.25	3.375	0.300	0.05	0.259 (CHF_3)
2	4	4	4.57	0.2916	0.042	0.242 (аммиак)

47 pairs of numbers. There we were limited to half-integer and integer numbers (in most sets). Choose from them pairs that satisfy condition (20) for the parameter λ. They are given in Table 4.

Using the same table from [18], we find for the EOS singled out by each pair of numbers in Table 4, the characteristic values of the EOS IPC – CCF characteristic (given in the 5th column). In the IPC model, thus, a consistent set of parameters of the selected EOS is determined. We define their meaning in the Martin model, where three different CCF values are used — experimental in the EOS and two — Z_C^{bt}, Z_C^{at}-before and after the translation. Since the predicted, determined by the value of the λ CCF parameter, is not equal to ¼, i.e. CCF value before the translation, and is not equal to the experimental CCF value, it represents the CCF value after the translation, i.e. takes place—$Z_C^{at} - 0.25 = $ t. The specific translation values found by this ratio are presented in the same table. An analysis of the predicted characteristics will be linked with the results of Martin. We note immediately that the value λ = 1/2 leads to such a IPC EOS, which by the value of the CCF (0.3333) and the controlling parameter (1.06) almost coincides with the Redlich-Kwong EOS (CCF = 1/3, χ = 1).

We considered one value from Table 4. If we turn to the article by Martin [6], it turns out that other values of the translation from Table 4 was also used by him in the construction of the EOSs: for ammonia t = 0.04, for trifluoromethane t = 0.055 and for isopentane t = 0.064. Empirical formulas are obtained that link CCF substances with the translation value. The values were determined by fitting the calculation results and comparing the parameter introduced by Martin for the EOS, equal to the sum of the CCF and the second virial coefficient at the critical point. Martin found that for different substances this value is almost constant and equal to 0.62.

As an example, consider the equation for the critical argon isotherm (after translation), selected by Martin as the best

$$P_r = \frac{T_r}{z_C V_r - 0.082} - \frac{27/64}{(z_C V_r + 0.043)^2}, \tag{21}$$

In [6], Martin gives the results of the calculation of the reduced pressure for this EOS, selected for reference, and for a number of well-known EOSs. The comparison shows that the best calculations for the EOS (21) coincide with the results for the Redlich-Kwong EOS. Calculation errors coincided in the entire area of density change (the reduced volume varied from 100 to 2/3).

We calculate the values of the parameters λ and χ for the Eq. (21): $\lambda = 0.043/0.082 = 0.524$, $\chi = 2 * 1/2 + 1/4 * 0.082 = 1.02$. The closest integer value of the control parameter $\chi = 1$ and it meets the EOS Redlich-Kwong. How the configuration contributions of these two EOSs, related to the IPC-type of the EOS, are described above (see previous section). From the foregoing it follows that the EOS of Martin and the EOS of Redlich-Kwong are EOS-analogues.

8.3 On the Physical Meaning of Translation by Volume in the IPC Model

When deriving the EOS in the IPC model, we used the concepts of two effective spherical volumes b. One of them, b^h, shows a point center when attraction does not work in the system, and the second, b^{ES}, is the resulting effective volume, which manifests itself a center as a result of repulsion and attraction. That it is included in the resulting EOS (6) IPC. Consequently, the difference between the two effective volumes is the result of the manifestation of attractive forces; it determines the value of the parameter c

$$c = b^h - b^{ES}.$$

The definition of the parameter c gives for the control parameter a relationship with the effective dimensions of the PC

$$\chi = b^h / b^{ES} - 1$$

We write the EOS IPC in the above form, which follows from the found meaning of the parameter c

$$P_r = \frac{T_r}{z_C V_r}\left[1 + \frac{\beta^{ES}}{V_r - \beta^{ES}} - \frac{\alpha}{V_r + \beta^h - \beta^{ES}}\right], \tag{22}$$

We will give the translated EOS Martin in the form of a three-term EOS IPC

$$P_r = \frac{T_r}{z_C V_r}\left[1 + \frac{B}{V_r - B} - \frac{A}{(V_r + 1/8 - B)^2}\right], \tag{23}$$

What is the difference between two EOS?

In EOS (22) IPC, all parameters are defined by one quantity, in which IIM forces manifest themselves, forming a consistent set. There are no fitting parameters; $\beta = b/V_C$.

In EOS (23), Martin—parameters: A = 27/64 (this is the reduced value of the parameter a of the van der Waals EOS, $a = (27/64)\, R^2T_C^2/P_C$) does not change, B shows translation, $B = tP_C/RT_C$ is a fitting parameter, 1/8 is the given value of the parameter b, $1/8 = b^W P_C/RT_C$.

Since we have shown that the Martins EOS are a certain sample from a family of EOSs obtained on the basis of the IPC model, the parameters of this EOSs should have a similar meaning. This means that B and β^{ES} have the same meaning—one and the same value, but expressed in different units.

Also, 1/8 and β^h are the effective size of an object when the forces of attraction do not manifest themselves (recall that van der Waals, making an assumption about the forces of attraction, considered them so weak that they do not affect the translation of objects, and this is in the IPC-model is equivalent to the fact that the parameter b EOS is determined only by repulsive forces, and the parameter $c = 0$).

Consequently, the Martin's EOS, reformulated to the type of the IPC EOS, is a physically based EOS (molecular model of point centers), which determines the proportion of its success.

In our opinion, the success of the selected IPC EOS compared to the "best" EOSs of Martin's, is determined by the fact that in the more general case of the IPC we have a consistent set of parameters and each value of the control parameter χ found from the reference value of the CCF substance corresponds values α, β, σ. While in Martin's EOS, two of the three parameters remain constant and only the translation changes, which leads to a coarsening of the result.

9 Conclusion

Conducted research and new results obtained in the framework of a simple molecular thermodynamic model related to the problem of choosing the optimal state-analog equations of different families show the need for further work. The next stage can become application on the basis of a new technique of more realistic the EOS in which the configuration contribution by repulsion determined is also optimized—just as it is made in known Carnahan-Starling EOS.

References

1. Batalin, O.Yu., Brusilovsky, A.I., Zakharov, M.Yu.: Phase Equilibria in Natural Hydrocarbon Systems, 272 p. Nedra (1992). [in Russian]
2. Wales, S.: Phase Equilibria in Chemical Technology: In 2 parts. Mir (1989). [in Russian]
3. Anderko, A.: Equation of state methods for the modeling of phase equilibria. Fluid Phase Equilib. **61**, 145–225 (1990)
4. Redlich, O., Kwong, J.N.S.: On the Thermodynamics of Solutions. V. An Equation of state. Fugacities of gaseous solutions. Chem. Rev. **44**, 233–244 (1949)
5. Vukalovich, M.P., Novikov, I.I.: The Equation of State of Real Gases. Energoizdat, Moscow (1948). [in Russian]
6. Martin, J.J.: Cubic equation of state-which? Ind. Eng. Chem. Fundam. **18**(2), 81–97 (1979)
7. Petrik, G.G.: Problems of low-parametrical equations of state. J.Phys.: Conf. Ser. **891**, 012328 (2017)
8. Petrik, G.G.: On the systematic approach to the search for an adequate equation of state and the first non-standard results. Protsessy v geosredah, no. 3, pp. 255–266 (2016). [in Russian]
9. Petrik, G.G., Gadzhiyeva, Z.R.: The one-parameter family of equations of state based on the point center model and its connection to the one-parameter law of the corresponding states. Monitoring. Sci. Technol. **1**, 67–78 (2010)
10. Martin, J.J.: Equations of state. Ind. Eng. Chem. **59**(12), 34 (1967)
11. Lin, H., Duan, Y.-Y., Zhang, T., Huang, Z.-M.: Volumetric property improvement for the Soave-Redlich-Kwong equation of state. Ind. Eng. Chem. Res. **45**, 1829–1839 (2006)
12. Hakayati, J., Roosta, A., Javanmardi, J.: Volumetric properties of supercritical carbon dioxide from volume-translated and modified Peng-Robinson equation of state. Corean J. Chem. Eng. **33**(11), 3231–3244 (2016). 10.1007/s11814-016-0176-5
13. Clausius, R.: Ann. Phys. Chem. **IX**, 337 (1881)
14. Petrik, G.G., Gadzhiyeva, Z.R.: In search of adequate models. On the new approach to obtaining thermal equations of state and its capabilities. Vestnik DSC of RAS. **27**, 5–12 (2007). [in Russian]
15. Petrik, G.G.: The equation of state of van der Waals and its modifications in the model of interacting point centers. Interuniversity collection. Scientific papers "Physico-chemical aspects of the study of clusters, nanostructures and nanomaterials." Tver State University, no. 4 (2014). [in Russian]
16. Petrik, G.G.: On a new approach to obtaining physically sound equations of state. 1. Model of interacting point centers. Monitoring. Nauka i technologii, no. 1, pp. 45–61 (2009). [in Russian]
17. Petrik, G.G.: On a new approach to obtaining physically sound equations of state. 3. The search for the optimal form of the repulsive contribution. Monitoring. Sci. Technol. **3**, 84–97 (2010). [in English]
18. Petrik, G.G. O:n a new approach to obtaining of physically substantiated equation of the state. 2. A search of optimal functional form of attractive term. Monitoring. Nauka i technologii, no. 2, pp. 79–92 (2010). [in Russian]
19. Petrik, G.G.: To the possible solution of «the problem of the third parameter» for low-parametric equations of state. Monitoring. Nauka i technologii, no. 2, pp. 70–81 (2017). [in Russian]
20. Petrik, G.G.: A new thermal equation of state—a physically based mathematical model. In: Proceedings of the II International School of Young Scientists "Physical and mathematical modeling of processes in geomedian", IPM RAS, Moscow, 19–21 October, pp. 130–134 (2016)

21. Vera, J.H., Huron, M.J., Vidal, J.: On the flexibility and limitations of cubic equations of state. Chem. Eng. Commun. **26**, 311–318 (1984)
22. Petrik, G.G.: On some new analytical and design capabilities of a simple molecular thermodynamic model. Monitoring. Sci. Technol. **3**, 55–64 (2018). https://doi.org/10.25714/MNT.2018.36.007. [in English]
23. Petrik, G.G.: On low parameter EOS—unnoticed anniversary, missed and realized opportunities. Bulletin of Novgorod State University Yaroslav the Wise **5**(103), 36–43 (2017). [in Russian]
24. Petrik, G.G.: Boyle—Bachinsky curve and its parameters in the model of interacting point centers. Monitoring. Sci. Technol. **1**, 87–98 (2011). [in English]
25. Petrik, G.G.: On the hierarchy of bonds of the critical molar volume of substances with effective intrinsic volumes of molecules at specific points of intermolecular curves. Monitoring. Sci. Technol. **4**, 80–92 (2012). [in English]
26. Petrik, G.G., Todorovsky, B.E.: Потенциал сферических оболочек. Общие соотношения между параметрами потенциалов взаимодействия свободных и связанных атомов. Jurnal fizicheskoi himii. **62**(12), 3257–3263 (1988). [in Russian]

Nuclear Reactions in Living Nature: The Possibility of Biological Processing and Deactivation of Liquid Radioactive Waste

A. A. Kornilova[1], V. I. Vysotskii[2], S. N. Gaydamaka[1(✉)], and M. A. Gladchenko[1]

[1] Lomonosov Moscow State University, Leninskye gory str. 1, 119991 Moscow, Russia
s.gaidamaka@gmail.com
[2] Shevchenko Kiev National University, Vladimirskaya str. 64, Kiev 01601, Ukraine

Abstract. The extent of contamination of the biosphere by industrial radionuclides requires the development of new deactivation methods. We have shown that one of the most optimal methods is related to the biotechnology of cleaning of territories and radioactive water, which consists in stimulating the process of nuclear transmutation of hazardous chemical elements and radioisotopes at low energy into stable and nontoxic elements and isotopes. In experiments on the disposal of radioactive cesium synthrophic associations (SA) of several thousand different types of microorganisms capable of withstanding a strong radiative forcing. A hypothesis was put forward and theoretically substantiated that the SA, which is put on the verge of survival by the absence of potassium in the nutrient medium, synthesizes barium nuclei, which is a biochemical analogue of potassium, from cesium nuclei by attaching to them the protons present in the liquid nutrient medium. It is assumed that the mechanism of nuclear transformations (transmutation) in biological systems proceeds according to the laws of quantum mechanics and nuclear physics in nanoscale nonstationary cavities of microbial cells. For protons, nanoscale cavities in growing biological cells are potential wells with dynamically changing walls that form coherent correlated states of quantum particles. Being in these states, protons are able to overcome the Coulomb barrier and enter into a nuclear reaction with cesium nuclei, as a result of which barium nuclei arise, which are required for the realization of biochemical processes in microorganisms.

Keywords: Caesium · Barium · Potassium · Radioactive waste · Syntrophic associations · Coherent correlated states · Nucleosynthesis

1 Introduction

Currently, there are about 400 operating nuclear power plants in 31 countries of the world, more than half of which are located in Europe [1]. During the operation of nuclear reactors, a very large number of hazardous radioactive wastes are generated.

V. I. Karev et al. (Eds.): *Physical and Mathematical Modeling of Earth and Environment Processes (2018)*, SPEES, pp. 213–230, 2019.
https://doi.org/10.1007/978-3-030-11533-3_21

Radioactive waste will emit different kinds of radiation for tens, hundreds and many thousands years. Therefore, the world is constantly looking for a solution to the problem of the optimal method of deactivating these wastes and the method of their safe storage [2].

Radioactive waste are all radioactive or contaminated radiation materials that are the product of human use of radioactivity and are not in further use. Deactivation is a very complicated and expensive process: water contaminated by radiation should not be released into sewage, contaminated soil must be removed and disposed of. The waste dump must be conserved so that radioactive substances do not enter the environment, and the radiation emanating from them does not reach people and the biosphere;

Before burial in the repository, nuclear waste must first be brought into the appropriate form. The form of pre-conservation depends on the level of radioactivity of the particular waste.

Liquid high-level radioactive waste is poured with molten glass in steel molds and chill molds. If these containers become brittle or even collapse under the action of radiation exposure and high temperature, then environmental contamination may occur. In this case, radioactive gases are able to enter the atmosphere. The greatest threat will arise in the event of radioactive substances entering the groundwater. Such global accidents have repeatedly occurred in violation of strict standards. The possibility of such a catastrophe depends on the geological stability of the burial site - the salt stock, on the groundwater streams surrounding it, on possible changes in the structure of the salt under the influence of radiation and strong heat transfer from radioactive waste, and also in violation of the relevant rules and regulations.

The dumping of radioactive waste into the sea has been stopped all over the world, as the containers used for this purpose provided safety only for several decades [3].

Containers with radioactive waste should be disposed of in places where they cannot be damaged by water, an earthquake, or any other forces. It should be noted that due to the high thermal energy produced during the decomposition of high-level waste, it is necessary to very effectively cool this waste for many decades.

Thus, the disposal of radioactive waste in burial grounds is an intractable global problem. None of the countries that exploit nuclear energy has yet found a safe, durable, and politically acceptable solution to the problem of the destruction of nuclear waste.

Any nuclear power plant has at its disposal only an intermediate repository of spent fuel, which allows temporary storage of nuclear waste.

In addition, at nuclear power plants there are hundreds of thousands of tons of highly active water containing many radioactive isotopes. Such water is formed during the operation of water-cooled nuclear reactors and other types of reactors both due to the long-term storage of fuel elements in water after the end of their operation, and through the use of water as a coolant circulating through the reactor core.

Thus, the problem of utilization has not yet been solved by anyone, and for spent fuel elements the decision to store them for an indefinitely long period of storage after primary processing is currently no alternative.

This article will briefly present the results of a series of studies conducted from the middle of the 90's of the twentieth century to the present, and aimed at the implementation of a nature-like technology for the disposal of liquid radioactive waste.

2 Background and Related Work

2.1 The Possibility of Using Nuclear Fusion for Disposal of Radioactive Waste

All the latest programs for the creation of recycling systems based on the methods of "burning out" spent fuel by intense neutron fluxes are extremely costly, each several tens of billions of dollars [4–7]. The analysis shows that in this case a large number of new radionuclides will be formed.

An alternative solution to this problem, both from an economic and from an environmental point of view, may be "low-temperature nuclear fusion."

It is necessary to note the extremely ambiguous attitude of scientists to such processes. Most "classical" experts in nuclear physics believe that the probability of nuclear transformations, if they are considered by the method of direct approximation of the physics of pair wise nuclear reactions from high energy to low energy, is very small and such effects a priori cannot be of practical importance.

There is another point of view. It is based on a number of fundamental positions of quantum mechanics. Its essence lies in the fact that direct approximation is incompetent, since many processes of nuclear physics with low energy are collective (not paired) and coherent, are influenced by the environment and can proceed with a sufficiently high probability, which makes them quite suitable for practical use. These results, in particular, directly follow from the specificity of the behavior of a particle located in a nonstationary potential well (in particular, in a nonstationary oscillator). In recent years [8–12], it was shown that one of the most effective methods to significantly increase the transparency of a potential barrier at low particle energy is associated with the use of coherent correlated states (CCS) of particles interacting with atoms (nuclei) that form this barrier. The most characteristic feature of CCS is the possibility of forming controlled giant fluctuations of the particle kinetic energy, the amplitude of which can be thousands and millions of times greater than the average (thermal) energy of the particle and far exceed $\delta E \approx 10\ldots50$ keV. In a concentrated form, this effect is reflected in modified uncertainty relations, called Schrödinger-Robertson uncertainty relations

$$\delta p \delta q \geq \hbar/2\sqrt{1 - r_{pq}^2}, \quad \delta E \delta t \geq \hbar/2\sqrt{1 - r_{Et}^2},$$

in which the product of fluctuations of the corresponding dynamic variables (coordinate, momentum, energy, time) is determined by the corresponding correlation coefficients, the value of which is limited by the interval $0 \leq r_{pq}^2 \leq 1, \ 0 \leq r_{Et}^2 \leq 1, \ldots$ [8–12].

In the stationary state in any system $r_{pq}^2, r_{Et}^2 \sim 1$ and these relations take the form of well-known Heiserberg uncertainty relations.

The situation is different in dynamic (non-stationary) systems (including in a living object). It is well known that the growth front of any biological object is never ideally homogeneous—local heterogeneities are always formed, which are leveled and eliminated in the process of growth. Each of these heterogeneities are, in fact, potential nanowells and is similar to a non-stationary oscillator for particles that are localized in it. In the process of dynamically changing the parameters of these wells, the CCS for these particles can be formed with a large value of the correlation coefficient $r_{pq}^2, r_{Et}^2 \rightarrow 1$ and, accordingly, with unlimited increase of impulse and energy fluctuations. Such the fluctuations are sufficient for passing through the potential barrier [8–12].

From this point of view, the growth front zone of any biological object is a set of potential nanoreactors, in each of which a reaction is possible with the participation of both these particles and atomic nuclei forming the "walls" of the dynamic nanowells. Similar processes can occur in the space between cells in the process of division, in the mitochondria, at the entrance to biological membranes, etc.

The effectiveness of using such states is easy to illustrate with a simple example.

Imagine that the considered particle (for example, a proton with a mass m) is in a small-sized nanowell width L in a coherent correlated state with a correlation coefficient $|r_{pq}|$ close to 1. In this case, the fluctuation of the kinetic energy $\delta T = (\delta p)^2/2m$ is determined by the expression

$$\delta T \geq \hbar^2/8m(\delta q)^2(1 - r_{pq}^2) = \hbar^2 G^2/2mL^2$$

In [8–12], it was shown that with a certain mode of compression of the nanowell, the parameter $G \equiv 1/\sqrt{1 - r_{pq}^2}$ can reach a very large value $G \geq (1\ldots3) * 10^4$.

If we take L = 3 Å as an estimate, then in the case of a proton, the fluctuation of its kinetic energy in this reaches very large values $\delta T \geq 16\ldots160$ keV, which are quite sufficient for effective nuclear fusion. It is very important to note that the average temperature of the whole medium can be arbitrarily small (including in the range that corresponds to biological systems).

The correct interpretation and possible use of the experiments discussed below can change the view on the relationship between living nature and the diversity of chemical elements and isotopes existing on Earth. The question concerns the possibility of nuclear transformations in biological systems under "non-reactor" conditions (moderate (e.g. room) temperature, normal (atmospheric) pressure, absence of neutrons and high-energy charged particles). These issues, in particular, are considered and formalized in the patent [13].

The application of the method of nuclear transformation of chemical elements in biological cultures to deactivate long-lived radioactive isotopes requires a serious study of the effect of these elements on microbiological organisms. From the biological point of view, the most dangerous component of spent nuclear fuel in nuclear power plants is the long-lived isotope cesium-137. Consider this element in more detail.

2.2 Accumulation of Cesium by Microorganisms

As is known, cesium (Cs) is a chemical element of the first group of the main subgroup of D.I. Mendeleev Periodic System, called alkali metals, which in addition to cesium include lithium, sodium, potassium and francium. Sequence number is 55, relative atomic mass is 133 a.m.u. [14]. Cesium is a silver-white metal with a density of 1.87 g/cm^3. Cesium atoms are easily excitable due to the small difference in the energies of atomic orbitals, and have low ionization potentials, have good electrical conductivity. Cesium stands out among the elements of the Periodic System with its low ability to complex and an electropositive character [15].

In nature, cesium is found only in the form of the stable Cs-133 isotope. In addition to the natural nuclide, more than 40 artificially synthesized radioactive isotopes with mass numbers from 114 to 145 are known. Isotopes with an atomic mass of less than 133 emit positrons, and with mass numbers above 133, are β-radioactive.

Trace amounts of cesium are present in many living organisms [16]. Many scientific groups have been studying the accumulation of cesium in flora and fauna. A detailed review of the results of research in this area can be found in [17]. Here we briefly present the most important aspects for this study. The danger of radioactive cesium is that it is a biochemical analogue of stable potassium and, in the absence of potassium, replaces it in a living organism. This leads to a sharp increase in the local radiation dose.

The authors [18–20] note the high solubility of cesium salts in water and the physicochemical similarity of cesium with the biogenic element potassium. In addition, cesium tends to remain in a form that is available to biota for several decades after being released into the environment [20, 21].

The chemical similarity of cesium to potassium ions and some other monovalent cations is a key factor determining the high mobility and availability of cesium in biological systems [22]. The atomic and ionic radii of potassium, ammonium, and cesium are sufficiently similar so that less selective transport systems located on the cell membranes of microorganisms can distinguish singly charged ions from each other [23]. Therefore, cesium probably acts as an analogue of potassium or ammonium cations and enters the cell through energy-dependent transport systems [18, 23–26].

The intracellular localization of cesium in a microbial cell is largely affected by its high solubility and weak coordinating abilities. Compared with many widely studied metal ions, the Cs^+ cation is slightly acidic (according to Lewis), that is, an ion having a low charge/radius ratio, and, as a result, a weak polarizing force. The Cs^+ ion weakly binds to ligands, where nitrogen or sulfur atoms act as donors. In this case, most likely, electrostatic interactions take place with groups containing oxygen [24], which are less stable than covalent bonds [24, 27, 28]. Therefore, cesium exhibits a low affinity for the functional groups of the cell surface responsible for the adsorption of metals [28].

It is believed that most of the cesium accumulated by the cells of microorganisms is maintained in soluble form in the cytoplasm or in vacuoles (if they exist) [24]. At the same time, weak coordination ability, demonstrated by cesium ions, first of all, causes its low toxicity against microorganisms compared with other toxic metals [27, 29]. It is believed that the main mechanism of cesium toxicity is associated with the displacement (usually stoichiometric) of intracellular potassium and the inability of cesium ions to replace potassium in cellular metabolism [22, 24, 30]. Note that potassium is an element accumulated by living cells to high intracellular concentrations [25] and is an indispensable element for living organisms [31]. Potassium, accumulating in its free form to high intracellular concentrations, creates the necessary osmotic pressure inside the cells. Potassium can be involved in the regulation of intracellular pH [24], act as a source of energy in the form of a transmembrane potassium gradient [32]. Potassium ions are necessary as a stimulator of enzyme synthesis [32], a catalyst of the enzyme function, in which they act as a cofactor, for example, malate dehydrogenase and pyruvate kinase [33]. Potassium is involved in the stabilization of cellular structures of the microbial cell, such as ribosomes [25], their maintenance in a functionally active state, and the specific binding of t-RNA to ribosomes.

Thus, the penetration of cesium into the cells of microorganisms usually occurs through the energy-dependent transport systems of monovalent cations. Accumulated cesium does not have a strong toxic effect on living organisms, which is probably due to the fact that cesium is a weak chelating agent and, penetrating into the cell, does not cause structural modifications, denaturation, blocking functional groups on biologically important molecules, and impair the integrity of cellular structures. At the same time, the toxicity of cesium correlates with the content of accumulated cesium.

As you can see, the determination of factors influencing the start of a nuclear fusion reaction (in particular, the transition of cesium to barium) by a biological culture includes the study of the mechanisms by which cesium affects living objects, the adaptation of microorganisms to the aggressive effects of the external environment and the determination of optimal conditions for their growth.

In our previous works, various groups of microorganisms that could exist under extreme conditions were considered: at low and high temperatures (psychrophiles and thermophiles), high salinity (halophiles), high pressure (barotolerans), and also at high levels of radiation under aerobic and anaerobic conditions. More than 500 experiments on the growth of microbiological cultures on a nutrient medium, including both stable and radioactive cesium, were carried out. The results of many experiments have shown that the effectiveness of individual types of microorganisms (e.g. E-coli etc.) for nuclear fusion is about 50-100 times lower than the effect of synthrophic associations - communities in which there are more than three thousand different types of microorganisms. This is due to many reasons, among which are self-intoxication processes, low resistance to radiation and much more. Briefly consider the results of some of our previous experiments.

2.3 Experiments with the Radioactive Isotope of Cesium ^{137}Cs and Anaerobic Synthrophic Association

The first experiments on the stimulated utilization of ^{137}Cs were conducted in 2002–2004 based on radioactive isotopes extracted from the reactor at the Chernobyl nuclear power plant [8]. These experiments were preceded by a study of the effect of synthrophic associations on the activity of reactor water extracted from the primary circuit of the reactor at the Kiev Institute for Nuclear Research [34]. In these experiments, we first discovered the process of transmutation and deactivation of the life expectancy of the radioactive isotope ^{140}Ba, which is part of the reactor water.

After testing the initial technology, experiments with ^{137}Cs isotope were performed.

Equal amounts of the concentrated biomass of the anaerobic syntrophic association (anSA) (Fig. 1) and 10 ml of a solution of cesium-137 salt in distilled water were placed in glass cuvettes.

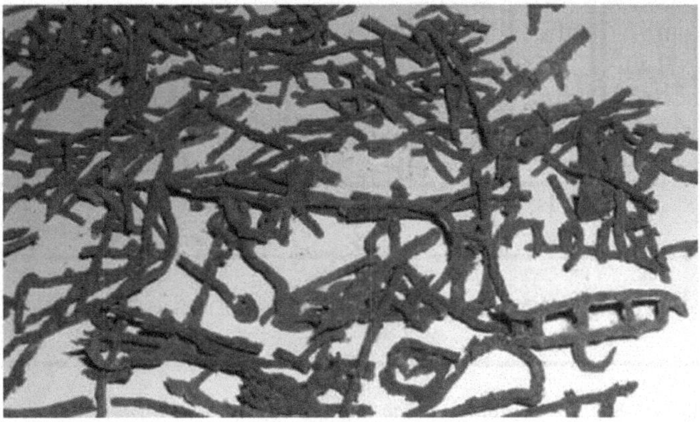

Fig. 1. Appearance of granular anaerobic syntrophic association

The initial gamma activity of the solution was 20,000 becquerels. The research scheme is shown in Fig. 2. The same amount of anSA granules was placed in 7 cuvettes. In 6 cuvettes, purified salts K, Ca, Na, Fe, Mg, and P were added to the active water. These chemical elements are among the vital for any living system. The main purpose of using such additives was to find ways to block possible transmutation channels (if the system contains a specific chemical element from among the vital ones, then assimilation of its biochemical analogue during transmutation becomes unlikely). In addition, such replacements were carried out in order to find the optimal composition of microelements for the rapid growth of microorganisms. The results below confirm the importance of such replacements.

Two additional cells were used for control: one contained active water and anSA (but did not contain additional salts), and the other - only active water.

All cuvettes were closed and kept at a temperature of 20 °C. The amplitude spectrum of the gamma-radiation cuvette was measured every 7 days on the same detector that used a Ge crystal. Special attention was paid to reducing the influence of errors associated with the measurement process. For this purpose, cells with a small height were used, and a detector with a large crystal size Ge. In each measurement, the cells were set at the same position in the center of the detector crystal. The results of the change in the relative activity of the Cs-137 isotope are presented in Fig. 3 and in Table 1.

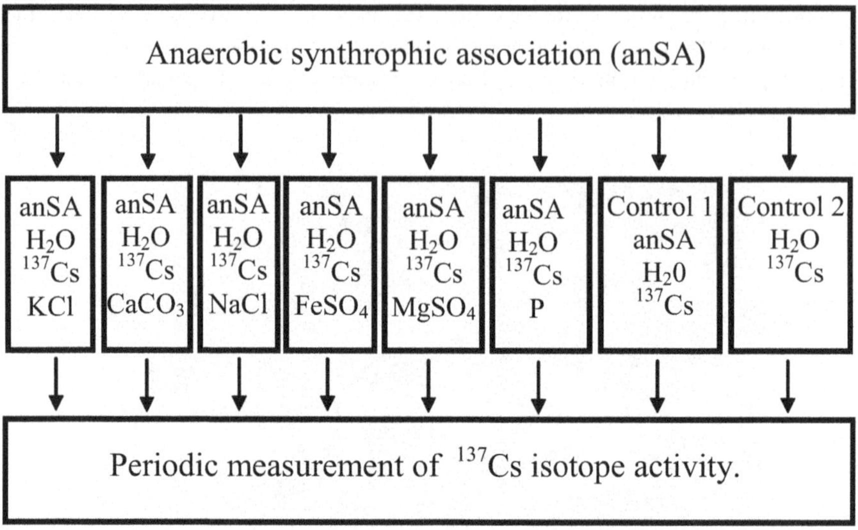

Fig. 2. Diagram of studies of the disposal of active isotopes using anSA

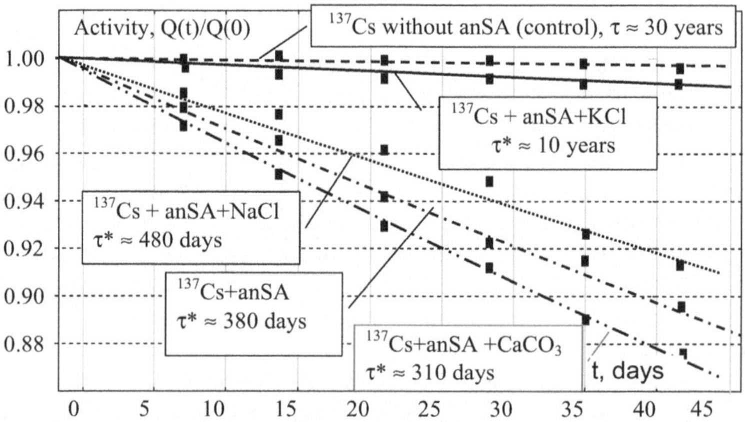

Fig. 3. Kinetics of accelerated utilization (deactivation) of the ^{137}Cs isotope in "biological cells" in the presence of anSA and various chemical elements.

Table 1. Changes in the activity of ^{137}Cs in the "optimal" environment (anSA + active water + $CaCO_3$)

Isotope	Energy, keV	Start experiment. N1, the number of counts in 10^3 s	The end of the experiment (100 days). N_2, the number of counts in 10^3 s	Spontaneous decay	Change in activity
^{137}Cs	661,7	266900	216800	−0,6%	−24%

From the data obtained, it follows that within 100 days of observation, a steady decrease in the activity of the ^{137}Cs isotope was recorded in all cells containing active water and anCA. In the control cell containing only active water, the change in the activity of the ^{137}Cs isotope corresponded to the standard spontaneous decay with a lifetime of about 30 years.

The fastest decrease in activity (it was equivalent to a decrease in the lifetime of ^{137}Cs by 35 times to $\tau^* \approx 310$ days) was observed in a cell containing a salt of calcium. This decrease in activity was not associated with accelerated decay, but was the result of the reaction of utilization of the radioactive isotope ^{137}Cs into a stable isotope of another element.

It is worth emphasizing that the analysis of possible transmutation paths should be based on several fundamental logical premises:

- this process should be energy efficient, i.e. transmutation should be characterized by a positive reaction energy $\Delta E \geq 0$;
- based on the fact that the reaction must be adapted in a certain sense to the biological system, the result of the reaction must be an isotope corresponding to one of the vital elements or its biochemical analogue;
- ceteris paribus, the original isotope of the transmutation reaction should reflect (correlate) the quantitative composition of the medium;
- among the different initial isotopes participating in the reaction, preference should be given to the lightest isotopes, for which the influence of potential barriers should be the smallest among the possible.

Based on these conditions, it is safe to say that the utilization of ^{137}Cs in these experiments was related to the reaction

$$^{137}\mathrm{Cs} + {}^{1}\mathrm{p} = {}^{138}\mathrm{Ba} + \Delta\mathrm{E},$$

flowing with the participation of protons of water. The result of the reaction is a stable isotope 138Ba. The reaction energy is positive and is equal to $\Delta E = 5.58$ meV.

Regarding the "biological feasibility" of this hypothesis, it can be noted that the Ba^{2+} and K^+ ions are biochemical analogues: they have approximately the same ionic radii in the bivalent state ($R_{Ba} \approx 1.4$Å, $R_K \approx 1.33$Å). Since the replaced element (potassium) is one of the vital microelements, the probability of such a replacement appears to be quite large, and barium synthesized ions can replace potassium ions in metabolic processes during crop growth. Such a replacement seems to be more effective than the "direct" replacement of potassium by cesium in the event of potassium deficiency (this is evident from the large difference in the ionic radii of cesium $R_{Cs} \approx 1.65\ldots 1.69$Å and potassium $R_K \approx 1.33$Å). It should be noted that a similar replacement of ions was previously observed and analyzed in experiments with the microbiological culture of Blastocladiella emersonii [35]. In these experiments, a sufficiently effective replacement of K^+ ions with Rb^+ and Ba^+ ions was recorded. These ions can replace each other in the processes associated with ion transport through the membrane into the cell.

Another issue is related to the reason for the increase in utilization efficiency when using an additional calcium salt. Apparently, this effect is associated with the general pattern of metabolism of microbiological cultures: the optimal growth of culture corresponds to the necessary balance of all micro and macro elements. It is possible that exactly calcium deficiency was the "bottleneck" that hampered the growth process and the concomitant transmutation in a specific growing microbiological system.

2.4 Experiments with a Radioactive Isotope ^{137}Cs and Aerobic Syntrophic Association

Further improvement of bio- and nuclear technologies has led to significant progress in the processes under consideration using more optimal biological substances and their growth regimes. The experiments completed in 2016 [36] showed that the utilization process can be accelerated many times using specially prepared aerobic synthrophic associations and an optimized set of macro- and microelements.

The experimental method was similar to the method for the growth of aerobic syntrophic associations in solution with a stable cesium isotope. The activity of ^{137}Cs in experimental and control bioreactors was about 10 kBq. The ambient temperature was about 25 °C. Aeration was carried out using microcompressors, evaporation was compensated by the addition of distilled water. Glucose was used as a substrate for aerobic syntrophic association.

At certain time intervals, gamma-spectroscopic measurements of gamma-radiation with energy of 661.65 keV were performed (was used NaI detector with a diameter of 50 mm).

The results of the measurements are presented in Table 2 and Fig. 4.

Table 2. Average measurement results of ^{137}Cs activity

The duration of the experiment, the day	0	5	12	20
^{137}Cs activity to the initial solution,%	100	82	78	77

In particular, the average (according to parallel experiments) decrease in the concentration of ^{137}Cs for 14 days was 23%, which corresponds to an acceleration of deactivation by more than 200 times. In some cases, a decrease in the concentration of ^{137}Cs in these experiments reached 70% (Fig. 4). These results correspond to an increase in the efficiency of transmutation (as compared to anaerobic associations—Fig. 3) by a factor of 20–30. It should be noted that with appropriate controlled correction of the composition of the medium, a further decrease in activity (and hence the concentration of ^{137}Cs) is possible at the same pace.

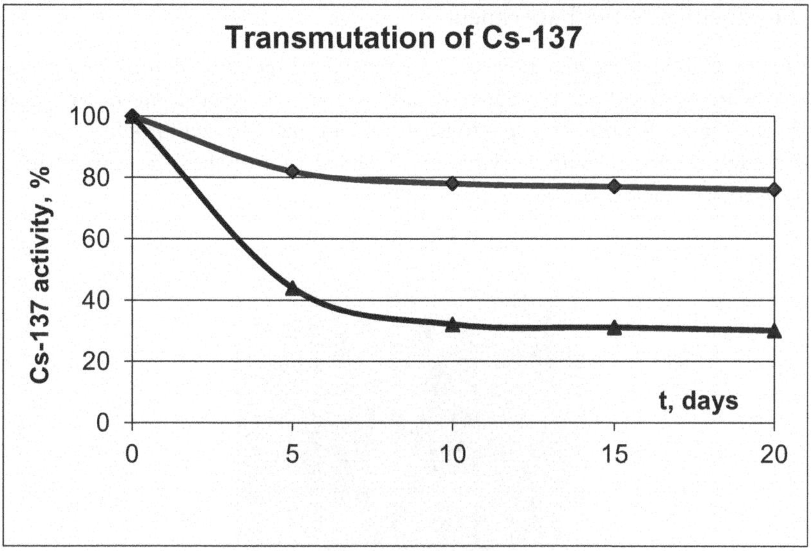

Fig. 4. Reduction of the activity of ^{137}Cs aqueous solution in optimized syntrophic association (the upper graph is the average data of a series of experiments, the lower one is the result of the most optimal series of experiments).

It should also be noted that in experiments on the study of the biosorption of cesium on aerobic and anaerobic SA, it was possible to achieve a decrease in the concentration of ^{137}Cs by more than 30 times over a period of 1 to 14 days. At the same time, the SA activity by the end of the experiment ranged from 10 kBq to 180 kBq per gram of absolutely dry biomass.

3 Experiments with a Stable Cesium Isotope 133Cs and an Aerobic Syntrophic Association

In experiments with ^{137}Cs, it was relatively easy to control changes in the content of radioactive cesium in the dynamics, but it is very important to prove the occurrence of barium in the structure of biomass. Such an analysis requires mass spectrometry studies of SA samples. In addition, the actual content of radioactive cesium in the biomass was estimated in nanograms per sample of the sample being analyzed, which made it impossible to carry out such measurements in situ on radioactive cesium. Therefore, to study changes in the content of cesium and barium in the process of transmutation, it was decided to use the stable isotope ^{133}Cs. Below are the most interesting results from a series of recent experiments with a stable cesium isotope and an aerobic syntrophic association.

3.1 Formulation of the Experiment

Experiments with stable cesium did not require special radiation control and personnel access to work with radioactive elements and were carried out in the microbiological laboratory. Figure 5 shows a photograph of an aerobic synthrophic association obtained from a special natural source and adapted to growth on a nutrient medium containing heavy metals (in particular, cesium).

Fig. 5. Lyophilized aerobic syntrophic association.

The preparation of the synthrophic association for the experiment consisted in the maximum possible removal from its composition of chemical elements that could potentially interfere with the performance of the target reaction. To transfer cesium to barium, you need to choose the right composition and maintain the necessary concentration of macro- and micronutrients in a nutrient medium. Among other things, a very important condition of the experiment was the fact that nuclear fusion occurs only with the active growth of culture.

The essence of the experiment was as follows. In bioreactors with a nutrient medium containing a salt of cesium ($CsNO_3$) introduced a wet centrifuged biomass of SA. Aeration and mixing was carried out by air using compressors. The ambient temperature was 20–23 °C. Glucose was used as a substrate. The duration of the experiment was 43 days. During the experiment, the liquid from the bioreactors evaporated, the initial volume was maintained by the periodic addition of deionized water to the bioreactor. In order to assess the influence of culture growth on the effect of the transition of cesium to barium, a control experiment was carried out, which consisted in the fact that the synthrophic association practically did not grow. In this case, there should be no change in the content of chemical elements. To this end, in a series of control experiments, the aeration was halved and the nutrition was reduced three times from the optimal level, while the biomass was not mixed. The duration of the control experiment was halved to prevent the death of aerobic microorganisms in the syntrophic association and not to increase the population of anaerobic bacteria or fungi.

During the experiment, the parameters important for the vital activity of the aerobic syntrophic association were monitored: pH of the medium, the content of organic matter, mobile forms of nitrogen and phosphorus. In the course of current analyzes, no more than 5% of the original medium was lost.

4 Results and Discussions

By the end of cultivation, the increase in biomass in experimental bioreactors was 3–5 times, while in control bioreactors no increase in the weight of SA was recorded. At the end of the experiment, samples of the original nutrient medium containing the cesium salt, the initial biomass washed from the accompanying elements, the supernatant and the biomass from the experimental and control bioreactors were analyzed on an inductively coupled plasma mass spectrometer in an independent laboratory for the presence of Cs, Ba, K. Error mass spectrometry measurements were 10%.

According to the results of analyzes presented by the analytical laboratory, the average content of Cs, Ba, K in the control and experimental bioreactors was calculated, and the balance of the elements was estimated (Tables 3 and 4).

Table 3. Balance of average content of elements in control bioreactors, mg

Element	0 days	20 days	Decrease/Increase
Cs	8,7825	8,8328	Without changes
Ba	0,0095	0,0093	Without changes
K	0,6099	0,5639	Without changes

Table 4. Balance of the average content of elements in experimental bioreactors, mg

Element	0 days	43 days	Decrease/Increase
Cs	8,7825	6,2270	Decrease on 29%
Ba	0,0095	0,0152	Increase on 60%
K	0,6099	1,0008	Increase on 65%

Mass spectrometry analysis confirmed that the original nutrient medium contained K and Ba only in trace amounts, but potassium and barium were contained in the biomass cells.

At the end of the experiment, a significant level of cesium biosorption by synthrophic microorganisms was noted, which is consistent with the work of other researchers [37–39]. The average content of cesium in the cells ranged from 20 to 40 mg per gram of absolutely dry biomass.

The calculation of the balance by elements showed that the content of cesium, barium and potassium in the control bioreactors did not change by the end of the experiment. In the experimental bioreactors, a decrease in the cesium content by 29% was observed, but an increase in barium by 60% and potassium by 65% was observed.

The results obtained require further detailed study; nevertheless, the fact of a significant change in the elemental composition in experimental bioreactors convincingly indicates the presence of nuclear transformations in the course of the growth of living culture.

Thus, the materials presented by us substantiate the reality of biological technology for the processing of radioactive waste, which consists of the following.

The waste is placed in a special bioreactor, to which microbiological syntrophic associations, a set of macro/micro elements and a substrate for SA growth are then added. Further two processes can be used: biosorption and nuclear fusion.

Biosorption

The applied SA selectively extracts and concentrates the specified radionuclides from the liquid phase. SA significantly reduces the volume of liquid radioactive waste (LRW) by transferring the bulk of radioactive elements into the SA phase. At the final stage of the process, it is expected to obtain safe "technical" water and a concentrate of microorganisms with radioactive elements accumulated in them. Extracted biomass of SA with a humidity of about 98–99% is subjected to soft ashing with a multiple decrease in its volume for disposal. If there is no possibility of burial, the recovered biomass of SA continues to grow under controlled conditions of nuclear fusion (in particular, the transition of cesium to barium).

Nuclear fusion

The applied SA, while observing certain conditions of its growth, contributes to the implementation of low-temperature nuclear fusion, the purpose of which is to convert radioactive elements into stable elements. In the case of a successful operation, a multiple decrease in the content of radioactive isotopes (in particular, cesium) is achieved.

5 Conclusion

In experiments with stable and radioactive cesium in the presence of microbiological syntrophic associations, it has been convincingly shown that when all the necessary conditions are fulfilled and if the appropriate components of the nutrient medium are present, low-temperature nuclear fusion reactions can be performed (in particular, by the transmutation of cesium to barium). The implementation of the nuclear fusion process requires the fulfillment of a number of mutually balanced conditions, the deviation from which drastically slows down the synthesis process or makes it impossible.

To achieve the effect of biosorption and nuclear transmutation, it is possible to use both individual types of microorganisms and synthrophic associations. Obviously, under conditions of radioactivity, the use of synthrophic associations is more likely to succeed. At the same time, in the reaction system there must be a certain set of elements, and microorganisms must grow using different groups of elements in a certain order and under certain conditions of the external environment. It is difficult to manage such synthrophic associations, many preliminary experiments are needed to identify the general patterns inherent in a particular type of association (especially when working with radioactive substances, where monitoring the composition of the medium and its characteristics is rather difficult).

The results presented in the article testify to the reality of the transmutation of cesium isotopes into barium isotopes accompanying the growth of microbiological cultures. However, the biochemical aspects of the observed reactions have yet to be understood.

Some stereotypes are firmly fixed in the consciousness of modern society, among which the question of the "industrial" nature of nuclear transformations is one of the most stable. Based on previous experience in the creation and operation of large-scale nuclear power, it is difficult to imagine that such processes can take place in a living organism. However, by and large, there is no significant opposition. The whole question is only in the expediency of such transformations and in their very low efficiency caused by the mutual repulsion of the nuclei due to the presence of the Coulomb barrier. At the molecular level, the specificity of the interaction and motion of microparticles is fully described by the laws of quantum mechanics and electrodynamics for both living and nonliving nature. From this point of view there is no difference between them. In non-stationary systems of the potential well type, an interference short-term suppression of such a repulsion is possible due to the formation of large fluctuations of the pulse and kinetic energy in the process of the formation of coherent correlated states. Such processes can be successfully implemented in any systems, if the necessary prerequisites are fulfilled.

These processes are ordinary nuclear reactions, but occurring in growing biological systems and under the catalytic influence of dynamic electric fields accompanying atomic-molecular processes that take place during the growth and development of these systems! In any case, the experiment is crucial. Such a process can be called "non-stationary dimensional nuclear catalysis."

In our understanding, the influence of the biological object itself on the process of nuclear transformations consists in the following circumstances:

- the dynamic spatial structure of a growing bioobject at the molecular and atomic level plays the role of a "spatial catalyst" of a nuclear reaction, ensuring the short-term elimination of the Coulomb barrier of the reaction,
- The specific development of a specific biological object provides selective functions that characterize the effectiveness of a particular reaction.

In conclusion, we can say a few words about the biological specificity of the observed phenomena. The very phenomenon of the transformation of chemical elements and isotopes in biological systems is based primarily on the repeatedly verified condition of the constancy of the elemental composition of microorganisms. From this fact, it follows a heuristic assumption that when some of the required macro or microelements are missing in the system, there are prerequisites that they will be synthesized and fixed in the final form as a result of physically sound nuclear reactions.

Acknowledgement. The work was done in the framework of the the Russian Foundation for Basic Research (grant 18-29-05064) and "ErisCom" company for scientific and technical support in carrying out research.

References

1. The World Nuclear Industry Status Report 2016. http://www.worldnuclearreport.org/IMG/pdf/20160713MSC-WNISR2016-S%26C-LR-EN.pdf
2. Krymskii, V.V., Balakirev, V.F., Plotnikova, N.V.: Deactivation of radioactive waste. In: Smirnov, L.A. (ed.) Publishing center of SUSU, Chelyabinsk, 70 p. (2018)
3. Antipov, S.V., Vysotsky, V.L., Sivintsev, Yu.V.: Risk assessment of radionuclide release for flooded radioactive waste taking into account the corrosive wear of protective barriers. NSI RAS, Moscow, 178 p. (2010)
4. Minato, K., Ikegami, T., Inoue, T.: Recent research and development activities on partitioning and transmutation of radioactive nuclides in Japan. In: Actinide and Fission Product Partitioning and Transmutation. Eight Information Exchange Meeting. OECD PUBLICATIONS, Las-Vegas, pp. 29–47 (2004)
5. Savage, B.: Overview of United States P&T programme. In: Actinide and Fission Product Partitioning and Transmutation. Eight Information Exchange Meeting. OECD PUBLICATIONS, Las-Vegas, pp. 49–56 (2004)
6. Ponomarev, L.I.: Overview of Russia P&T programme. In: Actinide and Fission Product Partitioning and Transmutation. Eight Information Exchange Meeting. OECD PUBLICATIONS, Las-Vegas, pp. 79–90 (2004)
7. Stanculescu, A.: IAEA activities in the area of partitioning and transmutation. In: Actinide and Fission Product Partitioning and Transmutation. Eight Information Exchange Meeting. OECD PUBLICATIONS, Las-Vegas, pp. 101–108 (2004)
8. Vysotskii, V.I., Adamenko, S.V., Vysotskyy, M.V.: Acceleration of low energy nuclear reactions by formation of correlated states of interacting particles in dynamical systems. Ann. Nucl. Energy **62**, 618–625 (2013). https://doi.org/10.1016/j.anucene.2013.02.021
9. Vysotskii, V.I., Kornilova, A.A.: Transmutation of stable isotopes and deactivation of radioactive waste in growing biological systems. Ann. Nucl. Energy **62**, 626–633 (2013). https://doi.org/10.1016/j.anucene.2013.02.008
10. Vysotskii, V.I., Vysotskyy, M.V.: Coherent correlated states of interacting particles – the possible key to paradoxes and features of LENR. Curr. Sci. **108**, 524–530 (2015)
11. Vysotskii, V.I., Vysotskyy, M.V.: The formation of correlated states and optimization of the tunnel effect for low-energy particles with nonmonochromatic and pulsed action on the potential barrier. J. Exp. Theor. Phys. **121**, 559–571 (2015). https://doi.org/10.1134/s1063776115100222
12. Vysotskii, V.I., Vysotskyy, M.V.: Formation of correlated states and tunneling at low energy and controlled pulsed exposure to particles. J. Exp. Theor. Phys. **125**, 195–209 (2017). https://doi.org/10.1134/s106377611707024x
13. Kornilova, A.A., Vysotskii, V.I.: Method for purifying water of radionuclides. Intern. patent WO 2015156698 A1 (2014)
14. Wieser, M.E., et al.: Atomic weights of the elements 2011. IUPAC Technical Report. Pure and Applied Chemistry. 85:1047-1078 (2013)
15. Pokrovskaya, L.I.: Cesium. In: Zefirov, N.S. (ed.) Chemical encyclopedia, vol. 5, pp. 654–657. Great Russian Encyclopedia, Moscow (1999)
16. Ghosh, A., Sharma, A., Talukder, G.: Effect of cesium on cellular systems. Biol. Trance Elem. Res. **38**, 165–203 (1993)
17. Peshkur, T.A.: Accumulation of cesium by actinobacteria of the genus Rhodococcus. Thesis for the degree of Ph.D. (biological sciences). Perm, Russian Federation (2002)

18. Tomioka, N., et al.: Kinetic studies on cesium transport in Rhodococcus erythropolis CS98 and Rhodococcus sp. strain CS402. Biosci. Biotechnol. Biochem. **59**, 2219–2222 (1995). https://doi.org/10.1271/bbb.59.2219
19. Tomioka, N., et al.: Recovery of Cs by a bioaccumulation system using Rhodococcus erythropolis CS98. J. Ferment. Bioeng. **85**, 604–608 (1998). https://doi.org/10.1016/S0922-338X(98)80013-5
20. Gadd, G.M.: Influence of microorganisms on the environmental fate of radionuclides. Endeavor **20**, 150–156 (1996). https://doi.org/10.1016/S0160-9327(96)10021-1
21. Mamikhin, S.T., Tikhomirov, F.A., Scheglov, A.I.: Dynamics of Cs content in forest biogeocenoses subjected to radioactive contamination as a result of the Chernobyl accident. Ecology **2**, 43–49 (1994)
22. Hoptroff, M.J., Thomes, S., Avery, S.V.: Influence of altered plasma membrane fatty acid composition on cesium transport characteristics and toxicity in Saccharomyces cerevisiae. Can. J. Microbiol. **43**, 954–962 (1997). https://doi.org/10.1139/m97-137
23. Bossemeyer, D., Schlosser, A., Bakker, E.P.: Specific cesium transport via the Escherichia coli Kup (TrkD) K + uptake system. J. Bacteriol. **171**, 2219–2221 (1989)
24. Avery, S.V., Codd, G.A., Gadd, G.M.: Salt-stimulation of caesium accumulation in the euryhaline green microalga Chlorella salina: potential relevance to the development of a biological Cs-removal process. J. Gen. Microbiol. **139**, 2239–2244 (1993). https://doi.org/10.1099/00221287-139-9-2239
25. Perkins, J., Gadd, G.M.: Caesium toxicity, accumulation and intracellular localization in yeasts. Mycol. Res. **97**, 717–724 (1993). https://doi.org/10.1016/S0953-7562(09)80153-6
26. Avery, S.V., Codd, G.A., Gadd, G.M.: Caesium transport in the cyanobacterium Anabaena variabilis: kinetics and evidence for uptake via ammonium transport system(s). FEMS Microbiol. Lett. **95**, 253–258 (1992). https://doi.org/10.1111/j.1574-6968.1992.tb05375.x
27. Rodriguez-Navarro, A., Ramos, J.: Dual system for potassium transport in Saccharomyces cerevisia. J. Bacteriol. **159**, 940–945 (1984). doi:002i-9193/84/090940-06$02.00/0
28. Gadd, G.M.: Metals and microorganisms: A problems of definition. FEMS Microbiol. Lett. **100**, 197–204 (1992). https://doi.org/10.1111/j.1574-6968.1992.tb14040.x
29. Howlett, N.G., Avery, S.V.: Induction of lipid peroxidation during heavy metal stress in Saccharomyces cerevisiae and influence of plasma membrane fatty acid instauration. Appl. Environ. Microbiol. **63**, 2971–2976 (1997)
30. Avery, S.V.: Microbial interaction with caesium - implication for biotechnology. J. Chem. Technol. Biotechnol. **62**, 3–16 (1995). https://doi.org/10.1002/jctb.280620102
31. Madigan, M.T., Martinko, J.M., Parker, J.: Brock Biology of Microorganisms, 986 p.. Intematinal Inc., Prentice-Hall (1997)
32. De la Pena, P., et al.: The elecfrochemical proton gradient of Saccharomyces - the role of potassium. Eur. J. Biochem. **143**, 183–191 (1982)
33. Rhodes, P., et al.: The regulatory properties of yeast pyruvate kinase - effect of N H / and K^+ concentrations. Biochem. J. **234**, 705–715 (1986)
34. Vysotskii, V.I., Kornilova, A.A.: Nuclear Transmutation of Stable and Radioactive Isotopes in Biological Systems, 208 p. India, Pentagon Press (2010)
35. Van Brunt, J., Caldwell, J.H., Harold, F.M.: Circulation of potassium across the plasma membrane of Blastocladiella emersonii: K + channel. J. Bacteriol. **150**, 1449–1461 (1982). doi:0021-9193/82/061449-13$02.00/0
36. Vysotskii, V., et al.: Biotransmutation of Cs133 and biodeactivation of Cs137 by aerobic microorganisms of sea ooze. In: 20th International Conference on Condensed Matter Nuclear Science (ICCF-20), Japan, Sendai. Book of Abstracts, p. A91 (2016)
37. Park, D., Yun, Y.-S., Park, J.M.: The past, present and future trends of biosorption. Biotechnol. Bioprocess Eng. **15**, 86–102 (2010). https://doi.org/10.1007/s12257-009-0199-4

38. Sayyadi, S., Ahmady-Asbchin, S., Kamali, K.: Biosorption of Cd(II) and Cs(I) from aqueous solution by live and dead cells of Saccharomyces carlsbergensis PTCC 5051. Environ. Technol. **39**, 450–456 (2018). https://doi.org/10.1080/09593330.2017.1302003
39. Sivaperumal, P., Kamala, K., Rajaram, R.: Adsorption of cesium ion by marine actinobacterium Nocardiopsis sp. 13H and their extracellular polymeric substances (EPS) role in bioremediation. Environ. Sci. Pollut. Res. **25**, 4254–4267 (2018). https://doi.org/10.1007/s11356-017-0818-0

Evolution of a Streaming Potential Near Measuring Borehole at Seismic Event Preparation

Ivan Panteleev[1(✉)] and Valery Gavrilov[2]

[1] Institute of Continuous Media Mechanics, Ural Branch, Russian Academy of Science, Perm 614014, Russia
pia@icmm.ru
[2] Institute of Volcanology and Seismology, Far Eastern Branch, Russian Academy of Sciences, Petropavlovsk-Kamchatskii 683006, Russia

Abstract. In this work we have presented results of numerical simulation of streaming potential and electrokinetic current near the measuring well G-1 during the preparation of a strong earthquake. The problems of evolution of volumetric strain (on the first spatial scale), the change in the fluid filtration rate and evolution of the electrokinetic current (on the second spatial scale) near the measuring well related to the heterogeneity of the shear modulus are solved successively for the specific seismic event. It has been shown that change in the electrokinetic current near the G-1 correlates with the change in the geoacoustic emission parameters recorded in this well at the depth of 1 km.

Keywords: Geoacoustic emission · Electrokinetic effects · Multiscale simulation

1 Introduction

The precursors of electromagnetic nature are currently considered as a separate class of earthquake precursors, primarily due to the difficulty of their direct measurements. Terada [28] was first to note the electrokinetic effect in relation to the preparation of an earthquake. The mathematical statement of the problem on calculating the electrokinetic effects in the Earth's crust was for the first time presented in [21]. The interest in the electrokinetic effect as a precursor of the earthquakes was rekindled by Mizutani and Ishido [17, 18], and a series of the works of Fitterman initiated the calculations of these precursors [6–8].

An electric precursor determined by the electrokinetic effect has the following physics. Deformation of the rock during the earthquake focus formation causes redistribution of pressure in a porous medium in the presence of a fluid. This gives rise to the emergence of the filtration flows and, in a consequence of the electrokinetic effect, the respective electric and magnetic fields are induced. In [5, 27, 29], the electrokinetic effects are the basis for the explanation of the generation of ultra-low-frequency (ULF) magnetic field.

V. I. Karev et al. (Eds.): *Physical and Mathematical Modeling of Earth and Environment Processes (2018)*, SPEES, pp. 231–243, 2019.
https://doi.org/10.1007/978-3-030-11533-3_22

The physical interpretation of anomalies in geoacoustic emission which are recorded in the deep wells in the seismically active regions of the Earth's crust during the strong nearby earthquakes is a separate application domain of the electrokinetic effect. The long-term multi-instrumental combined geophysical measurements conducted in the boreholes of the Petropavlocsk–Kamchatskii geodynamic site established a non-trivial effect of changes in the geoacoustic parameters emission near the borehole. In partition, it was established that the effect of modulation of geoacoustic emission (GAE) intensity by electromagnetic radiation (EMR) becomes more complicated during the sufficiently strong nearby earthquakes. Here, with the approximately constant level of external EMR, the level of the corresponding GAE variations can vary fairly rapidly and significantly [9–11] (Fig. 1). In [10, 11], the authors suggested the probable physical mechanisms of the discussed effects which are based on analysis of the interactions between the filtration and electric fields in the stressed geological medium. Inter alia, it was hypothesized that GAE amplitudes (the envelope) during the relatively strong seismic events are affected by streaming potential variations. This hypothesis relies on the works [7, 19] where the emergence of the electrokinetic current is considered as a short-term precursor of the earthquake and on the studies [13, 14] which demonstrated the direct dependence of the strength of the natural fluid-saturated dielectrics on the amplitudes of zeta-potential and ionic strength of the fluid.

This work is devoted to theoretical verification of this hypothesis and numerical simulation of streaming potential and corresponding electrokinetic current in rock (near the selected measuring well) during the preparation of a single seismic event. An

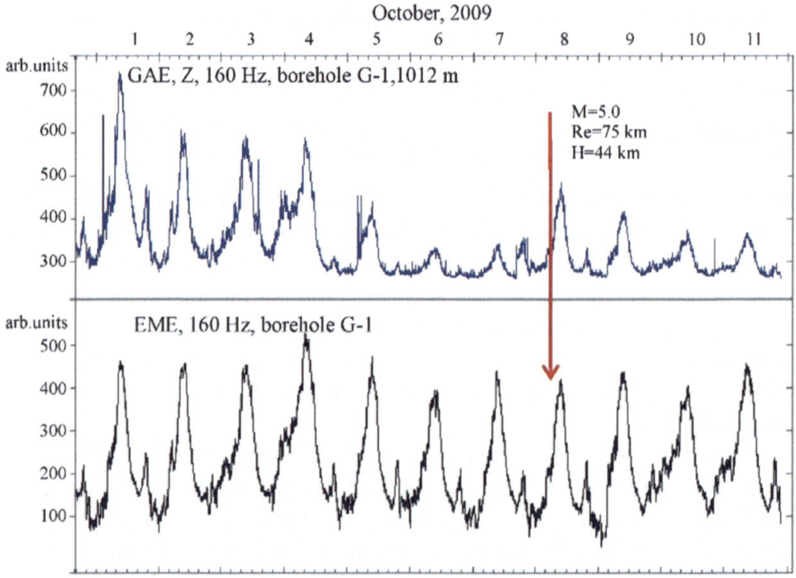

Fig. 1. The example of simultaneous geoacoustic and electromagnetic characteristics during the strong nearby earthquakes (the arrow denotes the time of the earthquake): (a) rms GAE variations at a depth of 1012 m; (b) EMR variations. Measuring methods of GAE and EMR are presented in [10].

important peculiarity of the developed model is a possibility to take into account the existing data on the parameters and structure of the geological medium near the selected measurement well. The measuring wells are located in the Kamchatskii geodynamic site. Approaches that are used for recording of the GAE and EMR are presented in [10].

2 Mathematical Model

The problem of the evolution of geophysical fields near the well has two characteristic spatial scales. The first spatial scale is the scale of perturbation of the stress-strain state induced by the formation of the future earthquake focus. Its characteristic size is hundreds and, in some cases, even thousands of kilometers. The second characteristic scale is the scale of the distortion of volumetric strain near the measuring well induced by its operation and by the initiation of the seismic source. Its characteristic size is hundreds of meters. The fact that the characteristic scales differ by more than an order of magnitude means that the mathematical statement of the problem should be formulated separately for each of the two scales.

2.1 The First Spatial Scale

As of now, the formalized models for the preparation of the tectonic earthquake source which would be suitable for calculation of stress-strain state perturbations of the geomedia during the initiation of the earthquake source are, in fact, absent. Of course, this is due to our incomplete understanding of the acts that take place in the Earth's crust under the preparation of the earthquake. Two pioneering models that describes the formation mechanism of the earthquake source on the physical, qualitative level have appeared in the second half of 20th century. These are the avalanche-unstable fracture model [20] and dilatancy–diffusion model [26]. One of the first formalized model of the seismic event formation was suggested by Dobrovolsky [1, 2]. This is the model, which considers the heterogeneity in elastic properties. The heterogeneity is understood as a domain in the geological medium, which perturbs the aseismic behavior of the tectonic processes and is a direct cause of the emergence and development of the precursors.

The heterogeneity vanishes when the earthquake occurs. One of the probable physical mechanisms of emergence of such heterogeneities in the geological medium is the cohesion of two blocks (structural units of a rock mass). In the field of the applied tectonic stresses, this cohesion leads to the formation of a consolidated domain (a heterogeneity) whose averaged properties differ from the properties of the medium before the cohesion (Reid's concept [23]). Loss of the heterogeneity corresponds to the slip along the boundary between these blocks leading to the generation of seismic waves. Clearly, this understanding of the initiation of a seismic source is very simplified and model; however, it allows direct calculation of the evolution of deformational, filtration, and electromagnetic fields induced by the preparation of a particular seismic event and can be used for estimation of the domain size accommodating the precursors of various nature. Considering these features of the Dobrovolsky's model, we selected it to describe the volumetric strain of the geomedia during earthquake focus formation in our study.

We consider the problem of the perturbation of the volumetric strain in the brittle elastic medium due to the change of a spherical heterogeneity in the elastic properties of the geomedia, which simulates the earthquake source. The calculation domain is a square prism with a 200 × 200 km base and a depth of 40 km (Fig. 2). At the center of symmetry, a sphere with radius R is located at the depth H. We assume that the source of a future earthquake (a spherical heterogeneity) only differs from the hosting rock by

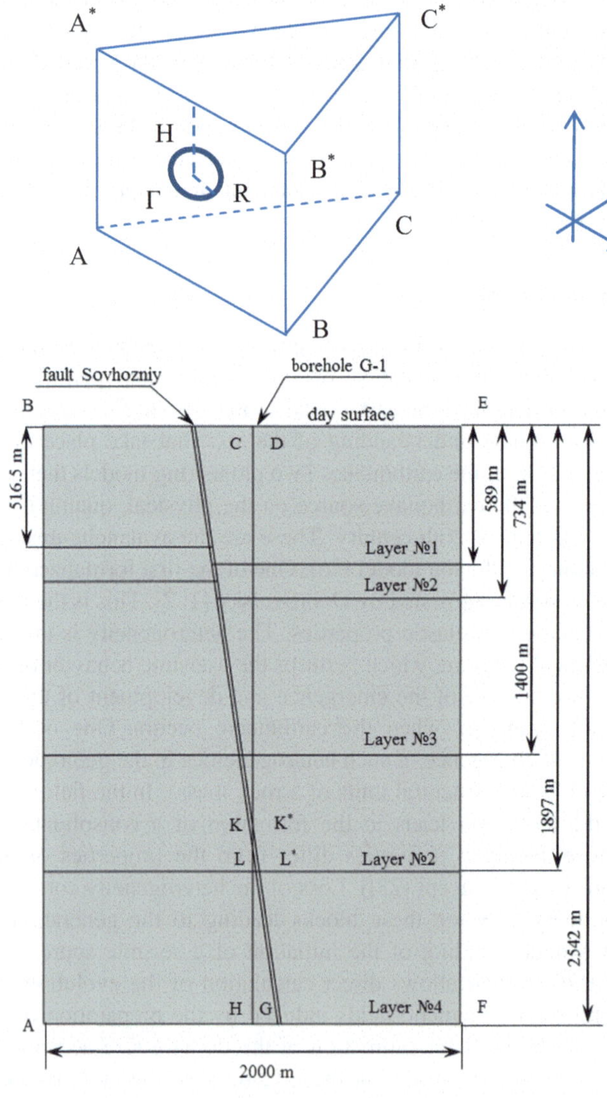

Fig. 2. The calculation domains for the first (top) and second (bottom) spatial levels.

the shear modulus. The preparation domain of the seismic source is specified in the form of a sphere rather than an ellipse because the average distance from the measuring well to the earthquake sources (according to the long-term measurements in wells) is more than 100 km. At this distance, the influence of the shape of the inclusion on the solution is negligible. In accordance with [25], the radius of the heterogeneity is determined by the magnitude of the preparing event:

$$\mathrm{R} = 10^{0.414\mathrm{M}-1.696}. \tag{1}$$

The prism is located in the Earth's gravitational field and is subjected to the pure shearing in the xOy-plane (the shearing stress is 200 MPa).

According to Dobrovolsky's model, preparation and implementation of an earthquake is determined only by the evolution of the physico-mechanical properties of the heterogeneity. The time when the heterogeneity of the elastic properties completely disappears corresponds to the occurrence of the earthquake.

The stress-strain state perturbation in geomedia that is caused by the initiation of heterogeneity in the elastic properties is specified by the well-known solution of the Eshelby problem for a spherical heterogeneity in an elastic medium [4]. The volumetric strain perturbations can be described by the following of equation:

$$\varepsilon = 2(1-2v)Gxy\left(\left[\frac{1}{r_1^5}\right] + \frac{3-4v}{r_2^5} - \frac{10H(z+H)}{r_2^7} + 2R^2\left(\frac{7(z+H)^2}{r_2^9} - \frac{1}{r_2^7}\right)\right), \tag{2}$$

where $G = \frac{\alpha\tau R^3}{2\pi\mu(1-v)}$, $r_1 = \sqrt{x^2+y^2+(z-H)^2}$, $r_2 = \sqrt{x^2+y^2+(z+H)^2}$, α is the relative change of the shear modulus in the heterogeneity, μ is the shear modulus of the medium (in Pa), v is the Poisson ratio, τ are the shear stresses at infinity (in Pa), H is the depth to the center of the heterogeneity in km, R is the radius of the heterogeneity in km, and the term in the square brackets only takes place beyond the heterogeneity.

2.2 The Second Spatial Scale

The model on the second spatial scale considers the evolution of various geophysical fields near the measuring well due to change in the stress-strain state calculated on the first scale level. Thus, the solution obtained at the first scale level is the governing action (control action) for the second-level model. In this work it is assumed that there is no feedback between these two spatial scales.

To numerical simulation of geophysical fields evolution near the operating measuring well was take into account the geological structure of the geomedia near the well [10]. Let us restrict ourselves to a plane case problem.

Based on the existing geological data, we constructed the calculation domain of the second scale level (spatial scale). This domain consists of four layers having different physico-mechanical properties. At an angle of 9.2^0 in the well's axis, there is the Sovkhozny fault, which intersects the well itself in the depth interval of 1700–1800 m. It is assumed that the width of the fault is constant along the entire length and that the diameter of the well also does not vary and is 146 mm. The physical properties of the rocks in each layer are determined by the laboratory studies of the cores. The hydrodynamic survey of the well established the flow rate of the well equal to 0.3 l/s.

The Problem on Fluid Migration in the Region of the G-1 Measuring Well. We describe the pore pressure changes in a fully fluid-saturated porous medium by the set of the stationary states each of which is characterized by the steady filtration regime at a given level of volumetric strains in the medium. According to the Brinkman model, the fluid is considered as incompressible and inertia effects are neglected. In this case, the filtration problem can be described by the following set of equations:

$$\begin{cases} \overline{\nabla} \cdot (\rho_f \bar{u}) = Q_{br} \\ -\overline{\nabla} p + \overline{\nabla} \cdot \left(\frac{\mu}{n}(\nabla \bar{u} + \nabla \bar{u}^T)\right) - \left(\frac{\mu}{k} + \frac{Q_{br}}{n^2}\right)\bar{u} - \rho_f \bar{g} = 0 \end{cases}, \quad \bar{u}|_{t=0} = 0;\ p|_{t=0} = 0$$
$$\begin{cases} p|_{BC,DE} = 0 \\ \bar{u} \cdot \bar{n}|_{CK,LH} = \bar{u} \cdot \bar{n}|_{DK^*,\, L^*G} \\ \bar{n}^T \cdot \left(-p\bar{I} + \frac{\mu}{n}(\nabla \bar{u} + \nabla \bar{u}^T)\right) \cdot \bar{n}|_{AB,AF,EF} = -(\rho_f h|y| + p_k) \\ \bar{n}^T \cdot \left(-p\bar{I} + \frac{\mu}{n}(\nabla \bar{u} + \nabla \bar{u}^T)\right) \cdot \bar{n}|_{KL,K^*L^*} = -\rho_f h|y| \end{cases} \tag{3}$$

where $\bar{u}$ is the Darcy velocity, ρ_f is the fluid density, $n = n_0 \frac{1 + \gamma\varepsilon_{vol}}{1 + \varepsilon_{vol}}$ is the porosity of the medium, n_0 is the initial porosity of the medium, γ is the coefficient of proportionality between the volumetric strains of the medium and the pores, Q_{br} is the mass source of a fluid, p is the pore pressure of a fluid, μ_f is dynamic viscosity of a fluid, p_k is pressure on the external boundary of the drainage area of a well, k is the permeability of the medium and ε_{vol} is the volumetric strain of the medium obtained from the solution of the problem at the first scale level. The expressions for the boundary conditions of the problem on fluid filtration near the G-1 well are presented by the relation (3) (the second bracket). The first boundary condition is the zero pore pressure on the Earth's surface; the second boundary condition is the periodic condition on the walls of the well (equal normal components of vertical velocities); the third boundary condition is a given flow rate that ensures the given pore pressure distribution at the boundaries of the calculation area; and the four boundary condition is the fluid discharge rate specified on the filtration part of the well (at the depths of 1700–1800 m) that ensures the known flow rate and bottomhole pressure.

Distribution of Fluid Salinity in the Region of the G-1 Measuring Well during the Aseismic Periods. The repeated chemical analysis of water in the G-1 well that was carried out immediately after the well's construction and during its operation as a measuring well [15] has shown that water is predominantly of the sodium-chloride type with average salinity of 9 g/l. For modeling of the variations in water salinity in the well according to the change in the regime of water inflow into the well during the preparation of the strong nearby seismic events, we assume the hypothesis that the degree of water salinity in the well is solely determined by the concentration of sodium chloride. In this case, the distribution of the concentration of sodium chloride in the fluid that fills the pore-fracture space of the geological medium near the well can be determined by following equation:

$$\bar{u}\cdot\overline{\nabla}C = \nabla\cdot\left(D\overline{\nabla}C\right)+R,\quad \begin{cases} C\big|_{ABCDEFGH} = C_0(y);\; C\big|_{CK,LH} = \big|_{DK^*,L^*G} \\ -\bar{n}\cdot\overline{N}\big|_{CK,LH} = \bar{n}\cdot\overline{N}\big|_{DK^*,\,L^*G};\; \overline{N} = -D\overline{\nabla}C+\bar{u}C. \\ -\bar{n}\cdot D\overline{\nabla}C\big|_{KL,K^*L^*} = 0 \end{cases} \quad (4)$$

where $\bar{u}$ is the Darcy velocity, C is the concentration of sodium chloride, D is the diffusion coefficient, and R is volumetric source. The diffusion coefficient of sodium chloride in porous medium is determined according to the Millington–Quirk model as $D = \frac{n}{n^{-1/3}}D_f$ where n is the porosity, D_f is the diffusion coefficient of sodium chloride in water (1.23·10–9 m^2/s). Based on the measurements of the degree of water salinity in the G-1 well, we assume the following initial distribution of NaCl concentration over the depth in the calculation area: the concentration is 6 g/l on the surface, 9 g/l at a depth of 1012 m, 11.7 g/l at a depth of 2510 m, and linearly varies on the intervals between these depth levels. For calculation of the sodium chloride concentration we assumed the following boundary conditions (in order of appearance in (4)): the first boundary condition is the given distribution of the sodium chloride concentration which corresponds to the initial distribution; the second and third equalities are the periodicity conditions on the walls of the well (equality of concentrations and equality of the concentration fluxes); the fourth equality is the flux of concentration; and the last equality is the outflow of sodium chloride together with the fluid in the filtration part of the well.

Calculation of the Streaming Potential Induced by the Fluid Filtration. The general form of the equations describing the electrokinetic effects of the first and second kind (the streaming potential and electroosmosis) follows from the approaches of equilibrium electrodynamics, specifically, from the Onsager principle and Curie theorem which make it possible to write out the relation between the thermodynamic forces and fluxes of the different ranks [22]. The system of electrokinetic equations for a porous fluid-saturated medium can be described by the following set of equations:

$$\begin{cases} \bar{j} = -\dfrac{\varepsilon_f\zeta}{F\mu_L}\nabla p(x,y,z,t) - \sigma_f\nabla\varphi(x,y,z,t) \\ \bar{W} = \dfrac{\varepsilon_f\zeta}{F\mu_L}\nabla\varphi(x,y,z,t) - \dfrac{k}{\mu_L}\nabla p(x,y,z,t) \end{cases}, \quad (5)$$

where $\bar{j}$ is the density of electric current, [A m^{-2}], $\bar{W} = m_0(\bar{V}-\bar{U})$ is the Darcy velocity, $\bar{V}$ and $\bar{U}$ are the motion velocities of the fluid and solid phase, [m/s], $\nabla p(x,y,z,t)$ is the pore pressure gradient, [Pa m^{-1}], $\nabla\varphi(x,y,z,t)$ is the gradient of the streaming potential, [V m^{-1}], k is permeability of the rock, [m^{-2}], σ_f is electrical conductivity of the fluid [S m^{-1}], μ_L is dynamic viscosity of the pore fluid, [Pa s], ε_f is the absolute permittivity of a fluid, [F m^{-1}], ζ is zeta potential, [V], $F = m_0^{-n}$ is the porosity parameter in the Dakhnov–Archie equation, n is the degree of cementation of the rock. The first equation describes the emergence of the electric current induced by the charge transfer in the fluid under the action of the applied pressure difference (the emergence of the streaming potential). Osmotic phenomenon arising due to the gradient of the streaming potential (the first term in the right-hand side of the second equation in system (5)) can be neglected because of the smallness of the electric field; however, in this case, the second equation will coincide with the Darcy's law, which has been already used in Eq. (3). The form of the first equation allows us to take into account the conduction current whose value is determined by the strength of the external electric field. The final problem on finding of the electric field (streaming potential *V*) induced by the variation in the pore pressure has the following form [8]:

$$\begin{cases} \bar{j} = -\dfrac{\varepsilon_{\phi}(C)\zeta(C,T)}{F\eta(C,T)}\nabla p - \sigma(C)\nabla V \\ \nabla \cdot \bar{j} = 0, j_z\,|_{z=0} = 0 \end{cases}. \tag{6}$$

One of the most important characteristics of the EDLs on the interfacial boundaries between the mineral skeleton of a rock and the fluid is zeta potential. Most published works devoted to the calculation of the electrokinetic precursors present a qualitative estimation for the coefficient in front of the pressure gradient [3, 7]. However, in our case, qualitative evaluations are insufficient for the comparison with the particular results of the measurements in the wells. The results of the experiments have shown that zeta potential for the specific rocks and fluids very strongly depends on the fluid temperature, degree of salinity, and *pH* [16, 24]. In this work, we have used the theoretical model that has been for the first time proposed in [12] for the calculation of the streaming potential during the preparation of the tectonic earthquake.

3 Evolution of Electrokinetic Current During the Preparation and Implementation of a Earthquake

The stationary solutions obtained for the problems of fluid filtration, distribution of fluid salinity, and emergence of the concomitant streaming potential near the G-1 measuring well will serve as the initial approximation in the modeling of precursory variations in the geophysical fields. For modeling of the streaming potential and corresponding electrokinetic current evolution at formation of a strong nearby earthquake, we selected the event with the magnitude of 5.2 that occurred in October 2009 at a distance of 114 km from the G-1 well [10]. We assumed that the diameter and the depth of the heterogeneity in the elastic properties of the geomedia do not vary with time except for the degree of heterogeneity. We assumed, that evolution of shear modulus in the heterogeneity consist of three intervals. Within the first interval, this parameter is constant, which corresponds to the aseismic period (the duration of the first interval is 4 days). On the second interval, which lasts two days, the parameter grows slowly. In fact, this segment reflects the process of strengthening of heterogeneity's consolidation. During the third interval which also has a duration of two days, the discussed parameter drops in the avalanche-like (power-law) manner to a certain value because of the vanishing of the heterogeneity and partial release of the accumulated elastic energy (Fig. 3). After the earthquake implementation, the permeability of rock increases [26].

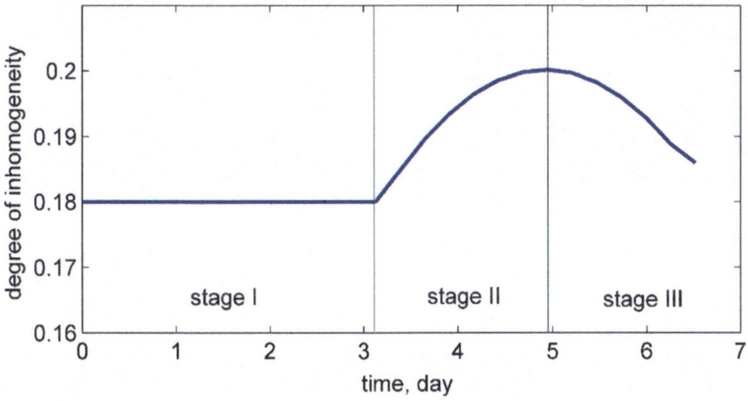

Fig. 3. The assumed evolution of the degree of heterogeneity of elastic properties.

Figure 4 shows evolution of volumetric strain of geomedia which is caused by the evolution of the degree of heterogeneity of the elastic properties in the future earthquake focus. The solution is presented for the vertical profile corresponding to the location of the well. It can be seen that the maximum changes in the volumetric strain is localized within the depth interval of 1200–2000 m.

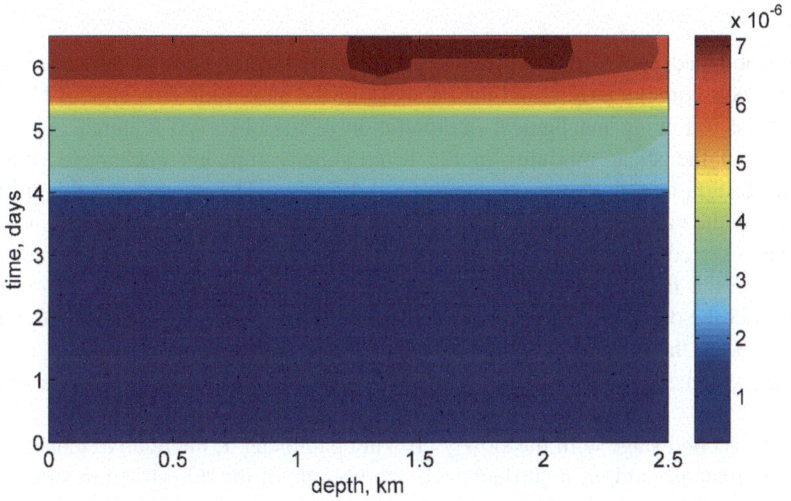

Fig. 4. The change of volumetric strain in the one of the well.

The electrokinetic current averaged over the spherical area with the center at the point of geophone location, which is obtained by successive solution to the problems at the second spatial level (3), (7), (9) is illustrated in Fig. 5. For comparison, also the envelope of the geoacoustic signals recorded in the G-1 well at a depth is 1 km is presented.

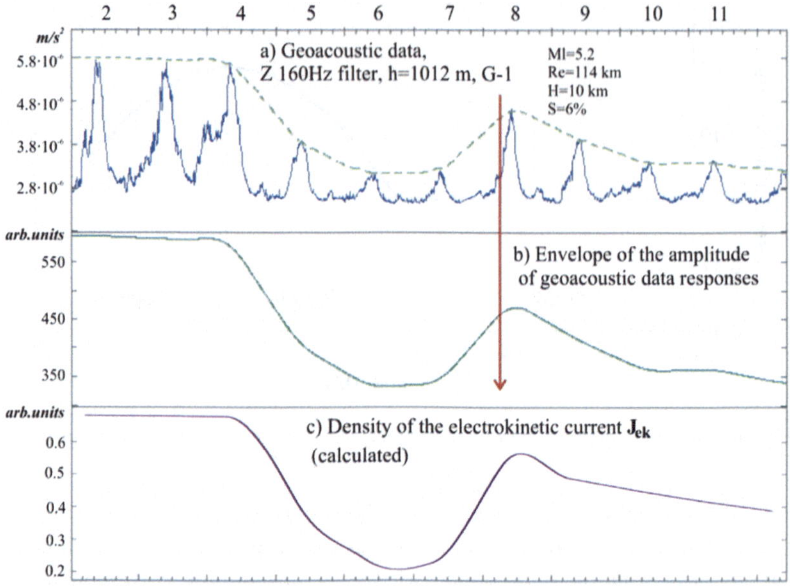

Fig. 5. The density of the electrokinetic current (c) compared to the changes in the envelope of the amplitude of GAE responses (b) during the nearby earthquake (the arrow is the time of earthquake).

The obtained results demonstrate a good qualitative agreement between the calculated electrokinetic current and the behavior of the amplitude envelope of GAE. We suggest the following mechanical interpretation of the observed features. At the first stage, when the shear modulus in the heterogeneity remains unchanged, the pore pressure insignificantly decreases due to fluid diffusion. The intensity of this diffusion depends on the value of the coefficient of piezoconductivity. This dependence is clearly seen when we compare the intensities of the reduction of electrokinetic current at the first and at the last (after the occurrence of the seismic event) stages. According to Fig. 5, after the occurrence of the earthquake, the electrokinetic current decreases much steeper than at the first stage. This dissimilarity is accounted for by the third assumed hypothesis, namely, by the sharp increase of the permeability of the medium near the source.

At the second stage with the slowly growing parameter α, the relative tensile strains gradually increase and, as a consequence, the volume of the pore-fracture space in the medium expands. In turn, this process leads to the drop of the pore pressure in the region of the measuring well and to the decrease of its gradient, which causes the relative reduction in the intensity of the electrokinetic current. The avalanche-like decrease of the parameter α (the third interval) results in the qualitative alteration of the fluid regime: the strain rate in the medium sharply grows (the relative compressive strains increase rapidly), the pore pressure builds up due to the decrease of the pore-fracture space and, as a consequence, the electrokinetic current also sharply increases. At the time of occurrence of the seismic event, variations in the parameter α

immediately stop, the changes in the stress-strain state of the medium is totally determined by the filtration processes in the region of the hypocenter of the earthquake, i.e. by the redistribution of the pore pressure.

Thus, the obtained results suggest that the character of changes in the pore pressure is largely determined by the second accepted hypothesis, i.e. the time variation in the degree of heterogeneity of the elastic properties in the area of future earthquake source. The close qualitative consistency between the results obtained in this study and the data yielded by the measurements in the well allows us to conclude that the proposed physical mechanisms adequately explain the set of the observed changes in GAE amplitudes during the nearby earthquakes. Hence, it can be asserted that the constructed theoretical model makes possible to qualitatively describe the perturbations in the geoacoustic emission amplitude near the large seismic event. This model can also be used for interpretation of the measuring well results in terms of identification of character of stress-strain state changes.

4 Conclusions

In the work, a two-scale model describing the evolution of the geophysical fields near the G-1 measuring well during the preparation and implementation of the nearby strong seismic event is developed. On the first (large) spatial scale, we solved the problem about the volumetric strain perturbations in the brittle elastic medium due to the changes of elastic properties in local volume - a prototype of the source of a seismic event. On the second spatial scale, we considered the model of the changes in fields of different nature near the measuring well that is induced by the perturbations in the strain filed calculated at the first scale level.

Under the assumption constraining the character of variations in the degree of heterogeneity of elastic properties within the spherical domain, the time dependence of the average (over the geophone sensitivity domain) electrokinetic current during the preparation of the selected seismic event is obtained on the first scale level. The character of variations of the current is qualitatively consistent with the envelope of GAE recorded by the geophone in the G-1 well at a depth of 1 km. Thus, the numerical simulation have established that the electrokinetic effects in the porous medium saturated by the mineralized fluid directly determine the intensity of the geoacoustic processes near the measuring borehole during the seismic events.

Acknowledgments. The work was carried out under the program of the RF President for state support of young scientists—Candidates of Sciences (project no. MK-2682.2017.5).

References

1. Dobrovolsky, I.P.: Theory of the Earthquake Preparation. UIPE RAS, Moscow (2000)
2. Dobrovolsky, I.P.: Strain and Stress Distribution Associated with the Preparation of a Tectonic Earthquake. Izvestiya Phys. Solid Earth **39**, 812–819 (2003)

3. Dobrovolsky, I.P.: Mathematical Theory of Preparation and Forecast of Tectonic Earthquakes. FIZMATLIT, Moscow (2009). (In Russian)
4. Eshelby, J.D.: The determination of the elastic field of an ellipsoidal inclusion, and related problems. Proc. Royal Soc. Lond. **A241**, 376–396 (1957). https://doi.org/10.1098/rspa.1957.0133
5. Fenoglio, M.A., Johnston, M.J.S., Byerlee, J.: Magnetic and electric fields associated with changes in high pore pressure in fault zones: application to the Loma Prieta ULF emissions. J. Geophys. Res. **100**, 12951–12958 (1994). https://doi.org/10.1029/95JB00076
6. Fitterman, D.V.: Electrokinetic and magnetic anomalies associated with dilatant regions in a layered Earth. J. Geophys. Res. **83**, 5923–5928 (1978). https://doi.org/10.1029/JB083iB12p05923
7. Fitterman, D.V.: Calculations of the self-potential anomalies near vertical contacts. J. Geophys. Res. **44**(2), 195–205 (1979). https://doi.org/10.1190/1.1440961
8. Fitterman, D.V.: Theory of electrokinetic-magnetic anomalies in a faulted half-space. J. Geophys. Res. **84**(B11), 6031–6040 (1979). https://doi.org/10.1029/JB084iB11p06031
9. Gavrilov, V.A., Bogomolov, L.M., Zakupin, A.S.: Comparison of the geoacoustic measurements in boreholes with the data of laboratory and in-situ experiments on electromagnetic excitation of rocks. Izv. Phys. Solid Earth **47**(11), 1009–1019 (2011). https://doi.org/10.1134/S1069351311100041
10. Gavrilov, V.A., Panteleev, I.A., Ryabinin, G.V., Morozova, Y.V.: Modulating impact of electromagnetic radiation on geoacoustic emission of rocks. Russ. J. Earth Sci. **13**(1), ES1002 (2013). https://doi.org/10.2205/2013ES000527
11. Gavrilov, V.A., Panteleev, I.A., Ryabinin, G.V.: The Physical basis of the effects caused by electromagnetic forcing in the intensity of geoacoustic processes. Izv. Phys. Solid Earth **50** (1), 87–101 (2014). https://doi.org/10.1134/S1069351314010042
12. Glover, P.W.J., Walker, E., Jackson, M.D.: Streaming-potential coefficient of reservoir rock: a theoretical model. Geophysics **77**(2), D17–D43 (2012). https://doi.org/10.1190/geo2011-0364.1
13. Ishido, T., Mizutani, H.: Relationship between fracture strength of rocks and ζ-potential. Tectonophysics **67**, 13–23 (1980). https://doi.org/10.1016/0040-1951(80)90161-4
14. Ishido, T., Nishizawa, O.: Effects of zeta potential on microcrack growth in rock under relatively low uniaxial compression. J. Geophys. Res. **89**(B6), 4153–4159 (1984). https://doi.org/10.1029/JB089iB06p04153
15. Khatkevich, Y.M., Ryabinin, G.V.: Hydrogeochemical studies in Kamchatka in connection with the search for earthquake precursors. J. Volcanol. Seismolog. **4**, 34–42 (2006). (In Russian)
16. Kosmulski, M., Dahlste, P.: High ionic strength electrokinetics of clay minerals. Colloids Surf. A **291**, 212–218 (2006). https://doi.org/10.1016/j.colsurfa.2006.06.037
17. Mizutani, H., Ishido, T.: A new interpretation of magnetic field variation associated with the Matsushiro earthquakes. J. Geomag. Geoelectr. **28**, 179–188 (1976). https://doi.org/10.5636/jgg.28.179
18. Mizutani, H., Ishido, T., Yokokura, T., Ohnishi, S.: Electrokinetic phenomena associated with earthquakes. Geophys. Res. Lett. **3**, 365–368 (1976). https://doi.org/10.1029/GL003i007p00365
19. Mizutani, H., Ishido, T.: Experimental and theoretical basis of electrokinetic phenomena in rock-water systems and its applications to geophysics. J. Geophys. Res. **86**(B3), 1763–1775 (1981). https://doi.org/10.1029/JB086iB03p01763
20. Myachkin, V.I., Kostrov, B.V., Sobolev, G.A., Shamina, O.G.: Basis of the source physics and earthquake precursors. In: Sadovsky, A.V. (ed.) Physics of an Earthquake Source, pp. 6–29. Nauka, Moscow (1975). (in Russian)

21. Nourbehecht, B.: Irreversible thermodynamic effects in inhomogeneous media and their applications in certain geoelectric problems. Ph.D. thesis, Massachusetts Institute of Technology, Cambridge, MA (1963)
22. Overbeek, J.T.: Thermodynamics of electrokinetic phenomena. J. Colloid Sci., 420–427 (1953). https://doi.org/10.1016/0095-8522(53)90026-9
23. Reid, H.F.: The California Earthquake of April 13 1906. V. 2. The Mechanics of the earthquakes. The Carnegie Institution, Washington (1910)
24. Revil, A., Pezard, P.A., Glover, P.W.J.: Streaming potential in porous media 1. Theory of the zeta potential. J. Geophys. Res. **104**(B9), 20021–20031 (1999). https://doi.org/10.1029/1999JB900089
25. Riznichenko, Y.V.: Dimensions of the core earthquake and seismic moment. In: Investigations on the Physics of the Earthquake Focus, pp. 9–18. Nauka, Moscow (1976). (in Russian)
26. Scholz, C.H., Sykes, L.R., Aggarwal, Y.P.: Earthquake prediction: a physical basis. Science **181**, 803–810 (1973). https://doi.org/10.1126/science.181.4102.803
27. Simpson, J.J., Taflove, A.: Electrokinetic effect of the Loma Prieta earthquake calculated by an entire-Earth FDTD solution of Maxwell's equations. Geophys. Res. Lett. **32**, L09302 (2005). https://doi.org/10.1029/2005GL022601
28. Terada, T.: On luminous phenomena accompanying earthquakes. Bull. Earthq. Res. Inst. Tokyo University **9**, 225–255 (1931)
29. Thomas, J.N., Lovea, J.J., Malcolm, J.S.J.: On the reported magnetic precursor of the 1989 Loma Prieta earthquake. Phys. Earth Planet. Inter. **173**, 207–215 (2009). https://doi.org/10.1016/j.pepi.2008.11.014

Analysis of Poroelastic Laboratory Experiments Using Numerical Simulation Techniques

Vitaly Borisov[1], Anton Ivanov[1], B. Kritskiy[1], I. Menshov[1], E. B. Savenkov[1], M. Trimonova[1,2](✉), S. B. Turuntaev[2], and E. V. Zenchenko[2]

[1] Keldysh Institute of Applied Mathematics of Russian Academy of Sciences, Moscow, Russia
Trimonova.ma@gmail.com
[2] Institute of Geosphere Dynamics of Russian Academy of Sciences, Moscow, Russia

Abstract. This research is aimed at the study of the numerical simulation of the laboratory experiment results. Namely, this work shows the results of numerical modeling the filtration process that takes place in the laboratory experiment on a gypsum/cement sample. Numerical simulator is three-dimensional and based on the Biot poroelastic theory. The laboratory experiments are aimed at the study of hydraulic fracturing with account of the fracturing fluid leak-off. This study is the first part of the total numerical-experimental simulation of the hydraulic fracturing and its aim is to test the assumptions in which both the laboratory model and the numerical one were created. As a result of the research, a great influence of the gas presence on the filtration was found. Because of the gas presence, the fluid filtration was much slower in the experiment than in the numerical simulation. Although the numerical model of the filtration process is single-phase, we have invented a method for counting the gas phase presence in the sample. Thanks to this, we were able to reproduce the results of the experiment.

Keywords: Pore pressure · Laboratory experiment · Numerical model · Elastic moduli

1 Introduction

Currently, mathematical modeling is one of the main tools used to analyze and optimize the development of oil and gas fields. The main attention is paid to mathematical modeling of filtration processes [1]. The mechanical state of the reservoir in filtration models is taken into account by setting the dependence of the porosity of the reservoir on the fluid pressure. To solve the filtration problems, such accuracy is usually sufficient. A number of situations are known, which analysis requires full account of the stress-strain state of the reservoir. In this case, to describe the reservoir-fluid system, a model of poroelastic medium is usually used. It allows to describe the fluid filtration in pores coupled with a mechanical model of the stress-strain state of the reservoir. Such

V. I. Karev et al. (Eds.): *Physical and Mathematical Modeling of Earth and Environment Processes (2018)*, SPEES, pp. 244–252, 2019.
https://doi.org/10.1007/978-3-030-11533-3_23

models were proposed by Biot [2]. The Biot model describes simultaneously the processes of elastic medium deformation and the fluid flow.

To confirm the correct choice of the mathematical model of the reservoir, it is necessary to check the model on real data. Obtaining data from real fields is a very complicated and expensive procedure. In addition, not all the necessary data can be obtained. For example, the maximum horizontal stress is one of the difficult-to-access parameters of the reservoir. But it may be necessary to set the boundary conditions for the elastic problem. Therefore, the best test of the numerical model is the verification of the model on the experiment, for which its complete formulation is known. This paper presents the experiment, its mathematical model, and a comparison of the results.

2 Mathematical Model

Here we consider a simplified version of the model, described in detail in [3]. This model will be used below in the description of the numerical algorithms and test calculations. The simplifications are not fundamental: the simplified model is isothermal, and the surrounding medium is homogeneous and isotropic.

Problem is solved in the spatial domain. The equations of the linear model of Bio are:

$$\nabla T + \rho g = 0 \tag{1}$$

$$\frac{\partial}{\partial t}(m_f) + \nabla(-\rho_f \frac{k}{\mu} \nabla p) = 0 \tag{2}$$

where T- stress tensor, ρ- sample density, m_f- fluid mass per unit volume, ρ_f- fluid density, k- sample permeability, μ- fluid viscosity, p- fluid pressure.

The problem formulation must be supplemented with the boundary conditions and initial conditions. For the elastic part of the problem (1), the components of the (total) normal stresses or components of the displacement can be specified on the boundaries of the region. For the filtration part of the problem (2), either the pressure values or the normal component of the fluid flux can be boundary conditions. The primary unknown variables of the problem are the displacements (u) and the fluid pressure in the reservoir (p).

3 Laboratory Experiment

3.1 Setup Description

The setup was developed at the Institute of Geosphere Dynamics of Russian Academy of Sciences [4]. Its main possibilities include: studies of hydraulic fracturing and related processes, the creation of a three-dimensional stress-strain state and pore pressure distribution.

Structurally, the setup for modeling the processes in a flat reservoir consists of upper and lower steel discs and a sidewall between them. A schematic view of the setup is shown in Fig. 1a. The sidewall height is 70 mm with an internal diameter of 430 mm.

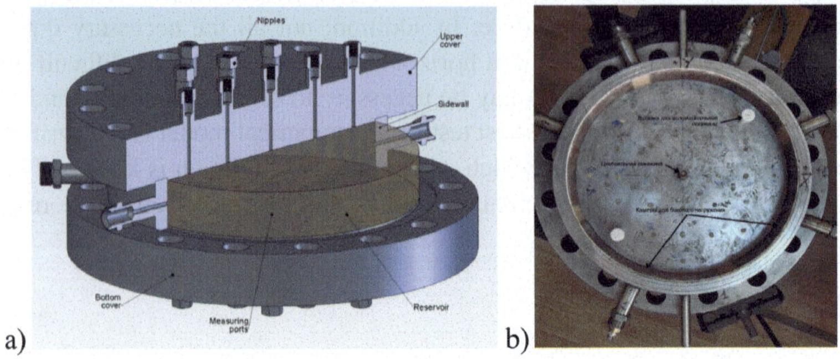

Fig. 1. Schematic view of the setup (a), top view of the lower cover of the installation (b)

Before the start of the experiment, the setup is filled with gypsum/cement/water mixture. Hardening, it forms model medium (a sample). The top disc is separated from the sample by a rubber membrane. A gap is created between the membrane and the disc. This gap is filled with the water under pressure, which allows to simulate lithostatic pressure in the sample. Horizontal loading of the model is ensured by the injection of fluid into opposing thin chambers located along the inner side of the setup sidewall (Fig. 1b). Several holes are made in both disc covers and in the sidewall. These holes are used either for mounting various sensors or for pumping the fluid in or out sample. The location map of the sensors is shown in the Fig. 2. The pore pressure in the sample is measured through the holes located in the bottom disc of the setup. Also the couple of the holes is used to imitate the wells for creation the pore pressure distribution in the sample.

This was a short description of the setup features used in our study.

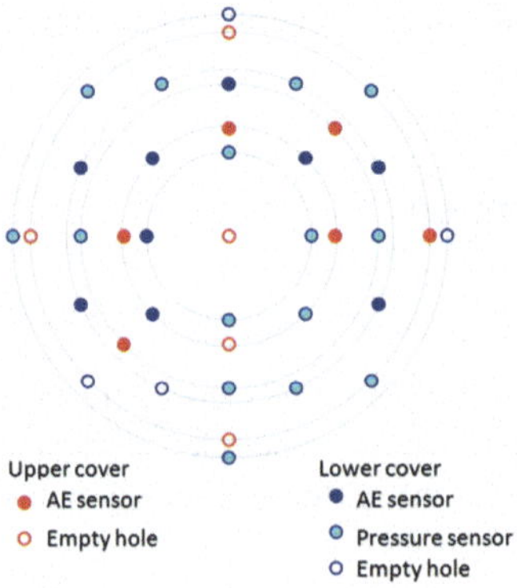

Fig. 2. Location map of the sensors

3.2 Model Medium

The choice of the medium modeling the reservoir is determined by the purpose and formulation of the experimental problems to be solved. The main purpose of the study is the experimental modeling of hydraulic fracturing with the possibility of transferring the results of the experiments to the real reservoir conditions. The choice of the sample material is based on the two main conditions:

1. similarity criteria, which are responsible for the possibility of transferring data from the experiment to real conditions. [5, 6];
2. technological factors associated with the possibility of manufacturing experimental samples.

From this point of view, the cement-based gypsum mix is a good choice. Good plasticity without shrinkage of this mixture allows to create homogeneous samples with necessary properties.

In the experiments considered below, the medium has the following properties.

$\upsilon_{dyn} = 0.25$ – dynamic Poisson coefficient,
$\upsilon_{stat} = 0.2$ – static Poisson coefficient,
$E_{dyn} = 7.5E9\ Pa$ – dynamic Young's modulus,
$E_{stat} = 3.7E9\ Pa$ – static Young's modulus,
$k = 2.4\ mD$ – sample permeability.

Description of the preliminary experiments on determination of the strength and filtration characteristics of the sample can be found in [4].

3.3 Experiment Description

To carry out this experiment, two wells were made at the points with coordinates [(0.057, 0.127), (−0.057, −0.127)]. The pressure sensors were mounted at the bottom of the setup according to Fig. 2. Then, the gypsum/cement/water mixture was poured into the setup and dried for a couple of days. After it, the vertical load was applied to the sample, as it was described above (20 atm, where 1 atm = 101335 Pa), simulating lithostatic pressure. The pressure in the side chambers was not set. After that, a gypsum water solution was pumped into the injection well with a constant pressure (14.5 atm). Another well was connected to the atmosphere (Fig. 3a). The sample saturation continued until a certain pore pressure distribution was created. After that, the injection was stopped. Accordingly, the pore pressure in the sensors began to decrease. The pressure distribution was recorded throughout the whole experiment (Fig. 3b). The duration of the experiment was approximately 220 min.

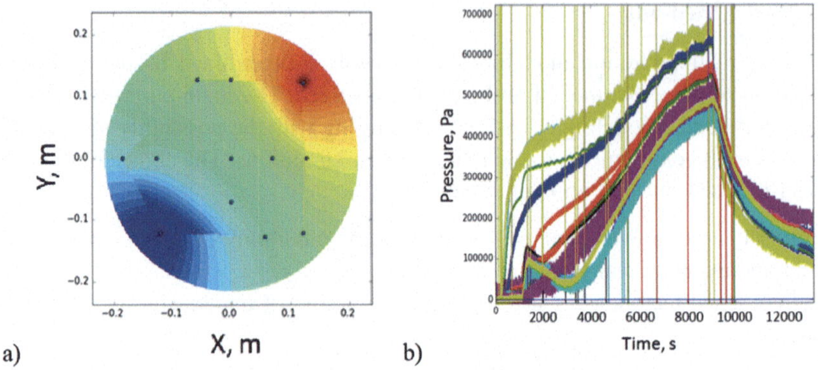

Fig. 3. Pore pressure distribution in the sample (a), pressure curves from the sensors (b)

4 Numerical Experiment

4.1 Boundary Conditions for the Elastic Problem

A vertical load was applied to the upper surface of the sample perpendicular to the surface. Therefore, the vector of the applied force is equal to the normal component of the force: $f = f_n = \sigma_n n$, where $\sigma_n = P_{top}$. Consequently, the Neumann boundary condition is set on the upper surface for a normal stress equal to the vertical load: $\sigma_n = P_{top} = 20\ atm$. The bottom disc cover of the setup is fixed rigidly vertically, hence the lower surface of the sample cannot move along the axis OZ. Thus, the Dirichlet boundary condition for displacement along the axis: $u_z = 0$ is placed on the bottom of the sample. To exclude the sample rotation around its axis and the sample shear in the plane OXY one can set the Dirichlet boundary condition on the whole side surface of the sample in the form of zero tangential displacement in the OXY plane: $-n_y u_x + n_x u = 0$.

Also it is necessary to add boundary conditions for the problem of elasticity to the wells. It follows from the fact that a fluid is pumped into the injection well under constant pressure and the production well is connected to the atmosphere. Thus, it is necessary to put the Neumann condition on the full normal stress on the wells: $\sigma_n = P_{inj} = 14.5\ atm$, $\sigma_n = P_{prod} = 1\ atm$.

4.2 Boundary Conditions for the Filtration Problem

On the top, bottom, side of the sample surface the condition of non-penetration is set. This condition means zero fluid flux through the given surfaces. But this fluid flux depends linearly on the pressure gradient. That is, the boundary condition is the Neumann pressure condition: $\partial p / \partial n = 0$. Constant pressures are first set on the wells: fluid is pumped into the injection well at a constant pressure, and the production well is connected to the atmosphere. After the pressure field in the sample has been established, pumping of the fluid into the injection well is stopped. Thus, the Dirichlet boundary condition is set on the whole surface of the injection well until the moment of switching ($P_{inj} = 14.5\ atm$). Further, the Neumann condition $\partial p/\partial n = 0$ is set on it until the end of the experiment. Also the Dirichlet boundary condition is set at the production well throughout the experiment ($P_{prod} = 1\ atm$).

4.3 Mesh Generation

To set the computational domain of complex geometry and generate an unstructured mesh, Salome is used. This platform provides the tools for the pre- and post-processing of the numerical simulation data. Salome allows you to build geometry and the mesh. The experimental sample is a cylinder with two holes symmetrical with respect to the center. This geometry is quite simply created in Salome (Fig. 4a). The mesh sizes were selected based on the requirements of the numerical model on the number of elements between the boundaries. The main advantages of Salome include the ability to divide the geometric and mesh regions into subregions. This opportunity is necessary to set the boundary and initial conditions. An example of the creation of the boundary region is shown below (Fig. 4b). The above features of the Salome make it possible to build a mesh, which is necessary for numerical simulation of the filtration experiment.

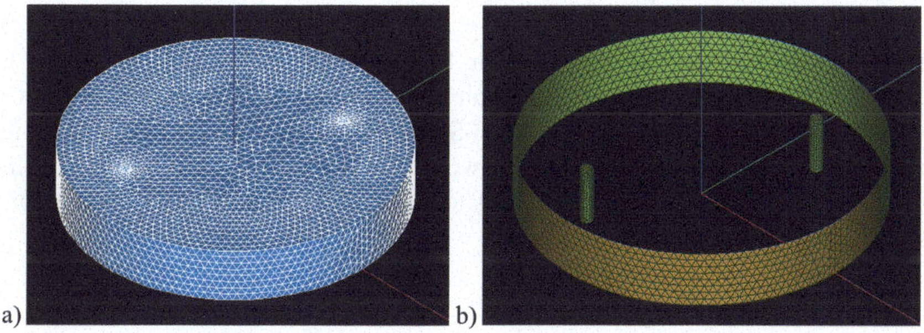

Fig. 4. Geometry and mesh of the sample (a), boundary regions (b)

5 Results

The results of the laboratory experiment are the pore pressure changes in the sample, namely, at the points where the sensors measure the pore pressure during the whole experiment. To verify the numerical model, the pressure change curves at these points were taken from the results of the numerical simulation and compared with the experimental curves.

From the formulation of the experiment described above, it was clear that the pressure in the sensors had to increase first, and then drop. During the experiment, it turned out that the initial process of maintaining constant pressure in the well was not stable. Therefore, to verify the numerical model, only the second half of the experiment was used, that is, the process of the pressure drop. To do this, the maximum pressure is determined at the points with sensors from the experimental curves. This pressure was specified in the numerical model as a condition for switching the injection mode.

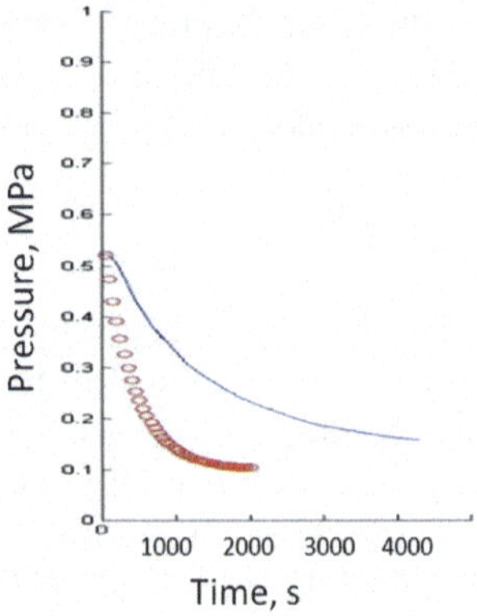

Fig. 5. Pressure curves from one of the sensors: experiment (blue), theory (red)

The first comparison of the results of numerical and experimental simulation showed their fundamental discrepancy in the rate of the pressure drop, i.e. in the filtration rate (Fig. 5). The reason for this, according to one of the assumptions, is the presence of air in the sample, which impedes the filtration process. Since the mathematical model of fluid filtration is single-phase rather than two-phase, it was decided to take into account the presence of air by "sticking" it to the matrix. For this, the matrix bulk modulus of the matrix + air mixture was recalculated according to the Hashin-Shtrikman method [7]. The Hashin-Shtrikman boundaries (3) are the most rigid boundaries of composite modules for two-phase material. Specifying the volume fraction of the constituent materials allows to calculate upper and lower bounds for the elastic moduli of any composite material.

$$K_{HS}^{\pm} = K_2 + \frac{\varphi}{(K_1 - K_2)^{-1} + (1 - \varphi)(K_2 + \frac{4}{3}G_2)^{-1}} \quad (3)$$

The upper bound is computed when $K_2 > K_1$. The lower bound is computed by interchanging the indices in the equations. Here K – is bulk modulus, G- shear modulus, φ- volume fraction.

The sequence of calculations was as follows:

- determination of the volume fraction of air in the mixture (matrix + air) by the known porosity of the sample and by the known air fraction in the pore space
- determination of the upper and lower bounds of new matrix bulk modulus ($\overline{K_s}^{\pm}$) of the material (matrix + air) using the Hashin-Shtrikman boundaries (3)
- determination of the upper and lower bounds of the new Biot coefficient ($\overline{M}^{\pm}$) of the material (matrix + air)

$$\frac{1}{\overline{M}^{\pm}} = \frac{b - \varphi_0}{\overline{K_s}^{\pm}} + \frac{\varphi_0}{K_f} \quad (4)$$

The Hashin-Shtrikman boundaries found for the matrix bulk modulus of a composite material (matrix + air) ($\overline{K_s}^{\pm}$) are 10^5 and 10^{10} Pa for the lower and upper bound, respectively. The boundaries of the Biot module ($\overline{M}^{\pm}$) are approximately of the same order. Since our experiment is mostly filtration, the Biot module has a greater influence on the filtration process.

It turned out that when the value of the Biot module is of the order of 10^7, the filtration rate in the numerical experiment almost matches with the rate determined in the laboratory (Fig. 6). It is possible that bulk modulus is not the only reason for slow filtration.

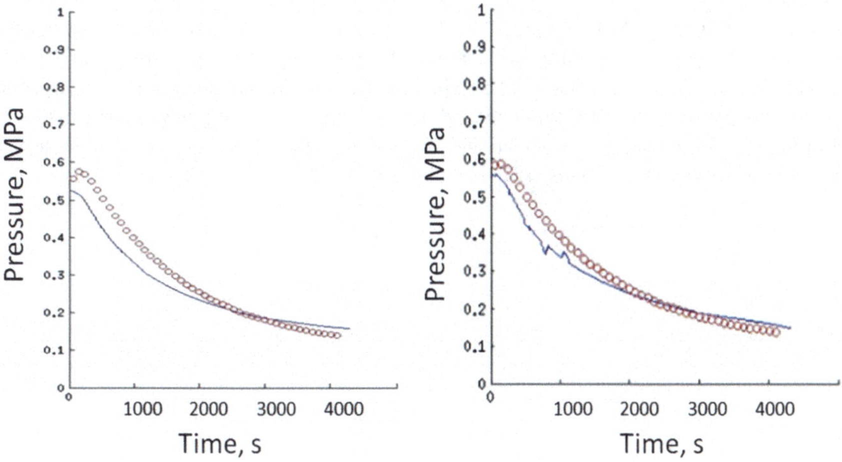

Fig. 6. Pressure curves from two sensors: experiment (blue), theory (red)

6 Conclusions

Verification of the numerical model in a quite simple laboratory experiment on the fluid filtration made it possible to reveal the enormous effect of air presence in the sample on the filtration process. Accounting for such air was produced by "sticking" it to the matrix and recalculating the matrix bulk modulus of the composite material. Since further laboratory and numerical experiments will be significantly complicated by the growth of a hydraulic fracture, the presence of air will need to be taken into account.

Laboratory modeling was carried out according to the State task (No. 0146-2017-0011), with the financial support of LLC "RN-UfaNIPIneft" and RFBR (project No. 16-05-00869).

Numerical modeling was funded by Russian Science Foundation according to the research project No. 15-11- 00021.

References

1. Aziz, K., Settari, A.: Petroleum Reservoir Simulation. Institute of Computer Research, Moscow (2004)
2. Biot, M.A.: General theory of three dimensional consolidation. J. Appl. Phys. **12**, 155–164 (1941)
3. Borisov, V.E., Ivanov, A.V., Kritskiy, B.V., Menshov, I.S., Savenkov, E.B.: Numerical modelling of poroelastic problems. In: Preprint of Keldysh Institute of Applied Mathematics, no. 81 (2017)
4. Trimonova, M., Baryshnikov, N., et al.: Estimation of the hydraulic fracture propagation rate in the laboratory experiment. In: Karev, V., et al. (eds.) Physical and Mathematical Modeling of Earth and Environment Processes. Springer International Publishing, Moscow, pp. 259–268 (2018)
5. de Pater, C.J., Cleary, M.P., Quinn, T.S.: Experimental verification of dimensional analysis for hydraulic fracturing. In: SPE Production & Facilities, pp. 230–238, November 1994
6. de Pater, C.J., Weijers, L., Savic, M.: Experimental study of nonlinear effects in hydraulic fracture propagation. In: SPE Production & Facilities, pp. 239–249, November 1994
7. Hashin, Z., Shtrikman, S.: A variational approach to the elastic behavior of multiphase minerals. J. Mech. Phys. Solids, 127–140 (1963)

Flows of a Viscous Incompressible Fluid in a Plane Convergent Channel

S. A. Kumakshev([✉])

Ishlinsky Institute for Problems in Mechanics of the Russian Academy of Sciences, Moscow, Russia
kumak@ipmnet.ru

Abstract. The problem of stationary flow of viscous incompressible fluid in a plane confusor is investigated. Solution in the form of one-mode and multi-mode (containing sectors of flow and outflow) velocity profiles of the liquid, depending on the confusor angle, is obtained numerically. For this purpose, a high-precision numerical-analytical procedure based on the method of accelerated convergence (Newton's algorithm type) and continuation with respect to a parameter (Reynolds number) was developed. For multi-mode solutions, the existence of symmetric and asymmetric (with respect to the bisector of the confusor angle) modes is established. The evolution of velocity profiles as the Reynolds number increases is studied.

Keywords: Jeffery-Hamel problem · Plane confusor

1 Problem Statement

We investigate the stationary flow of an incompressible fluid in a flat confusor (the angle between the walls is equal to 2β $(0 < \beta < \pi)$) with density ρ and outflow at the point of walls intersection with a power Q $(Q > 0)$. To describe the fluid motion we use the polar coordinates (r, θ), so that fluid region has the form $|\theta| < \beta$, $r > 0$. The medium is characterized by density ρ and kinematic viscosity ν. In the case of a flat confusor three quantities $\{\rho, Q, \nu\}$ are dimensionally dependent ($[Q] = [\nu]$); other dimensional quantities characteristic of this problem are absent, and the subsequent equations cannot be nondimensionalized completely. The dimensionless combination is the Reynolds number $Re = Q/\nu$.

Note, that for some distribution of pressure p with respect to the angle θ (which can be found on the basis of kinematics relations as $r \to \infty$) the velocity field (v_r, v_θ) is radial [1,3]

$$v_r = -\frac{Q}{r}V(\theta), \qquad v_\theta \equiv 0. \tag{1}$$

V. I. Karev et al. (Eds.): *Physical and Mathematical Modeling of Earth and Environment Processes (2018)*, SPEES, pp. 253–260, 2019.
https://doi.org/10.1007/978-3-030-11533-3_24

According to the expressions (1), guaranteeing the incompressibility condition, nonzero components of the stress tensor and the strain rate tensor are given by

$$v_{rr} = -v_{\theta\theta} = \frac{Q}{r^2}V(\theta), \qquad v_{r\theta} = -\frac{Q}{2r^2}V'(\theta), \tag{2}$$

$$\sigma_{rr,\theta\theta} = -p \pm \frac{2\rho Q^2}{r^2 Re}V(\theta), \qquad \sigma_{zz} = -p, \qquad \sigma_{r\theta} = -\frac{\rho Q^2}{r^2 Re}V'(\theta). \tag{3}$$

We can obtain an ordinary differential equation for the function $V(\theta)$ from (1)–(3) by substituting expressions (3) into two Navier-Stokes equations [1] and thus determine the pressure p

$$V'' + 4V - ReV^2 = C, \qquad C = \text{const}; \tag{4}$$

$$p = \frac{\rho Q^2}{2r^2 Re}(C - 4V). \tag{5}$$

The constant C has a mechanical meaning which can be obtained from the adhesion condition of the liquid on the walls $|\theta| = \beta$ of the confusor

$$V(\pm\beta) = 0, \qquad C = V''(\pm\beta). \tag{6}$$

To complete the problem of Jeffery-Hamel we have the constant flow condition in the integral form:

$$\int_{-\beta}^{\beta} V(\theta)d\theta = 1. \tag{7}$$

There are numerous special and educational publication (see the lists of references in [1–3]) dealing with analytical study of the problem (4), (6), (7). The order of Eq. (4) can be reduced. By multiplying both parts of this equation by V' and integrating the resulting equation we obtain the first integral

$$\frac{1}{2}V'^2 + 2V^2 - \frac{1}{3}ReV^3 - CV = \frac{1}{2}V'^2(\mp\beta). \tag{8}$$

From (8) it follows that if the function $V(\theta)$ is equal to zero at points θ_i, then the derivatives V' are equal in absolute value, i.e. $V'^2(\theta_i) = V'^2(\mp\beta)$. Equations (8), (4) allow integration in elliptic functions [1–3]. However, the analysis of the resulting elliptic integrals and the solution of the system of two transcendental equations with respect to the integration constants [1,2] is difficult. It is not possible to construct an explicit solution of the boundary value problem for arbitrary values of β and Re.

Now we have two-parametric boundary value problem (4), (6) with the condition (7). It allows variational interpretation [4]. For further numerical and analytical research, instead of a pair of parameters (β, Re), it is convenient to proceed to a pair (a, b), where $a = 4\beta$, $b = 2\beta Re$. The initial-value problem by transition from unknown $V(\theta)$ to a new desired function $y(x)$, where $V = y/(2\beta)$ and the argument $\theta = (2x-1)\beta$ $(0 < x < 1)$, can be represented in the form

$$y'' + a^2 y - by^2 = \lambda, \qquad y(0) = y(1) = 0, \qquad y'(0) = \gamma,$$
$$\int_0^1 y(x)dx = 1, \qquad \lambda = 8\beta^3 C = y''(0) = y''(1). \tag{9}$$

To characterize the flow rate, along with the variables y, y', we can introduce a variable z according to (8)

$$z' = y - 1, \qquad z(0) = 0, \qquad z(1) = 0. \tag{10}$$

The boundary value problem (9), (10) for variables y, z, is to be solved. The difficulty of this problem, among other things, is that the unknown parameter λ must be calculated; it can be eliminated by differentiating the Eq. (9), but the result is a nonlinear equation of the third order.

2 Numerical Solution

For certainty, a fixed value of the angle $\beta = 10°$ is considered, i.e., $a \approx 0.7$. This value is frequently encountered in applied problems. Also, for other a values, numerous calculations were made. We assume that the value b varies widely: $0 < b \leq 200$. High-precision solution of the problem (8), (10) with a relative error of $10^{-7} - 10^{-8}$ (absolute $O(10^{-5})$) is constructed using the procedure of continuation with respect to a parameter and a special modified method of accelerated convergence based on Newton's algorithm [4].

2.1 One-Mode Symmetric Flow

First, we apply this procedure to obtain one-mode symmetric flow [5] (similar to Poiseuille flow), for which as $b \to 0$ and any known a the exact limiting solution is known in analytical form. This solution will be used in the parameter continuation procedure as the first approximation of the solution for the next $Re \neq 0$ value. Then it will be updated and verified by an accelerated convergence method [4] of Newton's type algorithm.

As a result, the unknown constants $\gamma(b)$ and $\lambda(b)$ can be obtained. These curves are shown in Fig. 1: $\gamma(b)$ on the top side (above zero) and $\lambda(b)$ on the bottom side (below zero).

Now, using this figure we can fix some value b ($b = 2\beta Re$) and take the respective values $\gamma(b)$ and $\lambda(b)$. Then by substituting these values into (9) we can easily obtain the velocity profile $y(x)$. One-mode solutions of Jeffery-Hamel problem for some certain b are shown in Fig. 2.

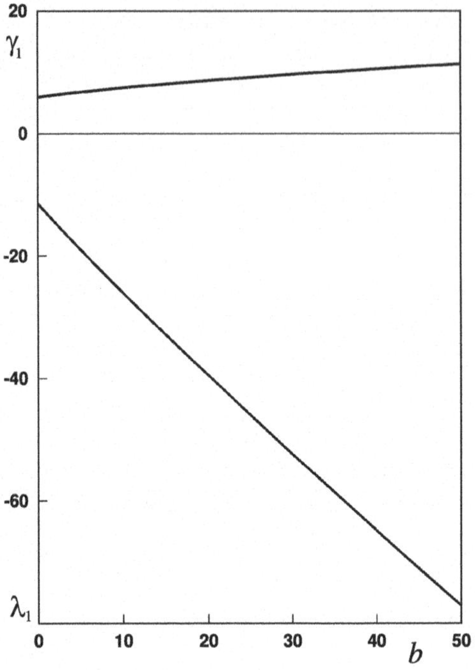

Fig. 1. Curves $\gamma_1(b)$ and $\lambda_1(b)$, $0 < b < 50$

2.2 Multimode Flows

Now let us obtain multimode profiles. The problem degenerates as $b \to 0$, which leads to significant difficulties in constructing multimode velocity profiles [6]. In this case, the values of the desired parameters $\gamma(b)$ and $\lambda(b)$ increase without limit and they are involved in the integration of the corresponding Cauchy problem. To avoid this circumstance we will fix the number of the selected mode ($n = 2, 3, 4, \dots$) and the value of the parameter $b = b_0$ (usually $b_0 \approx 10$). Then unknown γ, λ are searched for manually, looking for rough satisfaction of the boundary conditions. After this rough search we can verify γ and λ with high accuracy by method [4], providing low discrepancy in the boundary conditions. Then, using the continuation procedure with respect to the parameter b, the universal curves $\gamma(b)$, $\lambda(b)$ are constructed for $0 < b < b_0$ and $b_0 < b < \infty$. The difficulties of calculations are aggravated by the fact that the parameters $\gamma_n \gg \gamma_1$ and $\lambda_n \gg \lambda_1$ reach at $b = 1$ large values (measured by several orders of magnitude): for example $\gamma_5 \sim 10^4$, $\lambda_5 \sim 10^5$. These circumstances may explain the difficulties in the search, identification and analysis of multimode flows in Jeffrey-Hamel problem.

The results of analytical and numerical study and their graphical representations for the multimode flows (n = 2, 3, 4, 5) are presented and commented below. One can compare these results with the curves corresponding to the one-

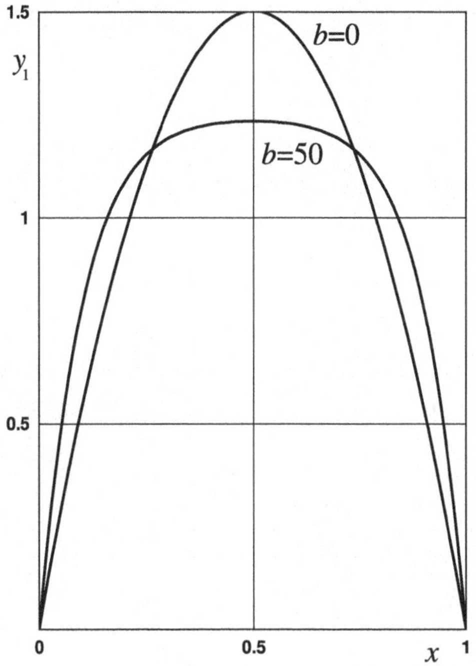

Fig. 2. Velocity profiles $y_1(x)$ for $b = 0$ and $b = 50$

mode solution, which is obtained above. Analysis shows that for $n \geq 2$ the odd modes correspond to symmetric (with respect to $x = 1/2$, or, which is the same, $\theta = 0$), and even n to asymmetric solutions.

We can establish some certain structural properties of multimode flows. Among other things, the (positive) maxima ($n \geq 3$) and (negative) minima ($n \geq 4$) of functions $y_n(x, b)$ have the same values, respectively. In addition, when n, b are fixed, the derivatives y_n' are equal in absolute value at all zeros x_i, $i = 1, \dots, n+1$, of the functions $y_n(x, b)$, i.e., $\gamma_n(b) = \pm y_n'(x_i, b)$. We can conclude in a certain sense that a multimode flow is a "combination" (set) of one-mode and two-mode flows. Highly likely that it follows from the general property of the radiality of the flows.

Functions $\gamma_n(b)$, $\lambda_n(b)$ on the interval $1 \leq b \leq 10$ are depicted in Fig. 3 on different scales above and below zero, respectively. For a fixed value of the parameter a (i.e., angle β) they define the solution of the boundary value problem (9), (10), which can obtain simply by integrating the Cauchy problem. It is interesting that the curves are grouped into pairs $n = 2, 3$ and $n = 4, 5$. Near $b = 0$, the curves have vertical asymptotes: $\gamma_n \to +\infty$, $\lambda_n \to -\infty$ as $b \to +0$, while the one-mode solution has finite values of γ_1, λ_1. Also, large absolute values of γ_n, λ_n for $b \sim 1$ should be noted, which leads to large values of the flow velocity in comparison with the one-mode solution. As the mode numbers increases, the differences $\gamma_n - \gamma_{n-1}$, $\lambda_{n-1} - \lambda_n$ increase without limit, as well

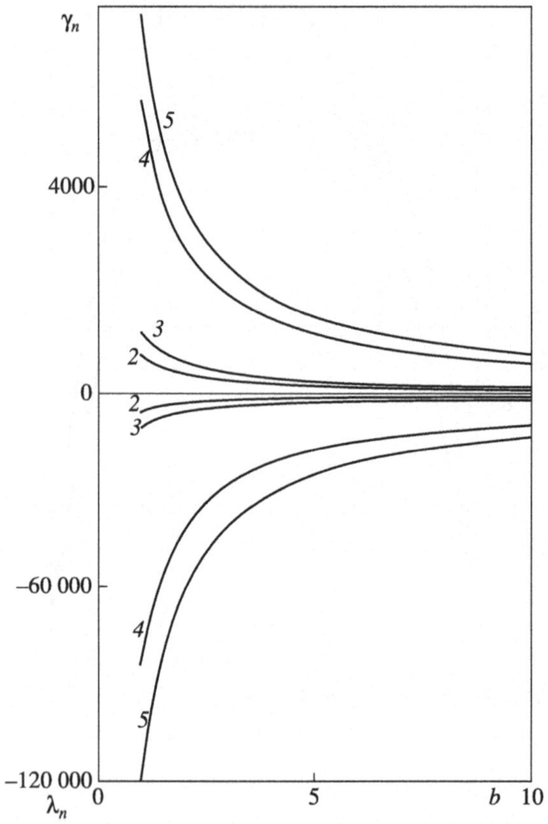

Fig. 3. Curves $\gamma_n(b)$ and $\lambda_n(b)$, $1 \leq b \leq 10$

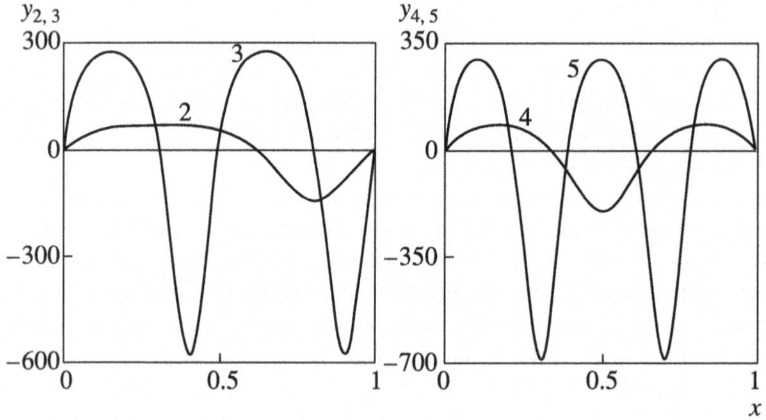

Fig. 4. Velocity profiles $y_n(x)$ for $b = 1$

as for $b \to +0$. Each curve γ_n fairly rapidly comes to relatively low values on the interval with small b and then, as b is growing, very slowly tends from above to the asymptote $\gamma_\infty = \sqrt{4b/3}$ [6]. Similarly, the curve λ_n very rapidly grows with increasing from zero and also tends very slowly from below to asymptote $\lambda_\infty = -b$ [6]. This implies an important conclusion in the mechanical aspect: all modes of stationary flows, including the main one ($n = 1$), have a common limit corresponding to a flow of an ideal fluid as $b \to \infty$ ($Re \to \infty$).

In Fig. 4 to illustrate multimode and asymmetrical flows for $n = 2, \ldots, 5$, velocity profiles $y_n(x)$ are given for relatively small $b = 1$. Here we can see the area of inflow ($y_n > 0$) and outflow ($y_n < 0$). There are large differences in absolute values of the positive maxima and negative minima of the velocity V, i.e., the functions $y_n(x)$ for small values of the parameter b (Re). As the parameter b (Re) increases, we observe a decrease in this difference and a decrease in return flows. Near the walls, $x = 0, 1$ ($\theta = \mp\beta$), typical boundary-layer phenomena are observed.

3 Conclusion

The main result of this study are the curves $\gamma_n(b)$, $\lambda_n(b)$. Using these curves, just by integration of the Cauchy problem for Eqs. (9), (10) we can obtained numerous characteristics of stationary flows in a convergent channel: velocity profile (1), the components of the strain rate tensor (2), stress tensor (3), and the pressure (5). The curves γ_n, λ_n have quite a simple shape, but their construction requires very high-precision calculations that are significantly more difficult for $b \to +0$ and $b \to \infty$. "Boundary-layer" effects appear in the problem. Computational algorithms based on known methods of functional analysis (Bubnov-Galerkin), finite-difference methods, and finite element method do not provide satisfactory results.

Acknowledgements. This study was performed according to the state assignment (AAAAA 17-117021310387-0) under partial financial support from RFBR (projects 17-01-00538, 17-08-00742, and 18-01-00812).

References

1. Kochin, N.E., Kibel, I.A., Roze, N.V.: Theoretical Hydromechanics. Interscience Publishers, New York (1965, 1964)
2. Landau, L.D., Lifshitz, E.M.: Fluid Mechanics (Volume 6 of A Course of Theoretical Physics). Pergamon Press, Oxford (2009)
3. Loitsyanskii, L.G.: Mechanics of Liquids and Gases. Elsevier Science (2014)
4. Akulenko, L.D., Kumakshev, S.A., Nesterov, S.V.: Effective numerical-analytical solution of isoperimetric variational problems of mechanics by an accelerated convergence method. J. Appl. Math. Mechanics **66**(5), 693–708 (2002)
5. Akulenko, L.D., Georgievskii, D.V., Kumakshev, S.A., Nesterov, S.V.: Numerical and analytical investigation of steady-state flow of viscous fluid in a plane confuser. Doklady Phys. **45**(9), 467–471 (2000)

6. Akulenko, L.D., Georgievskii, D.V., Kumakshev, S.A.: New asymmetric and multi-mode solutions of the problem on viscous-fluid flow inside a plane confusor. Doklady Phys. **47**(3), 219–223 (2002)

The Change in the Permeability of an Ultra-Low Permeable Limestone Sample Under the Influence of Confining Pressure During the Loading-Unloading Cycle

N. A. Baryshnikov(✉), E. V. Zenchenko, and S. B. Turuntaev

Institute of Geosphere Dynamics of Russian Academy of Sciences, Moscow, Russia
nabarysh@gmail.com

Abstract. The objective of this research was to study the long-term response of filtration properties of a porous rock sample to external confining pressure. An experimental dependence of the ultra-low-permeable limestone sample permeability on time during the loading-unloading cycle with duration of 29 days was obtained. As a result of the loading, the sample permeability decreased by more than one and a half times. The decrease was linear with time, while there was practically no visible permeability elastic reaction to an abrupt change in external load. An experimental study of the flow stabilizing process in the sample was also conducted. It is shown that the characteristic time of stabilization, both theoretically estimated and measured, is much less than the characteristic time at which the permeability change occurred during the main loading experiment. This allows us to conclude that the observed change in the permeability of the sample is related with its viscous deformation.

Keywords: Viscoelastic deformation · Filtration · Permeability · Long term loading · Experimental study

1 Introduction

An important feature of porous rocks and oil shale is the strong dependence of their filtration properties on the stress-strain state. Under external load such rocks may exhibit inelastic properties, which manifests itself in the form of a significant decrease in permeability with time, the absence of an elastic recovery of filtration properties after short-term loading [1–3]. Load-unloading cycles also affect the sensitivity of the rocks to the applied external pressure [4]. This is usually attributed to materials contained in the pore space of rocks such as clay. Previous laboratory studies have shown that the time-dependent deformation under external load can be characterized as linear viscoelastic deformation attended by the packing of clay and progressive collapse of the pore spaces [5]. The total creep deformation, which develops in time, can be comparable or even exceed the value of the instantaneous elastic deformation. It is obvious that such deformations should lead to significant change in the filtration properties of rocks. However, at present, there are few direct studies of the effect of stress on the

V. I. Karev et al. (Eds.): *Physical and Mathematical Modeling of Earth and Environment Processes (2018)*, SPEES, pp. 261–266, 2019.
https://doi.org/10.1007/978-3-030-11533-3_25

permeability of viscoelastic porous rocks, while experimental studies of the long-term change in the permeability of the low-permeability rocks under the external load are practically absent. Such studies are associated with a number of technical difficulties, primarily with significant time expenses: continuous measurements should be carried out over a long time (hours, days, weeks), which is not always possible. The purpose of this work was to study the long-term response of a porous rock sample to confining pressure. In the course of the work, an experimental time dependence of the permeability of the ultra-low-permeable sample of limestone during the loading-unloading cycle was obtained.

2 Experiment

In the figure (Fig. 1) a diagram of the laboratory setup that was used in our experiments to study filtration in cylindrical core samples is shown. The more detailed description of this setup is given in [6].

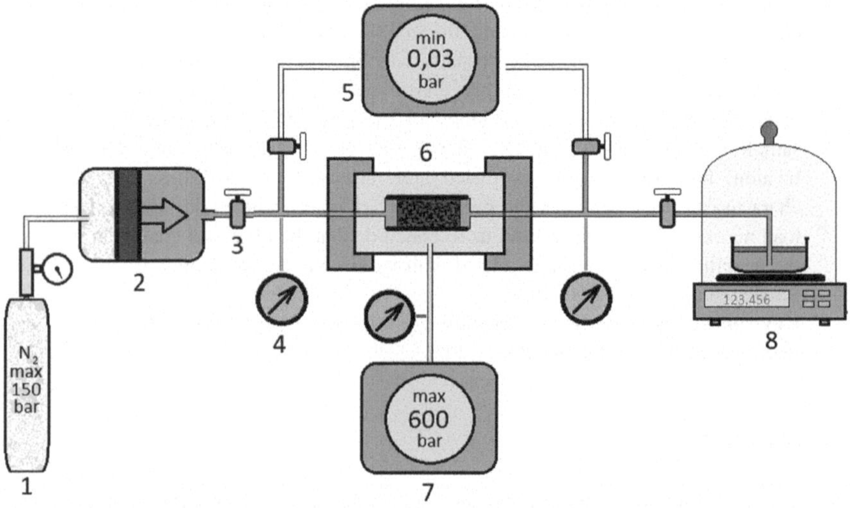

Fig. 1. Experimental setup. 1 - buffer tank; 2 - pneumohydraulic cylinder; 3 - valves; 4 - pressure sensors; 5 - vacuum pump; 6 - core holder; 7 - high pressure pump; 8 - scales with tank for accumulated fluid.

The tested sample in a rubber sleeve is placed in the core holder. The required all-round confining pressure is created in its shell using a high-pressure oil pump. Fluid filtering is produced by constant pore pressure gradients created by means of a pneumohydraulic cylinder located at the inlet of the core holder. The mass flow rate of the filtered fluid is measured using a laboratory scales by continuously weighing accumulated fluid at the outlet of the core holder. The accuracy of the scales is 0.001 g. The scales themselves and the fluid on them are protected from external influences and evaporation. The permeability of the sample under study is calculated according to the Darcy law in a stationary flow based on the rate of change of the fluid mass on the scales at a given pore pressure drop. Also we are using a correction for the temperature dependence of the viscosity and density of the fluid, as well as the deviation of the filtration flow in the sample from the flat one. This method allows investigating filtration in ultra-low permeability samples with permeability of the order of fraction of microdarsies.

In the described experiment the ultra-low permeable sample of limestone taken from the aquifer well from a depth of 27 m was studied. The sample was cylindrical 30 $\times$ 30 mm sized. The permeability of the sample, measured by gas flow, was 0.006 mD, and the porosity was 8.4%. Initially, a dry sample was placed into the core holder, in which an initial confining pressure of 3 MPa was created. After that, vacuuming was performed to remove air from the pore space. For the next three days, the sample was under vacuum (pore pressure of the order of $5 \cdot 10^{-3}$ MPa) with a constant confining pressure of 3 MPa.

Water was used as the filtering fluid in this experiment. The sample was saturated at a pore pressure 0.57 MPa at the inlet of the core holder. The sample saturation time was about 3 days. The duration of the main experiment (after saturation) was 29 days. All this time, continuous fluid filtration was performed through the sample at the pore pressure drop of 0.55 ± 0.02 MPa (with an average pore pressure about 0.33 MPa). The confining pressure increased up to values of 5, 10, 12, and 15 MPa, followed by a decrease down to 5 and 3 MPa. The solid line on the graph (Fig. 2) shows how the confining pressure P_c changed over time. The dots indicate the change in the sample permeability k over time. Each point corresponds to 0.05 mL of fluid passing through the sample, which is equal to about 3% of the total volume of the pore space of the sample. The increased point scattering in the left part of the permeability curve is related to the influence of air bubbles in the pipeline system at the outlet from the core holder. Bubbles interfered only in the process of measuring the fluid flow rate. The sample itself was not affected.

3 Experimental Results

An almost linear decay in the permeability with time throughout the entire loading stage with a slight recovery of permeability at the unloading stage (after $2 \cdot 10^6$ second) was observed (Fig. 2). The main feature of the graph is the almost complete absence of visible abrupt changes in the permeability of the sample at the moments when the confining pressure increased or decreased. Partially, this may be due to the forcing of fluid from the sample due to the poroelasticity. The forced fluid created an additional inflow at the outlet of the sample, which led to an apparent increase in its measurable permeability. With constant or slowly varying loads, the permeability of the sample varied linearly, while the rate of change was practically independent of the load created.

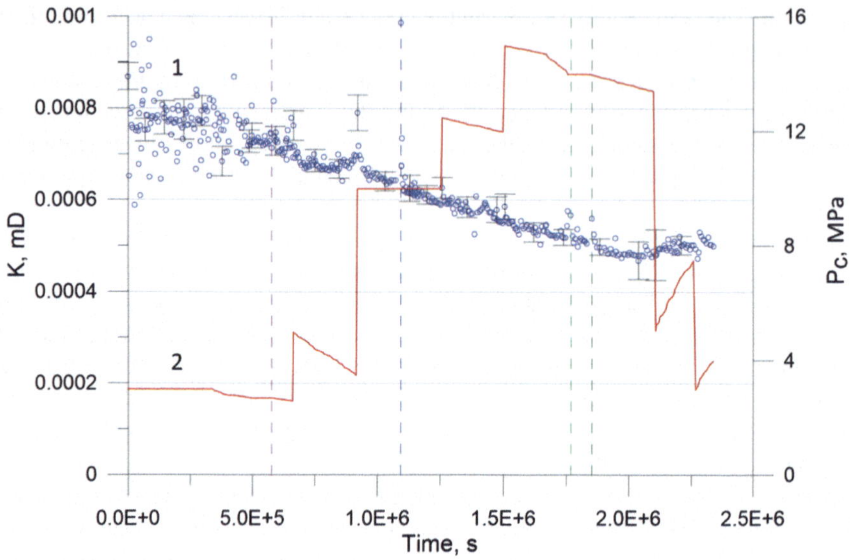

Fig. 2. Permeability 1 and confining pressure 2 with time. Dotted lines mark the moments when a flow stabilizing tests were performed.

Besides the main experiment, four additional tests were carried out at different stages of the loading cycle to study the process of the flow stabilizing in the sample. In the graph (Fig. 2) these moments are marked by dotted lines. The flow was stopped for a while by using a valve at the outlet of the core holder. When the pore pressure in the sample became constant, the valve reopened, and the flow stabilizing process was recorded. The graph (Fig. 3) shows how the flow varied over time from the moment of the valve opened. This experimental curve allows us to estimate the characteristic time of the flow stabilization in 1000 s. This time is the same for all four additional tests performed.

According to [7], in a poroelastic approximation the characteristic time of the flow stabilization in the sample with size L can be estimated by equation

$$\tau \approx L^2 \chi^{-1} \pi^{-2}$$

where $\chi = k/\beta\mu$ is the intrinsic diffusion coefficient, k - the sample permeability, μ - the fluid viscosity, β is the compressibility of the sample. By using permeability of the sample $k \approx 0.7$ mD and the compressibility of limestone $\beta \approx 10^{-4}$ MPa^{-1} we can estimate the characteristic time of the flow stabilization as 10 s. This is two orders of magnitude shorter than the time obtained from the flow decay curve analysis. Such a difference cannot be explained by the error in measuring the elastic and filtration properties of the sample and, apparently, is related to the inelastic nature of its deformation. At the same time, the characteristic time for the flow stabilization, both obtained theoretically and measured in additional tests, is much shorter than the characteristic times of the sample permeability changes during the main loading experiment (Fig. 2). This allows to conclude that the observed change in the permeability of the sample is related with its viscous deformation.

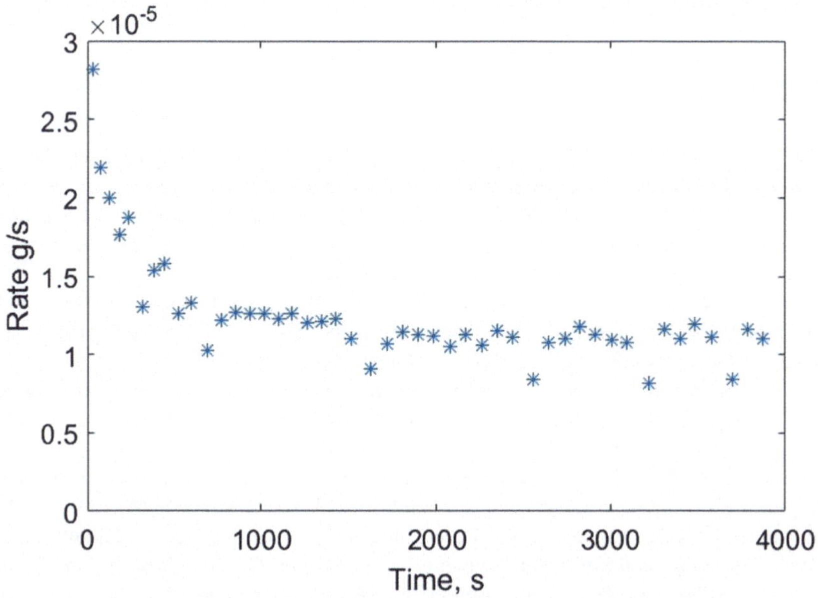

Fig. 3. Flow rate decay with time while stabilizing.

4 Conclusion

Previously, we carried out a series of experiments on the short-term loading of more permeable limestone samples [8]. Experiments have shown a relatively weak permeability response to external stress. During the stepped loading-unloading cycle lasting about two hours the permeability of the samples irreversibly decreased by the amount of 3–5% from the initial one. As a result of the long-term loading experiment we found that the change in permeability under the same loads over longer times can be much more significant: during the month, the permeability of the sample under study decreased by more than one and a half time. At the same time, there was practically no visible elastic response of the sample to abrupt increase in load, and the entire change in permeability occurred in the way of viscous deformation. This phenomenon requires further study, especially in connection with the increasing role of low-permeable reservoirs in the oil production. Such reservoir rocks may be subject to considerable compaction during development [9]. In most cases, this compaction is irreversible, and therefore cannot be described using the traditional theory of poroelasticity. This can lead to incorrect estimates of changes in the filtration properties of reservoirs during the development process. Understanding the mechanics of deformation is important for estimating the volume of recoverable hydrocarbon reserves, in development of hydrodynamic models of oilfields and oil production intensifying methods, in particular, hydraulic fracturing.

Acknowledgments. The work was supported by the RFBR project No. 16-05-00869 and the state task (project No. 0146-2017-00011).

References

1. Dobrynin, V.M.: Effect of Overburden Pressure on Some Properties of Sandstones. Society of Petroleum Engineers, 1 December 1962. https://doi.org/10.2118/461-pa
2. Gangi, A.F.: Variation of whole and fractured porous rock permeability with confining pressure. Int. J. Rock Mech. Min. Sci. Geomech. Abstr. **15**(5), 249–257 (1978). https://doi.org/10.1016/0148-9062(78)90957-9. ISSN 0148-9062
3. Lei, Q., Xiong, W., Yuang, J., Cui, Y., Wu, Y.-S.: Analysis of Stress Sensitivity and Its Influence on Oil Production from Tight Reservoirs. Society of Petroleum Engineers, 1 January 2007. https://doi.org/10.2118/111148-ms
4. Zhu, S.: Experiment Research of Tight Sandstone Gas Reservoir Stress Sensitivity Based on the Capillary Bundle Mode. Society of Petroleum Engineers, 30 September 2013. https://doi.org/10.2118/167638-stu
5. Sone, H., Zoback, M.D.: Time-dependent deformation of shale gas reservoir rocks and its long-term effect on the in situ state of stress. Int. J. Rock Mech. Min. Sci. **69**, 120–132 (2014)
6. Baryshnikov, N.A., Eliseev, S.V., Zenchenko, E.V., Lagutin, Yu.V., Turuntaev, S.B.: Laboratory study of the filtration properties of low-permeable porous media. In: Dynamic Processes in the Geospheres, no. **8**, pp. 35–42 (2016). (in Russian)
7. Barenblatt, G.I., Entov, V.M., Ryzhik, V.M.: Theory of Fluid Flows Through Natural Rocks, ix+395 pp. Kluwer Academic Publishers, Dordrecht (1990)
8. Arkhipov, Ya.D., Baryshnikov, N.A., Zenchenko, E.V., Turuntaev, S.B.: Changes in the filtration properties of low-permeable porous media under external load. In: Dynamic Processes in Geospheres, no. 9, pp. 68–75 (2017). (in Russian)
9. Yang, Y., Zoback, M.: Viscoplastic deformation of the Bakken and adjacent formations and its relation to hydraulic fracture growth. Int. Soc. Rock Mech. Rock Eng., 1 January 2015

Mathematical Modeling of the Internal Waves Emergence in the Stratified Viscous Fluid

Pavel Matyushin(✉)

Institute for Computer Aided Design, Russian Academy of Sciences, 19/18, 2nd Brestskaya str., Moscow 123056, Russia
pmatyushin@mail.ru

Abstract. The paper presents the results of the mathematical modeling and 3D visualization of the linearly density stratified 3D incompressible viscous fluid flows uniformly moving in a horizontal direction from the left to the right around a disk (along the axis of symmetry of the disk). If we will consider the fluid flows in the reference frame connected with fluid then we can investigate the nonlinear fundamental 3D mechanism of the formation of the 3D gravitational internal waves over the place M of the impulse start of the back side of the disk in the horizontal direction from the right to the left. At the present paper this 3D mechanism is analyzed in detail (for the first time). For the visualization of the 3D vortex structures of the fluid flows the isosurfaces of β were drawing, where β is the imaginary part of the complex-conjugate eigen-values of the velocity gradient tensor. During each half of the buoyancy period a small deformed vortex ring is generated over the point M (due to shear and gravitational instabilities), gradually grows in size and shifts down to the point M. The left half of ring is transformed in the internal half-wave. The right half of ring is compressed by next right halves of rings generated later. Thus the number of the internal waves between the back side of the disk and point M is always equal to the number of the buoyancy periods past since the disk start. This flows are described by the Navier-Stokes equations in the Boussinesq approximation. For the mathematical modeling the numerical method SMIF with an explicit hybrid finite difference scheme (second-order approximation, monotonicity) has been used.

Keywords: Density stratified viscous fluid · Disk · Mathematical modeling · Internal waves generation

1 Introduction

Understanding the physics of the complex processes occurring when the body is moving in the ocean is very important from both a practical and a theoretical point of view. Internal gravitational waves in the ocean depth in comparison with the ordinary surface waves have a number of surprising properties. For example, the group velocity of internal waves is perpendicular to the phase velocity and the wave energy propagates in radial directions from the place where the body starts with the group velocity parallel to the crests [1–3]. Let's try to understand how the first internal waves are generating after the impulse start of the body in the horizontal direction.

V. I. Karev et al. (Eds.): *Physical and Mathematical Modeling of Earth and Environment Processes (2018)*, SPEES, pp. 267–277, 2019.
https://doi.org/10.1007/978-3-030-11533-3_26

2 Equations and Boundary Conditions

Let us consider the linearly density stratified 3D incompressible viscous fluid flows uniformly moving with velocity U in a horizontal direction from the left to the right around a disk (with diameter d and thickness $H = 0.76 \cdot d$) (along the axis z of symmetry of the disk). Let ρ_0 is the initial density along axis z. Let x, y, z are the Cartesian coordinates starting from the disk center; z, x, y are the streamwise, lift and lateral directions non-dimensionalized by $d/2$. Let $\rho(x, y, z) = 1 - x/(2A) + S(x, y, z)$ is the non-dimensional (by ρ_0) density of the stratified viscous fluid, where $A = \Lambda/d$ is the scale ratio, Λ is the buoyancy scale, which is related to the buoyancy frequency N and period T_b ($N = 2\pi/T_b$, $N^2 = g/\Lambda$); g is the scalar of the gravitational acceleration; S is a dimensionless perturbation of salinity. The density stratified viscous fluid flows have been simulated on the basis of the Navier-Stokes equations (NSE) in the Boussinesq approximation (1)–(3) with four dimensionless parameters: internal Froude number $Fr = U/(N \cdot d)$, Reynolds number $Re = U \cdot d/\nu$, $A \gg 1$ and Schmidt number $Sc = \nu/\kappa = 709.22$, where ν is the kinematical viscosity coefficient, κ is the salt diffusion coefficient.

$$\frac{\partial S}{\partial t} + (\mathbf{v} \cdot \nabla)S = \frac{2}{Sc \cdot Re} \Delta S + \frac{v_x}{2A} \tag{1}$$

$$\frac{\partial \mathbf{v}}{\partial t} + (\mathbf{v} \cdot \nabla)\mathbf{v} = -\nabla p + \frac{2}{Re} \Delta \mathbf{v} + \frac{A}{2Fr^2} S \frac{\mathbf{g}}{g} \tag{2}$$

$$\nabla \cdot \mathbf{v} = 0 \tag{3}$$

In Eqs. (1)–(3) $\mathbf{v} = (v_x, v_y, v_z)$ is the velocity vector (non-dimensionalized by U), p is a perturbation of pressure (non-dimensionalized by $\rho_0 U^2$), t is time (non-dimensionalized by $t_0 = d/(2 \cdot U) = 1/(2 \cdot Fr \cdot N)$). It is convenient to introduce one more non-dimensional time $T = t \cdot t_0/T_b = [t/(2 \cdot Fr \cdot N)] \cdot N/2\pi = t/(4\pi \cdot Fr)$.

For solving of the Navier-Stokes Eqs. (1)–(3) the Splitting on physical factors Method for Incompressible Fluid flows (SMIF) with hybrid explicit finite difference scheme for approximation of the convective terms of NSE (second-order accuracy in space, minimum scheme viscosity and dispersion, capable for work in wide range of the Reynolds (Re) and the internal Froude (Fr) numbers and monotonous) has been used [4, 5]. In the numerical method SMIF the pressure is extracted from Eq. (2). The Poisson equation for the pressure is solved by the diagonal Preconditioned Conjugate Gradients Method. The numerical method SMIF has been successfully applied for solving of the different problems: 2D and 3D separated homogeneous and stratified fluid flows around a sphere [3, 6–17] and a circular cylinder; the flows with free surface including regimes with broken surface wave; the air, heat and mass transfer in the clean rooms.

In the present paper the cylindrical computational grid $[z, R, \varphi] = [240 \times 110 \times 40]$ is condensed to all surfaces of the disk and to the z axis so that there are 5 grid cells on the velocity boundary layer (where $x = R\ \cos\varphi, y = R\ \sin\varphi, z = z, \mathbf{v} = (v_z,\ v_R, v_\varphi)$). The condensation of the grid to the axis z is very important in order to correctly approximate the fine flow structures of the axial parts of the half-waves of the crests (see the right part of Fig. 1*i*). As a result, the horizontal plane of flow symmetry *y-z* is preserved during the entire calculation. As you move away from the disk along the z and R directions, the length of the corresponding side of the grid cell increases monotonically according to the polynomial law to a certain maximum value and does not change further. The minimum (maximum) dimensions of grid cells for $Re = 50$ along the z and R directions are 0.036 and 0.035 (0.41 and 0.67), respectively. The computational grid occupies the region $[-13 < z < 50,\ 0 < R < 30,\ 0 < \varphi < 2\pi]$.

The boundary conditions on the disk surface are given in Eq. (4).

$$v_z = v_R = v_\varphi = 0, \left.\frac{\partial \rho}{\partial z}\right|_{z=\pm H/d} = \left.\frac{\partial S}{\partial z}\right|_{z=\pm H/d} = 0, \quad \left.\frac{\partial \rho}{\partial R}\right|_{R=1} = \left.\left(\frac{\partial S}{\partial R} - \frac{\cos\varphi}{2A}\right)\right|_{R=1} = 0 \tag{4}$$

On the external boundary of the computational grid the following boundary conditions are used: (1) at $z = -13$: $\mathbf{v} = (1, 0, 0)$, $\partial S/\partial z = 0$; (2) at $R = 30$: $\mathbf{v} = (1, 0, 0)$, $\partial S/\partial R = 0$; (3) for $z = 50$: $v_z = 1$, $v_R = 0$, $\partial v_\varphi/\partial z = 0, \partial S/\partial z = 0$.

For $Re = Fr = 50$, $A = 9816$ the drag coefficient $C_d = 1.923$ and the length of the recirculation zone (**RZ**) $L/d = 0.741$ were obtained, which agrees well with experiment [18].

3 The Visualization Techniques

For the visualization of the 3D vortex structures of the fluid flows the isosurfaces of β (Fig. 1*e-f*) were drawing, where β is *the imaginary part of the complex-conjugate eigen-values* of the velocity gradient tensor **G**. β has a real physical meaning. Let us consider a local stream lines pattern around any point in a flow where $\beta > 0$ in a reference frame **x** starting from this point and moving with the velocity of this point ($\mathbf{v} = d\mathbf{x}/dt \approx \mathbf{G}\,\mathbf{x}$, where **v** is a velocity of a fluid particle in the considered reference frame **x** and t is time). It's easy to demonstrate that the local stream lines pattern in the considered reference frame **x** is closed or spiral, and β *is the angular velocity of this spiral motion* (see the theory of the ordinary differential equations). The good efficiency of this β-visualization technique has been demonstrated in [3, 6–17].

The steady flow of the stratified viscous fluid around a disk at $Fr = 1$, $Re = 500$, $A = 439$, $T = 16.45$ (with the time-average disk drag coefficient $< C_d > = 0.81$) is presented at Fig. 1. The skin friction patterns on the disk (Fig. 1*a–c*) in combination with the stream line patterns in the *x-z* and *y-z* planes (Fig. 1*h-i, k*) are also very useful for the understanding of the 3D stream line pattern of the flow around a disk in the reference frame, connected with disk. You can see the separation of the flow from the front edge of the disk and **RZs** near the lateral and back sides of the disk. The isolines of p, β, S in the *x-z* and *y-z* planes are also very useful for the understanding of the 3D flow around a disk (Fig. 1*d, g, m*). The isolines of S_z (Fig. 1*l*) are used for comparison with the experimental Schlieren images (received by the technique "Vertical slit - Foucault knife") [2].

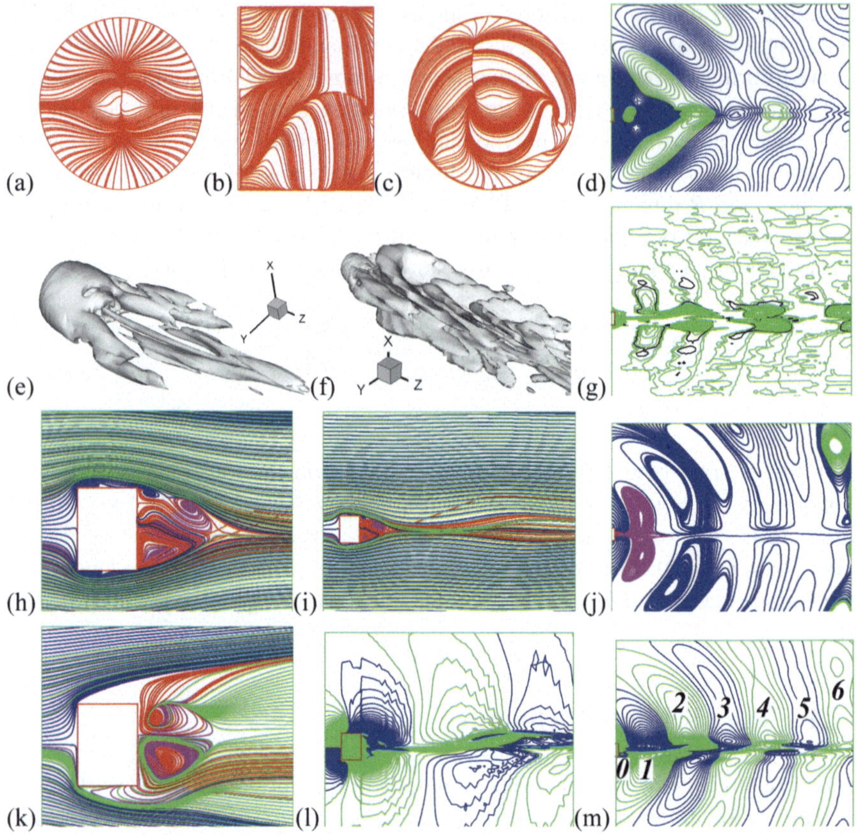

Fig. 1. Steady flow of the stratified viscous fluid around a disk at $Fr = 1$, $Re = 500$, $A = 439$, $T = 16.45$ ($< C_d > = 0.81$, $L/d = 0.947$): (*a-c*) the skin friction patterns on the front (a), side (at $y > 0$) (b) and back (c) of the disk; *e-f*) vortex structure of the flow (the isosurfaces of $\beta = 0.1$, 0.01); *h-k*) the stream lines in the planes *x-z* (h-j) and *y-z* (k) in the reference frames connected with disk (h-i, k) and with fluid (j); *d, g, l-m*) isolines of p ($\delta p = 0.005$) (d), β ($\delta\beta = 4{\cdot}10^{-3}$) (g), S_z ($\delta S_z = 2{\cdot}10^{-5}$) (l), S ($\delta S = 2{\cdot}10^{-5}$) (m) in the vertical plane *x-z* (the darker isolines correspond to negative p, S_z, S).

In some extent the velocity and pressure fields in Fig. 1 are characterized by the horizontal *y-z* and the vertical *x-z* symmetry planes. Therefore let us consider only the upper half-space ($x > 0$) at the last part of this paper. In the vicinity of the plane *x-z* the fluid flows in **RZ** (Fig. 1*h*). But in the vicinity of the plane *y-z* the fluid flows out **RZ** through two divergent foci (Fig. 1*k*). **RZ** is visualized at the vertical plane *x-z* not only in the picture of the streamlines in the reference frame connected with disk (Fig. 1*h-i*), but also in the picture of the isolines of the horizontal gradient of the density S_z (Fig. 1*l*). Moreover, the lines of troughs and crests in Fig. 1*i* coincide with the boundaries between the lighter ($S_z \geq 0$) and darker ($S_z < 0$) strips in the isolines S_z

(Fig. 1*l*). Each internal wave consists of one lighter and one darker strip in Fig. 1*l* at $x > 0$.

In the picture of the isolines *S* in the vertical plane *x*-*z* we also observe an alternation of darker ($S < 0$) and lighter ($S \geq 0$) strips (Fig. 1*m*). Let us call the darker strips in Fig. 1*m* (at $x > 0$) as half-waves of troughs, and lighter strips as half-waves of crests. The pattern of the streamlines in the vertical plane *x*-*z* at Fig. 1*j* (in the reference frame connected with fluid) is similar to the pattern of isolines *S* in Fig. 1*m*. Thus at Fig. 1*j*, *m* you can see the three internal waves and seven big cells No. 0, 1, 2, 3, 4, 5, 6 (at $x > 0$). In the zero convective cell adjacent to the disk in Fig. 1*j* the circulation of the liquid goes clockwise, in the next (first) convective cell (corresponding to the first half-wave of the troughs) the circulation goes counterclockwise, in the second convective cell (the first half-wave of the crests) the circulation goes again clockwise and so on. Thus, each internal wave in the picture of the streamlines in the reference frame connected with fluid (in the vertical plane *x*-*z*) consists of two convective cells with the opposite direction of the circulation of the liquid in them. The pattern of the isolines β in the vertical plane in Fig. 1*g* is to some extent similar to the pattern of the isolines *S* in Fig. 1*m*. The axial parts of the half-waves of the crests located in the vicinity of the *z*-axis and corresponding to the global maxima of *S* are distinguished in Fig. 1*g*. The vortex structure of flow in the wake of disk at Fig. 1*f* consists of the U-shaped structure the first half-wave of the troughs (cell No. 1) and the V-shaped structures of the first, second and third axial crests at $x > 0$ and the two side internal waves connected with the vortex shell surrounding disk. At Fig. 1*e* you can see connection of the first axial crest (at the right part of figure) with **RZ** (at the left part of figure).

4 Generation of the First Half-Wave of the Troughs Near Axis *Z* at $0.23 < T < 0.5$ ($Fr = 1$, $Re = 500$)

It is very interesting to understand how such complex structure of the internal waves (Fig. 1*f*) has been generated. Let us consider this generation process in the vertical plane *x*-*z* (Fig. 2) using the pictures of the instantaneous stream lines in the reference frame connected with fluid and the isolines of *S*. At $T = 0.01$ and $x > 0$ at the Fig. 2*a* you can see the convective cell No. 0 in the picture of stream lines and two cells in the picture of the isolines of *S*.

During clockwise motion of fluid in the right part of the zero convective cell (at $z > 1$ and $x > 0$) the lighter fluid particles descend along the current line curve (at some angle to the vertical axis) to the level of heavier fluid particles (Fig. 2*a*-*b*). The gravity forces tend to return these lighter particles to their upper level and raise these particles. As a result, at $T = 0.23$ the straight part of the current line turns into a wave-like line (Fig. 2*b*). By inertia, the amplitude of the wave of this section of the streamline increases with time (Fig. 2*c*–*f* at $0.26 \leq T \leq 0.5$).

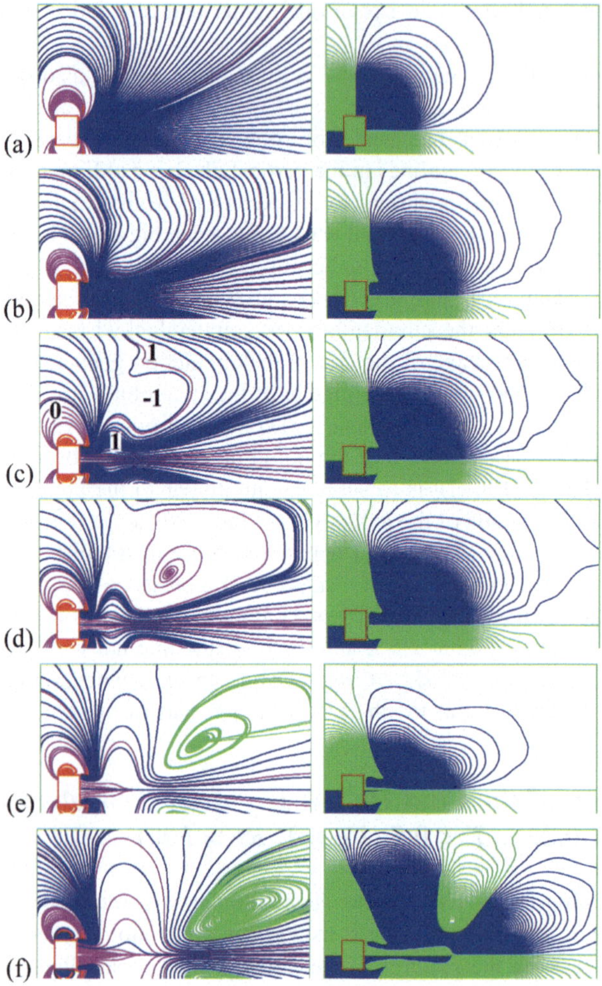

Fig. 2. The flows of the stratified viscous fluid around a disk at Fr = 1, Re = 500, A = 439 in the vertical plane x-z at T = 0.01, 0.2, 0.23, 0.26, 0.35, 0.5: a-f) the instantaneous stream lines in the reference frame connected with fluid (left figures); the isolines of S ($\delta S \cdot 10^6$ = 0.1, 1, 1, 1, 5, 1) (right figures).

Thus due to gravitational instability the first half-wave of troughs (convective cell No. 1 with counterclockwise motion of fluid) and the big convective cell No. -1 with clockwise motion of fluid are generated at $0.26 \leq T \leq 0.5$. Let us call this fundamental mechanism of the generation of the half-wave as **G_down-**mechanism because the fluid was moving down at $0 < T \leq 0.23$. The vertical line q passing through the point M of the impulse start of the center of the back side of the disk (or the start line) is situated approximately on the border between the cells No. 1 and No. -1 (Fig. 2*f*). At $0.35 \leq T \leq 0.5$ the fluid near the line q is already moving up and the right darker cell in the picture of the isolines of S is divided into two parts (cell No. 1 and cell No. -1) by new brighter cell No. N falling on it from above along the line q because the heavier fluid particles raise along the line q and therefore the value of S at the upper part of the line q is increased.

Thus at $T = 0.5$ the fluid near the line q is moving up. During $0.5 < T \leq 1$ two new convective cells No. 2 and No. -2 are generated around the line q (due to the **G_up**-mechanism of the gravitational instability which is similar to **G_down**-mechanism), gradually grow and fall on axis z from above along the line q and so on. Thus the number of the internal waves in the flow to the left of the line q at any time T is equal to T (Fig. 2*f*, 3).

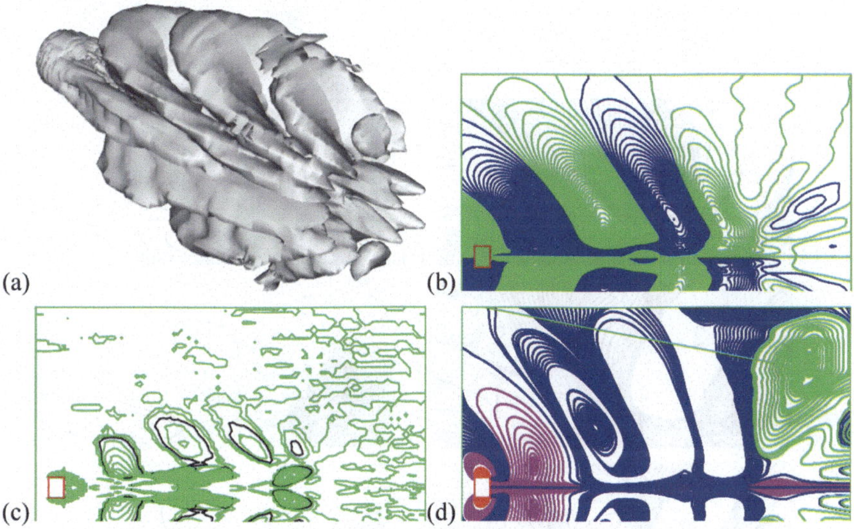

Fig. 3. The flow of the stratified viscous fluid around a disk at $Fr = 1$, $Re = 500$, $A = 439$, $T = 2$: (*a*) the isosurface of $\beta = 0.006$; (*b-d*) the isolines of S ($\delta S = 5 \cdot 10^{-6}$) (*b*) and β ($\delta\beta = 0.004$) (*c*) and the stream lines (*d*) (in the reference frame connected with fluid) in the vertical plane *x-z*.

At Fig. 3*a* the vortex structure of two internal waves generated at $0 < T < 2$ is demonstrated. At Fig. 3*b–d* you can see four big cells No. 1, 2, 3, 4 (corresponding to first and second internal waves) to the left of the line q and four small cells with the numbers –4, –3, –2, –1 to the right of the line q.

5 Generation of the Second Half-Wave of the Troughs at $0.9 < T < 1.5$ ($Fr = 0.5$, $Re = 50$)

It is known that the length λ of the internal waves in the vertical plane *x-z* is equal to $2\pi \cdot Fr \cdot d$. Thus at $Fr = 0.5$ (Figs. 4 and 5) it is possible to see at the same computational region in two times more waves than at $Fr = 1$.

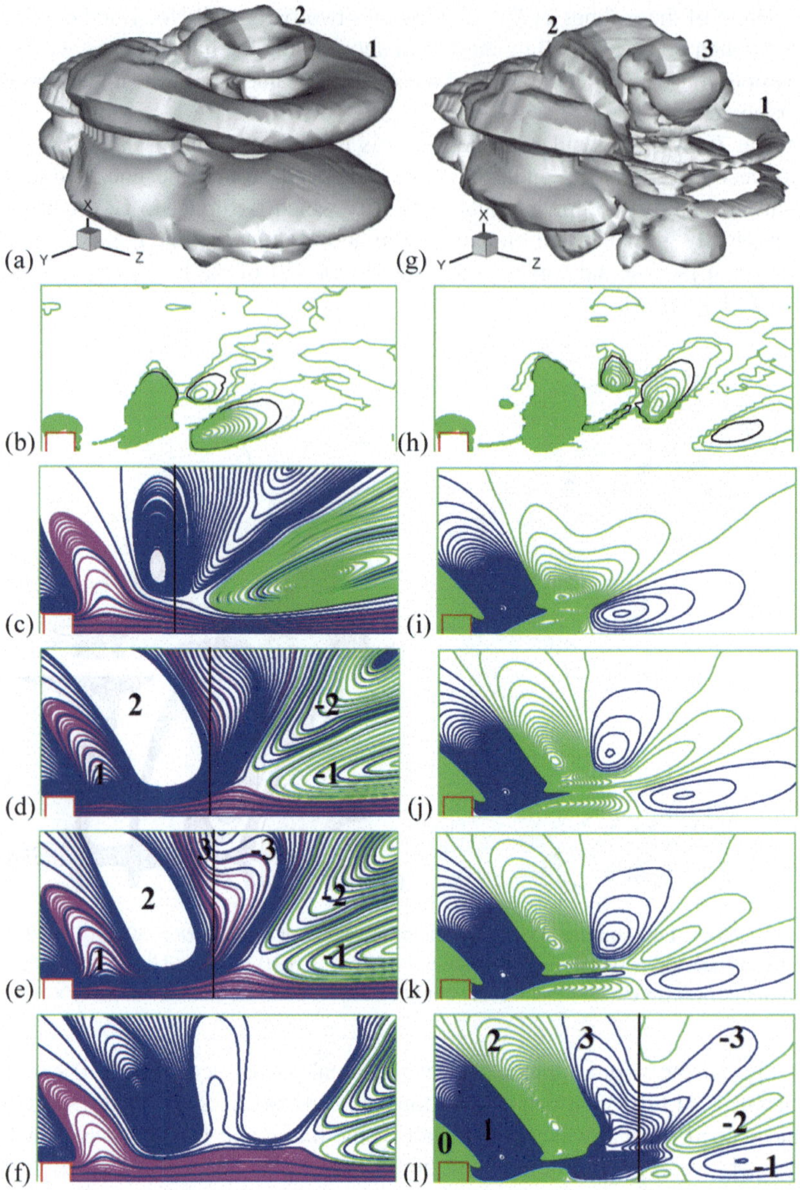

Fig. 4. Flow of the stratified viscous fluid around a disk at $Fr = 0.5$, $Re = 50$, $A = 981.6$, $0.9 \leq T \leq 1.5$: (*a, g*) the isosurfaces of $\beta = 0.005$ at $T = 0.9$, 1.5; (*b, h*) the isolines of β ($\delta\beta = 0.002$) in the vertical plane *x-z* at $T = 0.9$, 1.5; (*c-f*) the instantaneous stream lines (in the reference frame connected with fluid) in plane *x-z* at $T = 0.9$, 1.2, 1.25, 1.5; *i-l*) the isolines of *S* in vertical plane *x-z* ($\delta S = 3 \cdot 10^{-6}$, $3 \cdot 10^{-6}$, $3 \cdot 10^{-6}$, $2 \cdot 10^{-6}$) at $T = 0.9$, 1.2, 1.25, 1.5.

Let us consider the generation of the second half-wave of the troughs at $0.9 < T < 1.5$ in details (at $Fr = 0.5$, $Re = 50$) (Fig. 4). The vortex structure of the flow at $x > 0$ and $T = 0.9$ consists of the bottom and upper rings (Fig. 4*a*). The left halves of the ring No. 1 and the ring No. 2 are corresponding to first half-wave of the troughs (to the cell No. 1 at Fig. 4*c, i*) and to the first half-wave of the crests (to the cell No. 2 at Fig. 4*c, i*), respectively.

The right halves of the bottom and upper rings are corresponding to the cells No. -1 and No. -2 at Fig. 4*c, i*, respectively. The three vortex structures at the center of Fig. 4*b* at $T = 0.9$ are the cells No. 2, No. -2 and No. -1, respectively. During $1.0 < T \leq 1.5$ the new (third) vortex ring (Fig. 4*g*) is generated around the line q (the black vertical line at Fig. 4*c–e, l*) (due to the **G_down**-mechanism of the gravitational instability), grows and falls on axis z from above along the line q. The convective cells No. 3 and No. -3 (Fig. 4*f*) are situated in the cross section of the third ring by the vertical plane x-z. The **G_down**-mechanism of the generation of the third vortex ring at $1.2 < T \leq 1.25$ is demonstrated in details at Fig. 4*d-e*. At $T = 1.5$ cell No. 2 at Fig. 4*h* (the first axial crest) is situated at the same place in comparison with Fig. 4*b* ($T = 0.9$). But the cells No. -2 and No. -1 are shifted together from the center to the right end of the picture of the isolines of β during time interval $0.9 \leq T \leq 1.5$ (in the reference frame connected with disk). On the other hand in the reference frame connected with fluid during time interval $0.9 \leq T \leq 1.5$ the cells No. -2 and No. -1 are situated at the same place to the right of the line q. But the first axial crest and disk (during time interval $0.9 \leq T \leq 1.5$) shift from right to left and give free space for new cells No. 3 and -3 (the third vortex ring) (Fig. 4*h*).

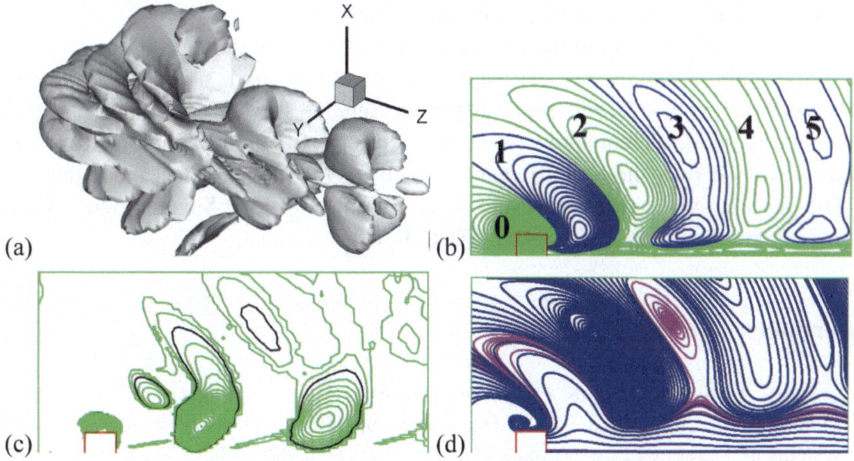

Fig. 5. Steady flow of the stratified viscous fluid around a disk at $Fr = 0.5$, $Re = 50$, $A = 981.6$, $T = 11.46$ ($< C_d > = 2.649$): (*a*) vortex structure of the flow (the isosurface $\beta = 0.02$); (*b-d*) the isolines of S ($\delta S = 10^{-5}$) (*b*) and β ($\delta\beta = 0.01$) (*c*) and the stream lines (*d*) (in the reference frame connected with fluid) in the vertical plane x-z.

The steady flow of the stratified viscous fluid around a disk at *Fr* = 0.5, *Re* = 50, *T* = 11.46 is presented at Fig. 5 in order to emphasize the difference of the vortex structures of the flows at *Re* = 50, *Fr* = 0.5 and at *Re* = 500, *Fr* = 1 (Fig. 1*f*). At Fig. 5*a* two side internal waves connected with the vortex shell surrounding disk are shorter than at Fig. 1*f*. The vortex structure of the flow at Fig. 5*a* consists of the large prominent U-shaped structures of the first half-wave of the troughs and the first half-wave of the crests and the large V-shaped structures of the second and third axial crests. The vortex structure of the flow at Fig. 1*f* consists of the smaller structures.

6 Conclusions

Some results of the mathematical modeling and 3D visualization of the linearly density stratified 3D incompressible viscous fluid flows (described by the Navier-Stokes equations in the Boussinesq approximation) around a disk (with diameter *d* and thickness *H* = 0.76•*d*) uniformly moving in a horizontal direction along its axis of symmetry *z* have been demonstrated. For the first time the nonlinear fundamental 3D mechanism of the formation of the gravitational internal waves over the place M of the impulse start of the centre of the back side of the disk (in the horizontal direction from the right to the left) is analyzed in detail at (*Fr*, *Re*) = (0.5, 50), (1, 500). During each half of the buoyancy period T_b a small deformed vortex ring is generated over the point M (due to shear and gravitational instabilities), gradually grows in size and shifts down to the point M. The left half of ring is transformed in the internal half-wave. The right half of ring is compressed by the next right halves of rings generated later. Thus the number of the internal waves between the back side of the disk and point M is always equal to the number of T_b past since the disk start.

The 3D vortex structure of the flow is visualized by the isosurfaces of the imaginary part of the *complex-conjugate* eigen-values of the velocity gradient tensor. The method SMIF with an explicit hybrid finite difference scheme for the approximation of the convective terms of the equations (second-order approximation, monotonicity), which has proved itself over the past 30 years, is used for the mathematical modeling.

This research was carried out using supercomputers at Joint Supercomputer Center of the Russian Academy of Sciences (JSCC RAS).

References

1. Lighthill, J.: Waves in Fluids. Cambridge University Press, Cambridge (1978)
2. Mitkin, V.V., Chashechkin, Y.D.: Transformation of hanging discontinuities into vortex systems in a stratified flow behind a cylinder. Fluid Dyn. **42**(1), 12–23 (2007)
3. Gushchin, V.A., Matyushin, P.V.: Simulation and study of stratified flows around finite bodies. Comput. Math. Math. Phys. **56**(6), 1034–1047 (2016)
4. Belotserkovskii, O.M., Gushchin, V.A., Konshin, V.N.: Splitting method for studying stratified fluid flows with free surfaces. USSR Comput. Math. Math. Phys. **27**(2), 181–196 (1987)
5. Gushchin, V.A., Konshin, V.N.: Computational aspects of the splitting method for incompressible flow with a free surface. J. Comput. Fluids **21**(3), 345–353 (1992)

6. Gushchin, V.A., Matyushin, P.V.: Numerical simulation and visualization of vortical structure transformation in the flow past a sphere at an increasing degree of stratification. Comput. Math. Math. Phys. **51**(2), 251–263 (2011)
7. Gushchin, V.A., Matyushin, P.V.: Vortex formation mechanisms in the wake behind a sphere for 200 < Re < 380. Fluid Dyn. **41**(5), 795–809 (2006)
8. Matyushin, P.V., Gushchin, V.A.: Transformation of vortex structures in the wake of a sphere moving in the stratified fluid with decreasing of internal froude number. J. Phys.: Conf. Ser. **318**, 062017 (2011)
9. Matyushin, P.V., Gushchin, V.A.: Direct numerical simulation of the sea flows around blunt bodies. In: AIP Conference Proceedings, vol. 1690, p. 030005 (2015)
10. Matyushin, P.V., Gushchin, V.A.: Direct numerical simulation of the 3D stratified separated viscous fluid flows. In: Fröhlich, J., Kuerten, H. (eds.) ERCOFTAC Series: Direct and Large-Eddy Simulation IX, vol. 20, pp. 459–466. Springer (2015)
11. Matyushin, P.V., Gushchin, V.A.: Direct numerical simulation of oceanic flows around blunt bodies. In: Proceedings of the Jointly Organized 11th World Congress on Computational Mechanics, 5th European Congress on Computational Mechanics, 6th European Conference on Computational Fluid Dynamics, pp. 7029–7040. Barcelona (2014)
12. Matyushin, P.V., Gushchin, V.A.: Direct numerical simulation of the 3D separated viscous fluid flows around the horizontally moving blunt bodies. In: Full Papers of European Congress on Computational Methods in Applied Sciences and Engineering, pp. 5245–5254. Vienna (2012)
13. Matyushin, P.V., Gushchin, V.A.: Direct numerical simulation of the 3D stratified viscous fluid flows around a sphere. In: Full Papers of European Congress on Computational Methods in Applied Sciences and Engineering, pp. 5232–5241. Lisbon (2010)
14. Gushchin, V.A., Matyushin, P.V.: The theory and applications of the SMIF method for correct mathematical modeling of the incompressible fluid flows. In: Dimov, I., Faragó, I. (eds.) FDM 2014, LNCS, vol. 9045, pp. 209–216. Springer, Heidelberg (2015)
15. Gushchin, V.A., Matyushin, P.V.: Mathematical modeling of the incompressible fluid flows. In: AIP Conference Proceedings, vol. 1631, pp. 122–134 (2014)
16. Gushchin, V.A., Matyushin, P.V.: Method SMIF for incompressible fluid flows modeling. In: Dimov, I., Faragó, I. (eds.) NAA 2012, LNCS, vol. 8236, pp. 311–318. Springer, Heidelberg (2013)
17. Gushchin, V.A., Matyushin, P.V.: Mathematical modeling of the 3D separated viscous flu-id flows. In: AIP Conference Proceedings, vol. 1487, pp. 22–29 (2012)
18. Bobinski, T., Goujon-Durand, S., Wesfreid, J.E.: Instabilities in the wake of a circular disk. Phys. Rev. E **89**, 053021 (2014)

The Features of Energy Dissipation of Spilling and Plunging Breaking Waves

V. V. Volkova[1(✉)] and Ya. V. Saprykina[2]

[1] Geography Faculty, Laboratory of the Geoecology of the North, Lomonosov Moscow State University, Moscow, Russia
kvendilll@yandex.ru

[2] Shirshov Institute of Oceanology of Russian Academy of Sciences, Moscow, Russia
saprykina@ocean.ru

Abstract. As a result of spectral analysis of experimental data obtained in laboratory and field conditions, this article points out the features of wave energy dissipation of wave breaking of different types. It was revealed, that the energy of plunging breaking mainly decreases at first and high nonlinear harmonics, and spilling breaking—within the span of first two harmonics. With the use of numerical modeling, the input of non-linear processes and dissipation into monochromatic waves transformation were separated, and it was confirmed that the dissipation rate significantly varies, depending on the type of breaking event.

Keywords: Wave breaking · Wave energy dissipation · Coastal oceanography · Numerical modeling

1 Introduction

The assessment of wave energy transformation in the surf zone is crucial for coastal engineering and numerical modeling of coastal processes, in particular—for defining the location of breaking point. Wave breaking leads to wave height decline, and eventually—to the wave energy dissipation. Among 4 existing type of breaking, the spilling (Sp) and plunging (Pl) are most common, and therefore most commonly used in hydrodynamic models. Prior researches have shown that the speed of wave energy dissipation depends on wave frequency [2], and the type of breaking depends on amplitude- phase-frequency content of wave [3].

The main tasks of this work were to observe the spectral wave energy dissipation of breaking of different types, as well as to define the most prominent features of this process. The results can then be used for numerical modeling of wave transformation in the surf zone.

V. I. Karev et al. (Eds.): *Physical and Mathematical Modeling of Earth and Environment Processes (2018)*, SPEES, pp. 278–286, 2019.
https://doi.org/10.1007/978-3-030-11533-3_27

2 Experiments

2.1 Laboratory Experiment

The laboratory experiment was carried out in 2004 in the wave flume of «Sea Shore» research centre in Sochi, Russia. The length of the flume was 22 m, the width—0.8 m, maximum depth—1 m. The waves were propagating above the underwater bar with a forward slope of 0.043° steepness, so that the depth of the flume varied from 0.165 to 0.59 m. The wave maker was located at 22 m mark, and the specially constructed wave absorber—at 0 m. The wave characteristics were synchronously measured by 8 string capacity-type wave gauges. The duration of record was between 1 and 4 min, with sampling frequency 200 Hz. The breaking point was being captured both visually and with the use of photo- and video assets, but mainly it occurred over the top of the bar at the end of the slope (Fig. 1).

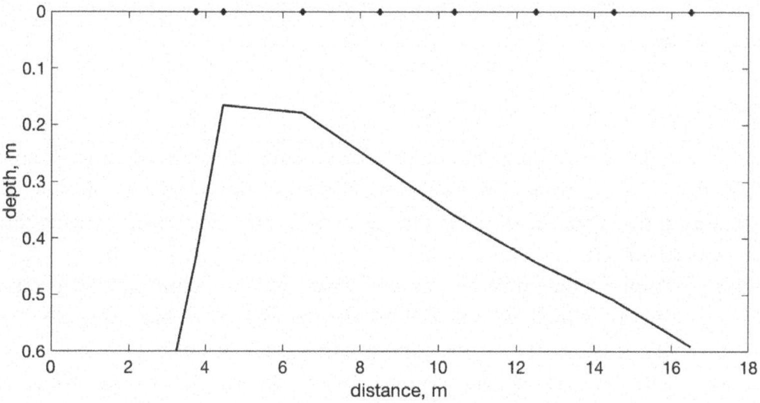

Fig. 1. Set-up of *laboratory experiment* and positions of wave gauges

2.2 Field Experiment

The experiment was carried out at the scientific base in Shkorpilovtsy (Bulgaria) in 2007. The waves were being measured by 15 string capacity-type wave gauges, positioned across the coast along the special trestle in order to minimize the effect of reflection on wave dynamics. The measurements were being made synchronously, with sampling frequency of 5 Hz. The record lasted from 20 to 60 min. The bottom cross-profile contained an underwater bar (Fig. 2). The breaking point was being captured both visually and with the use of photo- and video assets. For the analysis of experimental data, the methods of spectral analysis were applied.

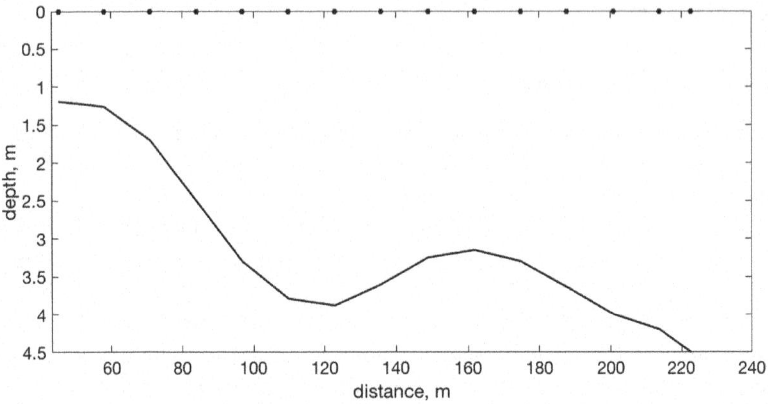

Fig. 2. Set-up of *field experiment* and positions of wave gauges

3 Discussion

3.1 Energy Decay of Breaking Waves

In order to estimate the common trends of wave energy distribution among non-linear harmonics during wave transformation and breaking, the spatial evolution of wave spectra, featuring the dynamic of wave energy of all harmonics before, during, and after breaking, was observed.

The Fig. 3 displays the most prominent series of monochromatic wave transformation measurements, which relate to different types of wave breaking, 48 (Spilling) and 44 (Plunging). It is clearly visible, that the biggest share of energy belongs to first two non-linear harmonics. It is also obvious, that different types of breaking evolve at different levels of energy transfer between the first and second harmonics—for spilling breaking the amplitudes of second harmonic reach their maximum, while for plunging breaking they still keep growing.

The energy transformation during the process of breaking was numerically estimated by the difference between spectra before and after the breaking point, at adjacent gauges:

$$dS = (S2 - S1))/dx, \text{where} \tag{1}$$

dS—difference between spectra before and after breaking, divided by the distance between gauges, m * c
S2—wave spectrum after breaking, m^2 * c

S1—wave of spectrum before breaking, в $m^2 * c$
dx—distance between gauges, m

Due to the uneven positioning of gauges, the formula (1) contains dividing by distance between gauges.

Figure 4 displays the share of dS for different harmonics for both investigated types of breaking. It can be seen that during the wave breaking of spilling type the energy mainly decays at first and second harmonics, and their proportion is 2:1. Meanwhile, the energy decay of high order harmonics is insufficient, or they may even grow in energy.

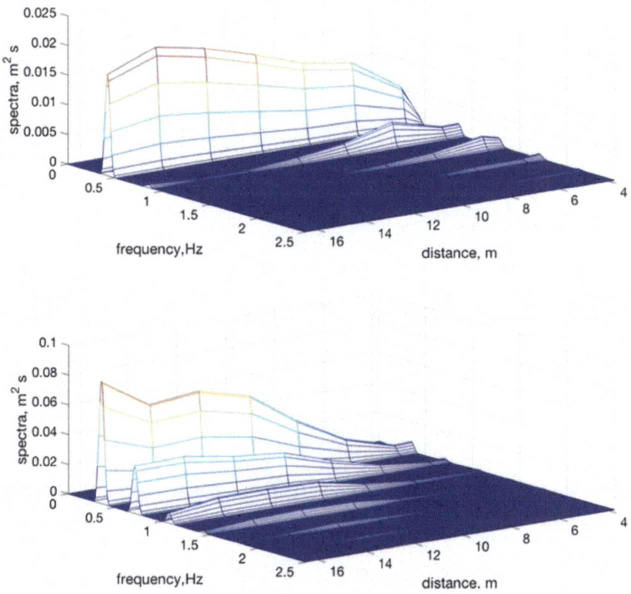

Fig. 3. Evolution of wave spectra for series 48, spilling breaker (above) and series 44, plunging breaker (under), *laboratory experiment*

During the wave breaking of plunging type, the energy decay of the first harmonic is prevailing over the others, and its proportion to the second harmonic decay is 3:1. In this case, the energy of high order harmonics decays evenly. In addition, the rate of energy decay of first harmonic for plunging breaking is 1.5–2 times bigger than for spilling ones.

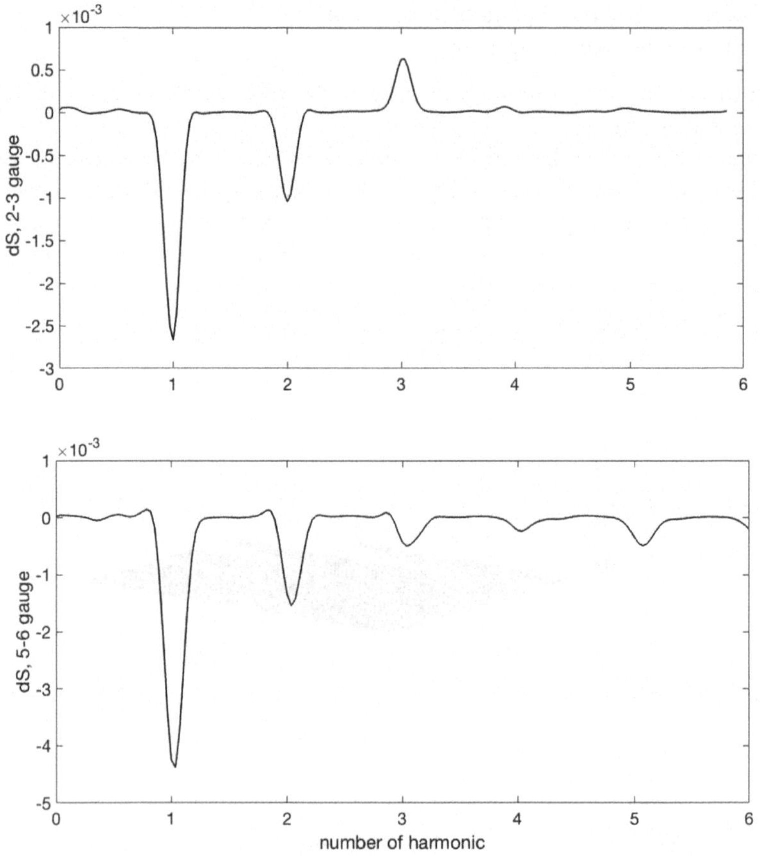

Fig. 4. The difference between spectra before and after breaking (dS), evaluated with the use of formula (1): series 48, spilling breaker (above), and series 44, plunging breaker (under), *laboratory experiment*

The above mentioned features were also discovered in natural waves. Figure 5 presents the evolution of wave spectra for series 38 of field experiment. For this series, different breaking events were observed at particular locations: spilling breaking at gauges 6, 8 and 11 (110, 135.9 and 174.8 m respectively), and plunging breaking at gauges 2, 7 and 13 (58, 122.9 and 200.8 m respectively).

Just like in case with monochromatic waves, right before the breaking point the first two harmonics of natural waves are on different levels of energy transfer. In spilling breaking, the energy of second non-linear harmonic is at its maximum, while it continues to grow in plunging breaking.

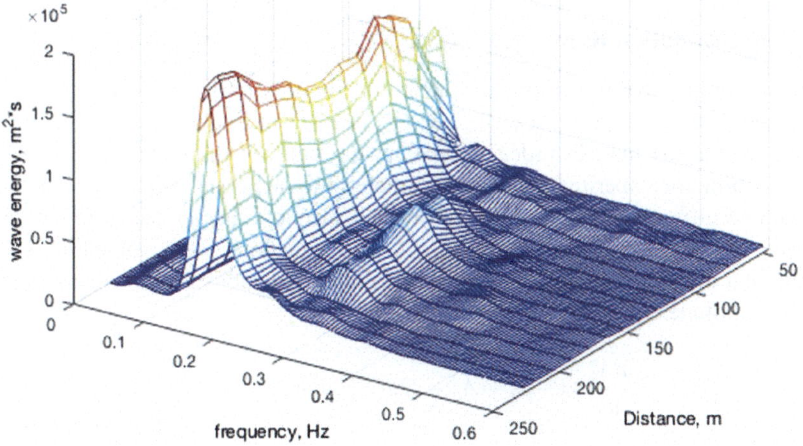

Fig. 5. The evolution of wave spectrum for series 38, *field experiment*

The spectral energy shift during the process of wave breaking is shown in Fig. 6. The same trends, that were pointed out for monochromatic waves during the laboratory experiment, were observed: (i) in spilling breaking, the energy decays within the span of first two harmonics, in proportion 3:1; (ii) in plunging breaking the energy decay of the first harmonic is significantly bigger than one of higher harmonics; (iii) the energy decay of first harmonic in plunging breaking is 3–4 times bigger than during spilling breaking events.

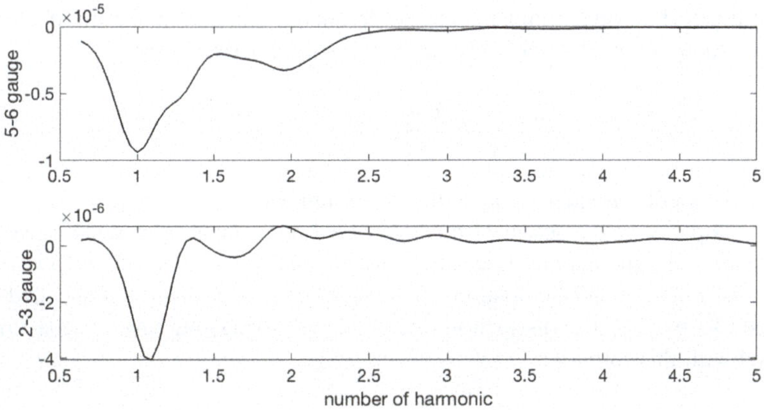

Fig. 6. The difference between spectra before and after breaking (dS), evaluated with the use of formula (1): series 38, spilling breaker (above), plunging breaker (under), *field experiment*

That way, during the process of wave breaking, the spectral energy decreases mainly within the span of first two harmonics in spilling breaking, while in plunging breaking the energy decays at all frequencies (Figs. 4 and 6).

3.2 The Dissipation Rate

The dynamic of wave transformation at the breaking point is affected both by dissipation and non-linear processes. The energy spectra evaluations, made in previous chapter, reflect just that. In order to separate the input of dissipation and non-linear transformation, the experimental data was imposed to numerical modeling with the use of Boussinesq-type equations with improved dispersion characteristics (Madsen and Sorensen [4]; Eldeberky and Battjes [1]). The equations were solved in spectral domain, and altogether they form a set of evolution equations for non-linear complex Fourier amplitudes (A_p):

$$\frac{dA_p}{dx} = -\beta_2 \frac{h_x}{h} A_p - i2g(F_p^+ + F_p^-) - \alpha_n A_p \tag{2}$$

The first term of the right side of (2) describes the linear increase of amplitude, when wave approaches the shallow water zone, and this term is proportional to slope steepness (h_x) and wave amplitude at the particular point. The second and third terms represent, respectively, the summary and incremental triad interaction of harmonics, described by

$$\begin{aligned} F_p^- &= \sum_{m=1}^{\infty} \frac{\alpha^-}{\beta_1} A_m^* A_{p+m} \exp[-i(\psi_{m+p} + \psi_m + \psi_p)] \\ F_p^+ &= \sum_{m=1}^{p-1} \frac{\alpha^+}{2\beta_1} A_m A_{p-m} \exp[-i(\psi_m + \psi_{p-m} + \psi_p)] \end{aligned} \tag{3}$$

where g is the gravitational force, β_1, α^+ and α^- are interaction coefficients, that depend on local depth (h), wave number (k_p), and frequency (ω_p); α_n is the dissipation rate, which is usually derived empirically, ψ_p is the phase, derived from

$$\frac{d\psi_p}{dx} = k_p, \tag{4}$$

where ω_p is angular frequency, k_p is the wave number.

The dissipation rate assessment method, based on comparing modeled (dissipation is negliged) and experimental (measurements, naturally including the effects of non-linear transformation and dissipation) of wave spectra, was suggested and laid out in (Mase and Kirby [5]). The dissipation rate of waves propagating towards coast may be estimated with this formula:

$$\alpha_n = \frac{\left(S(x_{start} + \Delta x)_{calc} - S(x_{start} + \Delta x)_{meas}\right)}{S(x_{start})_{meas} \times (2\Delta x)} \tag{5}$$

where S is wave spectra at specified location;

calc and meas indices apply to modeled and measured spectra, respectively;

Δx is the distance crossed by waves;

x_{start} is the coordinate of the starting point, where calculations begin.

Figure 7 represents the dissipation rate, calculated with the use of (5) for monochromatic waves during different breaking events. It can be observed, that in case with spilling breaking, the dissipation rate quadratically depends on wave frequency, while in plunging breaking, this dependency is virtually linear for all frequencies.

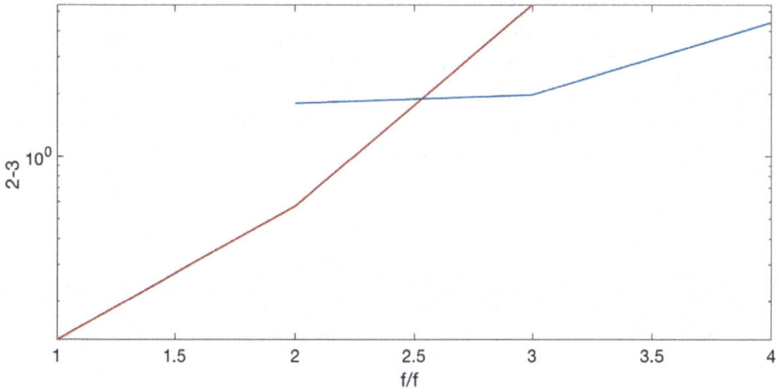

Fig. 7. The dissipation rate, calculated with the use of (5). Wave breaking occurred between gauges 2 and 3. Red line stands for series 48, spilling breaker, blue line stands for series 49, plunging breaker. *Laboratory experiment*

That way, the dissipation rate, related with the energy decrease during breaking, significantly differs, depending on the type of breaking event. The dissipation rate, which is equal to a constant, was suggested in (Eldeberky and Battjes [1]), and corresponds to plunging breaking, while for spilling breaking it requires quadratic parametrization (Mase and Kirby [5]).

4 Conclusion

The research has shown that during the process of wave breaking, the spectral energy decreases mainly within the span of first two harmonics in spilling breaking, while in plunging breaking the energy decays at all frequencies.

With the use of numerical modeling in order to separate the input of non-linear processes and dissipation into monochromatic waves transformation, it was confirmed that the dissipation rate significantly varies, depending on the type of breaking event—

in spilling breaking it quadratically depends on wave frequency, while in plunging breaking it equals to a constant.

Acknowledgements. This work was supported by the Russian Foundation for Basic Researches, project 18-55-45026.

References

1. Eldeberky, Y., Battjes, J.A.: Spectral modeling of wave breaking: application to Boussinesq equations. J. Geoph. Res. **101**(C1), 1253–1264 (1996)
2. Kuznetsov, S.Yu., Saprykina, Ya.V.: The dependency of wave energy dissipation of breaking non-regular waves on wave frequency. Water Resour. **31**(N4), 422–430 (2004). [in Russian]
3. Kuznetsov, S., Saprykina, Ya., Shtremel, M., Kuznetsova, O.: Spectral structure of breaking waves and its possible influence on the transport of sediments in coastal zone. In: Proceedings of Conference “Ocean's15 MTS/IEEE Genova”, Genova, Italy, 18–21 May 2015 (2015)
4. Madsen, P.A., Sørensen, O.R.: Bound waves and triad interactions in shallow water. J. Ocean Eng. **20**(4), 359–388 (1993)
5. Mase, H., Kirby, J.: Hybrid frequency-domain KdV equation for random wave transformation. In: Proceedings of 23D International Conference on Coastal Engineering, Venice, pp. 474–487 (1992)

Description of Ionosphere Disturbances Using Swarm Satellite Data Obtained During Hagupit Tropical Cyclone

V. I. Zakharov[1,2(✉)] and A. F. Khamidullin[1]

[1] Faculty of Physics, M.V. Lomonosov Moscow State University, Leninskie Gory, 1-2, 119991 Moscow, Russian Federation
zvi_555@list.ru
[2] A.M. Obukhov Institute of Atmospheric Physics Russian Academy of Sciences, Pyzhyovskiy pereulok, 3, 119017 Moscow, Russian Federation

Abstract. The paper discusses the allocation of disturbances in the ionosphere associated with large atmospheric events - tropical typhoons or cyclones. The experimental results of measurements of electron concentration by Langmuir probes carried out on ECA SWARM satellites were used. Selection technique for relative electron density perturbation in the selected frequency band were proposed and substantiated. The paper uses a frequency band equivalent to 15–180 s, which is associated with the characteristic spatial dimensions of the studied phenomenon in the atmosphere, this way allows us to distinguish disturbances from tropical cyclones in the ionosphere against the background of regular changes in the electron concentration determined by the dynamics of the Equatorial anomaly. The results of the calculations show a high correlation coefficient between the data from individual satellites of the constellation and a pronounced relationship between the spatial parameters of the cyclone and changes in the electronic content of the ionosphere.

Keywords: Ionosphere disturbances · SWARM · Electron density · Langmuir probe · Remote sensing · Tropical cyclone · Wave disturbances · Hagupit

1 Introduction

Earth's upper atmosphere and ionosphere are exposed to different processes occurring both in near-earth space and in the lower atmosphere. The mechanisms of interaction and energy transfer between the geospheres are not clear enough. For the time being such mechanisms are interpreted as a number of several assumptions, justified physically and in some cases confirmed experimentally. Consider the atmospheric-ionospheric interaction on the example of geophysical processes occurring with a large release of energy – typhoons and cyclones.

It is believed that the response from major atmospheric events is related to acoustic-gravity waves (AGW) and manifested as AGW generation by the atmospheric

V. I. Karev et al. (Eds.): *Physical and Mathematical Modeling of Earth and Environment Processes (2018)*, SPEES, pp. 287–294, 2019.
https://doi.org/10.1007/978-3-030-11533-3_28

turbulence in medium with natural frequencies spreading up to the ionosphere [1, 2]. Ionospheric response from the typhoons were detected by radiophysical techniques of the ionosphere monitoring: observation of the Doppler shift of high-frequency radio signals [3, 4], VLF transillumination of radio paths [5], measurements of ionospheric total electron content and its temporal variations using GPS signals [6, 7].

Purpose of this work is experimental study of ionospheric response from major tropical cyclone Hagupit 2014 on example of electromagnetic and plasma SWARM satellite measurement data.

2 Data and Method of Analysis

SWARM project consists of three satellites (A, B and C), located on two near-polar orbits: A and C are moving in the same plane with inclination of 87.4° actually at the same height, in similar orbits with a difference of eras of 2–10 s. Third satellite has other moving parameters and has a control character [8]. In this work we used data on the electron content spatial distribution, independently measured by onboard Langmuir probes with 2 Hz frequency and relative error not more than 1% [9].

Tropical typhoon trajectory data obtained from database [10, 11]. It contains information about cyclone evolution temporal dynamics and geolocation binding.

Study of tropical cyclone has one important feature despite all their power and the potential of their influence on the ionospheric heights. These events occur in the equatorial ionospheric anomaly localization area [12, 13], which makes it difficult to unambiguously interpret the possible manifestations of the influence of the atmospheric vortex. We can expect that there is no such problem in the middle latitudes from 30 to 60°, but in high latitudes influence of the ionospheric oval on the studied manifestations of atmospheric events in ionospheric plasma is possible.

Our idea of separating the effects of different processes on ionospheric response is related to different spatiotemporal characteristics of the studied processes. In fact, the EA region occupies a range of Equatorial latitudes of at least about 20…30°, while the influence of tropical cyclone should be localized in smaller latitude region, conventionally associated with the area of the storm or significant in the lower atmosphere. The spatial extent of the atmospheric storm region will be about 1500 km and less. The specified spatial size, taking into account velocity of the satellite, determines the choice of time filter parameters, which in our work are from 15 s and do not exceed 3–5 min. This approach allows us to consider the perturbation in the ionosphere as a wave, and the characteristics of the wave fall within the natural frequencies of the atmosphere.

According to authors of the technique and designers of SWARM satellites [http://directory.eoportal.org/web/eoportal/satellite-missions/s/swarm], the minimum value of the ion concentration, detected by Langmuir probes, are not less than 100 ions/cm^3, which allows to estimate the error in determining the relative electron concentration as not exceeding the relative value of 1%, under the assumption of plasma quasi-electroneutrality at altitudes of 450–500 km. The results of the selected effects everywhere exceed the specified value of experimental errors.

Typhoon Hagupit of the highest category 5 on the Saffir-Simpson scale [11] developed in the Equatorial Pacific ocean from tropical depression (on the "category" scale Fig. 1 value "−1") from 01 to 12 December 2014. The maximum wind speed on December 4 was almost 80 m/s. The geo-heliomagnetic situation during its operation was calm or weakly perturbed-the planetary Kp index did not exceed the value of 5 in some hours on December 6 and 7, and the Value of the DST index did not exceed 20 nT. The geo-heliomagnetic situation during its activity was calm or weakly perturbed, the planetary Kp index did not exceed the value of 5 in some hours on December 6 and 7, and the Value of the DST index did not exceed 20 nT.

Since the velocity of the measuring probe satellite-carrier is much higher than the velocity of the wave propagation, we obtain practically a section of modulation of the electron concentration in the direction of propagation of the satellite. The shape of the section can repeat the shape of the wave, but in general is the result of interference of waves of different origin. Therefore, calm conditions are important for interpreting observed effects.

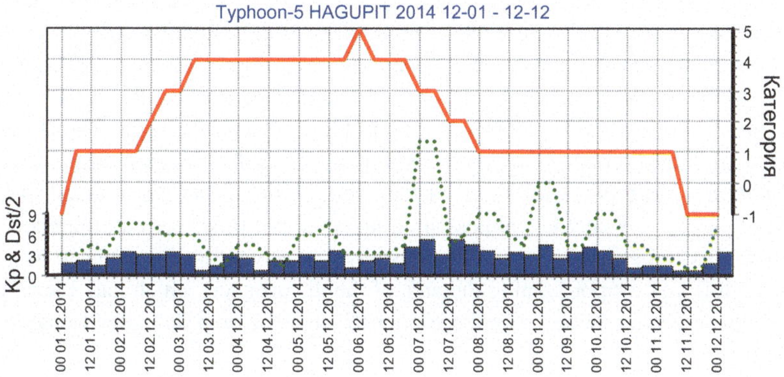

Fig. 1. Temporary dinamics of super Typhoon Hagupit, Saffir-Simpson scale (right axis, "−1" denote tropical depression, "5" - the highes step) and current geomagnetic indices Kp and Dst (left axis). Seen slight variations of Dst during the peak of the cyclone.

3 Analysis Results

Here is an example of the impact of the cyclone on the background of other processes acting during the passage of satellites.

Figure 2 shows typical electronic concentration registrations obtained by synchronous SWARM satellites-A and-C. The temporal behavior of the electron concentration in the Equatorial anomaly is seen. The software developed by us allows us to distinguish the responses in the electron concentration observed in the ionosphere against the influence of the Equatorial anomaly, which can be associated with the manifestations of tropical typhoons in the above-defined frequency band at ionospheric altitudes.

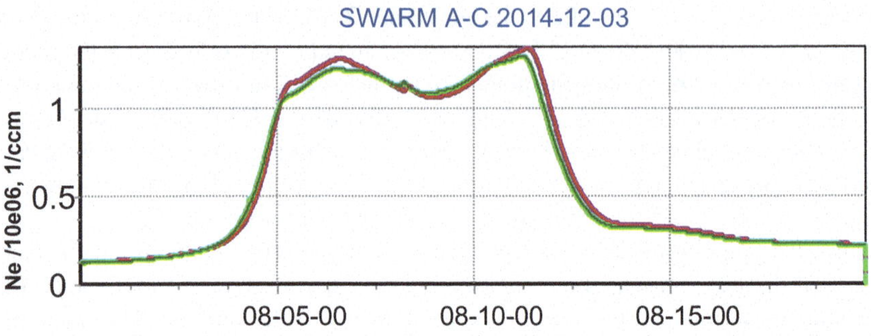

Fig. 2. Full registered evolution of the electron concentration during moving through the area of Equatorial magnetic anomaly approximately from 08:00 to 08:20 UTC, 3 December 2014

In Fig. 3, timestamps are given next to the tracks (bold almost vertical parallel rays) in the format hh:mm UTC. The track of movement of the Typhoon Hagupit for all the time of its existence is represented by a polyline starting at the point C coordinates (3.80 N, 154.0 E). The satellite tracks are color coded proportionally to the electronic concentration, the scale for which is shown in the figure on the right. The strength of the hurricane is also coded by color according to the Saffir-Simpson scale, a warmer color falls on a more powerful phase of the event, which is shown in detail in Fig. 1. It is seen that the two satellites almost simultaneously in time (with a difference of not more than 10 s) entering the region of the storm, is conventionally depicted as a circle of radius 1400 km, proportional to the area of the disturbance, which had 3 level on the Saffir-Simpson scale (panel a)) on December 02, and the next day goes to the level of 4th category.

The selected signal corresponding to the ionospheric response to the Typhoon is schematically presented in the form of curves relative to its track in Fig. 3(a), (b). Relative variations of the electron concentration magnitude are shown on these panels for convenience. Variations are refer to the band of natural frequencies, corresponding to periods of wave structures 15…150 s recorded by satellites A and C, respectively. The correlation coefficient of the selected variations amounted to more than 0.92, and the maximum corresponds to the time-shift in the movement of the SWARM -A and -C satellites.

A similar situation is observed for other analyzed spans, and the response value is proportional to the strength of the Typhoon. For example, the response is also observed on a SINGLE swarm satellite, when it flies into the region of ionospheric perturbation - see panel 3(в).

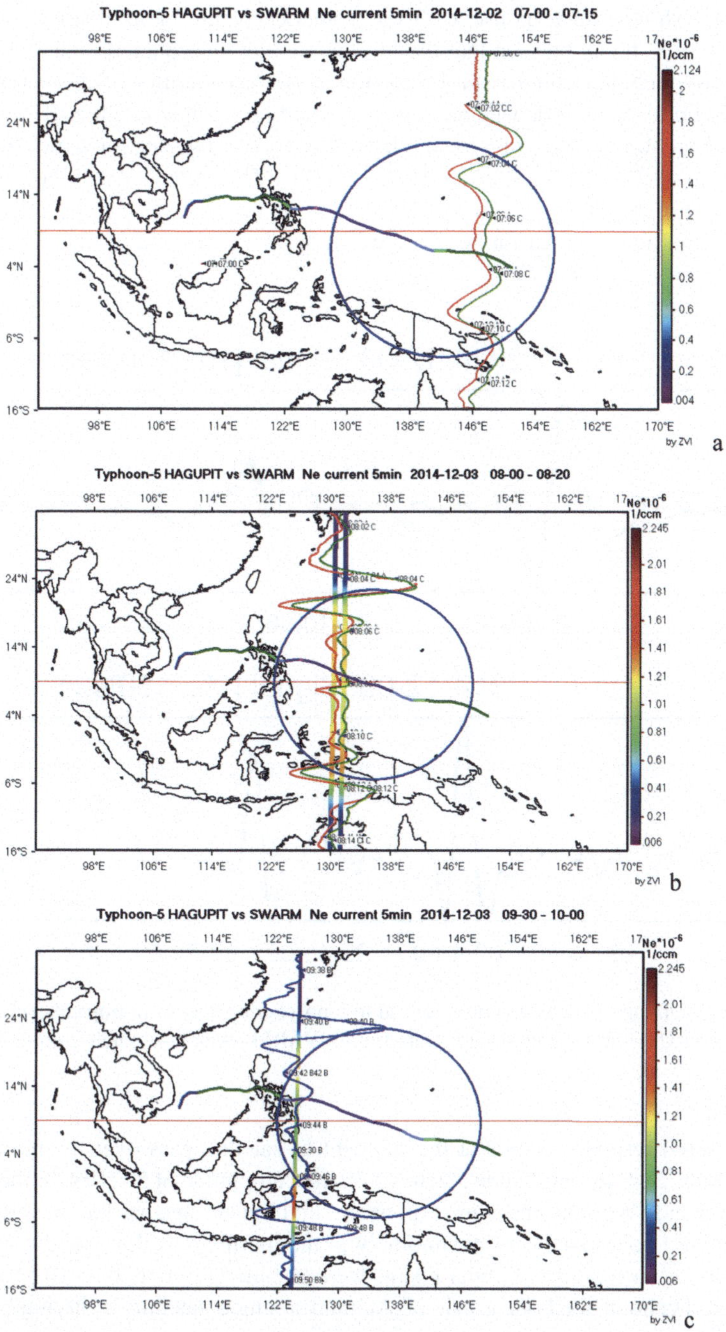

Fig. 3. Projections of SWARM satellite tracks on the earth surface. (A) satellites – A and –C, from 07 till 07:15 UTC, 2014, December 02. (B) satellites – A and –C, from 08 till 08:20 UTC, 2014, December 03. (C) satellite – B, from 09:30 till 10:00 UTC, 2014, December 03. The track of the Typhoon, the excitation region and the detected responses in the Ne ionosphere are presented. Panel (a) shows only the ionospheric response. The scale of responses is the same.

Figure 4 shows the response with the numerical value of its relative amplitude shown in Fig. 3. It can be seen that the response in the ionosphere to the atmospheric vortex is asymmetric, which is probably due to the peculiarities of the geomagnetic field of the Earth. Similar dependences are observed on a single satellite -B, see Fig. 4, panel (b). Note the presence of a local relative decrease in the electron concentration in the range up to 2%, observed at about 08:08 UTC. This behavior may be associated with the "eye of the Typhoon", in which variations in the atmospheric electrostatic field are significant, and requires further detailed study; it is observed in the primary data, see Fig. 2.

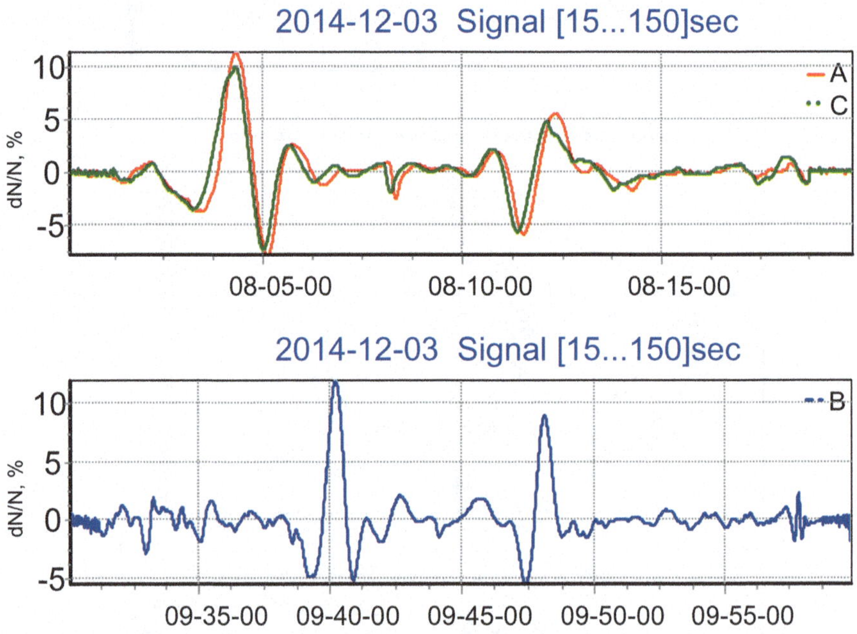

Fig. 4. Highlighting of wave structure, related to ionospheric effect from tropical cyclone, from 12:52 to 12:57 UTC, 2014, October 11. Data from SWARM –A and –C satellites, in range from 15 to 150 s period.

Discovered response occurs at the moment of satellite pass over the perturbation area, see Figs. 3–4, its magnitude reaches a relative amplitude of 5…10% in the studied frequency band. We note that the response is highlighted against the background of motion in the Equatorial electronic anomaly in the morning, as the spatial dimensions of the phenomena are differ. Similar results were obtained for more than 10 cases of the considered Typhoon for both a pair of AC and a single satellite B "flying" into the excitation region.

4 Main Conclusions

The review allows us to conclude that the proposed method of perturbation analysis in the selected frequency band corresponding to the spatial dimensions of atmospheric excitation is effective for isolating the contribution of the cyclone effect on the electron content in the ionosphere. This perturbation from a large cyclone reaches values up to 10% of the current values, which can be distinguished against the background of regular perturbations from other processes. It can be distinguished by the example of the considered large atmospheric event of the Hagupit tropical cyclone in more than a fifteen cases, mainly at the stages of development of 3–5 categories, using techniques of polynomial bandpass filtration, specially designed to solve this problem.

The obtained result can be wholly considered as a confirmation of the wave mechanism of perturbation transmission to ionospheric heights, strictly speaking, leaving out the question of the type of wave-agents of such transmission. We note that our estimates provide a spatial localization of perturbations "on the periphery", i.e. outside the area of the eye of the cyclone, which coincides with the findings [14] and the wave structure may correspond to acoustic-gravity branches of the natural oscillations of the atmosphere.

In conclusion, it is necessary to clarify that, most likely, the process of the impact of powerful atmospheric events on the ionosphere is not limited to wave effects, but is more complex. For example, our analysis techniques show the relative perturbation of the plasma, probably associated with the eye of a tropical Typhoon. In Fig. 4(a) this corresponds to a span in the immediate vicinity of the cyclone epicenter around 08:08 for SWARM-A and-C satellites and has a response rate of more than 2% and a duration of about 10…15 s. It is noteworthy that the track of the nearest flight of the satellite-B passes much to the West (almost 900 km, compare Fig. 3(b) and (c)) and has an amplitude at the measurement noise level (Fig. 4(b)). Estimation of the ionosphere perturbation field scale from the epicenter of the cyclone gives the value 70…100 km, which does not contradict the ground data on the spatial size of the eye of the cyclone.

Acknowledgements. SWARM data is provided by European Space Agency (https://earth.esa.int/web/guest/swarm/data-access). The work was partially supported by the grant 19-05-00941.

References

1. Danilov, A.D., et al.: Meteorological effects in ionosphere, Gidrometeolzdat., p. 267, in rus. (1987). (in Russian)
2. Hocke, K., Schlegel, K.: A review of atmospheric gravity waves and traveling ionospheric disturbances: 1982–1995. Ann. Geophysicae. **14**, 917–940 (1996)
3. Huang, Y.-N., Cheng, K., Chen, S.-W.: On the detection of acoustic-gravity waves generated by typhoon by use of real time HF Doppler frequency shift sounding system. Radio Sci. **20**, 897–906 (1985)
4. Xiao, Z., et al.: Morphological features of ionospheric response to typhoon. J. Geophys. Res. **112**, 1–5 (2007)

5. Rozhnoi, A., Solovieva, M., Levin, B.: Meteorological effects in the lower ionosphere as based on VLF/LF signal observations. Nat. Hazards Earth Syst. Sci. **14**, 2671–2679 (2014)
6. Zakharov, V.I., Kunitsyn, V.E.: Regional features of atmospheric manifestations of tropical cyclones according to ground-based GPS network data. Geomag. Aeron. **52–4**, 533–545 (2012)
7. Polyakova, A.S., Perevalova, N.P.: Comparative analysis of TEC disturbances over tropical cyclone zones in the North-West Pacific Ocean. Adv. Space Res. **52**, 1416–1426 (2013)
8. Olsen, N., et al.: The Swarm Satellite Constellation Application and Research Facility (SCARF) and Swarm data products. Earth, Planets and Space **65**, 1189–1200 (2013)
9. http://directory.eoportal.org/web/eoportal/satellite-missions/s/swarm
10. http://weather.unisys.com/hurricane/
11. http://www.nhc.noaa.gov/aboutsshws.php
12. Walker, G.O.: Longitudinal structure of the F-region equatorial anomaly - a review. J. Atmos. Terr. Phys. **43**(8), 763–774 (1981)
13. Balana, N.J., Souza, G., Baileyc, J.: Recent developments in the understanding of equatorial ionization anomaly: a review. J. Atmos. Solar Terr. Phys. **171**, 3–11 (2018)
14. Li, W., et al.: Ionospheric responses to typhoons in Australia during 2005–2014 using GNSS and FORMOSAT-3/COSMIC measurements. GPS Solut. **22**, 61 (2018)

Numerical Simulation of Heterogeneous Hyperelastic Model with RKDG

M. V. Alekseev(✉) and E. B. Savenkov

Keldysh Institute of Applied Mathematics of Russian Academy of Science, Moscow, Russian Federation
mikhail.alekseev@phystech.edu

Abstract. To describe shock wave propagation with large deformations when more than one material is presented, the first-order hyperbolic system of conservation laws in the Eulerian reference frame is considered. This model was developed by Godunov and Romenski [1] and differs from more traditional second-order formulations employed in elasticity. It has complex structure of contact, shock and rarefaction waves and their interactions. Even more complex models arise when multi-material medium is considered. Traditional approaches to describe them are based on the correct multi-phase extensions of the Godunov-Romensky model. In this paper different materials are described by the same equation of state (EoS) with spatially distributed parameters in material coordinates. A number of an efficient approaches were presented recently to handle hyperbolic hyperelastic problems. The most of them involve linearized HLL-type solvers and high-order WENO reconstruction techniques to obtain accurate fronts resolution and resolve complex solution structure. In this paper we consider discontinuous Galerkin method with third-order TVD Runge-Kutta scheme (RKDG) as an alternative to WENO schemes. Level-set method is used for multi-material description. It is implemented with non-conservative form of transport equation and is solved using the discontinuous Galerkin method based on Dal Maso, Le Floch and Murat framework. The presented results cover discussion of the problem statement, corresponding RKDG approximations and simulation results that demonstrate possibilities of RKDG approach applied for the hyperbolic hyperelastic formulations in the Eulerian reference frame.

Keywords: Hyperelasticity · Hyperbolic equations · Eulerian reference frame mechanics · Multimaterial · Level-set method

1 Introduction

Recently Eulerian formulations for solid dynamics with finite strain suitable to handle large deformations are paid substantial attention to. One of its major features is possibility to use Eulerian computational meshes which avoids a number

V. I. Karev et al. (Eds.): *Physical and Mathematical Modeling of Earth and Environment Processes (2018)*, SPEES, pp. 295–302, 2019.
https://doi.org/10.1007/978-3-030-11533-3_29

of problems typical for traditional arbitrary Eulerian-Lagrangian formulations. The latter ones suffer from lack of robustness in the case of extremely large deformations [2]. In this paper the Eulerian reference frame of first-order hyperbolic formulations of the problem is considered. The basic model of that type was originally developed by Godunov and Romenski [1]. The material is assumed to be hyperelastic. This means that material behavior is completely defined by a strain-energy (or stored-energy) function $\mathcal{U}$ which, in turn, can be used to obtain material thermodynamic properties including stress-strain relationship. The most common example of this kind of material is rubber, whose stress-strain relationship can be defined as non-linearly elastic, isotropic and generally independent of strain rate.

The more complex problem is description of multi-material medium. The most prominent approach in this case is utilization correct multi-phase extensions of the Godunov-Romenski model which is a complex task [2]. In this paper we use simple multi-material model which assumes that materials are described by the same equation of state (EoS) with spatially distributed parameters in material coordinates. In such way a multi-material medium is described by single phase medium with highly heterogeneous properties distribution. To handle hyperbolic hyperelastic problems, a number of an efficient approaches were presented recently. The most of them involve linearised HLL-type solvers and high-order WENO reconstruction techniques to obtain accurate fronts resolution and resolve complex solution structure.

In this paper we consider RKDG method as an alternative to WENO schemes. To make the approach as much "Riemann-free" as possible, a limiting procedure is applied to conservative variables and simple Lax-Friedrichs numerical fluxes are used. The multi-material treatment is implemented using a variant of level-set method. Non-conservative form of transport equation for level-set function is solved using the discontinuous Galerkin method based on Dal Maso, Le Floch and Murat framework [3].

2 Mathematical Model

2.1 Homogeneous Hyperelastic Model

To describe deformation of solid media in the Eulerian reference frame the model of Godunov and Romenski [1] is used. The complete three-dimensional system of momentum, strain and energy conservation laws is given respectively by:

$$\frac{\partial}{\partial t}(\rho \boldsymbol{u}) + \frac{\partial (\rho \boldsymbol{u} \otimes \boldsymbol{u} - \boldsymbol{T})}{\partial x} = 0, \tag{1a}$$

$$\frac{\partial}{\partial t}(\rho \boldsymbol{F}) + \frac{\partial \left(\rho \boldsymbol{F} \otimes \boldsymbol{u} - \rho \boldsymbol{u} \otimes \boldsymbol{F}^T\right)}{\partial x} = 0, \tag{1b}$$

$$\frac{\partial}{\partial t}(\rho E) + \frac{\partial (\rho \boldsymbol{u} E - \boldsymbol{u} \otimes \boldsymbol{T})}{\partial x} = 0. \tag{1c}$$

Here ρ is density given by equation $\rho = \rho_0 / \det(\boldsymbol{F})$, $\boldsymbol{T}$ is Cauchy stress tensor given by Murnaghan formula [4], ρ_0 is the initial density of unstressed medium, $E = \mathcal{U} + |\boldsymbol{u}|^2/2$ is the total energy. The solid media state is characterized by the elastic deformation gradient $\boldsymbol{F} = \nabla_{\boldsymbol{X}} \boldsymbol{x}$ ($\boldsymbol{x}$ and $\boldsymbol{X}$ denote the eulerian and material coordinates of the initial state respectively), velocity $\boldsymbol{u}$ and entropy S.

Equation of state for specific internal energy is used in the form $\mathcal{U} = \mathcal{U}(\boldsymbol{F}, S)$. It is shown in [5] that the internal energy must be expressed in terms of some symmetric strain tensor to satisfy frame indifference. In this paper the symmetric deformation Finger tensor defined as $\boldsymbol{G} = \boldsymbol{F}^{-T}\boldsymbol{F}^{-1}$ is used so internal energy has the form $\mathcal{U} = \mathcal{U}(\boldsymbol{G}, S)$. In that case the specific internal energy for a hyperelastic isotropic medium depends on principle invariants of Finger tensor [4]: $\mathcal{U} = \mathcal{U}(I_1, I_2, I_3, S)$. Here $I_1 = \mathbf{tr}(\boldsymbol{G})$, $I_2 = \left[\mathbf{tr}(\boldsymbol{G})^2 - \mathbf{tr}\left(\boldsymbol{G}^2\right)\right]/2$, $I_3 = \mathbf{det}(\boldsymbol{G})$.

Finally, the specific internal energy can be divided into hydrostatic potential part $\mathcal{U}^{\mathrm{h}}$ and shear deformation term $\mathcal{U}^{\mathrm{sh}}$:

$$\mathcal{U} = \mathcal{U}(I_1, I_2, I_3, S) = \mathcal{U}^{\mathrm{h}}(I_3, S) + \mathcal{U}^{\mathrm{sh}}(I_1, I_2, I_3, S). \tag{2}$$

Further development of this model can be found in [6].

In present paper the isotropic hyperelastic equation of state from [6] is used. The corresponding terms are given by:

$$\mathcal{U}^{\mathrm{h}}(I_3, S) = \frac{K_0}{2\alpha^2}\left(I_3^{\alpha/2} - 1\right)^2 + c_v T_0 I_3^{\gamma/2}\left(\exp\left[S/c_v\right] - 1\right) \tag{3}$$

and

$$\mathcal{U}^{\mathrm{sh}}(I_1, I_2, I_3) = B_0 I_3^{\beta/2}\left(I_1^2/3 - I_2\right)/2. \tag{4}$$

Here $K_0 = c_0^2 - (4/3)\, b_0^2, B_0 = b_0^2$ are the squared bulk speed of sound and the squared speed of shear wave, respectively, c_v is heat capacity at constant volume, α, β, γ are nonlinear coefficients of equation of state.

2.2 Heterogeneous Hyperelastic Model

We assume now that the medium properties (which are coefficients in EoS) can vary in media. Consider piece-wise constant distribution which is typical for applications. In that case one can utilize widely used level-set approach to describe their distribution in Eulerian reference frame. With this approach, the model is formally single-phase but allows to "simulate" real multi-phase setting [7] up to same accuracy. In traditional level-set method [8] level-set function Φ is given by:

$$|\Phi(\boldsymbol{x})| = d(\boldsymbol{x}) = \min_{x_I \in I}\left(|\boldsymbol{x} - \boldsymbol{x}_I|\right), \tag{5}$$

where I is interface, $\Phi(\boldsymbol{x}) > 0$ on one side of the interface, $\Phi(\boldsymbol{x}) < 0$ on the other. An interface I is fixed in material frame but naturally evolves in the Eulerian coordinates due to material motion. The Heaviside function $h(x)$, $h(x) = 0$

if $x < 0$ and $h(x) = 1$ if $x > 0$ is used as the weighting function defined in Eulerian reference frame. In this approach level-set function it can be reformulated as $H(\boldsymbol{x}) = h(\Phi((x)))$. For any piece-wise constant function f defined in material coordinates the corresponding parameters distribution in Eulerian frame is given by $f(\boldsymbol{x}) = (f_R - f_L)\, H(\boldsymbol{x}) + f_L$, where $f_{R,L}$ are values inside and outside domain bounded by interface I. To achieve numerical robustness we use smeared Heaviside function defined as:

$$H_{\rm sm} = \begin{cases} 0, & \Phi < \varepsilon, \\ \dfrac{1}{2} + \dfrac{\Phi}{2\varepsilon} + \dfrac{1}{2\pi}\sin\left(\dfrac{\pi\Phi}{\varepsilon}\right), & -\varepsilon \le \Phi \ge \varepsilon, \\ 1, & \Phi > \varepsilon, \end{cases} \tag{6}$$

where ε corresponds to half the thickness of the interface. If this smeared function replace level-set function $\overline{\Phi} = H_{\rm sm}$ we would not have to restore Φ from $H_{\rm sm}$. Assume the interface is advected with a given velocity field $\boldsymbol{u}$. The corresponding motion of the modified level-set function is given by the non-conservative transfer equation in one-dimensional case:

$$\frac{\partial\overline{\Phi}}{\partial t} + v\frac{\partial\overline{\Phi}}{\partial x} = 0. \tag{7}$$

Here $\overline{\Phi} \in [0, 1]$, v is one-dimensional velocity field. With modified level-set function the parameters weighting is formulated as $u(\boldsymbol{x}) = (u_R - u_L)\,\overline{\Phi}(\boldsymbol{x}) + u_L$.

In this paper the corresponding parameters to be weighted are b_0, c_0, c_v, T_0, α, γ. Thus, this approach formally sets a given inhomogeneous distribution of the properties of the medium in Lagrangian coordinates. In this case, the distribution of properties corresponds to two homogeneous subdomains defined by $\Phi = 0, 1$ and 1 and Lagrangian "diffuse boundary" with the width of 2ε, within which material properties change continuously, but rapidly.

3 Numerical Approach

The system of Eqs. (1a), (1b), (1c) and corresponding equation for level-set function advection (7) can be represented as:

$$\frac{\partial \boldsymbol{U}}{\partial t} + \frac{\partial \boldsymbol{F}(\boldsymbol{U})}{\partial x} = 0, \quad \frac{\partial\Phi}{\partial t} + \boldsymbol{v}\frac{\partial\Phi}{\partial x} = 0,$$

where $\boldsymbol{U}$ is vector of conservative variables, $\boldsymbol{F}(\boldsymbol{U})$ is corresponding flux, Φ is level-set function. System (3) is solved numerically. At first stage hyperelastic equations are solved with given distribution of level-set function from previous time step. At the second stage level-set equation is used to update level-set function with already calculated velocity field. In both stages the discontinuous Galerkin approach (DG) is used [9,10]. Since we do not introduce any problem specific features into the DG approximations, they are not described here.

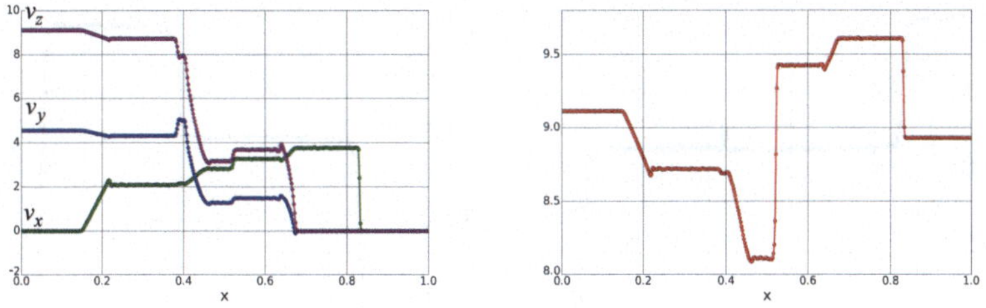

Fig. 1. Density (left) and velocity (right) profiles for the 1st case

For advection equation for Φ which contains non-conservative product the DG weak formulation can not be derived in its classical way. Different options can be suggested in that case. We use the one based on Dal Maso, Le Floch and Murat approach [3].

In order to reduce solution oscillations of high-order schemes the slope moment limiter introduced by Krivodonova [11] is used. It applies progressively limiting the high-order terms firstly if the solution starts to oscillate. Formally, it achieves the highest possible accuracy when some limiting is necessary. In this work it is conservative variables that are limited. Standard modified monotonicity preserving limiter [12] for level-set advection equation is used to obtain numerical solution with no over- and undershoots.

3.1 Implementation

The main purpose of developing software implementation is to create a test bed for analyzing various approximation methods and various models. For that reason, we tried to use the algorithms as close to physics-free approach as possible. In connection with this, the limiting of conservative variables and the simplest Lax-Friedrichs fluxes are used.

One of the key features of the class of models under consideration is the possibility of using various type of equations of state. For this reason, it is specified directly in the software implementation. At the same time, automatic differentiation is used to calculate all thermodynamic quantities. The results are obtained using the automatic differentiation library STAN [13] to calculate quantities derived from EoS such as $\boldsymbol{T}$ and diffusive term of flux with its magnitude $c_{\max}$. This allows for the flexibility of specifying the equation of state. Furthermore, Riemann-solver-free discontinuous Galerkin method considered in this paper could be implemented for systems without knowing the full eigenstructure. This also avoids a number of difficulties and adds variability of approach. The code is implemented with C++ and include template meta programming techniques.

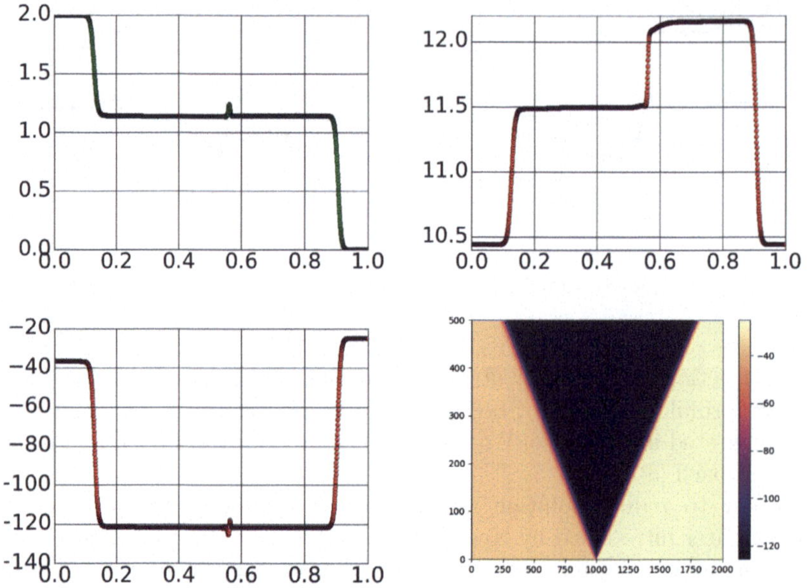

Fig. 2. Tets case 2: velocity (top left), density (top right), stress tensor component σ_{11} plot (bottom left) and x-t plot (bottom right)

4 Results

4.1 Homogeneous Material

The five wave example in [14] is taken as a basis for further validation of the numerical scheme presented in this paper. The initial conditions are

$$X_L = \left\{ \boldsymbol{u}_l = \begin{pmatrix} 0 \\ 0.5 \\ 1 \end{pmatrix} \mathrm{km\,s^{-1}}, \quad \boldsymbol{F} = \begin{pmatrix} 0.98 & 0 & 0 \\ 0.02 & 1 & 0.1 \\ 0 & 0 & 1 \end{pmatrix}, \quad S = 10^{-3}\,\mathrm{kJ\,g^{-1}\,K^{-1}} \right\},$$

$$X_R = \left\{ \boldsymbol{u}_r = \begin{pmatrix} 0 \\ 0 \\ 0 \end{pmatrix} \mathrm{km\,s^{-1}}, \quad \boldsymbol{F} = \begin{pmatrix} 1 & 0 & 0 \\ 0 & 1 & 0.1 \\ 0 & 0 & 1 \end{pmatrix}, \quad S = 0\,\mathrm{kJ\,g^{-1}\,K^{-1}} \right\}.$$

An additional degree of shear deformation as described in [6] is added in order to obtain full seven wave structure to further comparison of different schemes approximation order. There are three left travelling rarefaction waves, one right travelling contact wave, right travelling shock wave and two right travelling rarefactions waves. Figure 1 shows velocity (left hand side plot) and density (right hand side plot) profiles obtained using RKDG scheme with 500 cells for time 0.6 µs.

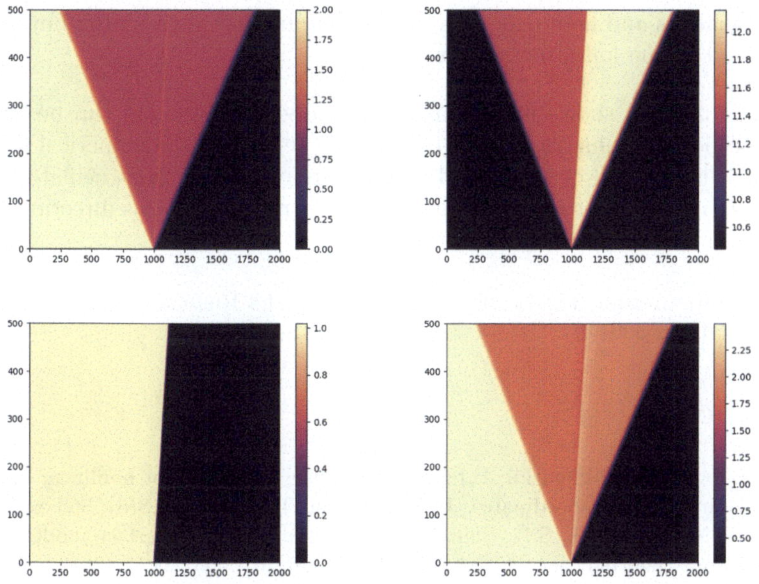

Fig. 3. x-t plots for velocity, density, level-set function and internal energy (left to right, top to bottom)

4.2 Heterogeneous Material

In this second example both materials are considered to be copper with different equation of state parameters corresponding to different types of material behavior. The level-set function is used to obtain material interface. The initial conditions are

$$X_L = \left\{ \boldsymbol{u}_l = \begin{pmatrix} 2 \\ 0 \\ 0.1 \end{pmatrix} \mathrm{km\,s^{-1}}, \quad \boldsymbol{F} = \begin{pmatrix} 1 & 0 & 0 \\ -0.01 & 0.95 & 0.02 \\ -0.015 & 0 & 0.9 \end{pmatrix}, \quad S = 0\,\mathrm{kJ\,g^{-1}\,K^{-1}} \right\},$$

$$X_R = \left\{ \boldsymbol{u}_r = \begin{pmatrix} 0 \\ -0.03 \\ -0.01 \end{pmatrix} \mathrm{km\,s^{-1}}, \quad \boldsymbol{F} = \begin{pmatrix} 1 & 0 & 0 \\ 0.015 & 0.95 & 0 \\ -0.01 & 0 & 0.9 \end{pmatrix}, \quad S = 0\,\mathrm{kJ\,g^{-1}\,K^{-1}} \right\}.$$

Figure 2 shows velocity and density profiles obtained using RKDG scheme and level-set method with 2000 cells for time 0.5 μs. Figures 3 show velocity and density x-t diagrams for the same conditions. The value of the diffuse interface parameter is $\varepsilon = 0.005$ cm which corresponds approximately to 10 mesh cells.

5 Conclusion

The purpose of the present study was to apply existing high-order RKDG method with simple Lax-Friedrichs numerical fluxes and Krivodonova slope limiter to the non-linear hyperelastic model in Eulerian reference frame for the case of homogeneous and heterogeneous thermodynamic properties distribution. The results lead to the following conclusions:

- RKDG method shows improvement in resolving waves and can be used for such types of problems in spite of simplest version of the numerical fluxes.
- Results using high-order RKDG method are essentially non-oscillatory when Krivodonova limiter is implemented. Further research in this direction will be put to improve the current situation.

Acknowledgements. This work is supported by the Russian Science Foundation, project $N^{\underline{o}}$ 17-71-30014.

References

1. Godunov, S.K., Romenskii, E.I.: Nonstationary equations of nonlinear elasticity theory in Eulerian coordinates. J. Appl. Mech. Tech. Phys. **13**(6), 868–884 (1972)
2. Favrie, N., Gavrilyuk, S.L., Saurel, R.: Solid fluid diffuse interface model in cases of extreme deformations. J. Comput. Phys. **228**(16), 6037–6077 (2009)
3. Dal Maso, G., Le Floch, P., Murat, F.: Definition and weak stability of nonconservative products. Journal de mathématiques pures et appliques. **74**, 483–548 (1995)
4. Godunov, S.K., Romenskii, E.I.: Elements of Continuum Mechanics and Conservation Laws. Springer Science and Business Media, London (2003)
5. Miller, G.H.: Minimal rotationally invariant bases for hyperelasticity. SIAM J. Appl. Math. **64**, 2050 (2004)
6. Barton, P.T., et al.: Exact and approximate solutions of Riemann problems in non-linear elasticity. J. Comput. Phys. **228**(18), 7046–7068 (2009)
7. Miller, G.H., Colella, P.: A high-order Eulerian Godunov method for elastic-plastic flow in solids. J. Comput. Phys. **167**, 131–176 (2001)
8. Olsson, E., Kreiss, G.: A conservative level-set method for two phase flow. J. Comput. Phys. **210**, 225–246 (2005)
9. Cockburn, B., Karniadakis, G.E. Shu, C.-W.: The development of discontinuous Galerkin methods. In: Discontinuous Galerkin Methods, pp. 3–50. Springer, Heidelberg (2000)
10. Shu, C., Osher, S.: Efficient implementation of essentially non-oscillatory shock-capturing schemes. J. Comput. Phys. **77**, 439 (1988)
11. Krivodonova, L.: Limiters for high-order discontinuous Galerkin methods. J. Comput. Phys. **226**(1), 879–896 (2007)
12. Owkes, M., Desjardins, O.: A discontinuous Galerkin conservative level-set scheme for interface capturing in multiphase flows. J. Comput. Phys. **249**, 275–302 (2013)
13. Carpenter, B., Hoffman, M., Brubaker, M., Lee, D., Li, P., Betancourt, J.M.: The Stan Math Library: Reverse-Mode Automatic Differentiation in C++. arXiv 1509.07164 (2015)
14. Titarev, V.A., Romenski, E.I., Toro, E.F.: MUSTA-type upwind fluxes for non-linear elasticity. Int. J. Numer. Meth. Eng. **73**, 897 (2008)

Localization of the Vortices and Road Sings in Images

E. N. Terentiev[1(✉)], I. N. Prikhodko[1], I. I. Farshakova[1], I. D. Kuznetsov[1], and N. E. Shilin-Terentiev[2]

[1] Faculty of Physics, Department of Mathematical Modeling and Informatics, M.V. Lomonosov Moscow State University, Moscow 119991, Russia
en.teren@physics.msu.ru
[2] Google, San Francisco, CA, USA
nikolay.terentyev@gmail.com

Abstract. An image in one color is a scalar field. The image is associated with its vector gradient field. The analysis of vector fields with the help of single-length vector templates made it possible to indicate in the image such objects as vortices, Road Signs (RS), Facial Features (FF), and so on. The proposed method of indicating local objects using gradient fields turned out to be highly precision and noise-immunity. The presented advantages of the proposed method recommend it as a promising tool in targets designation problems.

Keywords: Image analysis · Localization of object in the image · Computer vision · Designation problems of targets · Road signs · Facial features · False vortices

1 Introduction

The task of localizing objects in an image is one of the basic tasks of computer vision. The task can be considered as consisting of two tasks: finding the candidate's location in the desired object class and its subsequent classification. Problems of this type include: problems of recognition of RS, described in many articles [1–9] problems search for vortices [10], detection of person's face in article [11], etc.

Popular approaches to the localization of objects, for example, RS, based on methods of contour analysis [1], or combined with color segmentation of the image, as in [2], have problems in the accuracy and stability of localization. Because of poor illumination of the objects, the areas of the candidate for RS in the image are inaccurate, with the loss of a part of the RS, or vice versa, the area of the RS candidate is chosen too large. What makes it impossible to further classify the object? Estimation of the location of the object is given by morph lets, which are a special modification of the methods of morphological analysis [12]. Used in various articles [3–8] methods of classification of RS, like neural networks [3, 4, 6], contour descriptors [5], color

V. I. Karev et al. (Eds.): *Physical and Mathematical Modeling of Earth and Environment Processes (2018)*, SPEES, pp. 303–316, 2019.
https://doi.org/10.1007/978-3-030-11533-3_30

histograms [9], the search for key points [7] have their problems in the accuracy and stability of classification in their specific tools, terms. The use of precise localization methods with the estimation of the sizes of objects that are stable to the conditions of registration will make it possible to improve the accuracy of classification, reading, for example, RS.

2 Finite-Dimensional Sampling Theorems with Operations of Differentiation and Integration

Finite-Dimensional Sampling Theorems (FDST) realize interpolation operations, field theories with sample arrays with a mantissa accuracy of the order of $\sim 10^{-13}$ without using finite difference schemes and integral sums.

2D FDST (Terentiev)

Let three arrays of samples D, H and *H* be given. D is a two-dimensional array of samples (the image on the original grid NxN) D = f(x0,y0). H is a two-dimensional orthonormal matrix of the discrete Fourier harmonics H(x0) = {Hi(x0), i = 1:N, x0 = 0:N-1}. $H^{(n)}(x)=\left\{H_i^{(n)}(x),\ i = 1{:}N, x=0 : dx{:}N - dx\right\}$ is the "continuous" Fourier harmonics, with a smaller step dx < 1. All Fourier harmonics are numbered first - i index and in the discrete matrices H(x0), dx = 1 and in the interpolated matrices $H^{(n)}(x)$, dx < 1. The Fourier harmonics are arranged in rows.

The subscripts in parentheses (n) are the n-th (for positive n > 0) derivatives and the –n-th integral for negative values [13].

Then the function:

$$f^{(n,m)}(x,y) = \sum_{i,j=1}^{N} c_{i,j}\, H_i^{(n)}(x)\, H_j^{(m)}(y) \tag{1}$$

$$\begin{aligned} c_{i,j} &= \left(f(x0,y0),\ H_i(x0)\, H_j(y0)\right) \\ &= \sum_{x0,\,y0=1}^{N} f(x0,y0)\, H_i(x0)\, H_j(y0),\ i,j = 1{:}N \end{aligned} \tag{2}$$

passes through the sample points f (x0, y0) for n = m=0.

The scalar products (2) realize the two-dimensional Direct Fourier Transform (DFT). The Fourier series f (n, m) (x, y) (1) gives us an interpolated two-dimensional result for n = m=0, dx < 1. If there is no interpolation: dx = 1 and n = m=0, then the Fourier series f (n, m) (x, y) (1) realizes the Inverse FT. If n and m are not equal to zero, then we implement the operations of differentiation and integration from the sample arrays [13]. Note that the orders of the operations n, m are not necessarily integers [13].

1D FDST: ***Suppose*** we have a one-dimensional array of samples D = f (x0), ***then*** the result of interpolation with the n-th operation is represented by the formula in one line:

$$f^{(n)}(x) = (H(x0)^{*}D')'^{*}H^{(n)}(x) \tag{3}$$

An asterisk * in (3) denotes the rule "string * column with summation" when multiplying matrices and a prime 'is the transpose operation.

The first asterisk realizes a one-dimensional DFT (2), and the second asterisk is the sum of a discrete Fourier series (1). Note that the matrices of digitized Fourier harmonics H(x0) are unitary in construction and are intended for the realization of one-dimensional FTs. If n = 0 and x = x0, then the original equation $f(x0) = f^{(0)}(x0) = D$ follows from the fact that the matrix H(x0) is unitary (the transpose operation reverses the unitary matrix). The FDSTs give good results if there is no difference in image boundary. Otherwise, it is necessary to apply special methods protected from boundary effects in the FDST [13].

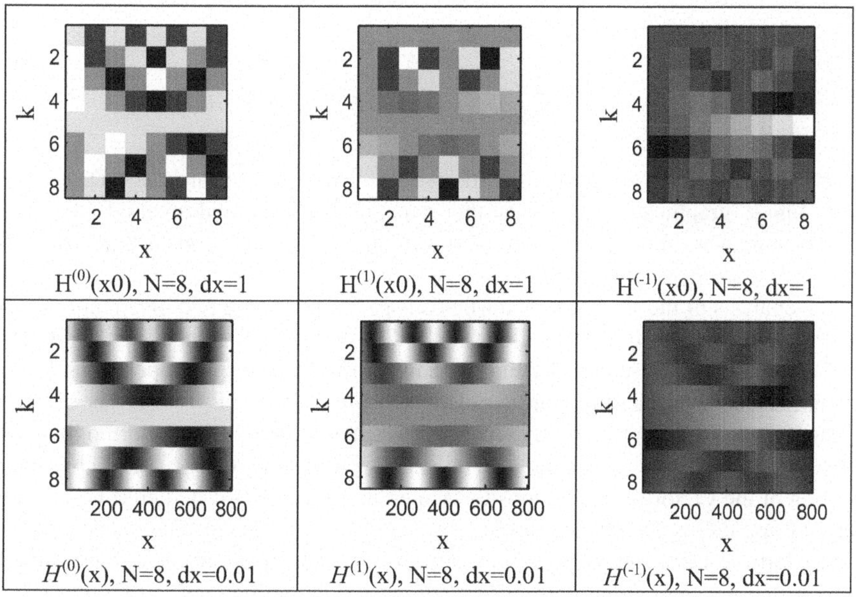

Fig. 1. Example of discrete and interpolated differentiated and integrated Fourier harmonics in the FDST.

Below, we give formulas with operations of differentiation and integration of the n-th order for sines and cosines:

$$H_k^{\cos^{(n)}(\rho_m x)} = \sqrt{2/N}\left\{(\rho_m)^n \cos\left(\rho_m x + n\frac{\pi}{2}\right)\right\}, \rho_m = \frac{2\pi}{N}m,\ m = \frac{N}{2} + 1 - k, \quad (4)$$

$$H_k^{\sin^{(n)}(\rho_m x)} = \sqrt{2/N}\left\{(\rho_m)^n \sin\left(\rho_m x + n\frac{\pi}{2}\right)\right\}, \rho_m = \frac{2\pi}{N}m,\ m = k - \frac{N}{2} + 1, \quad (5)$$

In (4, 5), the index k = 1: N indicates the line numbers in the matrix of continuous Fourier harmonics $H^{(n)}(x)$, see Fig. 1 for even N with $\sqrt{2/N}$ normalization. For the lines of the unpaired cosine and the constant in $H^{(0)}(x0)$, another normalization is $\sqrt{1/N}$.

The normalizations from the discrete case are necessary for the orthonormality of the rows of the matrix $H^{(0)}(x0)$ [13].

3 Gradient Fields, Direction Templates and Vector Operations

According to the 2D FDST (1, 2), the gradient from the sample array D = f (x0, y0) with interpolation at the points (x, y) is:

$$\text{gD} = \text{grad D} = \{\frac{\partial}{\partial x}\text{D}, \frac{\partial}{\partial y}\text{D}\} = \{f^{(1,0)}(x,y), f^{(0,1)}(x,y)\}. \quad (6)$$

gD (6) – there is a vector gradient field associated with the scalar image D.

In Fig. 2 graphically represents the steps of the proposed method of "vector analysis of images." For image analysis, it is often convenient to consider orthogonal or rotated in $\pi/2$ to a vector field ogD.

The analysis of the shape of vector fields is performed using two given templates of vector fields with unit directions along the radius and circumference [diR], [diC].

The second and third cells of Fig. 2 shows how the projection operations $P = P_X$ are implemented, on the direction template X = [diC].

The COS projection f =>f_C, f_C = (f,X)*X is called a signal, and the orthogonal SIN part $f_S = f–f_C$ – by noise. There is a revers operation for vectors f, i.e. change the direction of the vector f by the opposite –f operation if (f, diC) < 0, then if between the vectors f and diC is an obtuse angle.

Similarly, vectors can rotate in templates: from the template of directions along the radius [diR], we can obtain a template of directions along the circle [diC], see Fig. 2.

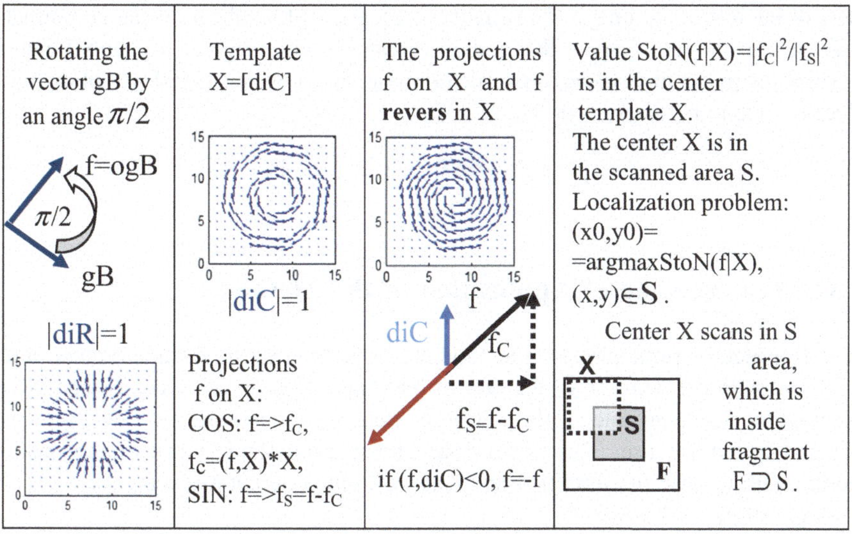

Fig. 2. Operations on vector fields, templates, projections, StoN function - "Signal/Noise".

Suppose that in image B – a bright ring is a scalar array B. Then ogB will represent two vector fields rotating in opposite directions. Of course, either we design a similar template to define the ring, or we allow the revers operation and remain with the simple direction template X = [diC], as in Fig. 2. The advantage of this approach is that we can have a template of larger sizes than the required ring and steadily realize the localization, without loss of the object.

4 Object Localization Problem

Let us write the signal-to-noise ratio StoN:

$$\mathrm{StoN}(f|X) = \|fc\|^2/\|fs\|^2 \tag{7}$$

Note that StoN (7) can have an infinite value if the part with noise $\|f_S\|^2 = 0$, for example, for $f_S = X$. Therefore, we will use the Mat Lab function atan2 (arctg):

$$\mathrm{aStoN}(f|X) = \pi/2^{*}\mathrm{atan2}(\|fs\|, \|fc\|) \tag{8}$$

The function aStoN with normalization on $\pi/2$ (8) will be used later in estimating the parameters of real vortices.

The center of the template X scans in the region S in the image fragment F of the image. The values of functions of the StoN, aStoN type (7, 8) are associated with the coordinate (x, y) in the S part F of the image fragment. The point (x0,y0) of the maximum of the function StoN (x,y), (x,y) ∈ S ⊂ F determines the positions of the object of interest in the image. The quality or accuracy of localization can be estimated by the size of the light part or the "radius of spot" in the image of the StoN function.

Thus, the problem of localizing or determining the coordinates (x0, y0) of an object reduces to the maximization problem:

$$(x0, y0) = \mathrm{argStoN}(f|x)(x,y), (x,y) \in S \tag{9}$$

5 General Scheme of Application of the Method

Let us consider the principle of the method in stages. In fact, this scheme is presented in Fig. 2, but with the addition of the task of filtering images in the second stage. When filtering, the High Frequency (HF) part of the image is cleared and the Low Frequency (LF) part is left for use. This is necessary for a more stable work of the method, together with the use of simplified templates with the operation revers.

Stages of general scheme:

1. The device measures or reads the input image B;
2. It sets to zero HF part of the image;
3. The device's processor calculates the vector gradient field gB = grad B;
4. It rotates all vectors of the vector field gB by an angle $\pi/2$ to obtain an orthogonal vector field f = ogB
5. It projects a vector field f = ogB in the directions of X: COS: f =>f_C = (f,X)*X, to obtain the signal vector part f_C and the noise vector part f_S = f–f_C.
6. It calculates the signal-to-noise ratio StoN (7, 8) at each point of the scanning region of the template X in S. The object sought is presumably in S;
7. It finds the coordinates (x0,y0) of the desired object corresponding to the pattern X, solving the problem to the maximum (9).

So, the scheme or algorithm of realization of a sequence of vector operations over the gradient field of the image is given, which gives the coordinates of the location of the object of the sought class in the images. Of course, all this is the content of the general problem of target designation.

6 Example of Indication of Vortices

Next, examples of the localization of objects of different classes will be demonstrated: vortices, RS, chamomile flowers and Pupil of the Eye (PE). If it is possible to say about RS, PE that the forms of RS, PE are standard, they are known. Vortices can not be attributed to objects whose form is known.

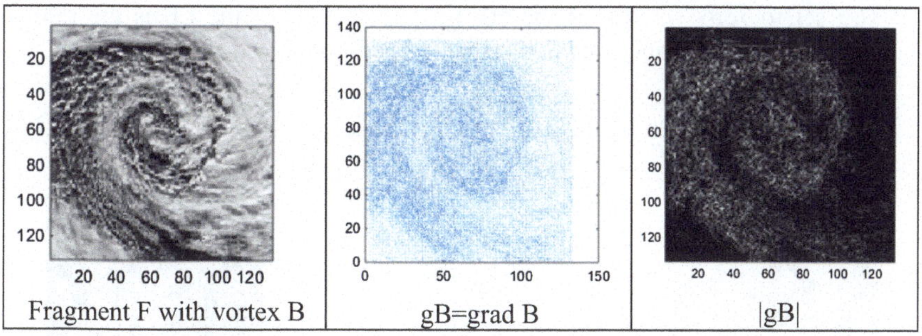

Fig. 3. An example of a fragment F with a vortex B, a gradient field gB and | gB |.

Let us take a fragment of a space image containing an atmospheric vortex, see Fig. 3. The vortex is identified by a vector template X = [diC], indicating the location by the maximum of the function StoN (7) in the scanning region S shown in Figs. 3, 4, 5 and 6.

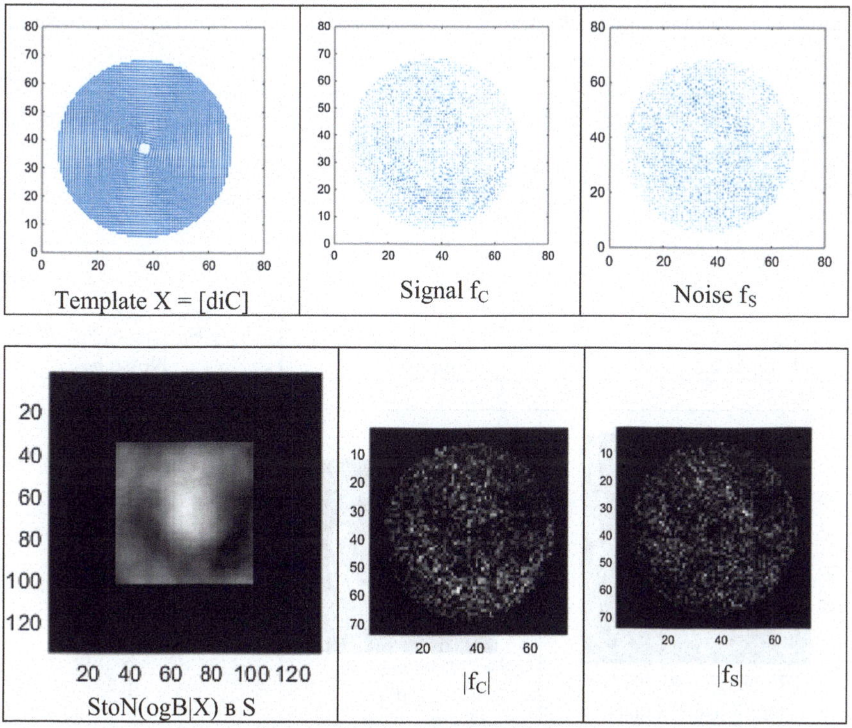

Fig. 4. Components of the StoN function for the vortex and its value in the scanning region S.

If you set to zero ~ 70% of the HF image, ~ 30% of the LF is the reminder image; the vortex localization method remains operational, see Figs. 5 and 6.

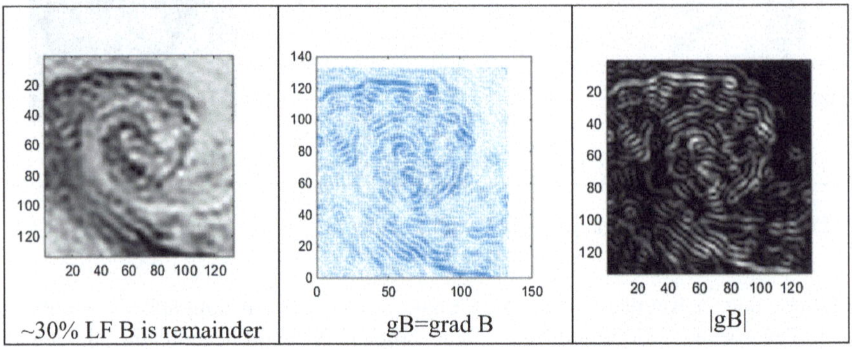

Fig. 5. An example of a 30% LF image with a vortex B, a gradient field gB, and | gB |.

In the construction of the functions StoN (7), in essence, the ratio of the energy of the circular motion is taken - the signal part to the radial motion energy - the noise part.

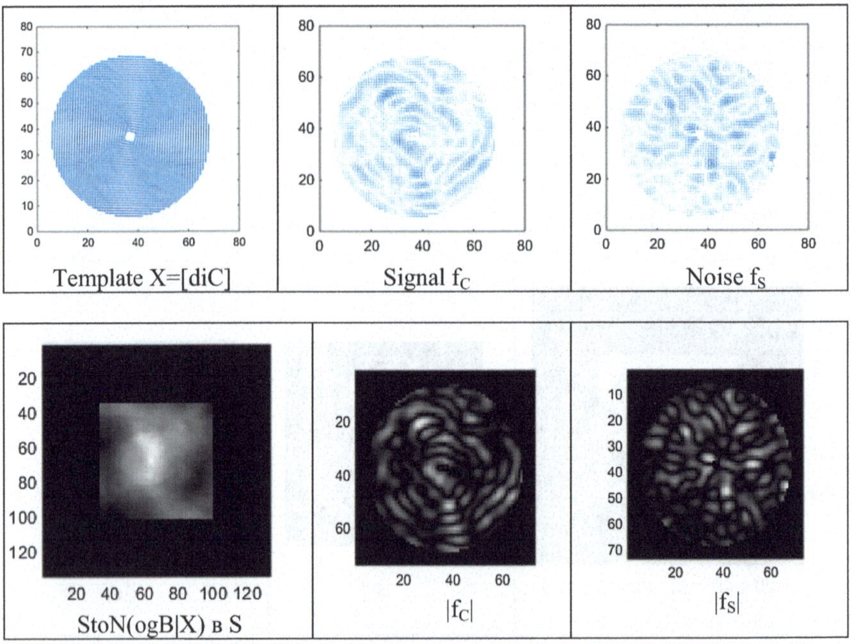

Fig. 6. Components of the StoN function for a vortex with 30% LF image and its values in the scanning area S.

7 False Vortices and Their Applications

If as an input image to take images of a chamomile flower, then the method will have the same reaction to the chamomile as to the vortex. Therefore, it can be said that the image of a daisy flower is a false vortex for a real vortex (Fig. 7).

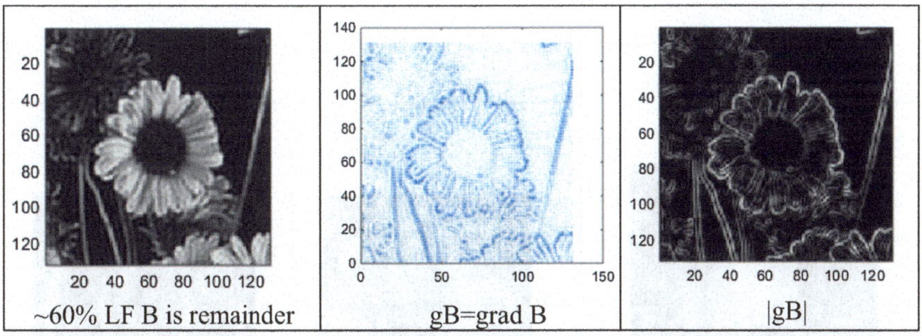

Fig. 7. An example of an image area with chamomile B, a gradient field gB and | gB |.

In space images, gradients or pressure drops, in the direction of which the air masses move, are not fixed. Therefore, the vortex localization method not only fixes vortices. By the way, to allocate flowers of a chamomile and a cornflower in a bouquet of flowers, as false vortices with an eye and without an eye, it appears possible.

On scalar space images, it is required to monitor the state of the atmosphere with the indication of vortices in the atmosphere. The proposed method allows you to do this automatically.

In addition, in the case of chamomile flower, the zeroing of the HF image improves the result. In 60% of the LF part, the result is even better than in the entire frequency band (Fig. 8).

A false vortex "around the perimeter of a triangle" allows you to localize the RS even if 30% of the LF remains from the image. In Blue (B), as an additional to the Red R, the picture | gB | looks more contrast than |gR|.

Another round RS, which is located lower prevents the localization of triangular RS, see Figs. 9 and 10.

In the problem of recognizing faces or in the problem of the Facial Features (FF), the distance between the Pupils of the Eyes (PE) in the image is the unit for measuring distances. All other elements of the FF can be found with the distance from the PE and with the help of this vector method for the analysis of other FF.

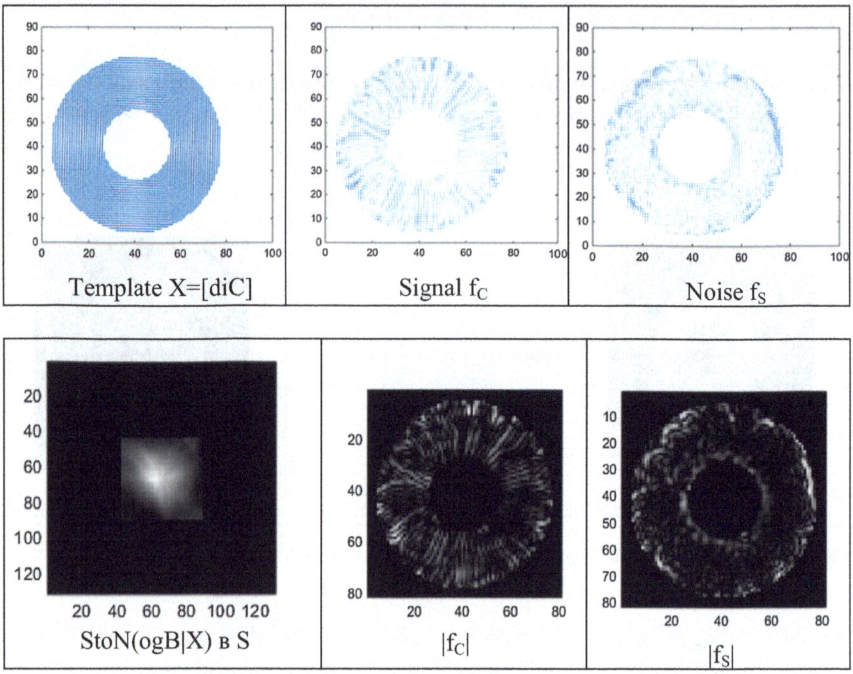

Fig. 8. Components of StoN function for chamomile flower and its values in the scanning area S.

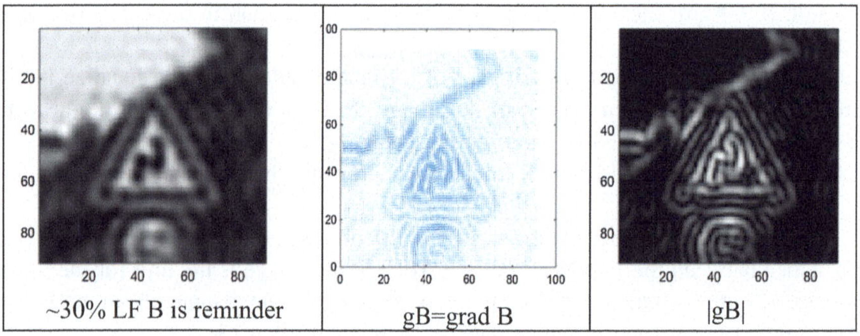

Fig. 9. An example of the image region B (Blue) with RS, the gradient field gB and |gB|.

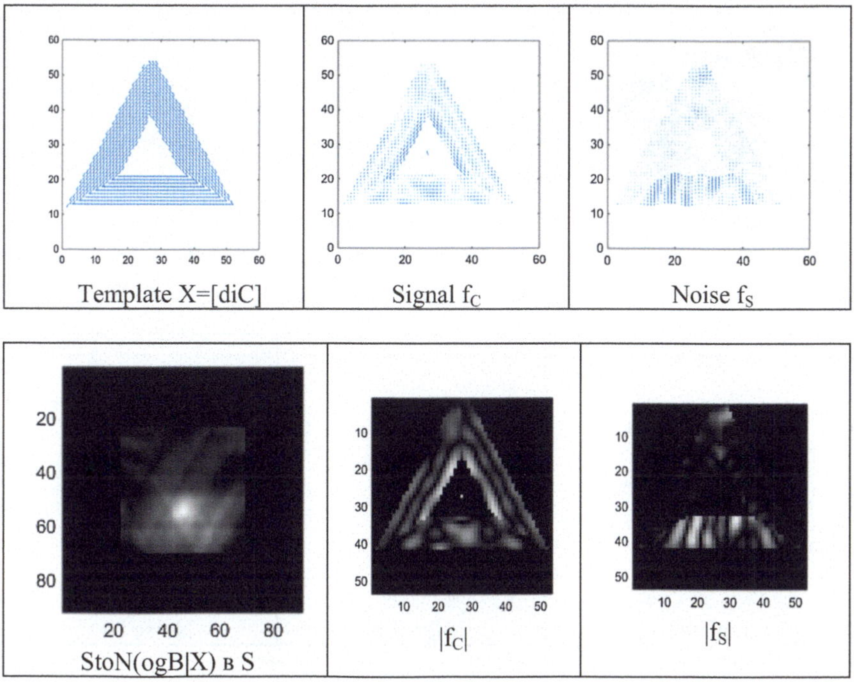

Fig. 10. Components of the StoN function for RS and its values in the scanning area S.

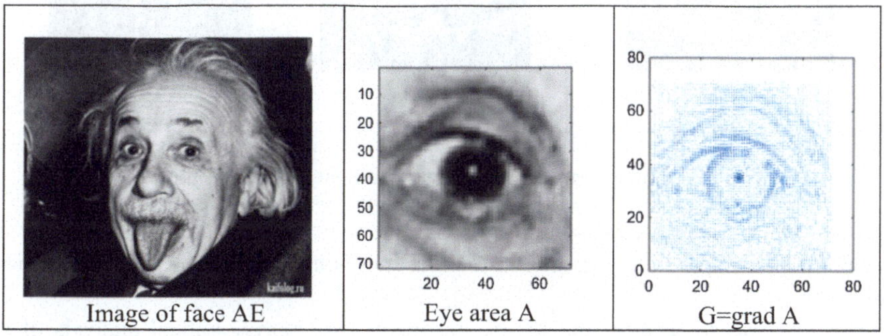

Fig. 11. Initial face image AE, eye area A, gradient field G

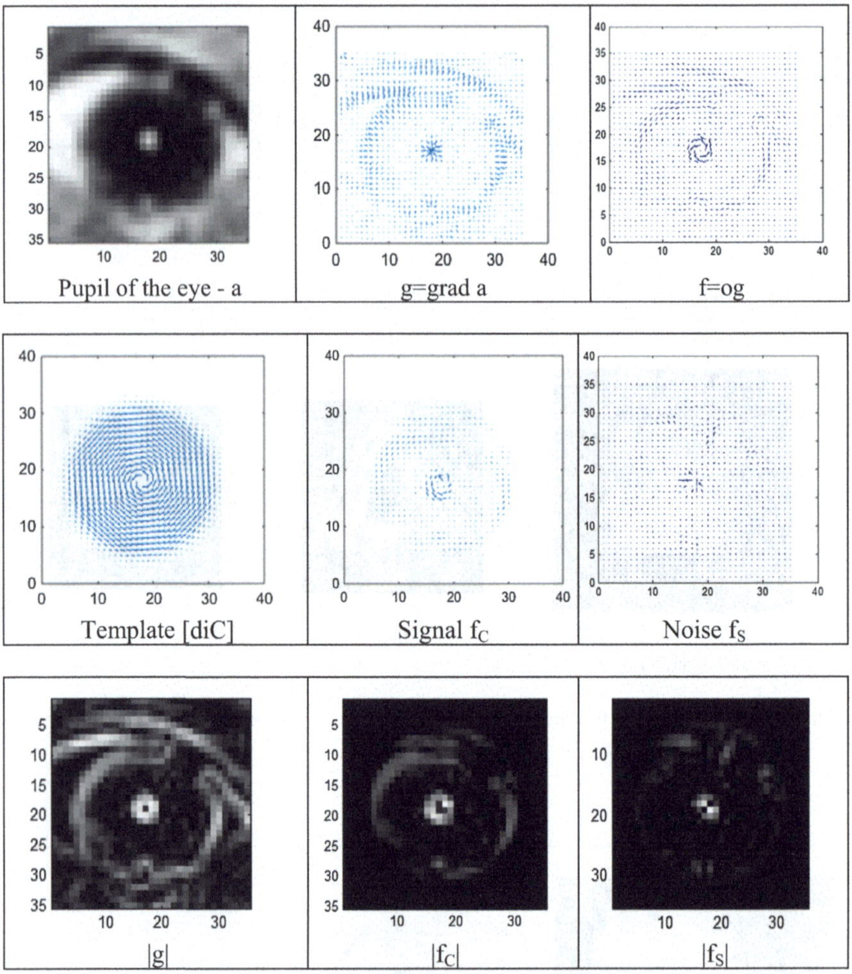

Fig. 12. Components of vector operations in the area of the PE.

The quality of the localization of objects in the region S (7) can be estimated by the size of the bright spot in the image of the StoN function. We calculated StoN functions in three ways.

The first was to localize the PE by the method of morphological analysis of the images (by areas) and to the input image of the fragment of eye A, see Figs. 11 and 13. The second method is the modernized morphological analysis of images, in which the "PE contours" were distinguished by a gradient method (see Fig. 12, | g |) and then StoN was constructed in the area of the contours in | G |. .

The smallest spot was obtained in the third compared method StoN (oG | X) (see Fig. 13), which is represented in this paper. We believe that this method is applicable to the solution of the complex problem of face recognition from images or specifically from the FF. There are ideas to realize the semantic analysis of images of the FF: nose, lips, PE region, ears, etc. using gradient fields and vector transformations.

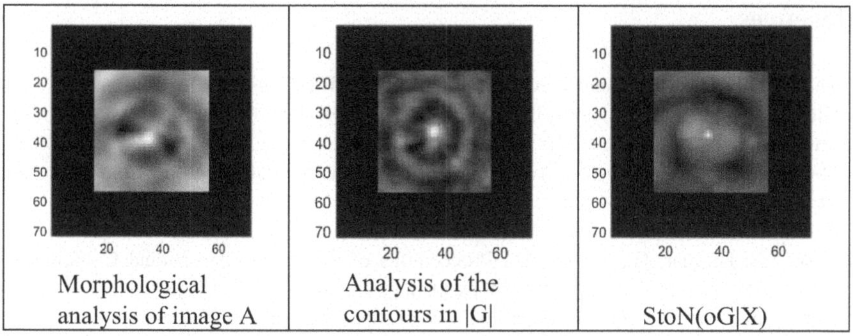

Fig. 13. Comparison of StoN functions in the scanning area S by different methods.

8 Conclusions

With the proposed approach, we were able to determine the location of objects with irregular and regular properties such as vortices and RS, which suggests that the proposed method is universal for the search for a wide class of objects. An unexpected result of the investigation is that gradient methods with projections on given templates of directions made it possible to indicate quite accurately with the aid of false vortices various objects of the type of RS. In real problems of reading RS in the preliminary task, together with the evaluation of the place of the RS, it is of course necessary to evaluate the size of the candidate in the RS. The ability to accurately assess the localization of the RS opens new avenues for solving the problem of semantic analysis of images such as reading the RS by the robot-navigator.

Estimating the exact distance between the PE makes it possible to formulate new semantic problems in estimating the parameters of the FF type for the realization of methods for recognizing face images of individuals. In the plans of the problem of estimating the parameters of atmospheric vortices and dipole eddies of sea currents [15] in space images of the Earth.

In the plans to implement the diagnostic task for the presence of tumors, ulcers, etc. in medical images, including in tomography by the method of false vortices.

Using the well-known sampling theorem of Wiener-Kotel'nikov [16], it is impossible to repeat the results of the FDST [13].

In our studies, it turned out that the realized accuracy was excessive and, to improve the result, we reset the HF images - coarsened images.

It is necessary to investigate, for example, double anti-vortices in templates X, see Fig. 2, to determine the position of the RS with a maximum accuracy of one pixel and less.

The limiting accuracy of localization can be achieved if done with interpolation and less to zero the HF images. We believe that this can happen on RS images and in the designation problems.

References

1. Thanh, H.N.: Electrical and Electronic Engineering, **4**(2), 36–44 (2014)
2. Шемарулин, И.А., Карпычев, В.Ю.: Труды НГТУ им. Р.Е. Алексеева, № 2, p. 60 (2016). (in Russian)
3. Shustanov, A., Yakimov, P.: Proceedings of the 14th International Joint Conference on e-Business and Telecommunications, vol. **5**, p. 42 (2017)
4. Krsak, E., Toth, S.: Acta Electrotechnica et Informatica, vol. **11**, №,4, p. 31 (2011)
5. Keser, T., Kramar, G., Nozica, D.: Proceedings of the IEEE International Conference on Smart Systems and Technologies (SST), Osijek, Croatia, pp. 12–14 (2016)
6. Xiong, B., Izmirli, O.: International Workshop on Image Processing and Optical Engineering, International Society for Optics and Photonics, vol. 8335, p. 83350B (2012)
7. Tun, H.M., Kyu, H.W., Maw, L.: International Journal of Scientific and Technology Research, **5**(06) (2016)
8. Приходько, И.Н.: XXV Международная конференция студентов, аспирантов и молодых учёных по фундаментальным наукам "Ломоносов - 2018". Секция "Физика" (2018). (in Russian)
9. Daraghmi, Y.-A., Hasasneh, A.M: International Journal of Signal Processing Systems, **4**(5), 417–421 (2016)
10. Алексанин, А.И., Загуменнов, А.А: Современные проблемы дистанционного зондирования Земли из космоса, В.5. Т.2. С, 17–21 (2008). (in Russian)
11. Liao, S., Anil, K.J., Stan, Z.L.: IEEE Transaction on Pattern Analysis and Machine Intelligence (2015)
12. Визильтер, Ю.В., Горбацевич, В.С., Вишняков, Б.В., Сидякин, С.В: Компьютерная оптика, **41**(3) (2017). (in Russian)
13. Терентьев, Е.Н., Терентьев, Н.Е.: ПРОЦЕССЫ В ГЕОСРЕДАХ, **4**(9), с.355–362 (2016). (in Russian)
14. Пытьев, Ю.П., Чуличков, А.И.: Методы морфологического анализа изображений. М.:Физматлит (2010) (in Russian)
15. Терентьев, Е.Н., Терентьев, Н.Е., Фаршакова, И.И.: Труды школы-семинара « Волны-2017 » . Математическое моделирование в радиофизике и оптике, с.56–58 (2017) (in Russian)
16. Кузнецов, Н.А., Синицын, И.Н.: Развитие теоремы отсчётов Котельникова, *УФН*179, 216–218 (2009) (in Russian)

Measurement of Gas Dynamic Parameters in the Gap at the Evaporation of Its Walls

A. A. Tairova[1,2(✉)] and G. V. Belyakov[1]

[1] Institute of Geosphere Dynamics (IDG RAS), Leninskiy prospect 38 bld. 1, 119334 Moscow, Russian Federation
moscouposte@gmail.com

[2] Moscow Institute of Physics and Technology (State University), Institutskiy pereulok 9, 141701 Dolgoprudnyi, Moscow Oblast, Russian Federation

Abstract. The results of a study aimed at optimization of heat methods of hydrocarbon extractions. These methods give the possibility of mining the shale solid minerals at great depths. This paper reviews the heat flow effect on the unconventional hydrocarbon reservoir model. The mechanism of filtration and propagation of exothermic reaction in a gap is considered. The results of measuring the gas-dynamic parameters in the gap during the melting and evaporation of its walls, obtained during laboratory studies are presented. The fracture of a shale layer was simulated by a gap with constant cross section area which was created in poly methyl methacrylate (PMMA or Plexiglas) previously. The gap was blown through the gas flow; its speed was measured by experiments. The gas flow was created by continuously working turbine driven pump. The pressure in the flow was determined by manometer. The values of flow rates, evaporation and entrainment of the mass of material from the surface of the gap are recorded.

Keywords: Exothermic reaction · Evaporation · Flow rate · Flow density

1 Introduction

The problem of mining and processing of shale solid minerals at great depths is very relevant, since open-pit mining is unprofitable. It is known that under the action of high temperature (>100 °C) shale solid hydrocarbons are transferred to the gas and liquid phases, which can be extracted by the well method. The thermal effect on hydrocarbon shales can be either an exothermic reaction of the combustion of the shales themselves or the combustion of the combustible gas supplied to them from the daylight area.

To maintain combustion, in both cases, the delivery of oxygen to the formation through the inlet well is necessary. To carry out the blowing and heating of the shale material with hot combustion products, it is possible to use the applied technology of hydraulic fracturing (hydrofracturing). As a result of hydraulic fracturing, a fracture is formed connecting the inlet and outlet wells. Combustion products and hydrocarbon agents decompose the heated mass of shale through the output or production well.

To optimize the process of decomposition of the material of the reservoir, it is necessary to know how the process of interaction of heated gas flows and hydrocarbon

V. I. Karev et al. (Eds.): *Physical and Mathematical Modeling of Earth and Environment Processes (2018)*, SPEES, pp. 317–322, 2019.
https://doi.org/10.1007/978-3-030-11533-3_31

phases of the formation material occur in the fracture. There is an important practical task: to determine the amount of flow of hydrocarbon material carried away by the co-current of hot gas acting on it.

2 Experiments

The study was conducted on the original experimental setup (see Fig. 1) [1]. The flat gap of the cell (1), the length $l = 30$ sm and opening $h = 0,2$ sm, had an initial clearance area $S = 1,2\ \mathrm{sm}^2$. In order to prevent the closure of the gap when heated, along the cell, four steel spokes with a diameter of 0.2 cm were equally spaced from each other. The differential pressure measured by the pressure gauge (6) over the cell length from the turbopump (8) was measured $\Delta p = p_0 - p = 50 \cdot 10^3$ g/(sm $\cdot$ s^2).

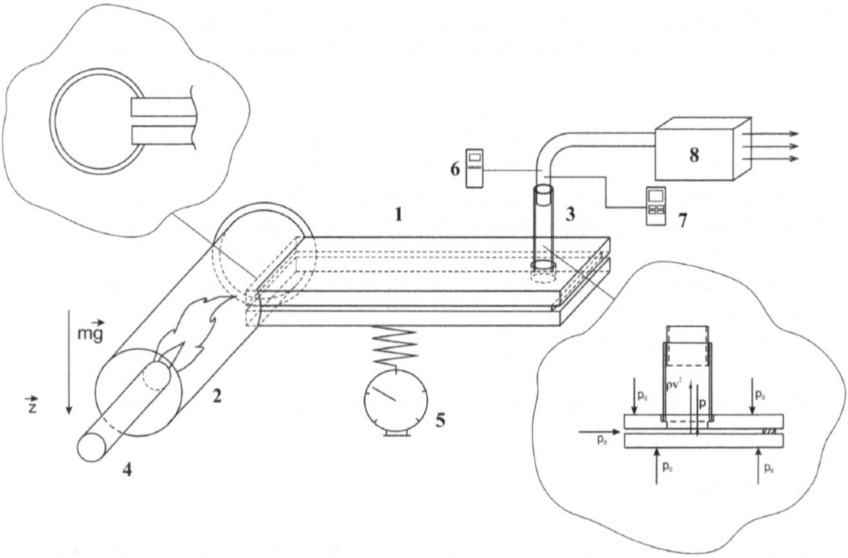

Fig. 1. The experimental setup scheme: 1 – the cell, 2 – combustion chamber, 3 – vertical tube, 4 – gas burner, 5 – precision balance, 6 – manometer, 7 – thermometer, 8 – turbopump.

The gas flow created by the pressure drop Δp passed along the cell and then went out into a vertical tube (3) with a cross-sectional area $S = 0.38\,\mathrm{sm}^2$, without friction connected to the tube and going to the turbo pump. The air flow was heated by a gas burner (4) using propane-butane mixture which was placed into the combustion chamber (2). The temperature was gauged by thermometer (7). In the course of the experiment, the measurement of the vertical force caused the effect on the gas flow cell was carried out with weights (5), the measurement accuracy of which was 0.01 g (the Shinko AJH 4200 balance).

3 Results and Discussion

If for estimating possible air flow rates arising in the cell gap with pressure drops $\Delta p \simeq (p_0 - p) \ll p_0$, when gas can be considered incompressible and viscosity independent of temperature, the exact solution of the Navier-Stokes equation is used, then the flow velocity along the streamline ($y = const$)

$$v_x = -\frac{1}{2\eta}\frac{dp}{dx}\left[\frac{h^2}{4} - (y - \frac{h}{2})^2\right] \tag{1}$$

where $y \leq h$. In this case, the average laminar velocity over the slit section $\bar{v}_x = \frac{h^2}{12\eta}\frac{4p}{l} \simeq 300\,\text{m/s}$. However, it is known [2] that if two layers of incompressible fluid or gas with a viscosity close to zero moves relatively to each other, then their motions cease to be laminar. Instabilities arise and the flow becomes turbulent. The experiment shows that the flow becomes turbulent, when a dimensionless number $\text{Re} = \frac{\bar{v}_x \cdot h}{(\eta/\rho)} > 3 \cdot 10^3$.

In this case, the determination of speed becomes possible only experimentally. For turbulent flow, we can assume that the front of a constant average flow velocity is perpendicular to the surface of the slit, and the near-wall gas layer moving at a different speed than the average velocity is thin and does not significantly affect the value of the flow velocity.

According to the scheme of the experiment (see Fig. 1), the cell is placed horizontally on the scales. The sum of the forces acting on the cell in the direction of the z axis is balanced by the flow of the gas pulse through the output vertical tube connected without friction along the z axis with the pipeline from the turbo pump. For a steady-state flow with a uniform ablation from the surfaces of the gap, the equilibrium equation takes the form:

$$-N + \sigma_{fr} + \dot{m}gt = -(p_0 - p)S + \rho v^2 S \tag{2}$$

where, N – is the support reaction along the z axis of the cell lying on the scales; σ_{fr} – friction force of the gas flow along the inner wall surface of the tube moving along the z axis; $\dot{m}gt$ – loss of cell weight by the time t due to evaporation and ablation of the material of the walls; $\dot{m}$ – mass rate of evaporation of cell surfaces; $(p_0 - p)S$ – "lifting" the power of the cell; $\rho v^2 S$ – the force of "thrust" along the z axis; S – output section of the tube; p_0, p – pressure, respectively, at the inlet and outlet of the cell.

Balance measures the response of the cell support $N = m \cdot g$, where m is the balance reading. The pressure and temperature of the gas at the outlet of the cell are recorded by a pressure gauge and a thermocouple respectively. The result of the preliminary weighing of the cell (at) before turning on the turbo pump is taken as starting point of the forces arising in the direction of the z axis. The maximum vacuum given by the turbo pump at zero flow rate $p \simeq 100\,\text{mbar}$. Records of the balance readings, pressure difference and temperature at the exit from the cell are shown in Fig. 2. The characteristic points of time for the experiment are shown also in Fig. 2. The range of numbers from 0 to 7 corresponds to the full time of the experiment

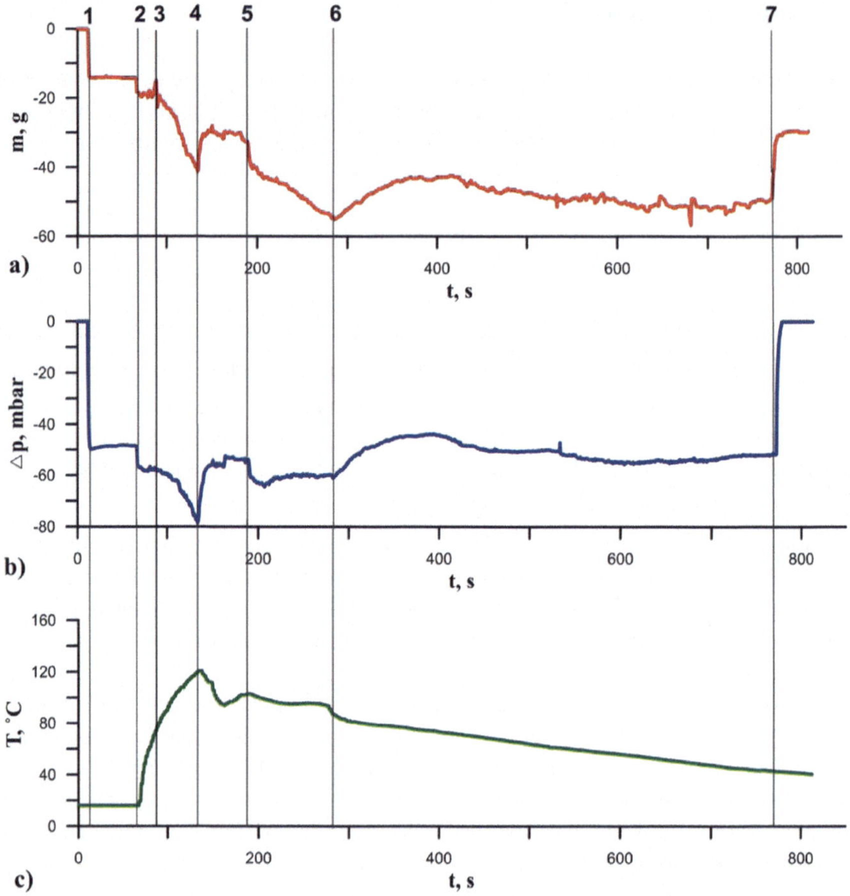

Fig. 2. Instrument readings: (a) the balance, (b) the manometer, (c) the thermometer. Stages of experiments: 1 – pump start, 2 – turn the burner, 3 – the beginning of the material evaporation, 4 – turn off the burner, 5 – turn the burner, 6 – stopping the burner, 7 – switch the pump.

$\Delta t = 0 \div 800$ s. The segment of the record from moment 0 to moment 1 over time $t \approx 10$ s corresponds to the initial recording of mass, pressure and temperature by instruments before turning on the turbo pump. Then, from moment 1 to moment 2, a turbo pump runs, which for about 1 min (until moment 2) causes a constant mass air flow $\dot{m} = \rho_0 v_3 S = 2,37$ g/s, at a pressure drop of 50 mbar. The flow rate is calculated based on the readings of the weights and the manometer using the equilibrium Eq. (1):

$$v_3 = \sqrt{\frac{\frac{-N}{S} + (p_0 - p)}{\rho}} = 52\,\text{m/s} \tag{3}$$

where $\rho = \frac{p\mu}{RT}$, from the equation of state of ideal gas, μ – molecule mass.

When calculating the flow rate, the friction of the gas against the walls of the exit tube is neglected, because its relative value is too small, estimated as

$$\frac{\sigma_{\text{fr}}}{(p_0 - p)S} \simeq \frac{\delta}{l} \simeq 3 \cdot 10^{-2} \tag{4}$$

here $l = 30\,\text{sm}$ is the length of the cell, $\delta \sim 1$ sm is the length of the tube. The reduced value of the flow velocity in comparison with the laminar velocity $v_x \approx 300\,\text{m/s}$, is explained by the increased gas-dynamic resistance arising due to the turbulization of the flow. Then, according to Fig. 2, at the moment of time 2, the gas burner is turned on, in which the propane-butane mixture burns with the specific heat of combustion $Q = 4 \cdot 10^{11}\text{erg/g}$. The fuel consumption in the burner is constant and amounts to $\widehat{m} = 5 \cdot 10^{-2}\,\text{g/s}$. This value is many times less than the airflow mass rate ($\dot{m} = \rho v_3 S \approx 2,37\,\text{g/s}$) moving through the cell. The jet of the air-propane mixture is formed and burned in the burner nozzle with a diameter of 2 cm. The nozzle is inserted at a small depth of 2 cm in a quartz glass with a diameter of 4 cm, forming a combustion chamber between the nozzle and the entrance to the cell. The side surface of the glass has a cut-out below the end cell size (Fig. 1). The mass flow rate of gas burner corresponds to the rate of heat

$$q = \frac{Q \cdot \widehat{m}}{S^*} = \frac{2 \cdot 10^{10}}{S^*} \tag{5}$$

where S^* is the area of effective heat release. The increment of the flow velocity in the cell from the heat release in the combustion chamber is estimated by solving the equations of mass, momentum, and energy fluxes through the control surfaces entering and exiting the combustion chambers [2, 3]:

$$\frac{d}{dt}m = \dot{m} = S\rho v = Sj = const \tag{6}$$

$$\Delta(p + jv) = 0 \rightarrow \Delta p \tag{7}$$

$$\Delta[j(\frac{v^2}{2} + \frac{\gamma p}{(\gamma - 1)\rho}] = q \tag{8}$$

The joint solution of the Eqs. (6–8) determines the flow rate increment

$$\Delta v = \frac{(\gamma - 1) \cdot q}{\rho(c^2 - v^2)} \tag{9}$$

where $c^2 = \gamma \frac{p}{\rho}$ - sound speed. For a short period of time, when the value of the mass carry-over from the cell surface is small, the velocity of the heated flow, determined from Eq. (2), equals to 64 m/s, and the speed increment from heating (9) is equal to $\Delta v \sim 12\,\text{m/s}$.

The small end part of the cell protruding into the volume of the combustion chamber, where the temperature $T_c \simeq 700\,^{\circ}\mathrm{C}$ is much higher than the boiling point of the Plexiglas $T_b \simeq 200\,^{\circ}\mathrm{C}$, boils (volume) and quickly turns into steam (from the moment 3 to the moment 4). The boiling vapor mass is slowly accelerated to the existing gas flow rate. As a result, there is a partial blockage of the flow at the entrance of the cell. A decrease in the flow rate with the turbo pump running will increase the vacuum in the cell to 80 mbar. After the burner is turned off in the time interval 4 – 5 the vacuum is reduced to a value almost equal to the initial vacuum pressure from the turbo pump operation in the absence of heat generation in the combustion chamber.

When the burner is switched on again, from the moment 5 to the moment 6, the increase of vacuum occurs. The vacuum remains constant at 60 mbar, corresponding to the previous density of the gas stream $\dot{m} \simeq 2,37$ g/s.

4 Conclusion

As the result of the flow of hot gases, the surface material of the gap evaporates and then it is carried away by the flow. Measured in stationary mode from the moment 5 to the moment 6, the removed mass takes the value $m \simeq 15$ g. The rate of evaporation and entrainment of mass by the stream during this period of time remains constant: $\frac{dN}{gdt} = const = \dot{m} = 0.16$ g/s

During the burner operation time, the mass of the burnt propane-butane mixture $\widehat{m}\Delta t = 0,05 \cdot 93 = 4,65$ g. If we assume that the material of the cell wall “corresponds to” the oil shale material, then the efficiency of its “production” when burning or burning the propane-butane mixture is close to $\eta = \frac{15-4,65}{15} \cdot 100\% = 70\%$.

Acknowledgements. The work was supported by the RFBR project No. 16-05-00605.

References

1. Tairova, A.A., Belyakov, G.V., Chervinchuk, S.Y.: Ablation in the slit in combustion. In: Proceedings of the International Conference on Advanced Materials with Hierarchical Structure for New Technologies and Reliable Structures 2017 (AMHS 2017), AIP Conference Proceedings, vol. 1909, p. 020216 (2017). https://doi.org/10.1063/1.5013897
2. Landau, L.D., Levich, E.M.: Hydrodynamics, 3d edn. Nauka, Moscow (1986). (in Russian)
3. Sedov, L.I.: A Course in Continuum Mechanics, vol. 2, 4th edn. Nauka, Moscow (1984). (in Russian)

Poroelastic Hydraulic Fracture Simulation Using X-FEM/CPP Approach

Vitaly Borisov, Anton Ivanov, Mukamay Ramazanov, and E. B. Savenkov(✉)

Keldysh Institute of Applied Mathematics of Russian Academy of Sciences, Moscow, Russian Federation
e.savenkov@gmail.com

Abstract. In this paper we describe self-consistent mathematical model for large scale hydraulic fracture development in poroelastic medium is proposed together with appropriate simulation techniques. The model includes Biot poroelasticity equations, fracture fluid flow model based in Reynolds lubrication equation, criteria for fracture development based on vector valued form of Rice-Cherepanov J-integral and bulk/fracture coupling conditions defined at fracture mid-surface. The proposed numerical algorithms is a new variant of the widely used eXtended Finite element (X-FEM) method which utilize closest-point projection technique for both fracture geometrical representation and fracture flow solver. A robust and efficient algorithm is proposed based on direct evolution of closest point projector. Numerical results which illustrate the developed numerical techniques are presented.

Keywords: Poroelasticity · Geomechanics · Hydraulic fracturing · XFEM · Closest point projection method

1 Introduction

The hydraulic fracturing is one of the most commonly used methods increasing oil recovery applied in the development of oil and gas fields. The essence of the fracturing technology consists in pumping a special liquid into the oil reservoir to create an artificial crack with a length of ~ 100 m, height of ~ 10 m, and an average opening of 5 10 mm. As a result, an artificial channel is created that is connected with a well and has a large inflow area and a high permeability (by orders of magnitude greater than the one of the oil reservoir). This provides a significant increase in the inflow of the formation fluid to the well.

Physical and mathematical description of the crack dynamics during hydraulic fracturing process lead to the solution of complex multiphysics problem which includes the following equation systems:

- a system of poroelasticity equations describing the evolution of stress-strain state of the formation and the fluid pressure fields in it during the development of the crack;
- equations of fluid flow in the crack;

V. I. Karev et al. (Eds.): *Physical and Mathematical Modeling of Earth and Environment Processes (2018)*, SPEES, pp. 323–333, 2019.
https://doi.org/10.1007/978-3-030-11533-3_32

- mechanical conditions for the development of the crack determining the direction of its development at each point of its front;
- dynamic interface conditions between the pressure field in the crack and in the formation defined on the lateral surfaces of the crack as well as kinematic conditions connecting the crack opening and the displacement of the formation.

Solution of such a problem is a complex task from numerical analysis and an efficient computer implementation point of view. In this work we present a mathematical model and simulation technique for the analysis of hydraulic fracture propagation (HF) in fully 3D setting and accounting for all major flow and geomechanical effects. The reservoir is described by linear poroelastic model [1]. The fracture is assumed to be as an arbitrary sufficiently smooth surface with boundary. The fracture flow is described by 2D lubrication layer equation. Reservoir and fracture flow models are coupled by means of appropriate interface conditions defined at the fracture mid-surface. The numerical approach is based on finite elements method and accounts for discontinuous displacement and pressure fields using an eXtended Finite Element (X-FEM) method [2, 3]. The fracture surface representation is based on closest point projection method [4]. The same approach is used for solution of lubrication layer equations for fracture flow. These implies two major benefits of the proposed approach. First, there is no need to introduce an independent computational for fracture flow simulation since CPP method for solution of surface PDEs is an embedding method which utilize bulk finite element mesh. Second, this allows for an efficient implementation of the coupled approach since the same data structures are utilized for X-FEM and fracture flow solver implementation. Moreover, an efficient algorithms can be developed for computation of fracture mid-surface geometrical evolution.

Below the main components of the proposed techniques are described as well together with numerical examples which demonstrates applicability of developed techniques.

2 Model of the Media

Mechanical equilibrium equations in matrix medium are given by:

$$\nabla \cdot \mathbf{T} + \rho \mathbf{g} = 0, \tag{1}$$

with $\rho = \rho(\mathbf{x})$ being a density of fluid-saturated medium, $\mathbf{g}$ beings gravity acceleration, $\mathbf{T}$ – Cauchy stress tensor. In the general anisotropic case and in linear approximation the constitutive relation is given by [1] $\Delta\mathbf{T} = \mathbf{C} : \mathbf{E} - \mathbf{B}\Delta p$, $\Delta\varphi = \mathbf{B} : \mathbf{E} + \Delta p/N$. Here p is fluid pressure, $\mathbf{E} = \left[\nabla \otimes \mathbf{u} + (\nabla \otimes \mathbf{u})^T\right]/2$ – deformation tensor, $\mathbf{u}$ – displacements field, $\mathbf{C} = \left[C_{ijkl}\right]$ – symmetric tensor of elastic coefficients, $\mathbf{B} = \left[B_{ij}\right]$ – symmetric Biot tensor, φ – porosity, N – Biot coefficient that describes dependence of porosity on fluid pressure, $\Delta f = f - f_0$ for any variable f,$\rho = (1-\varphi)\rho_s + \varphi\rho_f$, where $\rho_{s,f}$ is matrix and fluid density respectively.

Fluid mass conservation equation is given by:

$$\frac{\partial m_f}{\partial t} + \nabla \cdot (\mathbf{w}_m) = q, \quad \mathbf{v} = \frac{\mathbf{w}_m}{\rho_f} = \mathbf{K} \cdot (-\nabla p + \rho_f \mathbf{g}), \tag{2}$$

where m_f is fluid mass in a unit volume of a saturated medium, $\mathbf{w}_m = \rho_f \mathbf{v}$ – mass flux density vector, $\mathbf{v}$ – the filtration velocity determined by Darcy's law, $\mathbf{K}$ – symmetric tensor of permeability coefficients, q – mass density of sources.

Further we assume that $\Delta\rho_f / \rho_f^0 = \Delta p / K_f$, $\Delta m_f = \rho_f^0 v_f, v_f = \mathbf{B} : \Delta\mathbf{E} + \Delta p / M$, where p is fluid pressure, ρ_f – fluid density, K_f – bulk fluid compression module, v_f is relative change of fluid volume in an elementary volume of a saturated medium.

3 Geometrical and Flow Model for Fracture

Geometrically fracture is considered as of zero thickness with mid- surface being a smooth and simply-connected manifold $\mathcal{F} : D_\mathcal{F} \to \mathbb{R}^3$, $D_\mathcal{F} \subset \mathbb{R}^2$, with boundary (fracture front) $\partial\mathcal{F}\ominus$. At the time moment t fracture mid-surface is denoted by $\mathcal{F}_t$. The condition $\mathcal{F}_{t_1} \subset \mathcal{F}_{t_2}$ holds for any $t_{1,2}$ such that $t_1 \geq t_2$. At each time moment and point on $\gamma(t) = \partial\mathcal{F}_t$ the fracture front velocity field is given by $\mathbf{v} = \mathbf{v}(\mathbf{x}, t)$ so the fracture extension is described by the equation

$$\dot{\mathbf{x}} = \mathbf{v}(\mathbf{x}, t), \quad \mathbf{x}|_{t=0} = \mathbf{x}_0 \in \gamma(0), \tag{3}$$

where velocity field $\mathbf{v} = \mathbf{v}(\mathbf{x}, t)$ is smooth enough function of point $\mathbf{x} \in \gamma(t)$ at each fixed time moment and a smooth function of time for each fixed (Lagrangian) point of the boundary.

We assume that the fluid propagates along the fracture, trying to fill the its complete volume. Part of the middle surface $\mathcal{F}$ of fracture corresponds only to a part of the fracture occupied by the liquid. Crack opening is a function of the point of surface $\mathcal{F}, w = w(\mathbf{x})$, $\mathbf{x} \in \mathcal{F}$. Lateral fracture surfaces $\mathcal{F}^\pm = \mathbf{x} \mp \mathbf{n}(\mathbf{x}) w(\mathbf{x})/2$ are located at a distance $w/2$ from $\mathcal{F}$, where $\mathbf{n}(\mathbf{x})$ – unit normal to $\mathcal{F}$ at point $\mathbf{x}$ directed to $\mathcal{F}^-$ (thus, $\mathbf{n}^+ = \mathbf{n}$, $\mathbf{n}^- = -\mathbf{n}$). Function w is smooth, nonnegative and equals zero on $\partial\mathcal{F}$.

The equation of compressible fluid flow in a fracture has the form of Reynolds lubrication equation:

$$\frac{\partial \rho_f w}{\partial t} + \nabla_\mathcal{F} \cdot (\rho_f w \mathbf{v}_\mathcal{F}) = Q_f, \quad \mathbf{v}_\mathcal{F} = -D(\nabla_\mathcal{F} p_f - \rho_f \mathbf{g}_\mathcal{F}), \tag{4}$$

where $\mathbf{v}_\mathcal{F}$ is the velocity of fluid, D is an effective permeability, $\mathbf{g}_\mathcal{F}$ is the projection of the vector of gravity acceleration on the mid-surface of fracture, $\mathbf{g}_\mathcal{F} = (\mathbf{I} - \mathbf{n} \otimes \mathbf{n})\mathbf{g}$. In the case of a Newtonian fluid, $D = w^2/(12\mu_f)$, where μ_f is dynamic viscosity of fluid.

Source term Q_f can be represented as $Q_f = Q_f^{in} + Q_f^{out}$. In the latter the first term describes the inflow rate from the well to the fracture, the second one – the leakage of fluid into the poroelastic reservoir.

Equation (4) is considered in a bounded domain $\Omega_f = \Omega_f(t) \subset \mathcal{F}$ with boundary $\partial\Omega_f(t)$. At $\partial\Omega_f$ the appropriate boundary conditions are to be set, see Sect. 6.

4 Interface Conditions

Fracture mid-surface $\mathcal{F}$ is a surface of jump discontinuity for variables that describe reservoir state. Accordingly, the complete system of equations for poroelastic state and fracture fluid flow (see Sect. 2) has to be accompanied by a proper set of interface conditions defined at $\mathcal{F}$. Let $\mathbf{x} \in \mathcal{F}$ denotes point on fracture middle surface, $\mathbf{x}^{\pm}$ are the corresponding points on the fracture faces $\mathcal{F}^{\pm}$. The first group of interface conditions are kinematic conditions given by:$w(\mathbf{x}) = (\mathbf{u}^{+} - \mathbf{u}^{-}) \cdot \mathbf{n}, \quad \mathbf{u}^{\pm} = \mathbf{u}(\mathbf{x}^{\pm})$. Dynamic interface conditions are given by $\mathbf{T}^{\pm} \cdot \mathbf{n}^{\pm} = -p^{*}\mathbf{n}^{\pm}$.

Consider now interface conditions for the hydrodynamics part of problem. They can be given in different forms, depending on the assumptions imposed on the fracture flow.

In the first case fracture and medium are considered to be saturated with the same fluid that occupies entire volume of the fracture $(\Omega_f = \mathcal{F})$, and the hydrodynamic contact between fracture and the medium is ideal. In that case at each point $\mathbf{x} \in \mathcal{F}$ of the mid- surface the pressure on fracture faces are given by $p^{+} = p^{-} = p_f$ with $p^{\pm} = p(\mathbf{x}^{\pm})$. The continuity of the material flux conditions with the corresponding expressions for the flow in the medium (2) and the form of Eq. (4), expressing the material balance in the fracture, have the form $Q_f = Q^{+} + Q^{-}, \quad Q^{\pm} = \mathbf{w}_m \cdot \mathbf{n}^{\pm}|_{\mathbf{x}=\mathbf{x}^{\pm}}$ where p^{*} is defined as $p^{*}(\mathbf{x}) = p_f(\mathbf{x})$.

In the second case it is assumed that the fracture volume is isolated. In this case it is natural to assume that $\Omega_f \subset \mathcal{F}, \partial\Omega_f \cap \partial\mathcal{F} = \emptyset$. Interface conditions for the mass flux in domain Ω_f are given by $Q_f = 0, Q^{\pm} = 0, \mathbf{x} \in \Omega_f$. In other words, the interface conditions reduce to the Neumann boundary conditions for the flow equation in the medium. In this case conditions for pressure in the medium cannot be formulated, and the fracture pressure is given by (4) with non-zero right-hand side Q_f with p^{*} defined as $p^{*}(\mathbf{x}) = p_f(\mathbf{x})$.

Interface conditions in the domain $\mathcal{F} \backslash \Omega_f$ ("lag" domain) can be set in different ways depending on whether the fracture faces are permeable or not for the fluid in media. In the first case at points $\mathbf{x} \in \mathcal{F} \backslash \Omega_f$ the relations $Q^{\pm} = 0$ and $p^{*}(\mathbf{x}) = 0$ are satisfied. In the second case it is natural to consider that fluid in porous media fills fracture (sub)domain $\mathcal{F} \backslash \Omega_f$ and the pressure and mass flux are continuous at the fracture surfaces, i.e. $[p] = 0$, $[Q] = 0$.

5 Boundary Conditions at the Fracture Front

Let us assume that at each $\mathbf{x} \in \partial\mathcal{F}_t$ a vector valued function $\mathbf{J} = \mathbf{J}(\mathbf{x})$, can be defined such that energy released when fracture propagates in direction given by unit vector $\mathbf{l}$ ($\|\mathbf{l}\| = 1$) is equal to $J = \mathbf{J} \cdot \mathbf{l}$. In accordance to energy growth criterion fracture evolves in the direction which maximize value of J, $J = J_{\max} = J(\mathbf{l}_{\max})$. In the case when $J_{\max} \le J_*$, where J_* is some critical value fracture growth does not occur. When $J_{\max} = J_*$ the fracture is in equilibrium. In the case when $J_{\max} > J_*$ fracture develops in an unstable way. Consequently, if the fracture growth occur, the growth direction is given $\mathbf{J}$, i.e. one have $\mathbf{v} = \alpha \mathbf{J}$, $\alpha \ge 0$ in (3).

At the same moment, linear elastic fracture mechanics under quasi-equilibrium conditions does not provide a way to determine fracture growth velocity in the case $\alpha > 0$. Hence, (3) is a kind of "kinematic" equation and has to be supplemented by equation which determines fracture front profile under the given conditions: $J_{\max} \le J_*$. This equation is nonlocal (both in space and time) and introduce dependency of the future fracture configuration on the history of the process, its current configuration, problem (fracture and reservoir) solution at the given time in the whole domain of the interest, boundary conditions, etc.

The simplest option is to define Dirichlet boundary condition for fracture fluid pressure at the fluid front, $p_f(\mathbf{x}) = p_f^*(\mathbf{x}), \mathbf{x} \in \Gamma(t)$, with p_f^* being given function. Temporal evolution of the fluid domain id described by equation similar to (3). The value of p_f^* at the point of the fluid front can be set to zero when reservoir fluid does not enter fracture ("empty" lag zone). Otherwise it is natural to set to be equal to the reservoir pressure at the given point.

Finally let us note that (i) for practically important cases it is more important to account for lag itself but not for it precise value and (ii) consideration of lag is of the importance only in the relatively small neighborhood of the fluid front and at the initial stage of fracture development. This consideration usually allows obtaining lag values using some additional considerations or keeps being small but fixed in tome and space quantity (see, e.g., [5]).

6 X-FEM/CPP Approach

The core of the developed computational algorithm is an eXtended Finite Elements Method (X-FEM) [2, 3]. The method is based on the extension of a standard finite-element basis by special additional basis functions that allow one to correctly represent the discontinuous fields of the displacement, stresses and deformations on the middle surface of fracture. One important part of X-FEM method is an algorithm of fracture surface representation in a finite-dimensional problem, which should provide convenience of computing the all necessary quantities for the method implementation (for example, the stiffness matrix of the problem). In the modern version of X-FEM, the most commonly used is the level set method, see [6]. For the evolution of the level set functions Hamilton-Jacobi type equations are used, see [7, 8].

In this paper we use an approach based on the method of the nearest point projection to describe the geometry of a surface. In the context of fracture evolutions problems it is as general as the level set method, but it has a number of advantages when the boundary conditions on the lateral surfaces of a crack are determined by the solution of some equations (in our case, it is the lubricating layer equation). In addition, the closest point projection method is a convenient tool for developing computational algorithms for solution equations on surfaces of an arbitrary shape, see [9].

Thus, in this paper the CP projection method is used both to represent the fracture surface in X-FEM, and to solve the equation of the lubricating layer in the fracture. A complete description of the X-FEM/CP approach is presented in [10]. A complete description of CPP approach for midsurface evolution is given in [4]. The latter is a crucial issue when considering hydraulic fracture dynamics. For this reason it is briefly described below.

Let us consider the basic concepts of the closest point method. A rigorous mathematical justification of this method is given in [9]. Consider a smooth surface $\mathcal{F}$ with a boundary embedded in the three-dimensional space $\mathbb{R}^3$. Here and in what follows, the term "smooth" will mean that the set $\mathcal{F}$ is an image of a smooth mapping of a bounded closed subdomain of the two-dimensional plane $\mathbb{R}^2$ into the three-dimensional space $\mathbb{R}^3$. We will assume that the used mappings have any required number of derivatives. We also assume that the surface $\mathcal{F}$ is entirely located inside the spatial domain $\Omega \subset \mathbb{R}^3$.

Suppose that for an arbitrary $\mathbf{x} \in \Omega$, $\mathbf{x}_{cp}$ is the closest point to it on the surface $\mathcal{F} : \mathbf{x}_{cp} = \arg\min_{y \in \mathcal{F}} \|\mathbf{y} - \mathbf{x}\|$. Here and below, the Euclidean norms are used. The point $\mathbf{x}_{cp}$ will be called the projection of the point $\mathbf{x}$ onto the surface $\mathcal{F}$ in the sense of the shortest distance; the corresponding projection operator will be denoted by $\mathbf{P}$, $\mathbf{x}_{cp} = \mathbf{P}\mathbf{x}$.

The operator $\mathbf{P}$ is vector-valued; it maps the domain Ω onto the surface $\mathcal{F}$ considered as a subset of $\Omega \subset \mathbb{R}^3$. In the case that the surface and its boundary are smooth and the domain Ω is "sufficiently small", the operator $\mathbf{P}$ is uniquely defined, i.e., each point of Ω is uniquely projected onto a single point on the surface. It can be shown that the projections operator $\mathbf{P}$ uniquely describes the surface $\mathcal{F}$ as a set of its fixed points $\mathcal{F} = \{\mathbf{x} \in \Omega : \ \mathbf{x} = \mathbf{P}\mathbf{x}\}$. It should be noted that this method of description is quite general. It allows one to describe the geometry of a surface with a boundary and without a boundary, nonorientable manifolds or manifolds of codimension greater than one (i.e., curves (codimension is equal to 2) and points (codimension is equal to 3), and also the union of objects of different codimension.

Further, it will be necessary to distinguish the points of Ω that are projected into the interior points of $\mathcal{F}$ or onto its boundary $\partial\mathcal{F}$. With this purpose, let us consider the auxiliary operator $\tilde{\mathbf{P}}$ which we define as [4]: $\tilde{\mathbf{P}}(\mathbf{x}) = \mathbf{P}(2\mathbf{P}(\mathbf{x}) - \mathbf{x})$. It follows from the geometrical consideration that for the points $\mathbf{x}$ whose projections belong to the boundary of the surface, $\tilde{\mathbf{P}}(\mathbf{x}) \neq \mathbf{P}(\mathbf{x})$, while for the points whose projections belong to the interior points of $\mathcal{F}$, the values of the projectors $\tilde{\mathbf{P}}$ and $\mathbf{P}$ coincide: $\tilde{\mathbf{P}}(\mathbf{x}) = \mathbf{P}(\mathbf{x})$.

Together with the above presented method of describing the surface F, we will need a way of representation of the functions defined on this surface. Due to the fact that there are no local coordinates on the surface F, it is convenient to use the "implicit" way of their representation. Namely, a function on the surface will be defined as a trace on it of a function defined in the domain Ω.

The extension of the function given on the surface to the domain Ω can be constructed in various ways. A convenient and subsequently used method is its extension with the help of the operator $\mathbf{P}$. Namely, for an arbitrary function u given on the surface, its continuation $\varepsilon[u]$ in Ω is defined as $\varepsilon[u](\mathbf{x}) = u(\mathbf{P}\mathbf{x}), \mathbf{x} \in \Omega$. Thus, the constructed continuation is constant along the segments connecting a point of Ω and its projection. This fact allows us to calculate the derivatives of a function on the surface in a convenient way, as well as differential operators of higher order (see below in this section). Then, by virtue of the properties of the projector $\mathbf{P}$ and the extension operator ε, we have $\nabla\varepsilon[u](\mathbf{x}) = \nabla u(\mathbf{P}\mathbf{x}) = \nabla_{\mathcal{F}} u$.

Suppose that in a domain Ω containing a surface $\mathcal{F}$, a computational mesh composed of tetrahedral cells is defined. The resulting triangulation will be considered regular (i.e., the intersection of two different tetrahedra can be an empty set, a mesh node, its boundary, or its face). We denote the corresponding discrete domain by Ω_h. It is assumed that the mesh is sufficiently fine so that the characteristic mesh step h is much smaller than the radii of the principal curvatures of the surface and the radius of the curvature of its boundary.

Denote the set of mesh nodes by $\mathbf{x}_i$, $i = \overline{1, N}$, and N is the number of mesh nodes. We assign a basic function $\varphi_i = \varphi_i(\mathbf{x})$ to each node of the mesh. In the simplest case, continuous piecewise-linear basis functions can be used for which $\varphi_i(\mathbf{x}_j) = \delta_{ij}$, where δ_{ij} is the Kronecker symbol, and the cells of the mesh φ_i are continued linearly.

In the discrete case, the projection operator $\mathbf{P}^h$ is given by its values $\mathbf{P}_h^i$ at the mesh nodes. To calculate the projection operator $\mathbf{P}^h$ at an arbitrary point in the domain, the above introduced system of basis functions is used. If $\mathbf{x} \in \Omega_h$ is an arbitrary point of the domain Ω_h, ω is the tetrahedron in which it is located, $\{\xi_k\}_{k=1}^4$ is its barycentric coordinates with respect to the vertices of the tetrahedron ω, then the projection of the point $\mathbf{x}$ is the point $\mathbf{P}^h(\mathbf{x}) = \sum_{k=1}^4 \xi_i \mathbf{P}_k^h$. The constructed approximations of the projector $\mathbf{P}^h$ allow us to construct in a natural way the approximations of the projection $\tilde{\mathbf{P}}^h$. Now, let I be the set of all nodes of the mesh. Define its subsets $\mathrm{I}_{d,b}$, $\mathrm{I} = \mathrm{I}_d \cup \mathrm{I}_b$, $\mathrm{I}_d \cap \mathrm{I}_b = \emptyset$, corresponding to the nodes projected into the interior points of the surface and its boundary, respectively: $\mathrm{I}_d = \left\{\mathbf{x}_i \in \mathrm{I} : \left\| \mathbf{x}_i - \tilde{\mathbf{P}}^h(\mathbf{x}_i) \right\| < \varepsilon \right\}$, $\mathrm{I}_b = \mathrm{I} \backslash \mathrm{I}_d$, where $0 < \varepsilon \ll h$ is a small real parameter of the algorithm.

Obviously, in the course of the surface evolution, the set I_d can only increase (by inclusion). In other words, for $t_1 < t_2$ we have $\mathrm{I}_d(t_1) \subset \mathrm{I}_d(t_2)$, $\mathrm{I}_b(t_1) \supset \mathrm{I}_b(t_2)$. This allows, in particular, as the surface moves, to update information at each next time step of the discrete sampling only for the nodes from the set I_b.

The general idea of the algorithm for calculating the surface evolution is as follows: first, according to the known information, the coordinates of the projection points of the mesh nodes from the set I_b are calculated. Then, these points are shifted in a given direction in accordance with a given boundary speed and a given time step value. The resulting points belong to the new position of the front, but the vector connecting the original node of the spatial mesh and the new position of the boundary point is not the projection vector of the closest point. Therefore, the approximations of the projector are corrected. Then, we will assume that the surface evolves on the time interval $[0, T]$.. We introduce on this interval a computational mesh with steps $\{\Delta t_i\}$, $i = \overline{1, N_T}$, $T = \sum_{i=1}^{N_T} \Delta t_i$, where N_T is the number of mesh nodes.

At the initial time $t = 0$, the approximation of the projector $\mathbf{P}^h$ is assumed to be known. We also assume that during the time interval $[0, T]$ the surface remains inside the mesh domain.

```
1.  Input Data: Discrete representation of the surface at time t_n : sets I_{d,n}, I_{b,n}, projector
    approximations P^{h,n}, time step Δt, algorithm parameters: ε, l_min, R_f, R_max.
2.  Set I_{b,n+1} = ∅, I_{d,n+1} = I_{d,n}.
3.  Calculate the new positions of all projections onto the boundary:
    X = { P_i^{h,n} + Δt v(P_i^{h,n}), i ∈ I_{b,n} }.
4.  Forall i ∈ I_{b,n} // For all nodes projecting onto the boundary
5.      Determine the closest to x_i node from X :
        a = arg min_{x∈X} ||x_i − x|| = arg min_{i∈I_{b,n}} ||x_i − P_i^{h,n}||.
6.      Define (sub)set B ⊂ X such that:
        B = { b ∈ X : ||b − a|| > l_min, ||b − a|| < R_f, 0 < (x − a)·(b − a) < ||b − a||^2 }.
7.      if B = ∅ // Point a is a projection onto the boundary of the surface.
8.          Set q = a.
9.      else // Point a is not projection onto the boundary of the surface.
10.         Find such a point among the points in B that the length
            of the ab is minimal: b = arg min_{b̃∈B} ||a − b̃||.
11.         Set q = a + [(x_i − a)·ξ]ξ, ξ = (b − a)/||b − a||.
12.     end if
        // Determine the projection of the point x_i onto the segment P_i^{h,n}q :
13.     Set γ_min = ε/||P_i^{h,n} − q||, γ = (x_i − q)·(P_i^{h,n} − q)/||P_i^{h,n} − q||^2.
14.     if γ ∈ (γ_min, 1) // The point is projected onto the surface.
15.         Set P_i^{h,n+1} = q + γ(P_i^{h,n} − q), I_{d,n+1} := I_{d,n+1} ∪ {i},
16.     else if γ ≤ γ_min // The point is projected onto the boundary.
17.         Set P_i^{h,n+1} = q, I_{b,n+1} := I_{b,n+1} ∪ {i},
18.     else if γ ≥ 1 // The point is projected onto the surface.
19.         Set P_i^{h,n+1} = P_i^{h,n}, I_{d,n+1} := I_{d,n+1} ∪ {i}.
20.     end if
21. end forall
22. Output data: a discrete representation of the surface at t_{n+1}: the sets I_{d,n+1}, I_{b,n+1}, the
    projector approximations P^{h,n+1}.
```

Algorithm 1. Calculation of the evolution of the projector $\mathbf{P}^h$.

The complete algorithm for constructing the discrete operator $\mathbf{P}^{h,n+1}$ at time $t = t_n$ for given parameters of the method and the known approximation $\mathbf{P}^{h,n}$ is presented in the scheme 1.

The input data of the algorithm are the discrete sampling $\mathbf{P}^{h,n}$ of the projector at a time $t = t_n$ and the sets $\mathrm{I}_{d,n}$ and $\mathrm{I}_{b,n}$. As a result of application of the algorithm, the values of input data are replaced by the corresponding values of output data at time $t = t_{n+1}$.

The parameters of the method are the values $l_{\min}$, R_f, and $\gamma_{\min}$. The parameter $l_{\min}$ determines the conditionality of the projection of a point on the segment and the accuracy of the projection computation, $l_{\min} \ll R_f$, where R_f determines the radius of the sphere bounding the region in which the points are searched for in step 6 of algorithm 1. It is assumed that R_f is sufficiently large as compared with the mesh spacing, but does not exceed the maximum distance from the point to its projection onto the boundary, $R_f \leq R_{\max}$, $R_{\max} = \max\limits_{i \in \mathrm{I}_{b,n}} \left\| \mathbf{x}_i - \mathbf{P}_i^{h,n} \right\|$.

In the general case, it suffices to put $R_f = R_{\max}$. The parameter $\gamma_{\min}$ defines the threshold value that is used to determine whether the projection belongs to the boundary of the surface or to the set of its interior points. When calculating the value of $\gamma_{\min}$ (see step 5 of the algorithm).

7 Numerical Results

We present below some simulations results that illustrate the applicability of the presented algorithms.

On Fig. 1 an example of simulation with two fractures are presented. The left hand side plot shows pore pressure distribution and the right hand side σ_{yy} stress component. Blue markers at the fracture mid-surface shows close point projections of the bulk tetrahedral mesh nodes onto the fracture.

On the Fig. 2 (top) it is shown CPP representation of the fracture mid-surface. Red and green markers correspond to the nodes of the bulk FE mesh. These points are connected with their projection onto the mid-surface. Green line segments correspond

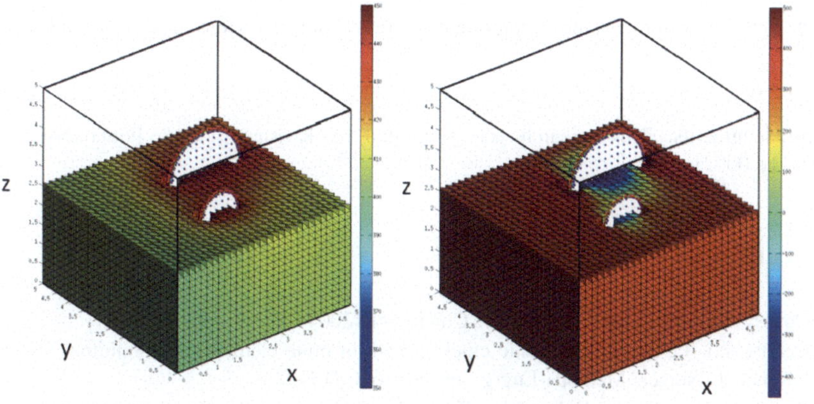

Fig. 1. Pore pressure distribution (left) and σ_{yy} stress component (right) in reservoir.

to the projection onto the internal part of the fracture while the red ones to the fracture front. On the bottom plot of the Fig. 2 stress component σ_{xx} distribution is shown for lower half of the computational domain.

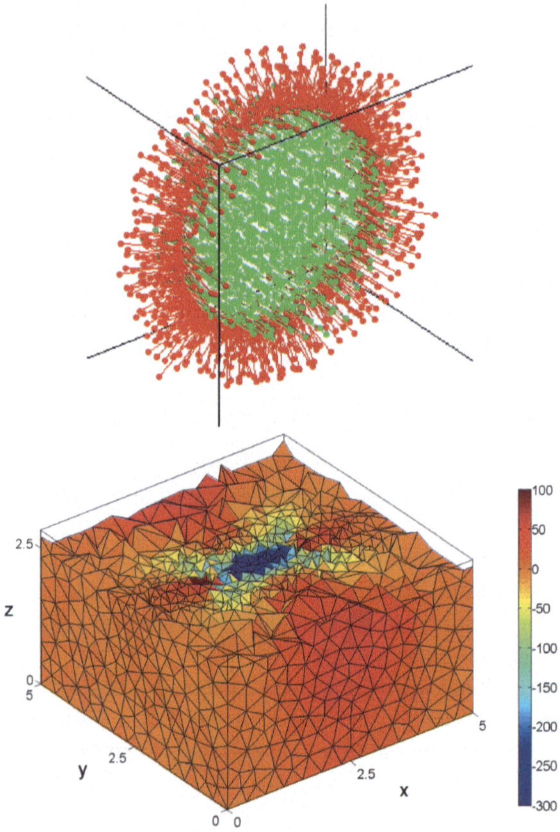

Fig. 2. CP projection fracture representation (top) and σ_{xx} stress component distribution (bottom)

Acknowledgments. This research was supported by Russian Science Foundation, project No. 15-11-00021.

References

1. Coussy, O.: Poromechanics. Wiley, Chichester (2004)
2. Belytschko, T., Black, T.: Elastic crack growth in finite elements with minimal remeshing. Internat. J. Numer. Methods Engrg. **45**, 601–620 (1999)
3. Moës, N., Dolbow, J., Belytschko, T.: A finite element method for crack growth without remeshing. Internat. J. Numer. Methods Engrg. **46**, 131–150 (1999)

4. Ivanov, A.V., Savenkov, E.B.: Simulation and visualization of the dynamics of a surface with a movable boundary on a stationary unstructured mesh. Sci. Vis. **9**(2), 64–81 (2017)
5. Linkov, A.M.: The particle velocity, speed equation and universal asymptotics for the efficient modelling of hydraulic fractures. J. Appl. Math. and Mech. **79**(1), 54–63 (2015)
6. Stolarska, M., Chopp, D., Moës, N., Belytschko, T.: Modelling crack growth by level sets in the extended finite element method. Int. J. Numer. Meth. Engng. **51**, 943–960 (2001)
7. Moës, N., Gravouil, A., Belytschko, T.: Non-planar 3D crack growth by the extended finite element and level sets – Part I: mechanical model. Int. J. Numer. Meth. Engng. **53**, 2549–2568 (2002)
8. Gravouil, A., Moës, N., Belytschko, T.: Non-planar 3D crack growth by the extended finite element and level sets–Part II: level set update. Int. J. Numer. Meth. Engng. **53**, 2569–2586 (2002)
9. Macdonald, C.B., Ruuth, S.J.: The implicit closest point method for the numerical solution of partial differential equations on surfaces. SIAM J. Sci. Comput. **31**, 4330–4350 (2009)
10. Savenkov, E.B., Borisov, V.E., Kritsky, B.V.: Utilization of closest point projection surface representation in extended finite element method. Math. Models and Computer Simulations (in press, 2019)

The Scheme Relating Frequency of Lightning Flashes to Statistical Characteristics of Convective Activity in the Atmosphere for the IAP RAS Climate Model

A. V. Eliseev[1,2,3(✉)], A. N. Ploskov[1], A. V. Chernokulsky[2], and I. I. Mokhov[1]

[1] Faculty of Physics, Lomonosov Moscow State University, Moscow, Russia
eliseev@ifaran.ru
[2] A.M. Obukhov Institute of the Atmospheric Physics, Russian Academy of Sciences, Moscow, Russia
[3] Kazan Federal University, Kazan, Russia

Abstract. A modification of the commonly used Price-Rind scheme for frequency of lightning flashes (FL), which can be used for calculations with large spatial and time steps, is developed. With such steps, the exponent in the relationship of LFF on convective cloud heights appears to be smaller by a factor of two over land and by one fourth over ocean in comparison to that in the original Price-Rind scheme. The modified version is implemented into the IAP RAS climate model (CM). The results of the lightning flash frequency simulations with the modified scheme agree better with the satellite data than those with the original one. In the IAP RAS CM, global warming (cooling) leads to FL increase (decrease) in all seasons. The sensitivity of lightning flashes frequency to the surface air temperature change at the global level is estimated equal to 10%/K.

Keywords: Lightning Flashes · Convective activity · Statistical gamma-distribution · EMIC · IAP RAS CM · Scenario RCP 8.5

1 Introduction

Lightning activity is associated with both dynamic processes in the atmosphere and atmospheric electricity processes [1]. It is important for environmental applications (lightning strikes often cause natural fires) and for problems of atmospheric chemistry (according to existing estimates, lightning flashes generate about 15% of the mass of nitrogen oxides in the atmosphere) [1]. As a result, parametrization of lightning activity is now being introduced into climate models, which connects the frequency of lightning flashes with the characteristics of convective clouds [2–5]. In particular, in [2] the frequency of lightning (FL) per unit area f is associated with the height of the upper boundary of convective clouds h:

V. I. Karev et al. (Eds.): *Physical and Mathematical Modeling of Earth and Environment Processes (2018)*, SPEES, pp. 334–343, 2019.
https://doi.org/10.1007/978-3-030-11533-3_33

$$f = C h^{\alpha} \tag{1}$$

and the values of the exponent α (as well as the coefficient C) are different for convective clouds of continental and marine origin: for the clouds of the first type $\alpha = 4.9$, of the second $\alpha = 1.7$.

2 Method and Problems

However, such parametrization cannot be used in Earth Models of Intermediate Complexity (EMICs) with a grid step in space of $\sim 10^8$ m and a time step of 10^5–10^6 s [6]; these values are several orders of magnitude greater than the characteristic spatial scale of convective cloud systems (104 m) and the characteristic time of their life (10^3–10^4 s). As a result, a set of convective systems is realized at every time step and in each horizontal model cell in the EMICs. In this case the "average" (reproducible by the model) characteristics of these systems are obtained by the space-time averaging of individual convective systems. Such averaging works as a smoothing operator, lowering the exponent in (1).

Climate model of the Institute of Atmospheric Physics RAS (IAP RAS CM) [7] is one of the EMICs and is the only model of this class in Russia. A specific feature of the IAP RAS CM is the explicit calculation of large-scale (with temporal and spatial scales exceeding the corresponding values for synoptic processes) and parameterized calculations of the statistical characteristics of synoptic processes. The spatial resolution of the IAP RAS CM is 4.5° in latitude and 6° in longitude, the time step is 5 days. Like the rest of the EMICs, it not allow to resolve clearly separate convective clouds or even separate systems such clouds.

The purpose of this work is to modify the scheme [2] for use in the EMICs, introduce this scheme into the IAP RAS CM and analyze the sensitivity of the characteristics of lightning activity during climate changes in numerical experiments with the IAP RAS CM.

A generalization of the FL calculation schemes can be carried out, if we consider not an individual convective system, but a statistical ensemble of such systems. In this case, it is necessary to introduce a probability distribution function for h. According to measurements over the European territory of the USSR [8], this function can be approximated by a gamma distribution with the shape parameter $a > 1$ and the scale parameter b. Similar can also be obtained for the maximum height of the radar echo from thunderstorm clouds according to [9]. For such a distribution, the average height of the clouds is $H = a\, b$, the standard deviation of the height of the clouds is $\sigma = a^{1/2}\, b$, and the coefficient of variation is $k = \sigma/H = a^{-1/2}$. From the condition $a > 1$ it follows that $k < 1$. Note that according to the sounding data for the maximum height of the radar echo from thunderstorm clouds, $k < 0.3$ [9].

When carrying out numerical calculations for h for different 1 km $\leq H \leq$ 20 km and $0 \leq k \leq 1$, a sample was constructed for the gamma distribution corresponding to each value of H and k. At the same time, in order to exclude very large and irrelevant observations (see, for example, [8]) of the values of h, it was assumed that $h \leq h_{max} = 20$ km. According to this sample, the ensemble average $\langle f \rangle = C \langle h^{\alpha} \rangle$ was calculated (triangular brackets mean ensemble averaging). Then, for each value of k, the exponent α was determined using linear regression ln $<f>$ on ln $<h>$ = ln H.

The results of the calculations revealed a monotonic decrease in α with increasing k both over land and on the ocean (Fig. 1). With a narrow distribution for h ($k << 1$), the values of the exponent in (1) are close to those recommended in [1]. With a wide distribution ($k \rightarrow 1$), these exponents decrease to 2.6 over land and to 1.3 over the ocean. For all $0 \leq k \leq 1$, the regression correlation coefficient ln $<f>$ on ln H exceeds 0.96. As a result, (1) with a modified value of α can actually be used to calculate the FL in EMICs.

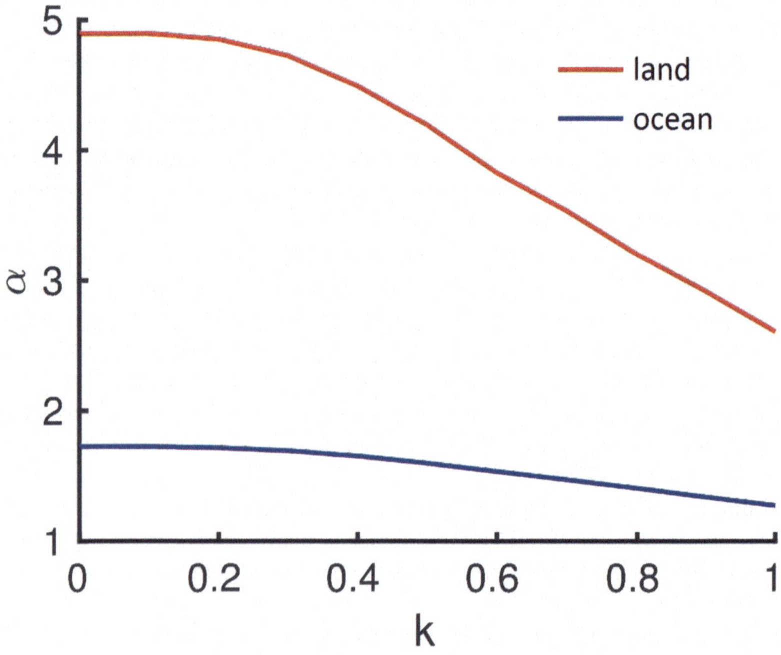

Fig. 1. The dependence of the exponent α in relation (1) on the coefficient of variation k for the probability distribution function of the height of convective clouds.

The scheme based on (1) was introduced in the IAP RAS CM. Additionally it was assumed that lightning flashes are possible only in the part of the cell occupied by convective clouds. As a result, instead of (1), a functional dependence was used in the IAP RAS CM:

$$f = C h^{\alpha} n_{co} \tag{2}$$

where n_{co} is convective cloud fraction in the model grid cell. The value of n_{co} was calculated in accordance with [10]. The tropopause height required for calculating H in the IAP RAS CM was set according to [11] as a third-order polynomial of the near-surface temperature of the atmosphere. It should be noted that the n_{co} multiplier was not taken into account when calculating the FL in accordance with the scheme [2] (or similar) in climate models (see, for example, [2–5]).

3 Results

With the IAP RAS CM, simulations were carried out for the period 1700–2100 AD under scenarios of anthropogenic impacts taking into account changes in the content of greenhouse gases (CO_2, CH_4 and N_2O) in the atmosphere, tropospheric and stratospheric volcanic sulfate aerosols, changes in solar constant and changes in agricultural areas. These impacts for the years 1700–2005 were set in accordance with the conditions of the project CMIP5 "Historical simulations" (Coupled Models Intercomparison Project, phase 5; http://www.Pasa.ac.at/web-apps/tnt/RcpDb). For the period 2006–2100, the corresponding impacts were specified in accordance with the RCP 8.5 scenario.

All simulations were carried out with the version of the model in which the exponents were set in accordance with the values obtained with the coefficient of variation k = 1 (version M), as well as with the values from [2] (version O). In connection with taking into account n_{co} in (2), simulations were carried out both with the full scheme (2) (version I) and with its variant, in which instead of the year-changing n_{co} used a multi-year average (taking into account the annual course) the number of convective clouds for the pre-industrial period (version P). Next, we use the notation X-Y, where X - or I, or P, and Y - or M, or O.

At the global level, for the modern period, all versions of the IAP RAS CM are in good agreement with the LIS/OTD (Lightning Imaging Sensor/Optical Transient Detector) satellite data [12]. If according to observations for 1995–2014, the global FL $F_g = 46 \pm 5\ s^{-1}$, then depending on the IAP RAS CM version for this time interval, F_g varies from 44 s^{-1} to 47 s^{-1} (Fig. 2). However, when separating land and ocean, it is clear that versions I-M and P-M are much better agreement with the LIS/OTD data than versions I-O and P-O. In particular, if in the first two versions for 1995–2014, when averaging over land $F_l = 35{-}37\,s^{-1}$ (in satisfactory agreement with the LIS/OTD data, in which the corresponding value is $41 \pm 4\ s^{-1}$), then the last two versions - $26{-}27\,s^{-1}$ (already significantly less empirical evaluation). When moving from versions O to versions M, the agreement of the FL averaged over the oceans with satellite data also improves significantly. If in versions I-M and P-M, the corresponding value for the period 1995–2014 equal to 8–9 s^{-1} (according to LIS/OTD - $5 \pm 1\ s^{-1}$), then in versions I-O and P-O it is 18–20 s^{-1}.

Improving the fit of f with satellite data during the transition from O to M versions

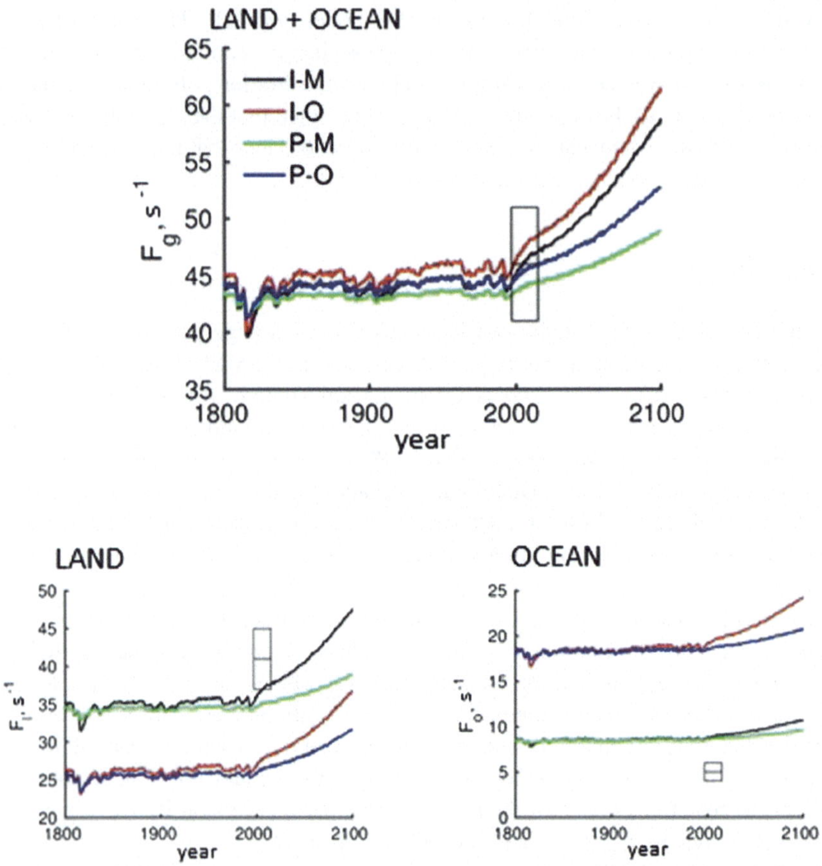

Fig. 2. Global (including separately over land and over the ocean) average annual frequency of lightning flashes in numerical experiments with the CM of the IAP RAS. The rectangles show the corresponding estimates for LIS/OTD satellite data (averages and uncertainty intervals).

is most significant for central Africa, the tropical oceanic areas, for southeastern North America, and central Eurasia (Fig. 3). It should be noted that, despite the improvement in f reproduction for central Africa, the FL here remains less than for central America. This discrepancy, however, is typical of all modern climate models with parametrization of the FL [3–5].

According to the obtained results, in the XIX-XX centuries climatic changes do not lead to significant changes in the frequency of lightning flashes in the IAP RAS CM both on the global (Fig. 2) and on the regional scale ($\leq 10\%$). This is consistent with the results of simulations with the ECHAM6 atmospheric general circulation model [4]. Short-term cooling after volcanic eruptions lead to an overall decrease in FL. After the eruption of the Tambor volcano (1815), the F^g value decreased from 44.1 s^{-1} to 39.6 s^{-1}, and after the eruption of the Pinatubo volcano (1991) - from 45.1 s^{-1} to

43.6 s^{-1}. At the same time, the decrease in f in the absolute magnitude in the tropics and subtropics.

Climate warming in the 21st century leads to an increase in the frequency of

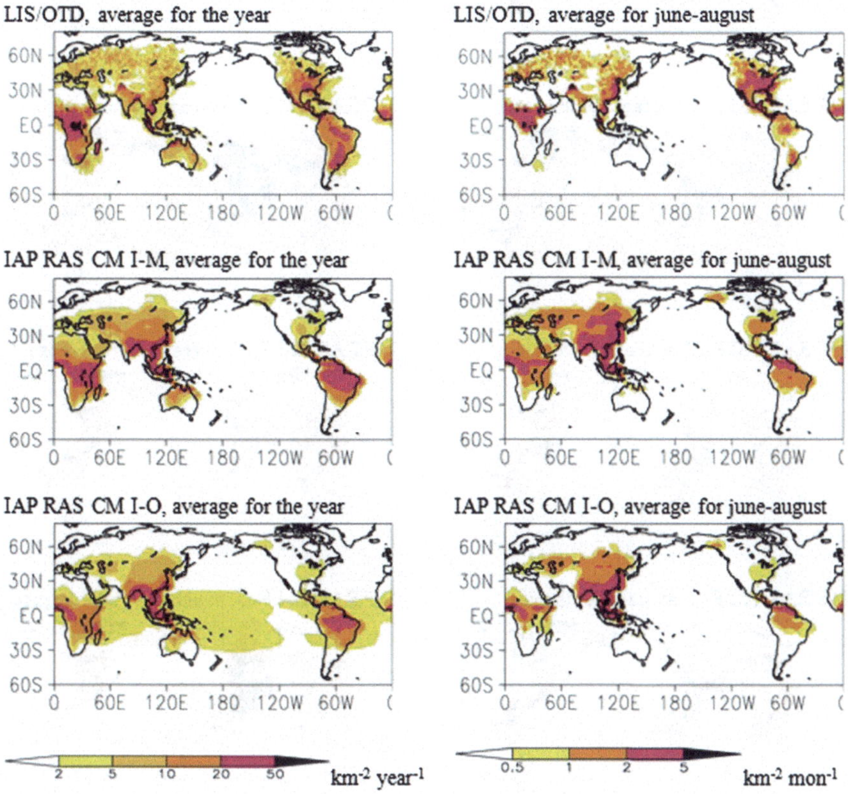

Fig. 3. The frequency of lightning flashes per unit area in 1995–2014 according to the LIS / OTD observations and in simulations with different versions of the IAP RAS CM.

lightning flashes. In the version I-M with the RCP 8.5 scenario, the Fg value increases by 2090–2100 to 58 s^{-1}, i.e. it increases by just over a quarter relative to the corresponding value for 1995–2014—close to the corresponding value obtained with the ECHAM6 model [4]. The frequency of lightning flashes per unit area varies in all seasons by no more than 10% in the tropics over the oceans between 1995–2014 and 2090–2100. In the regions of the maximum FL in the modern period (in south-east Asia, in the center of Africa and in Central America), by the end of the 21st century f increases by 10–20%; absolute changes in the frequency of lightning are maximum in these regions (Fig. 4). Over the continents in the middle latitudes, the increase in the FL by the end of the XXI century exceeds twofold relative to the modern period.

The decrease in the exponent α in (2) (the transition from versions O to versions M) leads to a sharp decrease in the response of the average annual values of f over the

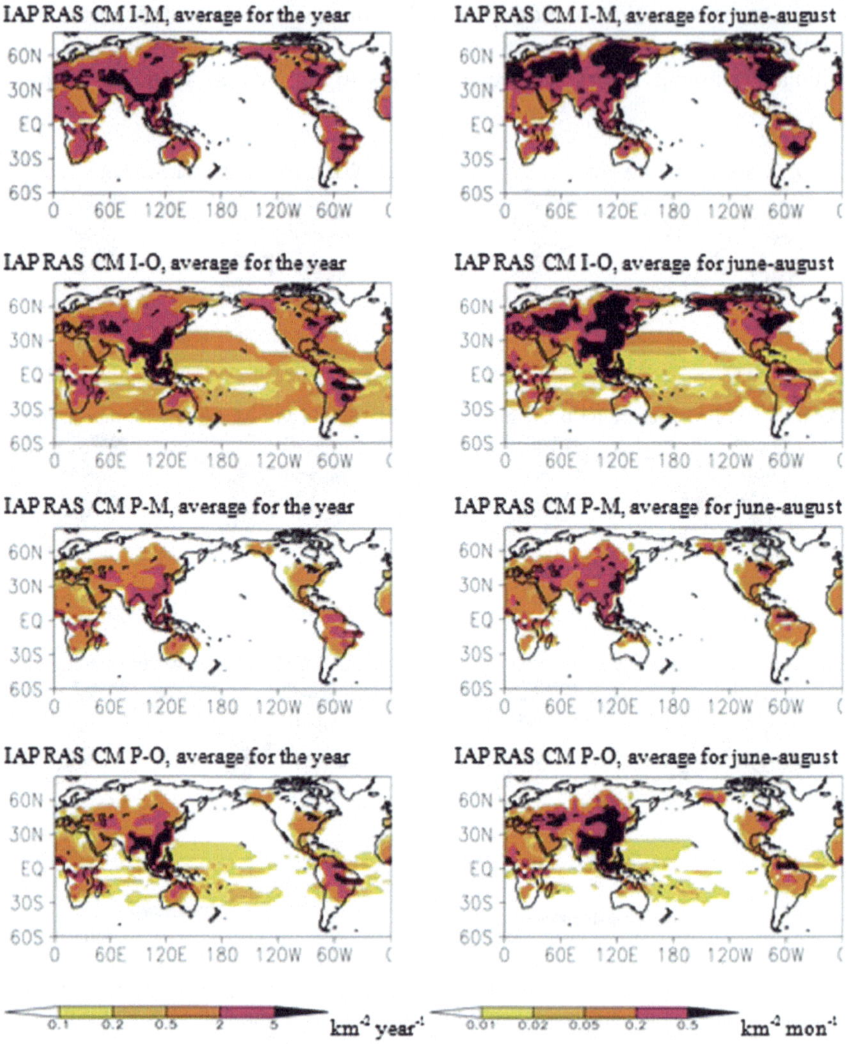

Fig. 4. Change in the frequency of lightning flashes per unit area from 1995–2014 to 2090–2100 in different versions of the IAP RAS CM.

oceans in the tropics (Fig. 4). At the same time, in other regions the average annual f response does not change very much. For both versions of model I, the spatial structure of the response of the f is fundamentally different for different seasons. If the average annual response is maximal in the tropics and subtropics, then in summer in these versions the climatic changes in f over the continents of middle latitudes in the 21st century with the RCP 8.5 scenario are similar in absolute value to those obtained for lower latitudes.

A comparison of versions I and P of the model suggests that the high sensitivity of f to climate change over mid-latitude continents is associated with corresponding changes in the number of convective clouds: nco in these regions in summer increases significantly in the 21st century with the RCP 8.5 scenario (Fig. 4). This is consistent with the increase of convective cloud fraction over Eurasia and North America, identified according to observations during climate warming in the last decades of the twentieth century [13] and is associated with changes in the static stability of the atmosphere [14]. It should be noted that the change in n_{co} obtained in simulations with the IAP RAS CM may not appear in other models. For example, in a climate model INMCM4 area of regions in which convective processes occur in the atmosphere decreases with warming [3]. In addition, trends in the number of convective clouds vary up to the sign for different data arrays for the end of the twentieth century [15].

For versions P of the IAP RAS CM, in which the change in the number of convective clouds during climate changes is not taken into account, the sensitivity of f to climate change is substantially less than for versions I, especially in mid-latitudes over the continents (Fig. 4). This is due to the neglect of a significant increase in the number of convective clouds in these regions in the warm period of the year. For the O version, this additionally shows a decrease in the sensitivity of the FL over the oceans of the tropics.

The development of lightning frequency parametrization in this work plays an essential role in determining the sensitivity K_g of the global FL to climate change. A decrease in the exponent in relations (1) and (2) (the transition from versions O to versions M) reduces this sensitivity. For the I-M version, it is equal to 9.7%/K, whereas for the I-O, it is 10.9%/K (Fig. 2). These values are close to those obtained with the ECHAM6 atmospheric circulation model [4]. According to the IAP RAS CM simulations, approximately half of the K_g value is associated with an increase in the number of convective clouds n_{co} during climatic changes. For versions P-M and P-O, in which the change in n_{co} during climate change does not affect f, the K_g value is 3.7%/K and 5.8%/K, respectively. These values are within the range of K_g values for different models (from −3%/K to + 14%/K [3–5]).

According to Fig. 2, the relative sensitivity of FL to climate change is greater over land than over the ocean (K_l and K_o, respectively). For versions I-M and I-O, K_l = 11%/K and K_l = 13%/K, respectively, and K_o = 7%/K and K_o = 9%/K, respectively. For versions P-M and P-O, K_l = 3.9%/K and K_l = 7.6%/K, respectively, and K_o = 3.7%/K and K_o = 3.8%/K, respectively. All these values are within the $-4\%/K \leq K_l \leq +17\%/K$ and $-6\%/K \leq K_l \leq +10\%/K$ intervals obtained with different parametrization of the FL in the CAM5 general atmospheric circulation model [5].

4 Conclusions

Thus, in this paper, a modification of the popular scheme [2] was developed to calculate the frequency of lightning, which can be used for models with large time and space steps, as well as in calculations for which only data with a time resolution are available that is insufficient to estimate the instantaneous values of the characteristics of convective activity in the atmosphere. Spatio-temporal averaging leads to a decrease in the exponent in the dependence of the frequency of lightning on the height of the upper of thunderstorm clouds. With a fairly wide (with a unit coefficient of variation) probability distribution for the height of thunderclouds, this exponent decreases with respect to the case of single convective systems by a factor of 2 over land and a quarter over the ocean.

A modified version of the scheme, taking into account that lightning activity is possible only in the part of the model cell covered with clouds, has been introduced in the IAP RAS CM, with which numerical experiments have been carried out in accordance with the conditions of the international climate comparison project CMIP5. In these simulations, it was found that the characteristics of lightning activity with a modified scheme are better consistent with satellite FL data than in the original version. At the same time, on the one hand, the decrease in the exponent of FL dependence on the average height of convective clouds reduces the sensitivity of the frequency of lightning flashes to climate change. On the other hand, a positive correlation of the number of convective clouds with the temperature of the troposphere increases the climate sensitivity of the FL. Global warming (cooling) of the climate in all seasons is accompanied by an increase (decrease) in the frequency of lightning. The sensitivity of the global frequency of lightning flashes to a change in the surface temperature of the atmosphere for the basic version I-M of a model obtained of 10%/K.

This work was supported by the RFBR grant 17-29-05098 using the results obtained in other projects of the RFBR and RAS programs. A.Eliseev's contribution was supported by the government program for federal support of Kazan (Volga region) Federal University for enhancing its competitiveness among the world's leading scientific and educational centers.

References

1. Rakov, V.A., Uman, M.A.: Lightning: Physics and Effects. Cambridge University Press, Cambridge (2003)
2. Price, C., Rind, D.: A simple lightning parameterization for calculating global lightning distributions. J. Geophys. Res. Atmos. **97**, 9919–9933 (1992). https://doi.org/10.1029/92JD00719
3. Mareev, E.A., Volodin, E.M.: Variation of the global electric circuit and Ionospheric potential in a general circulation model. Geophys. Res. Lett. **41**, 9009–9016 (2014). https://doi.org/10.1002/2014GL062352
4. Krause, A., Kloster, S., Wilkenskjeld, S., et al.: The sensitivity of global wildfires to simulated past, present, and future lightning frequency. J. Geophys. Res. Biogeosci. **119**, 312–322 (2014). https://doi.org/10.1002/2013jg002502
5. Clark, S.K., Ward, D.S., Mahowald, N.M.: Parameterization-based uncertainty in future lightning flash density. Geophys. Res. Lett. **44**, 2893–2901 (2017). https://doi.org/10.1002/2017GL073017
6. Claussen, M., Mysak, L.A., Weaver, A.J., et al.: Earth system models of intermediate complexity: closing the gap in the spectrum of climate system models. Clim. Dyn. **18**, 579–586 (2002). https://doi.org/10.1007/s00382-001-0200-1
7. Mokhov, I.I., Eliseev, A.V.: Modeling of global climate variations in the 20th–23rd centuries with new RCP scenarios of anthropogenic forcing. Dokl. Earth Sci. **443**, 532–536 (2012). https://doi.org/10.1134/S1028334X12040228

8. Mazin, I., Khrgian, A.: Handbook of Clouds and Cloudy Atmosphere. Gidrometeoizdat, Leningrad (1989). (in Russian)
9. Brylev, G.B., Gashina, S.B., Nizdoyminoga, G.L.: Radar Characteristics of Clouds and Precipitation. Gidrometeoizdat, Leningrad (1986). (in Russian)
10. Eliseev, A.V., Coumou, D., Chernokulsky, A.V., et al.: Scheme for calculation of multi-layer cloudiness and precipitation for climate models of intermediate complexity. Geosci. Model Devel. **6**, 1745–1765 (2013). https://doi.org/10.5194/gmd-6-1745-2013
11. Mokhov, I.I.: Diagnostics of Climate System Structure. Gidrometeoizdat, St. Petersburg (1993). (in Russian)
12. Cecil, D.J., Buechler, D.E., Blakeslee, R.J.: Gridded lightning climatology from TRMM-LIS and OTD: Dataset description. Atmos. Res. **135–136**, 404–414 (2014). https://doi.org/10.1016/j.atmosres.2012.06.028
13. Sun, B., Groisman, P.Ya., Mokhov, I.I.: Recent changes in cloud-type frequency and inferred increases in convection over the United States and the former USSR. J. Clim. **14**, 1864–1880 (2001). https://doi.org/10.1175/1520-0442(2001)014<1864:RCICTF>2.0.CO;2
14. Mokhov, I.I., Akperov, M.G.: Tropospheric lapse rate and its relation to surface temperature from reanalysis data. Izv. Atmos. Ocean. Phys. **42**, 430–438 (2006). https://doi.org/10.1134/S0001433806040037
15. Norris, J.R.: Trends in upper-level cloud cover and surface divergence over the tropical Indo-Pacific Ocean between 1952 and 1997. J. Geophys. Res. Atmos. **110**(D8), D08206 (2005). https://doi.org/10.1029/2005jd006183

Role of Electromagnetic Mechanisms in Downburst and Tornado Formation

Sergey Maslov[1,2](✉) and Vladimir Natyaganov[1]

[1] Lomonosov Moscow State University, Leninskiye gory 1, 119991 Moscow, Russia
sergm90@mail.ru, tenzor-home@yandex.ru
[2] Joint Institute for High Temperatures of RAS, Izhorskaya ul. 13, 125413 Moscow, Russia

Abstract. Basing on elecrohydrohynamic (EHD) equations, the article has researched role of thundercloud charge structure, atmospheric electric field (AEF) perturbations under it and giant dielectric permittivity effect (GDPE) in downburst and tornado formation. The authors have first suggested the formula qualitatively describing the process of thundercloud charge structure change from dipole to tripole one and simultaneously accounting the AEF presence even in case of fair weather. Thundercloud vorticity amplification during its recharging process, i.e. its transformation to the tornado-cyclone, is shown to be possible due to electromagnetic forces, not Coriolis ones. Besides that, dipole and tripole cloud charge distributions can create necessary conditions for generating downburst and tornado funnel correspondingly. The paper presents electrohydrodynamic model of forming downburst from a dipole cloud. The work considers total influence of tripole thundercloud charge structure and GDPE on tornado funnel formation and appearing number of accompanying specific phenomena. Using the formula of atmospheric electric field perturbations, the authors have theoretically explained frequently described "hill" ascent on the water surface toward the funnel descending. The authors have also shown that strong AEF perturbations can lead either to funnel descent or to its hanging above the ground.

Keywords: Atmospheric electric field · Thundercloud · Downburst · Tornado · Electrohydrodynamic pressure · Giant dielectric permittivity effect

1 Introduction

Model approaches to describing tornado and downbursts are often divided into pure hydrodynamic, thermohydrodynamic (THD) and hydroelectromagnetic ones. Hypotheses of the last type appeared more earlier that hydro- and thermohydrodynamic approaches. M. Lomonosov (1753, Russia) suggested one of the first electric hypotheses on tornado nature, which was supported by R. Hare (1837, the USA) and J. Peltier (1840, France). Only later pure hydrodynamic and thermohydrodynamic approaches to tornado modeling appeared, which at the beginning were used for typhoons research, and electric hypothesis was forgotten up to 1960-s.

V. I. Karev et al. (Eds.): *Physical and Mathematical Modeling of Earth and Environment Processes (2018)*, SPEES, pp. 344–358, 2019.
https://doi.org/10.1007/978-3-030-11533-3_34

The work [1] has explained (at energetic level) necessity of accounting important role of electromagnetic factors in tornado formation process. But even after that researches usually attempted to explain intriguing and often anomalous facts from numerous tornado behavior observations [2] basing on THD mechanisms [3].

At the same time, some of the natural facts [2] did not find adequate explanations in the limits of hydro- and thermohydrodynamic approaches. The main facts are high electric tornado activity (frequent lightning flashes, various funnel glows); its close connection with parent thundercloud; ascending dust or water droplet hill even before the funnel contact with underlying surface; forming still mysterious prester-tornado [1] descending from the thundercloud as a fiery column burning plants (grass and bushes up to root system); localizing vorticity in thin funnel walls which thickness can be by two orders less than tornado radius and can have a size about one meter. Hence a number of works after [1] suggest hydroelectromagnetic approaches to tornadogenesis research.

The work [4] showed determining character of electromagnetic factors for tornado funnel model in the mature stage and in the shape of coaxial cones. The article [5] attempted to explain important role of EHD-mechanisms at the initial tornado funnel formation stage before its contact with underlying surface accounting strong AEF perturbations under a parent thundercloud [2, 3, 6–9]. Concretization of this EHD-approach at model level for the beginning of tornado and downburst [10–12] formation suggested in [13] is continued in the present work.

2 Electric Field Perturbations Distribution Under the Thundercloud

The mature thundercloud usually has dipole or tripole charge structure [8, 9, 14]. As a rule, in the first case the upper layer is positively charged, the lower one is negative. The tripole cloud has comparatively small supplementary positive charge in its bottom center, which role and formation mechanisms raised questions of many researches, particularly Feynman [9]. Let us analyse role of cloud charge structure using simplified EHD-equations system [13, 15]:

$$\begin{cases} \rho(\partial \mathbf{u}/\partial t + (\mathbf{u}\nabla)\mathbf{u}) = -\nabla p_{\mathrm{e}} + \mu\Delta\mathbf{u} + \rho_{\mathrm{e}}\mathbf{E},\ \mathrm{div}\mathbf{u} = 0, \\ p_{\mathrm{e}} = p - \varepsilon_0(\varepsilon - 1)E^2/2, \quad \mathrm{div}(\varepsilon\varepsilon_0\mathbf{E}) = \rho_{\mathrm{e}}, \quad \mathrm{rot}\mathbf{E} = 0. \end{cases} \tag{1}$$

Here p_{e} is EHD-pressure, $\mathbf{E}$ – AEF value, ε_0 – dielectric constant, ε – relative dielectric permittivity, ρ_{e} – charge density, the rest of designations are standard for hydrodynamics. For convenience, let us work in cylindrical coordinate system (r, φ, z), where z-axis is directed straight up (perpendicular to the ground) and passes through the thundercloud center, r, φ – distance to z-axis and rotation angle around it correspondingly. In order to qualitative approximating vertical AEF component (which contribution is the most significant) under the cloud number of famous monographs [16, 17] apply the following formula with six defining parameters:

$$E_z = \frac{1}{2\pi\varepsilon_0}\left[\frac{Q_P H_P}{\left(H_P^2 + r^2\right)^{3/2}} - \frac{Q_N H_N}{\left(H_N^2 + r^2\right)^{3/2}} + \frac{Q_{LP} H_{LP}}{\left(H_{LP}^2 + r^2\right)^{3/2}}\right]. \qquad (2)$$

Here indexes «P», «N» and «LP» point to main positive, main negative and supplementary (lower) positive charges, Q and H – charge and center height of correspondent thundercloud part, each of them is approximated by a sphere in the given expression. Varying parameter Q_{LP} describes possible changing parent cloud charge structure from dipole to tripole one due to different mechanisms. However, thundercloud charge areas of different sign have not a spherical shape at all (according to indirect observations). Besides that, infinitely far from the cloud the expression (2) tends to zero, while in fair weather, i.e. in the absence of thunderclouds, AEF has the magnitude of about 100 V/m and is directed down (to the ground). So the work [5] qualitatively describes AEF perturbations under a tripole thundercloud with an approximative formula

$$\tilde{E}(\hat{r}) = \frac{E_z(\hat{r})}{E_0} = 1 - \frac{2(\hat{r}^3 - 8)(\hat{r}^4 - 16)}{(\hat{r}^7 + 64)}.$$

Here $E_0 \sim 100$ V/m – the fair weather field, $\hat{r} = r/L_1$ – distance to z-axis related to characteristic AEF changing scale L_1 under a tripole cloud. The article [13] first suggested a two-parameter formula accounting the AEF presence in fair weather, possibility of changing dipole cloud charge structure to tripole one and increasing electric field under severe thundercloud by more than one order:

$$\tilde{E}(\tilde{r}) = \frac{E_z(\tilde{r})}{E_0} = \frac{(4\tilde{r})^4 + \beta}{(4\tilde{r})^4 + 1}\left[N \cdot \exp(-\tilde{r}^2) - 1\right], \qquad (3)$$

where $\tilde{r} = r/L_2$ and the scale L_2 of AEF change is not equal to L_1 in general case. The value N corresponds to strong increasing electric field, sometimes up to $10^5 - 10^6$ V/m, i.e. under severe thunderclouds N can reach $10^3 - 10^4$. Varying parameter $\beta \leq 1$ describes the process of dipole thundercloud recharge to tripole one. The value $\beta = 1$ corresponds to a dipole cloud, $0.8 \leq \beta < 1$ – to the beginning of tripole structure formation, for $\beta \leq 0$ the zone where $\tilde{E} < 0$ appears near $\tilde{r} = 0$, and $\beta \leq -1$ corresponds to a mature tripole cloud (Fig. 1) with double reversal of AEF direction. For clarity and vertical scale convenience, graphs are plotted according to (3) for $N = 4$, as well as in the work [13].

One can conclude from (1), (3) that EHD-pressure p_e decreases significantly under the thundercloud center due to strong AEF perturbations for $N \gg 1$. Falling p_e is also caused by GDPE ε_e effect of the cloud [18] as a suspension of charged ice microparticles and supercooled water droplets with radius a. According to [8], these particles possess thin double electric layer (DEL) with the thickness $d \ll a$ on their surfaces:

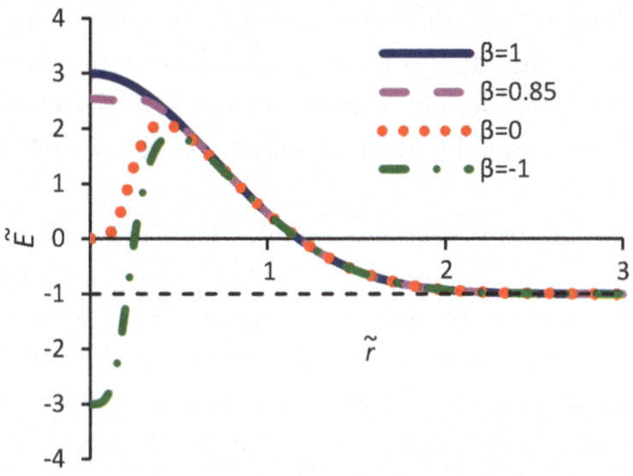

Fig. 1. Graphs $\tilde{E}(\tilde{r})$ for $N = 4$ and different β.

$$\varepsilon_e \approx c(\varepsilon_\Sigma - 1)[\mathbf{u} \times \mathbf{B}]/\langle\nabla\Phi\rangle_V \sim c(\varepsilon_\Sigma - 1)f \cdot a/d \gg 1. \quad (4)$$

Here ε_Σ – dielectric permittivity of the medium inside the DEL, $\langle\nabla\Phi\rangle_V$ – volume averaged electric potential gradient, **B** – magnetic field, c – volume particle concentration, f – dimensionless value [18] complicatedly depending on the medium characteristics in- and outside cloud particles. High **E** and ε_e values cause decreasing p_e and create appropriate conditions for Rayleigh – Taylor instability [12] realization. There exist two possible scenarios of developing this instability due to GDPE: changing charge structure to tripole with possible tornado funnels formation or jet downburst [10–12] of dense gas and droplet cloud substance usually differing from tornado by less vorticity and larger transverse size and by the presence of strongly pronounced divergent flow near the ground surface.

3 Forming Downburst from the Dipole Thundercloud Bottom

The work [10] assigns three main downburst stages. The first one is descent of thundercloud bottom up to its contact with the ground. At the second stage intense divergent flow appears near the ground. Maximal radial wind velocity at this step can reach 50 m/s which is comparable with a medium tornado. At the third phenomenon stage a so-called "cushion" of cold air descending from the cloud height generates near the ground surface. In case of vertical downburst the "cushion" prevents further thundercloud medium descent and divergent flow velocity near the ground decreases rather quickly. If the burst is oblique, the "cushion" forms only behind the downburst area and strong wind is observed in front of it. The downburst lifetime is usually from several minutes to hours (comparable with tornadoes).

As well as tornadoes, downbursts are very dangerous for forest, buildings and aviation. Therefore, researching this phenomenon is one of important physics and meteorology directions. A lot of works considered downbursts only in the terms of hydro- and thermohydrodynamics. However, strong AEF perturbations under a dipole cloud should also be discussed in the researching process of this dangerous natural phenomenon.

Let us analyze how the downburst descends from the dipole thundercloud to the underlying ground surface. In the simplest situation, when the cloud substance charge density is constant, AEF maximum and EHD-pressure minimum correspondingly are at the burst axis (Fig. 1). Hence, appearing a downburst is the most probable under the cloud center. In this approximation the given work considers the downburst motion at the initial stage – in its descent process to the ground believed infinitely remote from the burst ($z \to -\infty$). According to radar observations [10, 12], the downburst border with assumed DEL of constant surface charge density q_d = const is approximated by revolution paraboloid [19]. Besides that, because of small burst radius comparing with transverse thundercloud size AEF near the ground is believed as uniform and equal to $E_0(N-1)$, i.e. field under the dipole cloud center according to (3). For convenience let us work in the coordinate system connected with the burst and write the equations system for velocity $\mathbf{u}$ and electric field $\mathbf{E}$ in cylindrical coordinates (r, φ, z) outside the burst:

$$\begin{cases} u_r = -\frac{1}{r}\frac{\partial\psi}{\partial z}, \quad u_z = \frac{1}{r}\frac{\partial\psi}{\partial r}, \quad \mathrm{E}_r = \frac{1}{r}\frac{\partial\Psi_e}{\partial z}, \quad \mathrm{E}_z = -\frac{1}{r}\frac{\partial\Psi_e}{\partial r}, \\ \mathrm{rot}\mathbf{u} = \left(\frac{\partial^2\psi}{\partial z^2} + \frac{\partial^2\psi}{\partial r^2} - \frac{1}{r}\frac{\partial\psi}{\partial r}\right)\mathbf{e}_\varphi = 0, \ \mathrm{rot}\mathbf{E} = \left(\frac{\partial^2\Psi_e}{\partial z^2} + \frac{\partial^2\Psi_e}{\partial r^2} - \frac{1}{r}\frac{\partial\Psi_e}{\partial r}\right)\mathbf{e}_\varphi = 0, \\ \mathbf{u}_n|_\Sigma = 0, \ \lim\limits_{z\to-\infty} u_z = U_\infty, \ \lim\limits_{z\to-\infty} u_r = 0, \\ \lim\limits_{z\to-\infty} E_z = E_0(N-1), \ \lim\limits_{z\to-\infty} E_r = 0, \ \sigma\left(\frac{\partial\Phi}{\partial n}\right)_\Sigma = \mathrm{div}_\Sigma(q_\mathrm{d}\mathbf{u}_\tau). \end{cases} \tag{5}$$

Here Σ is a paraboloidal burst border with an equation $\Sigma = \sqrt{r^2+z^2} - (z+2b) = 0$, b = const; U_∞ = const – uniform flow velocity far from the paraboloid; ψ and Ψ_e – hydrodynamic and electric current functions accurate to sign; Φ and $(\partial\Phi/\partial n)_\Sigma$ – electric field potential and its derivative along the external normal to the paraboloid border (directed from the burst axis); σ – medium conductivity on the downburst border; $\mathrm{div}_\Sigma(q_\mathrm{d}\mathbf{u}_\tau)$ – convective surface current $q_\mathrm{d}\mathbf{u}_\tau$ divergence inside DEL. There is non-flow $\mathbf{u}_n|_\Sigma = 0$ and charge conservation $\sigma(\partial\Phi/\partial n)_\Sigma - \mathrm{div}_\Sigma(q_\mathrm{d}\mathbf{u}_\tau) = 0$ boundary conditions on the paraboloid border.

Let us introduce nondimensionalizing values U_∞, σ, $E_0(N-1)$, q_d. In this case dimensional and dimensionless (with the index "*") values are connected as

$$(r, z, b) = \frac{|q_\mathrm{d}|U_\infty}{\sigma E_0(N-1)}(r_*, z_*, b_*), \ \mathbf{u} = U_\infty\mathbf{u}_*, \ \mathbf{E} = E_0(N-1)\mathbf{E}_*,$$

$$\psi = U_\infty\left[\frac{q_\mathrm{d}U_\infty}{\sigma E_0(N-1)}\right]^2\psi_*, \ \Psi_\mathrm{e} = \frac{q_\mathrm{d}^2U_\infty^2}{\sigma^2 E_0(N-1)}\Psi_\mathrm{e}^* \text{ etc.}$$

The system (5) solution for dimensionless hydrodynamic ψ_* and electric Ψ_e^* current functions satisfy the expressions

$$\psi_* = \frac{1}{2}\left[r_*^2 - 2b_*\left(\sqrt{z_*^2 + r_*^2} + z_*\right)\right], \quad \Psi_e^* = \frac{1}{2}\left[r_*^2 - 2b_*\left(\sqrt{z_*^2 + r_*^2} + z_*\right)\right] + \Psi_1^*,$$

where the addition Ψ_1^* is determined by iterative method described in [20]. Figure 2 images dimensionless AEF value dependence on the distance to downburst axis. One can see near the burst border AEF value can grow significantly comparing with the field far from it.

Note that electric field calculation inside the downburst is a rather difficult problem. It is caused by sufficient number of parameters indefiniteness: charge density ρ_e distribution depending not only on electric, but also on hydrodynamic medium charac-

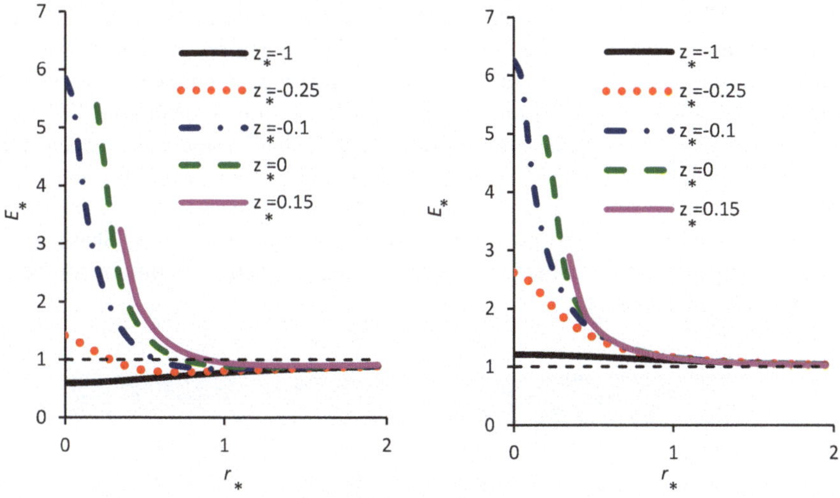

Fig. 2. Dependence of AEF magnitude on the distance to burst axis in case of $b_* = 0.1$ for $q_d > 0$ (left) and $q_d < 0$ (right).

teristics inside the burst (e.g. according to the scheme [19] of gas jet penetration into the incompressible fluid); electric potential bound inside the DEL; possible double layer breakdown or simple layer presence instead of it on the downburst border (in this case boundary conditions are significantly changed) etc. Similar facts are supposed to be analyzed later and here we only underline that rather simple downburst EHD-model described in the given section allowed to explain theoretically main observation data characterizing this dangerous natural phenomenon.

4 Specific Phenomena Accompanying Tornado Formation from a Tripole Thundercloud

Tripole charge structure of parent thundercloud creates favorable conditions for tornado funnel formation. It is caused by two main electromagnetic factors. First, there exists interaction of radial positive cluster ions current $\mathbf{j}_r$ with vertical geomagnetic component $\mathbf{B}_z$ at the parent thundercloud recharge stage from dipole to tripole one. Therefore, azimuthal magnetic force $\mathbf{j}_r \times \mathbf{B}_z$ together with Coriolis one generates parent cloud vorticity transforming it into a tornado-cyclone. The described mechanism does not act in a dipole cloud, so a downburst vorticity is usually 10^{-2} s^{-1} and a tornado one is more higher – up to 10 s^{-1}.

Second, strong electric field perturbations under the cloud in conjunction with GDPE cause EHD-pressure p_e decrease at the certain distance from the flow axis. A tornado funnel must form namely in local p_e falling region. A funnel descent is frequently accompanied by some specific features and even anomalous phenomena [2, 13], such as a "collar" – a lug of quasi-cylindrical shape in its base or so-called "devil horns" – two (or more) funnels forming in a wide "collar" corners. Appearing these effects in connected with peculiar AEF distribution at different stages of forming tripole structure. One can see from Fig. 1 at its initial formation stage ($\beta = 0.85$ in the formula (3)) profiles of AEF and EHD-pressure p_e have a plateau near z-axis. Shape of this plateau corresponds to the "collar" configuration (Fig. 3), hence conditions for the last one formation can occur.

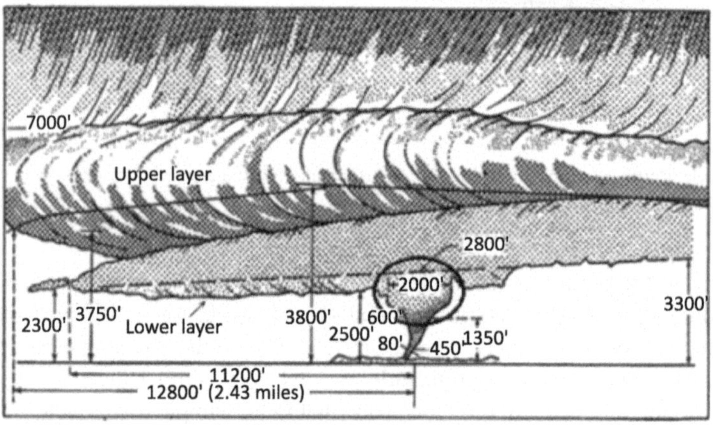

Fig. 3. Forming a tornado "collar" in the thundercloud bottom. Sizes are in foots, the "collar" is designated by the circle (Source: Nalivkin D.V. [2], p. 196)

At the beginning of appearing AEF double reversal zone near the cloud axis, where AEF is directed from the cloud to the ground, when in (3) $\beta \to 0$ and $\tilde{E} \to 0$ for $\tilde{r} \to 0$, $\tilde{E}$ and p_e reach their maximum and minimum correspondingly for some $\tilde{r} = \tilde{r}_m$. In similar cases "devil horns" can appear from a wide "collar" corners in the thundercloud bottom. Sometimes these "horns" can be observed in the presence of a central funnel reached the ground (Fig. 4). Due to initial thundercloud vorticity two or more funnels can fuse in a single "super-tornado" with multi-vortex flow structure. The article [21] notes that such tornadoes possess large destruction force.

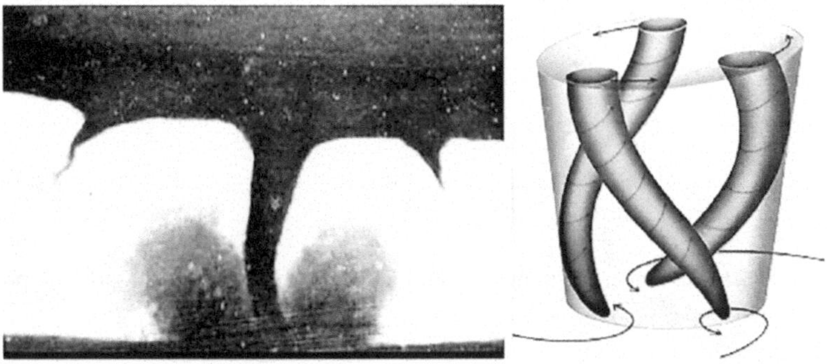

Fig. 4. Left: forming "devil horns" near the central funnel, one of the first tornado photo, 1884 (Source: Nalivkin D.V. [2], p. 218). Right: scheme of multi-vortex flow structure in the funnel (Source: Snow J.T. [21], p. 94)

A centrifugal force caused by parent thundercloud rotation can also take part in tornado funnels formation. In this case for small Rossby numbers $\mathrm{Ro} = U/(L\Omega) \ll 1$ [22], where U and L are characteristic velocity and size, Ω – cloud angular velocity, velocity $\mathbf{u}$ and pressure p fields can be approximately determined from a simple equations system:

$$\begin{cases} p_{\mathrm{el}} = p - \frac{\rho}{2}[\mathbf{\Omega} \times \mathbf{r}]^2 - \frac{\varepsilon_0(\varepsilon-1)}{2}E^2, & \frac{\partial p_{\mathrm{el}}}{\partial z} = 0, \\ \frac{\partial p_{\mathrm{el}}}{\partial x} = 2\rho u_y \Omega, \quad \frac{\partial p_{\mathrm{el}}}{\partial y} = -2\rho u_x \Omega, & \frac{\partial u_x}{\partial x} + \frac{\partial u_y}{\partial y} + \frac{\partial u_z}{\partial z} = 0. \end{cases} \tag{6}$$

Here p_{el} is modified EHD-pressure caused both by AEF and centrifugal force $\rho\Omega^2\mathbf{r}$. If $\mathbf{\Omega} = \Omega\mathbf{e}_z$, $\Omega = \mathrm{const}$, the last three equations of the given system lead to $\partial u_z/\partial z = 0$, $\partial u_x/\partial x + \partial u_y/\partial y = 0$, i.e. any slow motion in a fluid rotating as a solid body represents a superposition of axial flow parallel to z-axis and two-dimensional one in perpendicular plane. This result explaines possibility of forming Taylor column [22] analogue near the descending funnel. The system (6) can also be useful during researching anomalous landing of the whole rotating thundercloud on the ground [2] without a funnel formation (so-called "helicopter" landing).

High AEF values under the thundercloud also influence on a cascade appearing – dust or water droplet column frequently observed near a mature tornado funnel bottom or under a descending funnel still before its contact with underlying surface. Electric mechanism of cascade formation is the following [5]: radial force $F_e \sim \nabla E^2$ for certain β (e.g. $\beta = -1$) in formula (3) takes large negative values (Fig. 5). This force together with

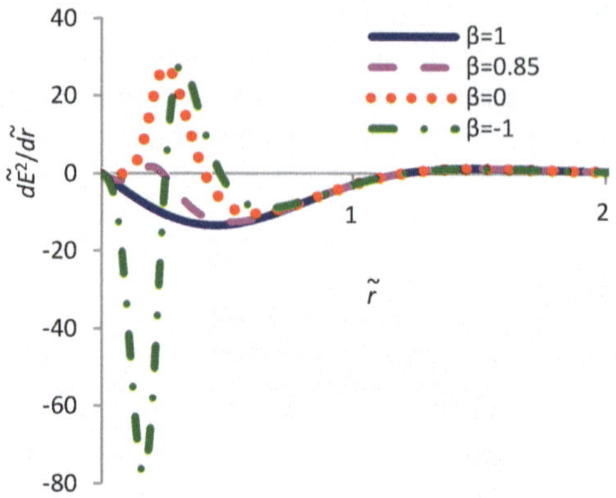

Fig. 5. Graphs $d\tilde{E}^2/d\tilde{r}$ under the thundercloud for different β corresponding to AEF profiles in Fig. 1

vertical Coulomb one $F_c = qE_z$, where q is a particle charge, leads to ascending "hill" of droplets or dust toward the funnel vertex long before its contact with the ground.

Water and dust particles in the cascade are comparatively large, relation of their radius a to DEL thickness d in (4) is usually much greater than is case of thundercloud bottom cluster ions, microdroplets or ice microcrystals. It follows from (4) that GDPE and strong p_e decrease can play significant role in cascade generation process. The electrical mechanism described explains in a certain degree possibility of appearing column of water droplets or dust with the height of 10 m and more [2] at the very beginning of tornado descent process and even earlier. This fact is not still explained in the limits of hydro- and thermohydrodynamic approaches.

AEF perturbations under the thundercloud can lead not only to descending tornado funnel to the ground, but also to its ascending into the cloud vice versa. The electric force $\mathbf{F}_z = \rho_e E_0 N \beta \mathbf{e}_z / \rho$ acting on the forming tornado funnel from the side of the thundercloud bottom [7] accelerate its descent when $\beta < 0$ and decelerate when $\beta > 0$. Therefore, in case of positive β action time of THD and EHD-mechanisms [6] increases in intensive vorticity generation process. It leads to growing azimuthal velocity of the funnel which can hang at the certain height or pulse periodically increasing or decreasing its length due to the noted force $\mathbf{F}_z$. Rotation of hanging vase-shaped funnel [2] can be described with the expression for vertical modification of combined Rankine vortex [7, 23]

$$u_r = u_z = 0, \quad u_\varphi = \omega_0 \left[rH(r_0 - r) + r^{-(d+1)} H(r - r_0) \right],$$

where $d \geq 1$, ω_0 are constants, r_0 – the vortex core radius, $H(x)$ – step Heaviside function, or by the Q-vortex [24] formula $u_\varphi = \Gamma/r \cdot [1 - \exp(-\lambda r^2)]$ where Γ, λ = const.

5 Conclusions

The article has analyzed the influence of thundercloud charge structure on forming intensive atmospheric jets and vortices basing on EHD-equations. Downbursts and tornado funnels are shown to represent the types of Rayleigh – Taylor instability realization for different thundercloud charge distribution. For a dipole cloud authors have solved the problem of forming downburst with paraboloidal shape. For thunderclouds with tripole charge structure the article has shown possibility of generating initial vorticity and subsequent tornado formation. A funnel can hang above the ground or go down to it. In certain cases the funnel descent is accompanied by appearing "collar", "devil horns" or oncoming ascent of droplet or dust "hill" near the underlying ground surface.

Appendix

The work [25] derived electromagnetic generalization of classic V. Levich [26] problem solution of electrocapillary drift (ECD) of a spherical drop with radius a and thin DEL with the thickness $d \ll a$ on its surface. As a result of this general Levich problem full solution in coordinate system connected with immobile drop center, one can obtain that axisymmetric deformation current is added to potential uniform flow around the drop, and toroidal Taylor vortex [25] appears besides spherical Hill one inside the drop.

Dimensionless electric potentials Φ' inside and Φ outside the drop is determined by the formulae in spherical coordinates (R,θ,φ)

$$\begin{aligned} R<1: \Phi' &= -\frac{2q_d u_0}{\breve{\sigma}} R\cos\theta - \frac{2q_d u_1}{\breve{\sigma}} R^2 P_2(\cos\theta) + C', \\ R>1: \Phi &= \left(-R+\gamma/R^2\right)\cos\theta + \frac{2q_d u_1}{3R^3} P_2(\cos\theta) + C \end{aligned} \tag{7}$$

correspondingly. Here the constant difference $\left|C' - C\right| \gg 1$ defines the potential bound across the DEL, $\breve{\sigma} = \sigma'/\sigma$ is the relation of drop and surrounding medium conductivities, $\gamma = q_d u_0 - 1/2$, q_d – DEL surface charge density, $P_2(\cos\theta)$ is the second degree Legendre polynomial. Values u_0 and u_1 are equal (accurate to small parameters $d/a \ll 1$ and $1/\breve{\sigma} \ll 1$) to

$$u_0 = \frac{3q_d}{2\left(2+3\breve{\eta}+q_d^2\right)}, \; u_1 = \frac{M\gamma(\gamma+4)}{20\left(1+\breve{\eta}+0.4q_d^2\right)} \tag{8}$$

and are connected with dimensionless tangential velocities on the drop surface by the conditions $u_\theta = u'_\theta = -u_0\sin\theta$ and $u_{1\theta} = u'_{1\theta} = -u_1\sin\theta\cos\theta$ correspondingly for ECD with Hill vortex inside the drop and deformation electro-vortex flow [4] outside and with toroidal Taylor vortex inside.

In the formulae (8) the value $\breve{\eta} = \eta'/\eta$ – relation of dynamic viscosities in- and outside the drop, M – Hartmann number [25] calculated using dimensional magnetic field induction proportional to external electric field value. This electric field provides ECD of a conductive drop with thin ($d/a \ll 1$) double electric layer on its surface. Underline that due to ECD oppositely charged double layer surfaces start to move which transform DEL into a double current layer (DCL) with convective surface currents $j = \pm q_d u_0 \sin\theta$ of meridional direction. These anti-parallel (due to different DEL surfaces charge signs) surface currents generate their own azimuthal magnetic field with induction

$$B = \frac{1}{2}\left(R + \frac{2\gamma}{R^2}\right)\sin\theta,\ B' = q_d u_0 R \sin\theta \tag{9}$$

in- and outside the drop correspondingly.

Moreover, there is a simple magnetic layer similar to surface Dirac δ-function with azimuthal induction $B_\Sigma = -q_d u_0 \delta(R-1)\sin\theta$ inside a thin DCL. As a result, a new complex structure, namely triple electromagnetic layer (TEML) with perpendicular to each other meridional surface currents directions and azimuthal magnetic field appears on a conductive drop surface in the presence of external electric field. The main TEML properties briefly described in [25] were used in [27] for creating electocapillary-vortex modeling one of the most mysterious natural phenomenon – a ball lightning.

Adding magnetic Lorentz force $[\mathbf{j} \times \mathbf{B}]$ in the motion equations with account of (9), one can obtain the second terms in formulae (7) and the expression for u_1. Then one should repeat the procedure described for calculating magnetic field of deformation electro-vortex current outside the drop and toroidal Taylor vortex inside it. As a result, the following azimuthal appendices for magnetic induction (correspondingly out- and inside the drop) can be derived:

$$B_1 = \frac{q_d u_1}{R^3}\sin\theta\cos\theta,\ B_1 = q_d u_1 R^2 \sin\theta\cos\theta.$$

However, there tangential induction components must break going through the current layer. Hence, an azimuthal simple magnetic layer like a surface δ-function with induction

$$B_\Sigma = -q_d \delta(R-1)(u_0 + u_1\cos\theta)\sin\theta \tag{10}$$

appears inside the DEL. Then, in the limits of generalized function theory the ECD problem of a single drop (more exactly, uniform flow with ECD velocity around it) is described by the equations system

$$\begin{cases} \mathrm{rotrot}\mathbf{u} + \nabla p = 8\pi A_1 \mathbf{U}_0 \delta(\mathbf{R}), \\ \mathrm{div}\mathbf{u} = -4\pi A_2 (\mathbf{U}_0 \cdot \nabla\delta(\mathbf{R})),\ \Delta\Phi = 4\pi\gamma(\mathbf{k}\cdot\nabla\delta(\mathbf{R})). \end{cases} \tag{11}$$

Here A_1, A_2 are constants, $\mathbf{U}_0$ – electrocapillary streamlining velocity (that of a drop ECD accurate to a sign) of uniform flow with collinear unit vector $\mathbf{k}$, $\gamma = q_d u_0 - 1/2$,

$\delta(\mathbf{R})$ – three-dimensional Dirac delta-function. It follows from the system (11) that a drop with a unit radius and thin DEL on its surface (thickness is much less than radius) can be considered for $R > 1$ as some combination of a singular force and a dipole with correspondent coefficients A_1, A_2 in terms of hydrodynamics and as an electric dipole in terms of electrodynamics.

Studying uniform suspension of such drops with thin DEL on their surfaces requires changing right parts of the system (11) equations to the correspondent sums:

$$8\pi A_1 \mathbf{U}_0 \sum_{i=1}^{m} \delta(\mathbf{R} - \mathbf{R}_i),\ -4\pi A_2 \sum_{i=1}^{m} (\mathbf{U}_0 \cdot \nabla\delta(\mathbf{R} - \mathbf{R}_i)),\ 4\pi\gamma \sum_{i=1}^{m} (\mathbf{k} \cdot \nabla\delta(\mathbf{R} - \mathbf{R}_i)).$$

Here $m + 1$ is the number of suspended particles, $\mathbf{R}$ and $\mathbf{R}_i$ – radius-vectors of the observation point and i-th particle center. Then one should make an averaging procedure of derived system (11) analogue using binary correlation function $g(R)$. Similar procedure is described in detail in the work [28] assuming that correlation function of volume drop concentration $c \leq 0.2$ equals to

$$g(R) = \begin{cases} 0,\ R < 2, \\ 1 + c(8 - 3R + R^3/16), 2 \leq R \leq 4, \\ 1,\ R \geq 4. \end{cases}$$

As a result, there can be derived [28] an equation system for such drops suspension instead of the system (11):

$$\begin{cases} \mathrm{div}\langle\mathbf{u}\rangle = -3cA_2\left[\frac{dg}{dR} + g(2)\delta(R-2)\right]u_e\cos\theta, \\ \mathrm{rotrot}\langle\mathbf{u}\rangle + \nabla\langle p\rangle = 6cA_1 g(R)\mathbf{u}_e, \\ \Delta\langle\Phi\rangle = 3c\gamma\left[\frac{dg}{dR} + g(2)\delta(R-2)\right]\cos\theta. \end{cases} \quad (12)$$

One should add to this system the equations for internal region of the test drop placed in the coordinates origin and corresponding conditions for $R = 1$ (the drop border) and $R = 2$ (spherical surface of drops center number break surrounding the test drop).

In this system $\mathbf{u}_e$ is averaged drops ECD velocity, angle brackets $\langle\rangle$ mean averaging over possible configurations ensemble, $g(2)$ is correlation function bound for $R = 2$. For explaining uniform suspensions GDPE of drops with thin surface DEL the most important equation is the last one of (12) for electric potential which is convenient to be written separately for the regions $1 < R < 2$ and $R > 2$; one should also add the Laplace equation for $R < 1$. As a result, the electric potential equations system

$$\Delta\langle\Phi'\rangle = 0,\ \Delta\langle\Phi\rangle = 0,\ \Delta\langle\Phi_*\rangle = 3c\gamma\frac{dg}{dR}\cos\theta \quad (13)$$

can be obtained correspondingly for $R < 1$ (inside the selected drop), $1 < R < 2$ (surrounding medium) and $R > 2$ ("effective" medium containing singular specialities, which model the presence of other drops). The index "*" in the last equation of (13) and further points to the region $R > 2$.

Skipping the derivation and explanation of the boundary conditions form of the general system like (12) and obtaining procedure of its general solution (see the details in [29]), let us give only subsystem (13) solution which can be found in the form

$$\langle\Phi'\rangle = \alpha' R\cos\theta + C',\ \langle\Phi\rangle = \left(\alpha R + \gamma/R^2\right)\cos\theta + C,$$
$$\langle\Phi_*\rangle = \left(\alpha_* R + \gamma_*/R^2\right)\cos\theta,$$

where the coefficients (necessary for the further explanation of drop suspension GDPE) determined in [29] from the general system (12) boundary conditions have the following form accurate to the terms linear in $c < 0.2$:

$$\alpha' = -\frac{2q_d u_0}{\breve{\sigma}},\ \alpha = -(1+4c\gamma),\ \alpha_* = -(1+3c\gamma),$$
$$\gamma_* = \gamma + c\left(1 - 4q_d u_0 + \frac{13}{3}q_d^2 u_0^2\right).$$

The work [29] derived a generalized Maxwell formula for effective electric conductivity σ_e of uniform spherical drops suspension with surface charge of thin DEL, when ECD takes place:

$$\sigma_e = \frac{\sigma[\alpha_* - c(\alpha+\gamma)] + \sigma' c\alpha'}{\alpha_* - c(\alpha+\gamma) + c\alpha'}. \tag{14}$$

This formula numerator corresponds to the current density $\langle\mathbf{j}\rangle_V = -\sigma\left(\nabla\langle\Phi_*\rangle - c\overline{\nabla\langle\Phi\rangle}\right) - \sigma' c\overline{\nabla\langle\Phi'\rangle}$ average over the volume (in electric approximation), denominator – to volume average electric potential gradient $\nabla\langle\Phi\rangle_V = \nabla\langle\Phi_*\rangle - c\overline{\nabla\langle\Phi\rangle} + c\overline{\nabla\langle\Phi'\rangle}$, and the line above the certain terms in these expressions means average over the unit radius test ball volume [28, 29].

Changing formally all conductivities σ_e, σ', σ values to correspondent ones of dielectric permittivities ε_e, ε', ε in (14), one can obtain a classic Maxwell formula [30] generalization for effective permittivity ε_e of uniform spherical drops (with DEL) suspension during ECD, but only in electric approximation. However, during the calculation of e volume average electric induction $\mathbf{D} = \varepsilon_0(\varepsilon_1\mathbf{E} + (\varepsilon_1 - 1)[\mathbf{u}\times\mathbf{B}])$ in electromagnetic approximation one should regard strongly pronounced surface, or Σ-phase forming a thin TEML with currents and magnetic field orthogonal to each other. As a result, calculated effective dielectric permittivity of uniform drop suspension with thin surface TEML during ECD, the following expression is the main one:

$$\varepsilon_e \approx \varepsilon_0 \frac{c(\varepsilon_\Sigma - 1)\overline{[\mathbf{u}_\Sigma \times \mathbf{B}_\Sigma]}}{\nabla\langle\Phi_*\rangle - c\overline{\nabla\langle\Phi\rangle} + c\overline{\nabla\langle\Phi'\rangle}} \sim \varepsilon_0(\varepsilon_\Sigma - 1)c q_d u_0 u_1 \frac{a}{d}. \tag{15}$$

Here ε_Σ is a relative dielectric permittivity of polarized medium inside a thin DEL and a line above the vector product $\overline{[\mathbf{u}_\Sigma \times \mathbf{B}_\Sigma]}$ is now meaning average over a thin sphere layer volume corresponding to TEML.

It follows from (8), (10) that after integrating $[\mathbf{u}_\Sigma \times \mathbf{B}_\Sigma]$ along this sphere layer only the term $2q_d u_0 u_1 \sin^2\theta \cos\theta \cdot \mathbf{n}$ ($\mathbf{n}$ is the external normal) gives non-zero $\mathbf{k}$-component (along ECD velocity and external electric field providing it). Then, averaging over this layer equals to

$$\overline{[\mathbf{u}_\Sigma \times \mathbf{B}_\Sigma]} = \overline{[\mathbf{u}_\theta \times \mathbf{B}_\varphi]} = \frac{4}{15} q_d u_0 u_1 \cdot \frac{a}{d} \mathbf{k},$$

which leads to the result (15) (accurate to numeric coefficient) with large dimensionless parameter $a/d \gg 1$ for a thin DEL ($d/a \ll 1$) exactly.

Let us underline as a conclusion there does not appear similar increasing effective electric conductivity of drops suspension with thin DEL, although Ohm's law includes the vector term $\mathbf{j} = \sigma(\mathbf{E} + [\mathbf{u} \times \mathbf{B}])$, but transverse conductivity $\sigma_\Sigma = 0$ inside the DCL generated during ECD. Hence the result derived in [29] remains true.

References

1. Vonnegut, B.: Electrical theory of tornadoes. J. Geophys. Res. **65**(1), 203–212 (1960). https://doi.org/10.1029/JZ065i001p00203
2. Nalivkin, D.V.: Hurricanes, Storms and Tornadoes. Geographic Peculiarities and Geological Activity (in Russian). Nauka, Leningrad (1969)
3. Bengtsson, L., Lighthill, J. (eds.): Intense atmospheric vortices. Springer-Verlag, New York (1982)
4. Shcherbinin, E.V. (ed.): Electro-vortex flow (in Russian). Zinatne, Riga (1985)
5. Natyaganov, V.L., Maslov, S.A.: Electromagnetic mechanisims of forming a tornado-like whirlwind. Moscow Universty Mechanics Bulletin **69**(2), 29–34 (2014). https://doi.org/10.3103/S0027133014020010
6. Maslov, S.A.: Electric mechanisms of vorticity amplification in the funnel of a tornado. Moscow Univ. Mech. Bull. **70**(6), 149–152 (2015). https://doi.org/10.3103/S0027133015060035
7. Maslov, S.A.: Effect of the atmospheric electric field under a thundercloud on tornado funnel formation. Moscow Univ. Mech. Bull. **72**(1), 23–27 (2017). https://doi.org/10.3103/S0027133017010058
8. Frenkel, Ya.I.: The theory of atmospheric electricity phenomena (in Russian). GITTL, Moscow – Leningrad (1947)
9. Feynman, R., Leighton, R., Sands, M.: The Feynman Lectures on Physics, Mainly Electromagnetism and Matter, vol. 2. Addison-Wesley, Reading (1964)
10. Fujita, T.T.: Tornadoes and downbursts in the context of generalized planetary scales. J. Atmos. Sci. **38**(8), 1511–1534 (1981). https://doi.org/10.1175/1520-0469(1981)038%3c1511:TADITC%3e2.0.CO;2
11. Fujita, T.T.: Downbursts: meteorological features and wind field characteristics. J. Wind Eng. Ind. Aerodyn. **36**(1), 75–86 (1990). https://doi.org/10.1016/0167-6105(90)90294-M
12. Gendugov, V.M., Natyaganov, V.L., Chaika, A.A.: Oblique impact of a cylindrical jet on a plane. Dokl. Phys. **55**(8), 405–408 (2010). https://doi.org/10.1134/S1028335810080094

13. Maslov, S.A., Natyaganov, V.L.: Influence of an electric thundercloud structure on forming the tornado-like vortices (in Russian). Prikladnaya Fizika **6**, 16–20 (2015)
14. Williams, E.R.: The tripole structure of thunderstorms. J. Geophys. Res. D **94**(11), 13151–13167 (1989). https://doi.org/10.1029/JD094iD11p13151
15. Saranin, V.A.: Equilibrium stability, charging, convection and interaction of liquids in electric fields (in Russian). Regulyarnaya i Khaoticheskaya Dinamika, Moscow – Izhevsk (2009)
16. Rakov, V.A., Uman, M.A.: Lightning: Physics and Effects. Cambridge University Press, New York (2003)
17. MacGorman, D.R., Rust, W.D.: The Electrical Nature of Storms. Oxford University Press, New York (1998)
18. Natyaganov, V.L.: Conventionality of the boundaries of EHD and MHD approximations in some problems of electromagnetic hydrodynamics (in Russian). In: 9th International Scientific Conference on Modern Problems of electrophysics and Electrohydrodynamics of Fluids, pp. 131–134. SOLO, Saint-Petersburg (2009)
19. Sagomonyan, A.Y.: Impact and Penetration of Solid Bodies into Liquid (in Russian). Moscow State University Press, Moscow (1986)
20. Vabishchevich, P.N., Pulatov, P.A.: Numerical solution of the Neumann exterior problem. USSR Comput. Math. Math. Phys. **27**(2), 141–146 (1987). https://doi.org/10.1016/0041-5553(87)90169-8
21. Snow, J.T.: The tornado. Sci. A. **250**(4), 86–96 (1984)
22. Batchelor, G.K.: An Introduction to Fluid Dynamics. Cambridge University Press, Cambridge (1967)
23. Lamb, H.: Hydrodynamics. Dover, New York (1945)
24. Alekseenko, S.V., Kuibin, P.A., Okulov, V.L.: Introduction to Theory of Concentrated Vortices (in Russian). Institute of Thermophysics, Novosibirsk (2003)
25. Natyaganov, V.L.: Electrocapillary vortex model for a Hill – Taylor spherical vortex. Dokl. Phys. **46**(11), 800–802 (2001). https://doi.org/10.1134/1.1424375
26. Levich, V.G.: Physical-Chemical Hydrodynamics (in Russian). Fizmatgiz Publisher, Moscow (1959)
27. Natyaganov, V.L.: An electrocapillary eddy model of ball lightning. Dokl. Phys. **48**(6), 319–322 (2003). https://doi.org/10.1134/1.1591322
28. Natyaganov, V.L., Oreshina, I.V.: Electrohydrodynamics of monodisperse emulsions: 1. Sedimentation of an emulsion of droplets with an electrical double layer. Colloid J. **62**(1), 79–85 (2000)
29. Natyaganov, V.L., Oreshina, I.V.: Electrohydrodynamics of monodisperse emulsions: 2. Electrophoretic motion of spherical emulsion droplets. Colloid J. **62**(1), 86–89 (2000)
30. Landau, L.D., Lifshits, E.M.: Electrodynamics of continuous media. Addison-Wesley, Reading (1960)

Linear Approximations of Turbulent Moments of Horizontal Velocity and Temperature Fluctuations Within a Forced Convection Sublayer of the Atmospheric Surface Layer

Alexander Vulfson[1,2](✉) and Petr Nikolaev[3]

[1] Institute of Oil and Gas Problems, Russian Academy of Sciences, Moscow 117701, Russia
vulfson@ipng.ru

[2] National Research University Higher School of Economics, Moscow 101000, Russia

[3] National University of Science and Technology MISIS, Moscow 119049, Russia

Abstract. In the convective surface layer a heavy forced convection sublayer is distinguished. The turbulent moments of this sublayer depend mainly on the buoyance flux. It is shown that "linear" approximations are effective for describing turbulent moments of this sublayer. These approximations correspond to truncated Taylor series expansions in a modified height, that include only two terms. The first-order expansion terms do not take into account the wind and represent the free convection limits of the Monin-Obukhov similarity theory. The second-order expansion terms take into account the wind and its effect on the convection. The proposed approximations are compared with the experimental data.

Keywords: Linear approximations · Monin-Obukhov similarity theory · Atmospheric surface layer

1 Introduction

The classical similarity theory of the atmospheric surface layer was first formulated in [1–3] for approximation of the first-order turbulent moments.

It is significant that the Monin-Obukhov similarity theory allows us to find turbulent moments of a higher order under no-wind conditions. In particular, second-order turbulent moments were calculated in [4–7]. Approximations of the turbulent moments of the surface layer, corresponding to no-wind conditions, are called free-convection limits. For them, the parameter $z/|L_*| = \infty$, where z is the height of the level above the underlying surface; $|L_*| = 0$ is the Monin-Obukhov length parameter.

The free-convection limits of the convective surface layer can be used as a first-order approach in the construction of linear approximations of turbulent moments in a sufficiently heavy region $\zeta_0 \leq z/|L_*| < \infty$, where $\zeta_0 \approx 4 \cdot 10^{-2}$. Linear approximations of turbulent moments complement the free-convective limits by the components that

V. I. Karev et al. (Eds.): *Physical and Mathematical Modeling of Earth and Environment Processes (2018)*, SPEES, pp. 359–367, 2019.
https://doi.org/10.1007/978-3-030-11533-3_35

take into account the presence of wind. Therefore, the region $\zeta_0 \leq z/|L_*| < \infty$ should be interpreted as a sublayer of forced convection.

In the present work, linear approximations are used to describe known observational data. A comparison with experimental data indicates the existence of a forced convection sublayer with a lower boundary $\zeta_0 \approx 4 \cdot 10^{-2}$.

2 Local Parameters of Friction Velocity and Buoyancy Flux

Let t be time and x, y, z be a Cartesian coordinate system placed at the underlying surface $z = 0$ so that the z axis is opposite to gravitational acceleration g.

We assume that $u = u(x, y, z, t)$ and $w = w\,(x, y, z, t)$ are the velocity vector components along the horizontal and vertical axes; $\bar{u} = \bar{u}(z)$ is the mean value of the horizontal wind velocity vector along the x axis; and $u'(x, y, z, t) = u(x, y, z, t) - \bar{u}$ is the fluctuation of the horizontal velocity.

It is supposed that $\Theta = \Theta(x, y, z, t)$ is the potential air temperature and Θ_0 is a constant mean value of the potential temperature at the upper boundary of the surface layer [8]. A potential temperature fluctuation $\Theta'(x, y, z, t) = \Theta(x, y, z, t) - \Theta_0$ and a dimensionless potential-temperature fluctuation $\theta(x, y, z, t) = \Theta'(x, y, z, t)/\Theta_0$ are introduced following [8]. The quantity $g\theta(x, y, z, t)$ denotes local buoyancy [9].

Let us assume that

$$U_* = \lim_{z \to 0} \left(-\overline{u'w'}\right)^{1/2} > 0, \quad gS_\theta = \lim_{z \to 0} \overline{g\theta w} > 0 \tag{1}$$

where U_* and gS_θ have dimensions $[U_*] = \mathrm{m/s}$ and $[gS_\theta] = \mathrm{m^2/s^3}$.

For application of the similarity theory in the convective surface layer, we use three key parameters: z, gS_θ and U_*. The parameter of height z is a variable; the buoyancy flux gS_θ and friction velocity U_* parameters are constants.

The parameters $U_* > 0$ and $gS_\theta > 0$ allow us to introduce constant parameters of length $L_* < 0$ and buoyancy $g\Theta_* > 0$ such that

$$L_* = -\frac{U_*^3}{k_v gS_\theta}, \quad g\Theta_* = \frac{gS_\theta}{U_*} \tag{2}$$

where $k_v = 0.4$ is the von Kármán constant.

We suppose that H is the mean heat flux from the underlying surface, ρ_0 is the mean air density at the underlying surface, c_P is the specific heat capacity of dry air and T_* is the temperature parameter of the Monin-Obukhov theory.

Then, taking into account (2), it is found that

$$gS_\theta = \left(\frac{g}{\Theta_0}\right)\frac{H}{c_P\rho_0}, \quad g\Theta_* = -\left(\frac{g}{\Theta_0}\right)T_*, \quad T_* = -\Theta_*\Theta_0 \tag{3}$$

Relationships (3) indicate the proportionality of the key parameters gS_θ and $g\Theta_*$ to the conventional parameters H and T_* of the Monin-Obukhov similarity theory.

3 Monin-Obukhov Similarity Theory and Free-Convection Limits of the Surface Layer

We examine a convective surface layer in conditions of free convection, $U_* = 0$. In this case, only two key parameters z and gS_θ are finite.

In accordance with [3, 6], the first and second moments of buoyancy and the second moment of horizontal velocity fluctuation have the forms

$$\frac{\partial}{\partial z} g\overline{\theta} = -\lambda_\theta z^{-4/3} (gS_\theta)^{2/3} \tag{4}$$

$$\overline{(g\theta)^2} = \lambda_{\theta\theta} (gS_\theta)^{4/3} z^{-2/3}, \quad \overline{(u')^2} = \lambda_{uu} (gS_\theta)^{2/3} z^{2/3} \tag{5}$$

where $\lambda_\theta = 1$ from [6], $\lambda_{\theta\theta} = 1.8$ from [10] and λ_{uu} is a positive constant.

According to [11–13], the second relationship of (5) is formed by a stochastic ensemble of convective jets.

It is supposed that that K_h is the coefficient of turbulent heat transport. Under the free convection regime, the buoyance flux does not change with the height. In accordance with the Boussinesq gradient relationships and the similarity equality (4), we obtain

$$-K_h \frac{\partial}{\partial z} g\overline{\theta} = \overline{g\theta w} = gS_\theta, \quad K_h = \frac{1}{\lambda_\theta} (gS_\theta)^{1/3} z^{4/3} \tag{6}$$

Let us consider the convective surface layer in forced convection with low winds $0 \neq U_* << 1$. In this case, all three key parameters z, gS_θ and U_* are finite.

To construct free-convection limits of the Monin-Obukhov similarity theory, it is suggested that the turbulent moments in low wind $0 \neq U_* << 1$ are the same as those when there is no wind, $U_* = 0$.

Substituting (2) and (3) into the relationship for the first buoyancy moment (4) leads to the equality

$$-\frac{k_v z}{g\Theta_*} \frac{\partial}{\partial z} g\overline{\theta} = -\frac{k_v z}{\Theta_*} \frac{\partial}{\partial z} \overline{\theta} = -\frac{k_v z}{|T_*|} \frac{\partial}{\partial z} \overline{\Theta} = \alpha_\theta (z/|L_*|)^{-1/3} \tag{7}$$

where $\alpha_\theta = k_v^{4/3} \lambda_\theta \approx 0.3$ is a positive constant (see [5, 6, 14]).

The transformation of (6), with (2) and (3) taken into account, leads to the equality

$$\frac{K_h}{U_* |L_*|} = \frac{1}{\lambda_\theta k_v^{1/3}} (z/|L_*|)^{4/3} \tag{8}$$

The second moments of buoyance and horizontal velocity fluctuations (5) are transformed by taking into account (2) and (3). Thus

$$\frac{\overline{(g\theta)^2}}{(g\Theta_*)^2} = \frac{\overline{\theta^2}}{\Theta_*^2} = \frac{\overline{(\Theta')^2}}{|T_*|^2} = \alpha_{\theta\theta}^2 (z/|L_*|)^{-2/3} \tag{9}$$

$$\frac{\overline{(u')^2}}{U_*^2} = \alpha_{uu}^2 (z/|L_*|)^{2/3} \tag{10}$$

where the constant coefficients $\alpha_{\theta\theta}^2 = k_v^{2/3}\lambda_{\theta\theta} \approx 1$ from the data of [7, 14] and $\alpha_{uu} =$ 5.5 from the data of [15].

It is assumed that K_m is the coefficient of turbulent transport of horizontal momentum; $P_t = K_m/K_h$ is the turbulent Prandtl number. According to [16, 17], it is assumed that $P_t = 0.76$.

Within the surface layer, the buoyance and momentum fluxes do not change with the height. In accordance with the Boussinesq gradient relationships, the relationship for momentum flux can be written as

$$K_m = K_h P_t, \quad K_m \frac{\partial}{\partial z} \bar{u} = U_*^2 \tag{11}$$

Substituting (8) into the relationships (11), we find the relationship for the free convection limit of horizontal velocity

$$-\frac{k_v z}{|U_*|} \frac{\partial}{\partial z} \bar{u} = \alpha_u (z/|L_*|)^{-1/3}, \quad \alpha_u = \alpha_\theta / P_t = \lambda_\theta k_v^{4/3} / P_t \tag{12}$$

When $P_t = 0.76$ and $\alpha_\theta = 0.3$, the relationship (12) gives $\alpha_u = 0.4$.

The relationships (7), (9), (10) and (12) are the free convection limits of the bouyance and horizontal velocity.

4 Linear Approximations of the Monin-Obukhov Universal Similarity Functions, the First and Second Moments of the Surface Layer

Length parameter $L_* \leq 0$ is used to construct the dimensionless height $z/|L_*|$. The region $\zeta_0 < z/|L_*| < \infty$ is denoted as a forced-convection sublayer with moderate wind. We suppose that $\zeta_0 \approx 4 \cdot 10^{-2}$. In the forced-convection sublayer a free-convection area $1 << z/|L_*|$ is also distinguished.

It can be assumed that the dimensionless forms of the equations for the first (7) and second (9) buoyancy moments in the forced-convection sublayer within $4 \cdot 10^{-2} < z/|L_*| < \infty$ can be written as

$$-\frac{k_v z}{|T_*|} \frac{\partial}{\partial z} \overline{\Theta} = \alpha_\theta (z/|L_*|)^{-1/3} F_\theta (z/|L_*|), \quad \lim_{z/|L_*| \to \infty} F_\theta (z/|L_*|) = 1 \tag{13}$$

$$\frac{\overline{(\Theta')^2}}{|T_*|^2} = \alpha_{\theta\theta}^2 (z/|L_*|)^{-2/3} F_{\theta\theta}^2 (z/|L_*|), \qquad \lim_{z/|L_*| \to \infty} F_{\theta\theta}^2 (z/|L_*|) = 1 \tag{14}$$

It should be noted that the coefficient $\alpha_\theta = k_v^{4/3} \lambda_\theta \approx 0.3$ from the data of [5, 6, 14] and the coefficient $\alpha_{\theta\theta}^2 \approx 0.95$ from the data of [18–20].

The first and second moments of horizontal velocity can be defined as the forms that are similar to the first (13) and second (14) moments of buoyance:

$$-\frac{k_v z}{U_*} \frac{\partial}{\partial z} \bar{u} = \alpha_u (z/|L_*|)^{-1/3} F_u (z/|L_*|), \qquad \lim_{z/|L_*| \to \infty} F_u (z/|L_*|) = 1 \tag{15}$$

$$\frac{\overline{(u')^2}}{|U_*|^2} = \alpha_{uu}^2 (z/|L_*|)^{2/3} F_{uu}^2 (z/|L_*|), \qquad \lim_{z/|L_*| \to \infty} F_{uu}^2 (z/|L_*|) = 1 \tag{16}$$

Linear approximations of the moments of horizontal velocity and potential temperature fluctuations are constructed in the forced-convection sublayer. The first-order expansion terms of linear approximations are chosen according to the free convective limits of the Monin-Obukhov similarity theory under no-wind conditions, $U_* = 0$. The second-order expansion terms of linear approximations will then describe profiles of turbulent moments with the presence of wind, $U_* \neq 0$.

Limiting ourselves to a linear Taylor expansion of $F_\theta(z/|L_*|)$, $F_u(z/|L_*|)$, $F_{\theta\theta}(z/|L_*|)$ and $F_{uu}(z/|L_*|)$ in parameter $(z/|L_*|)^{-2/3}$, we obtain

$$-\frac{k_v z}{|T_*|} \frac{\partial}{\partial z} \overline{\Theta} = \alpha_\theta (z/|L_*|)^{-1/3} \left\{ 1 - \beta_\theta (z/|L_*|)^{-2/3} \right\} \tag{17}$$

$$-\frac{k_v z}{U_*} \frac{\partial}{\partial z} \bar{u} = \alpha_u (z/|L_*|)^{-1/3} \left\{ 1 - \beta_u (z/|L_*|)^{-2/3} \right\} \tag{18}$$

$$\frac{\overline{(\Theta')^2}}{|T_*|^2} = \frac{\sigma_\theta^2}{|T_*|^2} = \alpha_{\theta\theta}^2 (z/|L_*|)^{-2/3} \left\{ 1 - \beta_{\theta\theta} (z/|L_*|)^{-2/3} \right\} \tag{19}$$

$$\frac{\overline{(u')^2}}{U_*^2} = \frac{\sigma_u^2}{U_*^2} = \alpha_{uu}^2 (z/|L_*|)^{2/3} \left\{ 1 + \beta_{uu} (z/|L_*|)^{-2/3} \right\} \tag{20}$$

where σ_θ, σ_u are variances of potential-temperature fluctuation and horizontal velocity; β_θ, β_u, $\beta_{\theta\theta}$, β_{uu} are constant coefficients.

Figure 1a presents a comparison of the linear approximation of Monin-Obukhov similarity theory (17) at $\alpha_\theta = 0.3$, $\beta_\theta = 3 \cdot 10^{-2}$ with the field data [21]. It shows that the approximation (17) slightly overestimates this field data within the free-convection sublayer.

Figure 1b presents a comparison between the field data [21] and the linear approximation of Monin-Obukhov similarity theory (18) at $\alpha_u = 0.4$ and $\beta_u = 3 \cdot 10^{-2}$.

A comparison between the field data [22] of the dimensionless potential temperature variance $\sigma_\theta/|T_*|$ and its linear approximation in the form of (19) with $\alpha_{\theta\theta}^2 = 0.95$, $\beta_{\theta\theta} = 0.06$ is shown in Fig. 2a.

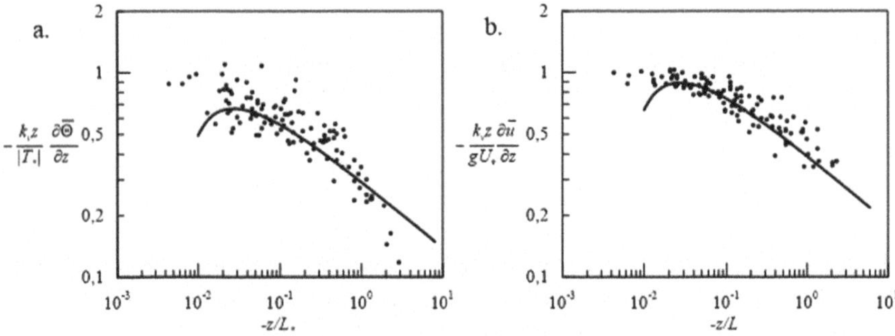

Fig. 1. Comparison of linear approximations (17) and (18) with empirical values of the function $-k_v z|T_*|^{-1}\partial\bar{\Theta}/\partial z$ and $-k_v z|U_*|^{-1}\partial\bar{u}/\partial z$ from [21]. The solid line (a) corresponds to approximation (17) with coefficients $\alpha_\theta = 0.3$ and $\beta_\theta = 3 \cdot 10^{-2}$. The solid line (b) corresponds to approximation (18) with coefficients $\alpha_u = 0.4$ and $\beta_u = 3 \cdot 10^{-2}$.

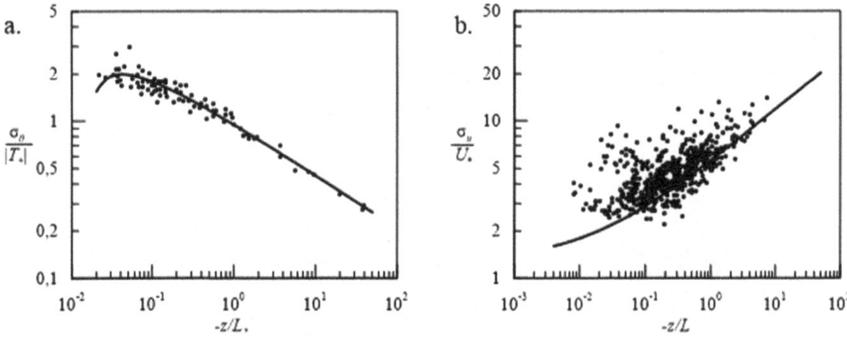

Fig. 2. Comparison of linear approximations (19) and (20) with empirical values of $\sigma_\theta/|T_*|$ from [22] and $\sigma_u/|U_*|$ from [20]. The solid line (a) corresponds to approximation (19) with coefficients $\alpha_{\theta\theta}^2 = 0.95$ and $\beta_{\theta\theta} = 0.06$. The solid line (b) corresponds to approximation (20) with coefficients $\alpha_{uu} = 5.5$ and $\beta_{uu} = 0.06$.

A comparison between the field data [20] of the dimensionless horizontal velocity variance $\sigma_u/|U_*|$ and its linear approximation in the form of (20) with $\alpha_u = 5.5$ and $\beta_{uu} = 0.06$ is shown in Fig. 2b.

The results, that are presented in Fig. 1, show that the linear approximations (17) and (18) are correct within the range of $2.7 \cdot 10^{-2} < z/|L_*| < \infty$.

Figure 2 shows that the linear approximations (19) and (20) describe the empirical values within the range $4 \cdot 10^{-2} < z/|L_*| < \infty$ fairly well.

We choose $\zeta_0 \approx 4 \cdot 10^{-2}$ as the lower boundary of the forced convection sublayer. Thus, the results of comparisons in Figs. 1 and 2 prove that the linear approximations (17)–(20) correspond to turbulent moments within the forced convection sublayer, $4 \cdot 10^{-2} < z/|L_*| < \infty$.

Another expansion, which is similar to (17)–(20), and its comparison with experimental data sets are shown in [23].

Alternative global approaches that are using the local similarity theory are presented in [24–27].

Modern comprehensive field data of atmospheric surface layer parameters and techniques of its processing are described in [28, 29].

5 Conclusion

The analysis of the known experimental data shows that in the convective surface layer there is a heavy sublayer of forced convection, with a clearly defined lower boundary. For describing turbulent moments of this sublayer, the Monin-Obukhov similarity theory is used. The universal functions of the similarity theory are expanded to truncated Taylor series expansions, that include only two terms.

The first-order expansion terms correspond to the free-convection limits of the Monin-Obukhov theory in the static atmosphere. The second-order expansion terms take into account the presence of wind and its effect on convection. The presented results can be used as a basis for construction of a layered structure of the atmospheric surface layer.

This study was supported by the Russian Foundation for Basic Research, grant 17-05-01116.

References

1. Obukhov, A.M.: Turbulence in thermally inhomogeneous atmosphere. Tr. Inst. Teor. Geofiz. Akad. Nauk SSSR. **1**, 95–115 (1946)
2. Monin, A.S., Obukhov, A.M.: The dimensionless characteristics of turbulence in the surface layer of the atmosphere. Dokl. Akad. Nauk SSSR **93**, 257–260 (1953)
3. Monin, A.S., Obukhov, A.M.: Basic laws of turbulent mixing in the surface layer of the atmosphere. Tr. Geofiz. Inst. Akad. Nauk SSSR. **151**, 167–187 (1954)
4. Monin, A.S., Yaglom, A.M.: Statistical Fluid Mechanics: Mechanics of Turbulence, vol. 1. MIT, Cambridge (1975)
5. Kader, B.A., Yaglom, A.M.: Mean fields and fluctuation moments in unstably stratified turbulent boundary layers. J. Fluid Mech. **212**, 637–662 (1990)
6. Priestley, C.H.B.: Turbulent Transfer in the Lower Atmosphere. University of Chicago Press, Chicago (1959)

7. Wyngaard, J.C., Cote, O.R., Izumi, Y.: Local free convection, similarity and the budgets of shear stress and heat flux. J. Atmos. Sci. **28**, 1171–1182 (1971)
8. Vulfson, A.N.: Equations of deep convection in a dry atmosphere. Izv. Atmos. Ocean. Phys. **17**, 646–649 (1981)
9. Turner, J.S.: Buoyancy Effects in Fluids. Cambridge University Press, Cambridge (2009)
10. Kaimal, J.C., et al.: Turbulence structure in the convective boundary layer. J. Atmos. Sci. **33**, 2152–2169 (1976)
11. Vulfson, A.N., Borodin, O.O.: An ensemble of dynamically identical thermals and vertical profiles of turbulent moments in the convective surface layer of atmosphere. Russ. Meteorol. Hydrol. **34**, 491 (2009)
12. Vulfson, A., Borodin, O., Nikolaev, P.: Convective jets: volcanic activity and turbulent mixing in the boundary layers of the atmosphere and ocean. In: Karev, V., Klimov, D., Pokazeev, K. (eds.) Physical and Mathematical Modeling of Earth and Environment Processes, PMMEEP 2017, pp. 71–83. Springer, Cham (2017)
13. Vulfson, A.N., Nikolaev, P.V.: An integral model of a convective jet with a pressure force and forms of vertical fluxes in the atmospheric surface layer. J. Phys: Conf. Ser. **955**, 012013 (2018)
14. Li, D., Katul, G.G., Bou-Zeid, E.: Mean velocity and temperature profiles in a sheared diabatic turbulent boundary layer. Phys. Fluids **24**, 105105 (2012)
15. Andreas, E.L., Hill, R.J., Gosz, J.R., Moore, D.I., Otto, W.D., Sarma, A.D.: Statistics of surface-layer turbulence over terrain with metre-scale heterogeneity. Bound.-Layer Meteorol. **86**, 379–408 (1998)
16. Businger, J.A., Wyngaard, J.C., Izumi, Y., Bradley, E.F.: Flux-profile relationships in the atmospheric surface layer. J. Atmos. Sci. **28**, 181–189 (1971)
17. Huo, Q., Cai, X., Kang, L., Zhang, H., Song, Y.: Effects of surface source/sink distributions on the flux–gradient similarity in the unstable surface layer. Theoret. Appl. Climatol. **119**, 313–322 (2015)
18. Katul, G.G., Hsieh, C.I.: A note on the flux-variance similarity relationships for heat and water vapour in the unstable atmospheric surface layer. Bound.-Layer Meteorol. **90**, 327–338 (1999)
19. Asanuma, J., Brutsaert, W.: Turbulence variance characteristics of temperature and humidity in the unstable atmospheric surface layer above a variable pine forest. Water Resour. Res. **35**, 515–521 (1999)
20. Choi, T., et al.: Turbulent exchange of heat, water vapor, and momentum over a Tibetan prairie by eddy covariance and flux variance measurements. J. Geophys. Res. D: Atmos. **109**, 1–12 (2004)
21. Wilson, D.K.: An alternative function for the wind and temperature gradients in unstable surface layers. Bound.-Layer Meteorol. **99**, 151–158 (2001)
22. Liu, X., Tsukamoto, O., Oikawa, T., Ohtaki, E.: A study of correlations of scalar quantities in the atmospheric surface layer. Bound.-Layer Meteorol. **87**, 499–508 (1998)
23. Vulfson, A.N., Nikolaev, P.V: Linear approximations of the second turbulent moments of the atmospheric convective surface layer in a forced-convection sublayer. Izv. Atmos. Ocean. Phys. **54**, 472–479 (2018)
24. Vulfson, A.N., Volodin, I.A., Borodin, O.O.: Local similarity theory and universal profiles of turbulent characteristics in the convective boundary layer. Russ. Meteorol. Hydrol. **10**, 1–10 (2004)
25. Shao, Y., Hacker, J.M.: Local similarity relationships in a horizontally inhomogeneous boundary layer. Bound.-Layer Meteorol. **52**, 17–40 (1990)

26. Wood, C.R., et al.: Turbulent flow at 190 m height above London during 2006–2008: a climatology and the applicability of similarity theory. Bound.-Layer Meteorol. **137**, 77–96 (2010)
27. Nadeau, D.F., Pardyjak, E.R., Higgins, C.W., Parlange, M.B.: Similarity scaling over a steep alpine slope. Bound.-Layer Meteorol. **147**, 401–419 (2013)
28. Liu, Y., et al.: Nondimensional wind and temperature profiles in the atmospheric surface layer over the hinterland of the Taklimakan Desert in China. Adv. Meteorol. **2016**, 1–8 (2016)
29. Kang, D., Wang, Q.: Optimized estimation of surface layer characteristics from profiling measurements. Atmosphere (Basel) **7**, 14 (2016)

Simulation of the Circulating Flow in the Vicinity of a Horizontal Well

P. E. Morozov(✉)

Institute of Mechanics and Engineering, Subdivision of FIC KazanSC of RAS, Kazan, Russia
morozov@imm.knc.ru

Abstract. In this paper we present new semi analytical and approximate analytical pressure transient and steady-state solutions for a horizontal circulation well in anisotropic reservoir. The pressure transient solutions incorporate the skin and wellbore storage effects for production and injection sections of a horizontal circulation well. For the steady state case the approximate analytical solution and engineering formula for estimating productivity of horizontal circulation well are derived The solutions presented in this paper are compared with the numerical simulation in KAPPA Saphir.

Keywords: Horizontal well · Production and injection sections · Circulation flow · Analytical solution · Geothermal energy · Productivity

1 Introduction

One of the promising ways to extract geothermal energy is to use horizontal circulation wells with the installation of a heat exchanger directly in the geothermal reservoir [1]. To this end, a horizontal wellbore is divided into the production and injection sections (Fig. 1). Passing through a heat exchanger installed in the production section of a horizontal well, the working fluid (formation fluid or brine) is cooled and pumped back into the reservoir, where it is again heated to reservoir temperature. Due to the occurrence of a circulation flow in the vicinity of a horizontal well, a significant area of the reservoir is involved in the heat exchange process.

The choice of pumping equipment for a horizontal circulating well requires a definition the pressure drop between the production and injection sections for a given flow rate. In this paper, a semi-analytical solutions of unsteady and steady-state fluid flow to a horizontal circulation well in an anisotropic reservoir, taking into account the wellbore storage and skin effects, was obtained. Based on the assumption of the uniform flux distribution along the length of production and injection sections, an approximate analytical solution of the problem is proposed.

V. I. Karev et al. (Eds.): *Physical and Mathematical Modeling of Earth and Environment Processes (2018)*, SPEES, pp. 368–373, 2019.
https://doi.org/10.1007/978-3-030-11533-3_36

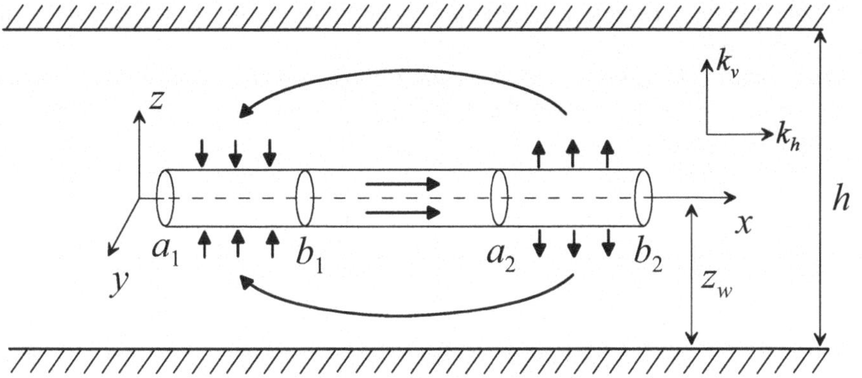

Fig. 1. Scheme of horizontal circulating well.

2 Semi-analytical Solution

Suppose that the anisotropic reservoir has an impermeable top and bottom, and a horizontal well is divided into two sections (Fig. 1). Fluid is pumped at constant flow rate Q from reservoir to the first section. Next, it is transferred by pump within the well to the second section, and subsequently returned to the reservoir from there. Neglecting the pressure loss on the friction, we will assume that the pressure distribution in both section of the horizontal well is uniform, i.e. $p(x,t) = p_i(t)$, $x \in [a_i, b_i]$, $i = 1,2$. To obtain a solution to the problem, we use the solution for a point source in an anisotropic reservoir bounded by two parallel planes. After integrating elementary sources with a flow density q and –q along production and injection sections in Laplace domain, we obtain a system of integral equations for determining pressures $p_i(t)$ and flux distributions $q_i(x,t)$ along the length of sections [2]:

$$\overline{p}_{id}(u) = \sum\nolimits_{j=1}^{2} \int_{\alpha_j}^{\beta_j} \overline{q}_j(\theta,u) \left[K_0\left(\sqrt{u}|x_d-\theta|\right) + 2\sum_{n=1}^{\infty} K_0\left(\sqrt{u}|x_d-\theta|\right)\Omega_n \right] d\theta \\ + S_i l_i \overline{q}_i(x_d,u) \tag{1}$$

$$\int_{\alpha_i}^{\beta_i} \overline{q}_i(\theta,u)d\theta = \frac{(-1)^{i+1}}{u} - C_{id}u\overline{p}_{id}(u),\ x \in [a_i,b_i],\ i = 1,2 \tag{2}$$

where u – Laplace transform variable; $p_{id} = \frac{2\pi k_h h(p_k - p_i(t))}{\mu Q}$ – dimensionless pressure in the i-th section of the horizontal well; p_k – reservoir pressure; $t_d = \frac{k_h t}{\mu \beta L^2}$ – dimensionless time; $L = b_2 - a_1$ – horizontal well length; h – reservoir thickness; r_c – wellbore radius; k_h, k_v – horizontal and vertical reservoir permeability; μ – fluid viscosity; β – compressibility; $l_i = \beta_i - \alpha_i = (b_2 - a_1)/L$ – dimensionless length of i-th section; C_i – wellbore storage of the i-th section; $r_{cd} = \frac{r_c}{2L}\left(1 + \sqrt{\frac{k_h}{k_v}}\right)$; $C_{id} = \frac{C_i}{2\pi h \beta L^2}$; $x_d = \frac{x}{L}$; $\xi_n = \frac{\pi n}{h_d}$; $h_d = \sqrt{\frac{k_h}{k_v}}\frac{h}{L}$; $z_{wd} = \sqrt{\frac{k_h}{k_v}}\frac{z_w}{L}$; $\Omega_n = \cos\xi_n z_{wd}\cos\xi_n(z_{wd} + r_{cd})$. Skin factor S_i characterizes additional filtration resistance in the near-wellbore zone of the i-th section of the horizontal well.

Assuming in (1), (2) $u \to \infty$, which corresponds $t_d \to 0$, we obtain the system of integral equations for determining the steady state pressures and flux distribution along the length of sections:

$$p_{id} = \sum_{j=1}^{2} \int_{\alpha j}^{\beta_j} q_j(\theta) \left[In\left(\frac{1}{|x_d - \theta|} \right) + 2\sum_{n=1}^{\infty} K_0(\xi_n |x_d - \theta|) \Omega_n \right] d\theta \\ + S_i l_i q_i(x_d) \tag{3}$$

$$\int_{\alpha_i}^{\beta_i} q_i(\theta) d\theta = (-1)^{i+1},\ x \in [a_i, b_i],\ i = 1, 2. \tag{4}$$

3 Approximate Analytical Solution

Suppose that the flux distribution along the length of the production and injection sections is uniform. Integrating (1) over x_d with regard to (2), we find of weighted average pressure in production and injection sections in Laplace domain:

$$\overline{p}_{1d} = \frac{U_1 I_2 - U_2 Y_1}{I_1 I_2 - Y_2 Y_1},\ \overline{p}_{2d} = \frac{U_2 I_1 - U_1 Y_2}{I_1 I_2 - Y_2 Y_1} \tag{5}$$

where

$$I_i = 1 + C_{id} u^2 \left(\overline{p}_{i,i} + \frac{S_i}{u} \right),\ Y_i = C_{id} u^2 \overline{p}_{1,2},\ U_i = (-1)^{i+1} \left(\overline{p}_{i,i} - \overline{p}_{1,2} + \frac{S_i}{u} \right),$$

$$\overline{p}_{ij} = \frac{1}{l_i l_j u} \int_{\alpha_i}^{\beta_i} \int_{\alpha_j}^{\beta_j} \left[K_0(\sqrt{u}|x_d - \theta|) + 2\sum_{n=1}^{\infty} K_0\left(\sqrt{u + \xi_n^2}|x_d - \theta| \right) \Omega_n \right] d\theta dx_d,\ i,j = 1,2$$

For steady state case the weighted average pressures in production and injection sections are equal to

$$p_{id} = \sum_{j=1}^{2} \frac{(-1)^{j+1}}{l_i l_j} \int_{\alpha_i}^{\beta_i} \int_{\alpha_j}^{\beta_j} \left[In\left(\frac{1}{|x_d - \theta|} \right) + 2\sum_{n=1}^{\infty} K_0(\xi_n |x_d - \theta|) \Omega_n \right] d\theta dx_d + \\ + (-1)^{i+1} S_i,\ i = 1, 2. \tag{6}$$

After simplifying expressions (6) we get

$$p_{id} = (-1)^{i+1} \left\{ S_i - \ln l_i + \frac{l_{12}^2}{2 l_1 l_2} \ln l_{12} - \frac{(l_1 + l_{12})^2}{2 l_1 l_2} \ln(l_1 + l_{12}) \right. \\ \left. - \frac{(l_2 + l_{12})^2}{2 l_1 l_2} \ln(l_2 + l_{12}) - \frac{h_d}{l_i} \ln\left[4 \sin\left(\frac{\pi r_{cd}}{2 h_d} \right) \sin\left(\frac{\pi (2 z_{wd} + r_{cd})}{2 h_d} \right) \right] \right\}, \tag{7}$$

where $l_{12} = (b_1 - a_2)/L = 1 - l_1 - l_2$ – dimensionless distance between the sections. If horizontal well is located at the same distance from the top and the bottom of the reservoir and both sections have the same length ($l_1 = l_2 = l$), then the expressions (7) for $r_c \ll h$ take the form:

$$p_{id} = (-1)^{i+1}\left\{S_i - \ln l + \frac{(1-2l)^2}{2l^2}\ln(1-2l) - \frac{(1-l)^2}{l^2}\ln(1-l) - \frac{h_d}{l}\ln\left(\frac{2\pi r_{cd}}{h_d}\right)\right\} \tag{8}$$

Formula (8) can also be used to estimate the productivity of a horizontal well in a reservoir with linear constant pressure boundary, by analogy with the Borisov's formula for a horizontal well in a circular reservoir with constant pressure boundary [3].

4 Result

Figure 2 shows the log/log pressure and pressure derivative plots for the production section of a horizontal well, obtained by a semi-analytical infinite-conductivity solution (1), (2) (solid lines), an approximate analytical uniform-flux solution (5) (puncture lines), and formula (8) (dash-dotted line) with the following parameters: L = 500 m, r_c = 0.1 m, a_1 = 0 m, b_1 = 200 m, a_2 = 300 m, b_2 = 500 m, h = 20 m, z_w = 10 m, Q = 500 m^3/d, k_h = 0.1 mkm^2, k_v = 0.01 mkm^2, $\mu = 10^{-3}$ Pa·s, $\beta = 2 \cdot 10^{-4}$ MPa^{-1}, S = 1, C = 0.1 m^3/MPa.

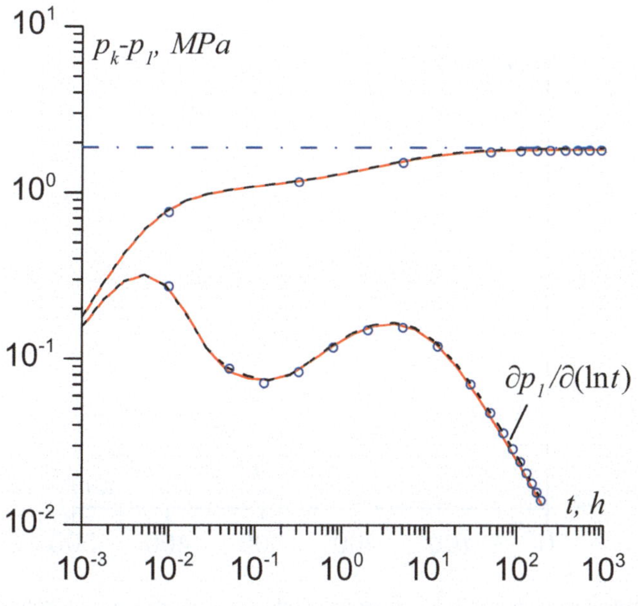

Fig. 2. Pressure response and pressure derivative in the production section of a horizontal well.

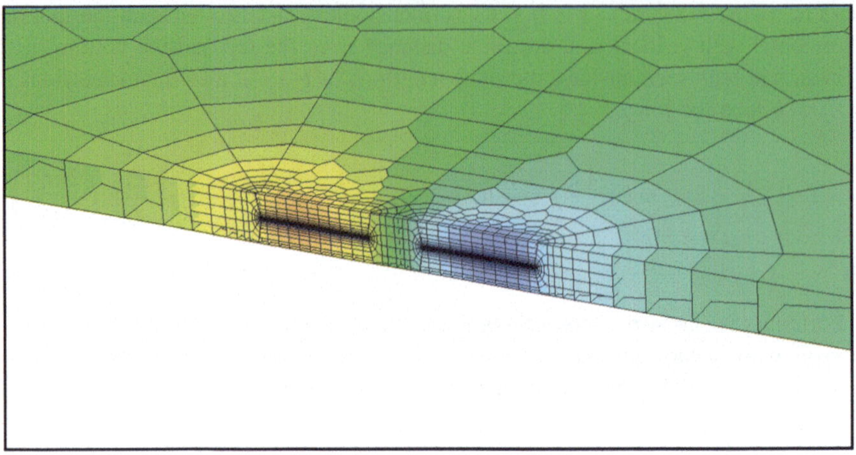

Fig. 3. Pressure field around horizontal circulation well simulated in *KAPPA Saphir*.

For comparison, the symbols in Fig. 2 shows the results of a numerical simulation performed in *KAPPA Saphir* [4]. The 3D mesh and pressure field in the vicinity of production and injection sections of horizontal circulation well is presented in Fig. 3.

Figure 4 shows the steady-state flux distribution along the length of the production and injection sections of a horizontal well, obtained by solving the system of integral Eqs. (3) and (4).

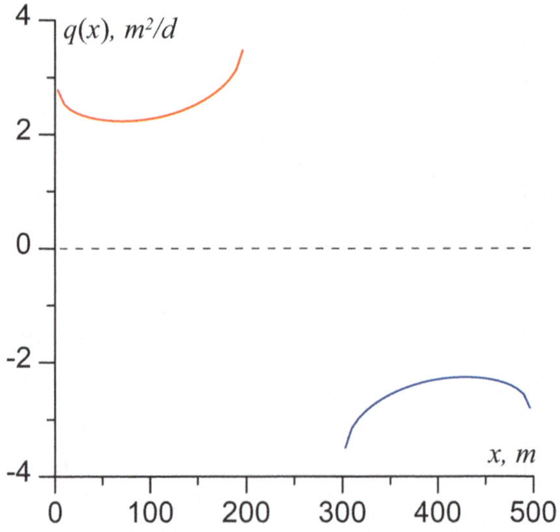

Fig. 4. Flux distribution along the length of the production and injection sections of a horizontal well.

5 Conclusions

This work presents new semi-analytical and approximate analytical solution of steady-state and unsteady fluid flow to a horizontal circulation well in an anisotropic reservoir. The engineering formula for estimating productivity of horizontal circulation well is obtained. The proposed solutions are validated by comparison with the numerical model in KAPPA Saphir.

References

1. Akhmadullin, I., et al.: Numerical analysis of downhole heat exchanger designed for geothermal energy production. Geotherm. Energy. **5**, 1–24 (2017)
2. Morozov, P.E.: Modeling of non-stationary fluid inflow to a multisectional horizontal well. Georesources **20**(1), 44–50 (2018)
3. Morozov, P.E.: Steady fluid flow to a radial system of horizontal wells. J. Appl. Mech. Tech. Phys. **59**(2), 99–107 (2018)
4. https://www.kappaeng.com/

Direct Numerical Simulation of Single and Two-Phase Flows at Pore-Scale

Vladislav Balashov(✉) and E. B. Savenkov

Keldysh Institute of Applied Mathematics of Russian Academy of Sciences, Moscow, Russian Federation
vladislav.balashov@gmail.com

Abstract. The paper is devoted to numerical techniques for an analysis of single and multiphase fluid flows in the pore space of core samples with direct resolution of the pore space structure. The core of the approach is quasi-hydrodynamic (QHD) regularization of classical Navier–Stokes (1-phase flow) and Navier–Stokes–Cahn–Hilliard (2-phase flow) equations. In essence QHD regularization consists in specific addition of physically based dissipative terms to initial equations. These terms act as stabilizers and allow to use logically simple explicit finite difference schemes with central difference approximations. Moreover, for moderately-rarefied gas flows these terms provide good agreement of obtained numerical results with the experimental data for microflows. A number of simulations demonstrating consistency of the model and algorithms as well as realistic flow simulations within realistic μ-CT models are presented.

Keywords: Two-phase flows · Moderately rarefied gas flows · Navier–Stokes–Cahn–Hilliard · Quasi-hydrodynamic regularization · Digital rock physics

1 Introduction

Analysis of the micro-flows in different natural and artificial porous materials is relevant and important problem. With the development of high performance computing and μ-CT technologies which allows to construct detailed geometrical description of material microstructure, it turns possible to use numerical experiments for direct evaluation of material properties. In geophysical applications such techniques are called digital rock physics and is meant to accompany and supply conventional laboratory measurements for evaluation of transport properties of the rock core samples.

The paper is devoted to numerical techniques for an analysis of single and two phase flows in the pore space of core samples with direct resolution of the pore space structure. The core of the approach is quasi-hydrodynamic (QHD) regularization of classical Navier–Stokes (1-phase flow) and Navier–Stokes–Cahn–Hilliard (2-phase flow) equations. As geometrical model voxel based binary 3D-

V. I. Karev et al. (Eds.): *Physical and Mathematical Modeling of Earth and Environment Processes (2018)*, SPEES, pp. 374–379, 2019.
https://doi.org/10.1007/978-3-030-11533-3_37

images of a core sample are used. A number of simulations demonstrating consistency of the used models and algorithms as well as realistic flow simulations within realistic μ-CT models are presented.

2 Mathematical Models

2.1 Single Phase Flows

QHD-regularization of Navier–Stokes equations describing isothermal flow neglecting bulk forces are [1,2]

$$\partial_t \rho + \nabla \cdot \boldsymbol{j}_m = 0, \tag{1}$$
$$\partial_t(\rho \boldsymbol{u}) + \nabla \cdot (\boldsymbol{j}_m \otimes \boldsymbol{u}) + \nabla p = \nabla \cdot \boldsymbol{\Pi}, \tag{2}$$

where ρ is a mass density, $\boldsymbol{u}$ is a velocity, $\boldsymbol{j}_m = \rho(\boldsymbol{u} - \boldsymbol{w})$ is mass flux, $p = c_s^2 \rho$ is a pressure, c_s is a sound speed, $\boldsymbol{\Pi} = \boldsymbol{\Pi}^{\mathrm{NS}} + \boldsymbol{\Pi}^{\tau}$ is a viscous stress tensor, $\boldsymbol{\Pi}^{\mathrm{NS}} = \eta\left[(\nabla \boldsymbol{u} + \nabla \boldsymbol{u}^T) - (2/3)\boldsymbol{I}\,\mathrm{div}\,\boldsymbol{u}\right]$ is a classical Navier–Stokes viscous stress tensor, $\boldsymbol{I}$ is a identity tensor, η is a dynamic viscosity, $\boldsymbol{\Pi}^{\tau} = \rho\,\boldsymbol{u} \otimes \boldsymbol{w}$ is a regularizing stress tensor, $\boldsymbol{w} = (\tau/\rho)\,(\rho \boldsymbol{u} \cdot \nabla \boldsymbol{u} + \nabla p)$ is a regularizing velocity, τ is a small relaxation parameter with dimensionality of time. For gas flows τ equals to mean free time up to coefficient of order of unity [1]. For moderately rarefied gas flows relaxation parameter is defined accordingly to $\tau = \eta/(p\,\mathrm{Sc})$, where Sc is a Shmidt number. On the solid boundary we impose classical Maxwell slip boundary conditions:

$$\partial_n \rho = 0, \quad u_n = 0, \quad \boldsymbol{u}_\tau = \frac{2-\sigma}{\sigma}\lambda \partial_n \boldsymbol{u}_\tau, \tag{3}$$

where σ is an accommodation coefficient, λ is a mean free path length, $\boldsymbol{n}$ is an inward boundary unit normal, $u_n = \boldsymbol{u} \cdot \boldsymbol{n}$, $\boldsymbol{u}_\tau = (\boldsymbol{I} - \boldsymbol{n} \otimes \boldsymbol{n})\boldsymbol{u}$. Here we emphasize that no-flux condition $\boldsymbol{j}_m \cdot \boldsymbol{n} = 0$ on the solid boundary follows from the first and second conditions in (3).

For dense gas flows (or for liquid) we use no slip boundary conditions. The QHD-terms are considered as numerical regularizers only. For relaxation parameter in this case we use $\tau = \alpha^* h/c_s$, where α^* is empirically chosen on the basis of numerical stability issues, h is space step of numerical grid.

2.2 Two Phase Flows

In [3] QHD-regularization technique was applied to equations of phase field model describing multiphase multicomponent nonisothermal flows with surface effects. For isothermal two-component mixture (Navier–Stokes–Cahn–Hilliard equations) the system is as follows:

$$\partial_t \rho + \mathrm{div}\,\boldsymbol{j}_m = 0, \tag{4}$$
$$\partial_t(\rho \boldsymbol{u}) + \mathrm{div}(\boldsymbol{j}_m \otimes \boldsymbol{u}) + \nabla p = \mathrm{div}\,\boldsymbol{\Pi}, \tag{5}$$
$$\partial_t(\rho C) + \mathrm{div}(\boldsymbol{j}_m C) = \mathrm{div}\left(M \nabla \mu\right). \tag{6}$$

Here $\rho > 0$ and $\boldsymbol{u}$ are mixture density and velocity; C is the mass concentration of mixture components: $C = C_1$, $C_2 = 1 - C_1$, $C_i = \rho_i/\rho$, where ρ_i is a mass density of i-th component, $i = 1, 2$.

Accordingly to [4,5], Helmholts free energy of mixture is

$$\Psi = \Psi_0(C) + \frac{\lambda_1}{2}|\nabla C|^2, \tag{7}$$

where first term depends only on local conditions and is responsible for phase separation. Often double-well form $\Psi_0(C) = A_\psi C^2(1 - C)^2$ is used [6]. The second term in (7) depending on ∇C, is a gradient energy and allows to account for surface effects.

On the foundation of (7) using the Colleman–Noll procedure constitutive relations for stress tensor and QHD-terms can be derived [3,7]: $\boldsymbol{\Pi} = \boldsymbol{\Pi}^{NS} - \boldsymbol{Q} + \boldsymbol{\Pi}^\tau$, $\boldsymbol{Q} = \lambda_1 \rho \nabla C \otimes \nabla C$, $\boldsymbol{w} = (\tau/\rho)\left[\rho(\boldsymbol{u} \cdot \nabla)\boldsymbol{u} + \nabla p + \operatorname{div} \boldsymbol{Q}\right]$. In (6) $M = M(C) > 0$ and μ are mobility coefficient and generalized chemical potential:

$$\mu(\rho, C) = \Psi'_{0C}(\rho, C) - \frac{\lambda_1}{\rho} \operatorname{div}(\rho \nabla C). \tag{8}$$

To set specific value of equilibrium contact angle θ we use following boundary condition [8]:

$$\partial_n C = \frac{6\sigma \cos\theta}{\rho \lambda_1} C(1 - C). \tag{9}$$

Other boundary conditions reads

$$\partial_n \mu = 0, \quad \boldsymbol{u} = \boldsymbol{0}, \quad \partial_n p = -\boldsymbol{n} \cdot \operatorname{div} \boldsymbol{Q}. \tag{10}$$

In the following simulations mixture of two liquids with same equations of state was considered. In this case $p = c_s^2 \rho$.

3 Numerical Results

In this section some numerical results of flow simulation in pore space of core samples are presented. Voxel images are freely available and are taken from Imperial College website [9].

3.1 Single Phase Flows

To simulate one phase flow in pore space we set on two opposite faces of numerical domain boundary conditions corresponding to pressure drop (~1 Pa). On other domain boundaries and boundary of pore space we set conditions, corresponding to solid wall: no-slip for liquid or dense gas and slippage for moderately rarefied gas flow.

Permeability coefficient k is obtained from Darcy law. If simulation was performed with no-slip boundary conditions then we consider permeability coefficient as absolute and denote it as k_∞. Several numerical simulations results with

Table 1. Coefficients k_∞ (mDarcy) and $b\,(10^5$ Pa) for core samples from collection [9]

Sample	$b(N_2)$	k_∞^{QHD}	$k_\infty^{Palabos}$	k_∞ [10]	Sample	$b(N_2)$	k_∞^{QHD}	$k_\infty^{Palabos}$	k_∞ [10]
S_1	645.75	2.168	2.613	1.729	S_6	544.84	13.215	14.588	10.543
S_2	838.57	4.293	4.928	3.931	S_7	658.87	9.410	10.387	7.495
S_3	869.09	0.9205	1.264	0.415	S_8	643.00	14.930	16.564	13.330
S_4	879.12	0.455	0.652	0.285	S_9	800.87	2.539	2.747	2.066
S_5	743.87	5.915	6.516	4.810	S_b	907.91	1.542	1.856	1.287

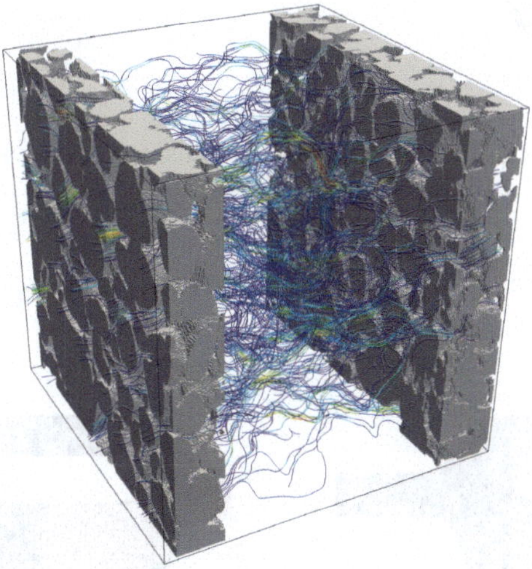

Fig. 1. Streamlines in S_8 core sample. Grey color indicates part of rock.

different average pressure of moderately rarefied gas flows can be used to obtain Klinkenberg coefficient b.

Values of k_∞ for some core samples obtained with QHD-regularized equations, with open source software package Palabos [11] and from [10] are presented in Table 1. Also there are values of Klinkenberg coefficients b for nitrogen obtained with QHD-regularized equations.

On the Fig. 1 streamlines obtained for sandstone S_8 sample [9] are presented. Grey color indicates part (for illustration purposes) of rock.

3.2 Two Phase Flows

Numerical simulation of two-phase displacement in core sample S_9 [9] with neutral contact angle ($\theta = 90°$) was performed. Phase distribution for certain intermediate time moment is presented on the Fig. 2 (left). Red color indicates phase with $C > 0.5$, and blue color — phase with $C \leq 0.5$. For illustration purposes only red fluid is presented. Dynamics of saturation of core sample by the red fluid is presented on the Fig. 2, where dashed line corresponds to inflow velocity

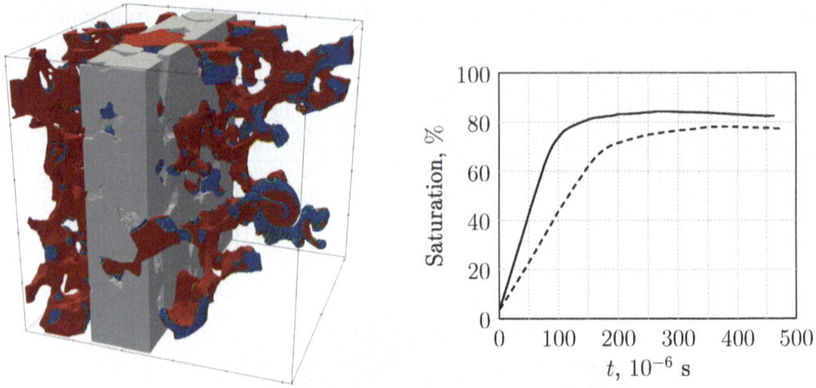

Fig. 2. Distribution of displacing phase (left) and saturation dynamics (right) during displacement process for different inflow velocities

$\boldsymbol{u}_l = (0, 5, 0)$ m/s, full line — to $\boldsymbol{u}_l = (0, 10, 0)$ m/s. Other details about this simulation can be found in [12].

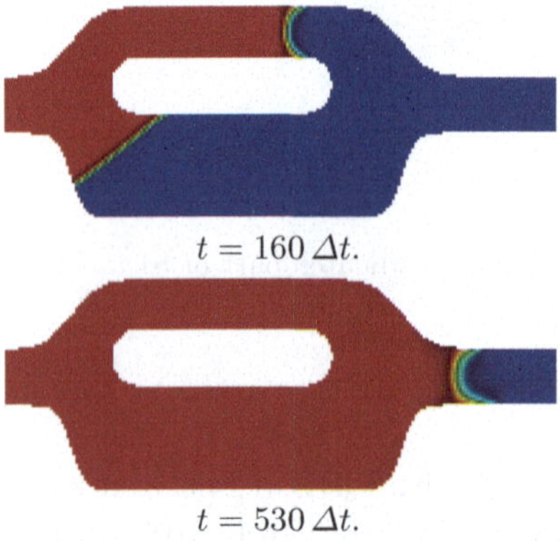

Fig. 3. Phase distribution for several consecutive time moments for $\theta = 150°$

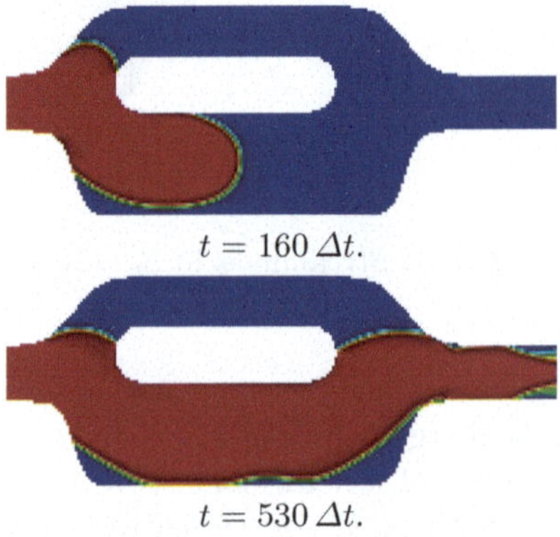

Fig. 4. Phase distribution for several consecutive time moments for $\theta = 30°$

Also two phase displacement was performed in system of two parallel connected pore throats (so called pore doublet model). Different wettability of solid boundary was considered: (i) $\theta = 150°$, i.e. wetting of the solid wall by displacing fluid is very favorable; (ii) $\theta = 30°$ i.e. wetting of the solid wall by displacing fluid is unfavorable. Domain sizes are $L_x = 0.02$ m, $L_y = L_x/2$. On the Figs. 3 and 4 distribution of phases are presented for several consecutive time moments, where $\Delta t = 8 \cdot 10^{-8}$ s is the time step. Space step is $h = L_x/200$. Qualitative difference between two cases is obvious: if wetting of the solid wall by displacing fluid (the red one) is unfavorable then considerable part of pore space is occupied by displaced phase (the blue one); in other case the whole pore space is occupied by displacing fluid. Other details can be found in [13].

Acknowledgments. This research was supported by Russian Science Foundation, project No. 17-71-30014.

References

1. Sheretov, Yu.V.: Continuum Dynamics Under Spatiotemporal Averaging. RKhD, Moscow-Izhevsk (2009). (in Russian)
2. Elizarova, T.G.: Quasi-Gas Dynamic Equations. Springer, Heidelberg (2009)
3. Balashov, V.A., Savenkov, E.B.: Quasi-hydrodynamic model of multiphase fluid flows taking into account phase interaction. J. Appl. Mech. Tech. Phys. **59**(3), 434–444 (2018). https://doi.org/10.1134/S0021894418030069
4. Novick-Cohen, A., Segel, L.A.: Nonlinear aspects of the Cahn-Hilliard equation. Phys. D Nonlinear Phenom. **10**(3), 277–298 (1984). https://doi.org/10.1016/0167-2789(84)90180-5
5. Cahn, J.W., Hilliard, J.E.: Free energy of a nonuniform system. I. Interfacial free energy. J. Chem. Phys. **28**(2), 258–267 (1958). https://doi.org/10.1063/1.1744102
6. Tierra, G., Guillén-González, F.: Numerical methods for solving the Cahn-Hilliard equation and its applicability to related energy-based models. Arch. Comput. Methods Eng. **22**(2), 269–289 (2015). https://doi.org/10.1007/s11831-014-9112-1
7. Liu, J.: Thermodynamically consistent modeling and simulation of multiphase flows. Ph.D. Dissertation, The University of Texas at Austin (2014)
8. Carlson, A., Do-Quang, M., Amberg, G.: Modelling of dynamic wetting far from equillibrium. Phys. Fluids **21**(12), 121701 (2009). https://doi.org/10.1063/1.3275853
9. Imperial College London. http://www.imperial.ac.uk/earth-science/research/research-groups/perm/research/pore-scale-modelling/micro-ct-images-and-networks/
10. Dong, H., Blunt, M.J.: Pore-network extraction from micro-computerized-tomography images. Phys. Rev. E **80**(3), 036307 (2009). https://doi.org/10.1103/PhysRevE.80.036307
11. Degruyter, W., Burgisser, A., Bachmann, O., Malaspinas, O.: Synchrotron X-ray microtomography and lattice Boltzmann simulations of gas flow through volcanic pumices. Geosphere **6**(5), 470–481 (2010). https://doi.org/10.1130/GES00555.1
12. Balashov, V., Zlotnik, A., Savenkov, E.: Numerical method for 3D two-component isothermal compressible flows with application to digital rock physics. Russ. J. Numer. Anal. Math. Model. **34** (2019, in press)
13. Balashov, V., Savenkov, E.: Numerical simulation of two-dimensional two-phase fluid flows accounting for wettability with quasi-hydrodynamic equations. Preprint of KIAM RAS, No. 131, Moscow (2018) (in Russian). https://doi.org/10.20948/prepr-2018-131

Modeling of Thermal Gas Treatment of Low-Permeability Reservoirs of Bazhenov Formation

S. E. Yakush[1(✉)], A. N. Galybin[2], A. M. Polishchuk[3], and S. A. Vlasov[3]

[1] Ishlinskiy Institute for Problems in Mechanics RAS, Ave. Vernadskogo 101 Bldg 1, Moscow 119526, Russia
yakush@ipmnet.ru
[2] Schmidt Institute of Physics of the Earth RAS, Bolshaya Gruzinskaya str., 10-1, Moscow 123242, Russia
a.n.galybin@gmail.com
[3] Reseach and Technology Organization ITIN, Moscow, Russia
itinmsk@gmail.com

Abstract. A coupled compositional and geomechanical model is presented which is applicable to simulation of thermal gas treatment of Bazhenov formation. The geomechanical module includes the criteria for rock fracturing caused by the increase in the internal pore pressure due to pyrolysis (decomposition) of solid organic material, kerogen, with formation of synthetic oil and gases, criteria for the formation of hydraulically connected system of cracks, as well as models for the permeability of fractured rock. Numerical simulations are performed for the scheme of air injection into high-permeability layers available in the impermeable kerogen-containing matrix. It is shown that straightforward technological scheme using the same permeable layer for air injection and oil production may not be viable because the cracked zone is lags behind the in-situ combustion front propagation, and synthetic oil may be burnt in the injected air, rather than reach the production well.

Keywords: Kerogen · Hydraulic fracturing · Thermal methods · Bazhenov formation

1 Introduction

Thermal gas treatment is considered one of the prospective methods for development of Bazhenov formation and involvement into the recovery process of vast hydrocarbon reserves contained as native oil and solid kerogen [1–3]. For the most part, these hydrocarbon substances are contained in practically impermeable rock matrix in closed pores or unconnected lens reservoirs, making traditional approaches for enhanced oil recovery ineffective.

The possibility of thermal gas treatment of Bazhenov formation reservoirs is based on the availablilty of high-permeability carbonate layers containing native oil and some amount of kerogen. By injecting oxidizer (air) into these high-permeability layers,

V. I. Karev et al. (Eds.): *Physical and Mathematical Modeling of Earth and Environment Processes (2018)*, SPEES, pp. 380–394, 2019.
https://doi.org/10.1007/978-3-030-11533-3_38

exothermal reactions between the hydrocarbons contained there and oxygen can be initiated (due to high enough temperature in Bazhenov formation rocks, self-ignition can be achieved without the need for any external thermal source). The in-situ combustion front, propagating in the permeable layers, can provide thermal energy for heating of the rock matrix, causing thermal decomposition of kerogen in the pores, its conversion to hydrocarbon fluids (synthetic oil) and gas (light hydrocarbons and combustion products). Due to the development of high internal pressure in closed pores, fracturing of rock matrix is going to occur, resulting in the appearance of a connected system of filtration channels (cracks) allowing filtration of hydrocarbon fluids towards the production well.

This general scheme of involvement of rock matrix kerogen and native oil into the development process and hydrocarbon recovery from Bazhenov formation requires, however, detailed study on the complex physical, chemical, geomechanical, and fluid-dynamical processes accompanying the thermal gas treatment. The main difference between Bazhenov formation and conventional oil deposits is that rock matrix permeability is practically absent, and hydrocarbons are contained in the form of solid kerogen. During thermal gas treatment, the processes of kerogen conversion and rock fracturing (leading to the appearance of rock permeability) are tightly coupled, and the main target of a successful development strategy would be to provide conditions under which the synthetic oil recovery is feasible and cost-effective.

Mathematical modeling of Bazhenov formation development by thermal gas treatment method is not straightforward, reflecting the complexity of the object. While traditional models can be applied to the description of fluid filtration and in-situ combustion in the permeable carbonate layer, special models are required for the processes in the initially impermeable rock matrix initiated by its heating. Only on the basis of such models, the fundamental aspects of kerogen transformation and rock fracturing can be studied, allowing development of appropriate technology.

The purpose of this paper is to present a coupled model for the processes involved in Bazhenov formation development on the basis of thermal gas method, and to demonstrate that simplistic approaches to the arrangement of thermal treatment will not prove successful. The difficulties of one such approach, consisting in injection of air into a high-permeability layer and in-situ combustion as a source of thermal energy, are elucidated both by considering a simpler thermal model, and by performing more comprehensive simulations on the basis of a full compositional model.

2 Mathematical Model

2.1 Compositional Model

The compositional model (e.g., [4–7]) considers all fluids present in the system as a set of N_c components present in the three phases (water, oil, gas) denoted by subscripts w, o, and g, respectively; the solid phase (rock) is denoted by subscript r. Conservation equations for i-th component and energy equation for the whole system take the form

$$\frac{\partial}{\partial t}\left[\varphi_f\left(\rho_w S_w w_i + \rho_o S_o x_i + \rho_g S_g y_i\right)\right] + \nabla\left(\rho_w w_i \mathbf{W}_w + \rho_o x_i \mathbf{W}_o + \rho_g y_g \mathbf{W}_g\right) = Q_i \quad (1)$$

$$\begin{aligned} &\tfrac{\partial}{\partial t}\left[\varphi_f\left(\rho_w S_w E_w + \rho_o S_o E_o + \rho_g S_g E_g\right) + \varphi_r \rho_r E_r\right] + \\ &\quad + \nabla\left(\rho_w H_w \mathbf{W}_w + \rho_o H_o \mathbf{W}_o + \rho_g H_g \mathbf{W}_g\right) = \nabla\left(\lambda_{eff} \nabla T\right) + Q_H \end{aligned} \quad (2)$$

Here, w_i, x_i, and y_i are molar fractions of i-th component in water, oil, and gas phases; ρ_k and S_k are phase saturations, φ_f is the volume fraction of fluids, $\varphi_r = 1 - \varphi_f$ is the volume fraction of solid material (rock, including solid kerogen). Also, E_k and $H_k = E_k + P_k/\rho_k$ are internal energies and enthalpies of fluids ($k = w, o, g$) and rock ($k = r$), λ_{eff} is the effective conductivity, $\mathbf{W}_k$ are superficial phase velocities, Q_i are source terms describing consumption or production of components in chemical reactions, Q_H is the corresponding energy equation source term.

The superficial phase velocities $\mathbf{W}_k$ ($k = w, o, g$) in (1), (2) are described by the generalized Darcy law

$$\mathbf{W}_k = -\frac{k_{rk} K_{r,z}}{\mu_k} \nabla P \quad (3)$$

where $K_{r,z}$ is permeability which is generally anisotropic (here, we assume that permeability is different in vertical and radial directions, taking the values K_z and K_r, respectively); k_{rk} is the relative permeability, μ_k is viscosity of k-th phase. The effective conductivity is weighted over the conductivities of all fluids and rock: $\lambda_{eff} = \varphi_f(S_w\lambda_w + S_o\lambda_o + S_g\lambda_g) + \varphi_r\lambda_r$.

The phase densities are described by the Peng-Robinson equation of state [8] (see also [9]), with specific volume correction [10]. The molar fractions of components in water, oil, and gas phases are related to each other by the equilibrium constants, $y_i = K_i^{go} x_i$ and $y_i = K_i^{gw} w_i$ evaluated by the Wilson correlation (see details in [6]). The molar fractions of all components in each phase sum up to unity $\sum w_i = 1$, $\sum x_i = 1$, $\sum y_i = 1$; also, saturations of all fluids satisfy the condition $\sum S_k = 1$.

The chemical reactions between different components are considered in the form

$$\frac{d[C]}{dt} = -A[C]^n [P_{O2}]^m \exp\left(-\frac{E}{RT}\right) \quad (4)$$

where где C is the concentration of reactant, P_{O2} is the partial pressure of oxygen, measured in atmospheres, A is the pre-exponential factor, E is the activation energy, $R = 8.314$ J/mol K is the universal gas constant, T is the absolute temperature, n and m are reaction order with respect to the current reactant and oxygen (for the pyrolysis reactions $m = 0$). The source term Q_H in the energy Eq. (2) is obtained as by summing up contributions of all individual reactions.

The equations of compositional model presented above are formulated in their general form, applicable to all kinds of problems, from 1D to 3D. However, in this work all simulations were performed in 2D axisymmetric set up, sufficient to show the main features of thermal gas treatment of rock matrix of Bazhenov formation. Extension of the analysis to 3D configurations is straightforward.

2.2 Geomechanical Model

Geomechanical module is the key element of present coupled model; it provides the conditions under which rock fracturing occurs due to the growth of internal pore pressure, as well as predicts the development of connected fracture system capable of sustaining the flow of synthetic oil and gas.

The criterion for the onset of rock fracturing is based on the comparison of internal pore pressure with compressing stresses in the rock. The in-pore pressure $P_{pore}(T, \chi)$ is calculated from the compositional model described above, albeit without the transport terms (zero permeability); it is a function of both the current temperature T and kerogen pyrolysis progress variable χ (defined as a fraction of kerogen which was converted to synthetic oil and gas). Since χ is a time integral of the kerogen conversion rate, the in-pore pressure is not a single-valued function of temperature; rather, it depends on the *history* of rock heating. This is an important point, because heating by in-situ combustion implies that the heating source is moving, and rock matrix temperature changes with time.

The minimum compressing stress component in the rock is obtained by summing the "natural" stresses available at a given deposit depth, and thermal stresses caused by rock heating. For the natural stresses, the vertical component was taken equal to the geostatic pressure, whereas the horizontal stress components were taken equal to the lateral pressure, evaluated as some fraction of the vertical component (in modeling of some particular oil field, these quantities must be taken from geological data). Stress components (vertical σ_z, radial σ_r, hoop σ_θ, and tangential σ_{rz}) arising due to heating are calculated from linear thermoelasticity theory.

The criterion for the onset of rock fracturing includes the following additional geomechanical parameters:

- Tear strength of matrix rock σ_t (set to 0 in simulations);
- Vertical stress (geostatic pressure) σ_V equal to the weight of overlying rock (if unavailable, can be evaluated as deposit depth/40, MPa).
- Minimum lateral compressing stress $\sigma_h = \lambda\sigma_V$, where $\lambda \leq 1$ is the non-dimensional lateral pressure coefficient (taken equal to 0.6 in the simulations);
- Maximum lateral compressing stress σ_H(assumed $\sigma_H = \sigma_h$ in the simulations).

For evaluation of crack development and merging into a connected crack system, also specified are:

- Characteristic distance between pores d, m;
- Crack resistance of the material K_{Ic}, MPa·m$^{1/2}$ (taken $K_{Ic} = 0$ in the simulations);
- Coefficient of crack interaction β (taken equal to 1 in the simulations).

The criterion for onset of fracturing with subsequent formation of a connected crack system is based upon stress analysis at each point, with checking for the following condition:

$$P_{pore}(T, \chi) + \sigma_{temp}(T) + \sigma_h = \sigma_t \tag{5}$$

where $\sigma_{temp}(T) = \max(\sigma_r, \sigma_\theta)$. This condition is checked only in the matrix rock containing kerogen. The minimum resistance to fracturing is determined by

$$\sigma_{resist} = \left[P_{pore}(T, \chi) + \sigma_{temp}(T)\right] - \left[\sigma_t - \sigma_h\right] \tag{6}$$

If $\sigma_{resist} > 0$, onset of cracks is possible at the current point. In this case, check is performed for the criterion of crack propagation, in the form $K_I(\sigma_{resist},\ d) = K_{Ic}$. First, the stress intensity factor is calculated from

$$K_I(\sigma_{resist}, d) = \sigma_{resist}\sqrt{\pi\frac{d}{2}}, \tag{7}$$

after which the condition is checked

$$\beta K_I(\sigma_{resist}, d) - K_{Ic} > 0. \tag{8}$$

Should the criterion (8) be satisfied, it is concluded that rock fracturing occurred, resulting in the development of a system of predominantly vertical cracks (this direction corresponds to the direction of strongest compression). The vertical rock permeability arising due to these induced cracks, is taken equal to

$$K_z = \frac{1}{24a} N_f \left[e_0 + a\Delta P \frac{\pi(1 - \nu^2)}{E}\right]^3, \tag{9}$$

where ΔP is the pressure excess in the crack, E is Young's modulus, ν is Poisson's ratio, N_f is the number of merged cracks having the length of the order of representative volume of rock, e_0 is some crack opening characterizing plastic (irreversible) fracturing, a is the characteristic size of the representative volume. For the reservoirs in question, it is assumed that most of the induced permeability will be provided by cracks with $a \approx 1$ m. A study on crack merging has shown that $N_f = 3 \div 5$.

The minimum crack opening e_0 in (9) is an uncertain value dependent on irreversible rock fracturing details, unavailable for Bazhenov formation. In (9), it is rasied to the power of three, augmenting significantly the uncertainty. Therefore, a reverse approach is taken: taking a typical permeability K_0 of tight reservoirs (10–50 mD), evaluate $e_0 = \left(24aK_0/N_f\right)^{1/3}$. For example, at $K_0 = 20$ mD, and $N_f = 5$ we obtain $e_0 = 4.6 \cdot 10^{-5}$ m.

The horizontal permeability is taken proportional to the vertical one: $K_r = \psi K_z$, where ψ reflects the effects of rock inhomogeneity and possible deviation of the main component of stress tensor from vertical; in the absence of any objective data, it is assumed that $\psi = 0.1$, maintaining the predominantly vertical crack direction.

2.3 Implementation of Coupled Model

The model was implemented numerically in the axisymmetric coordinate system, allowing simulations of vertical cross-sections of permeable reservoir and impermeable kerogen-containing rock matrix. The vertical coordinate axis Oz coincided with the axis of vertical injection well, the radial axis passing along the bottom rock boundary. Note that the purpose of numerical implementation of the composite simulator was to have in hand a research tool for testing the performance of the geomechanical module, unavailable in closed-code commercial simulators. Evidently, the two-dimensional framework is just optional, and all models can be implemented in 3D simulators for full-scale studies of Bazhenov formation development.

The compositional model is discretized on Cartesian non-uniform mesh by the finite-volume method; all variables are defined in the cell centers, while the filtration velocities defined on the corresponding cell boundaries. The non-linear system of equations is solved by fully implicit numerical scheme with Newton iterations at each step. The chemical kinetics equations are solved at a separate step by stiff ODE solver. The thermoelasticity problem is discretized by the finite-element method.

Coupling between the fluid dynamics (compositional) and geomechanics modules is performed on each time step. After updating the pressure and temperature fields, together with phase saturations and component mole fractions in all phases, the criteria for rock fracturing are checked in each mesh cell where cracks were not detected previously (i.e., fracturing is irreversible, and, once cracked, a mesh cell becomes permeable and remains such for the rest of the simulation). If the fracturing criterion is met, the cell is marked cracked, and fluids start to flow at the next time step in the compositional model.

Note that such tight coupling between the compositional and geomechanical models might be difficult to implement in the commercial simulators where a fixed number of variable-permeability models is available. Detailed study on the possibility of implementation of the current geomechanical model in commercial simulators, though, is outside of the scope of current paper.

3 Components and Reactions

On the basis of literature review, a set of pseudo-components was chosen for numerical simulations of thermal gas treatment of Bazhenov formation kerogen-containing deposits. The components are similar to those in SARA (Saturates, Aromatics, Resins, and Asphaltenes) approach (see [6]), their names and molecular masses are presented in Table 1.

The kinetic scheme, including pyrolysis, low-temperature oxidation, and high-temperature oxidation of components, is summarized in Table 2. For each reaction, the parameters of kinetic Eq. (4) are presented, including the reaction orders n and m in the reactant and oxygen, respectively, activation energy E, and pre-exponential factor A. The heat of reaction is also indicated in the second column, together with the stoichiometry of the reaction.

Table 1. Components used in the compositional model.

Pseudo-component	Abbreviation	Molecular mass
Kerogen	Kerogen	17034.75
Asphaltenes	Asph	1285.7
Pyrolysis coke	PyrCoke	617.28
Resins	Resins	745.36
Aromatics	Arom	410.14
Saturates	Sat	291.84
Oxidized saturates	OxdSat	165.84
Light oil	Lites	170.00
Oxygen	O_2	32.0
Nitrogen	N_2	28.0
Water	H_2O	18.0
Carbon dioxide	CO_2	44.0

Table 2. Kinetic scheme.

No.	Reaction	*n*	*m*	*E*, kJ/mole	*A*, h^{-1}
Pyrolysis					
1	Kerogen → 5.5 Asph + 7.6 Resine + 4.28 Sat + 1.2 PyrCoke + 6.34 Lite + 18H_2O + 21.5 $CO_{2-}$$_{7}$80 kJ/mole	1	–	222	$9.02 \cdot 10^{13}$
2	Asph → 0.38 Arom + 0.54 Sat + 0.19 Lites + 1.43 PyrCoke + 0.55 H_2O + 1.08 CO_{2-2}25 kJ/mole	1	–	222	$2.01 \cdot 10^{17}$
3	Resins → 0.17 Asph + 0.44 Arom + 0.42 Sat + 0.79 Lites + 0.48 H_2O + 0.09 CO_2 – 116 kJ/mole	1	–	184.45	$6.2 \cdot 10^{13}$
4	Arom → 0.09 Asph + 1.81 Lites – 74.3 kJ/mole	1	–	184.45	$3.1 \cdot 10^{13}$
Low-temperature oxidation					
5	Lites + 18.5 O_2 → 13.0 H_2O + 12 CO_2 + 7140 kJ/mole	1	1	138.281	$4.4394 \cdot 10^{11}$
6	Sat + 18.71 O_2 → 14.54 H_2O +10.65 O_2 + OxdSat +10729.2 kJ/mole	1	0.7	78.604	$6.0005 \cdot 10^{7}$
High-temperature oxidation					
7	PyrCoke + 68.3 O_2 → 21.54 H_2O + 54.89 CO_2 +28456.6 kJ/mole	1	1	68.405	$9.2968 \cdot 10^{3}$
8	OxdSat + 12.96 O_2 → 6.22 H_2O + 10.65 CO_2 +3555.6 kJ/mole	1	1	68.405	$9.2968 \cdot 10^{2}$
9	Asph + 133.58 O_2 → 59.56 H_2O + 100 CO_2 +46285.2 kJ/mole	1	1	68.405	$9.2968 \cdot 10^{1}$
10	Resins + 84.34 O_2 → 44.68 H_2O + 60 CO_2 + 28323.7 kJ/mole	1	1	68.405	$9.2968 \cdot 10^{1}$
11	Arom + 44.27 O_2 → 23.47 H_2O + 31.93 CO_2 + 17020.8 kJ/mole	1	1	68.405	$9.2968 \cdot 10^{1}$

4 Results

4.1 Parameters and Initial Conditions

Simulations were performed for the following parameters: rock density 2500 kg/m^3, density of kerogen $\rho_k = 1100$ kg/m^3, heat capacity of rock $C_r = 800$ J/kg K, rock conductivity $\lambda = 1.7$ W/m K, Young's modulus of rock $E = 2 \cdot 10^4$ MPa, Poisson's ratio $v = 0.2$, thermal expansion coefficient $\alpha_T = 1 \cdot 10^{-5}$ 1/°K.

The kerogen-containing rock matrix was of the thickness 10 m (in the vertical direction), of which the lower 1 or 2 m were occupied by the high-permeability layer into which air was injected at the reservoir temperature of 120°K. The bottom and upper boundaries of kerogen-containing rock are bounded by impermeable rocks for which the same mechanical and thermal properties were assumed.

The relative permeability in oil-water and liquid-gas systems taken in the simulations is plotted in Fig. 1. The same two-phase relative permeability curves were used in the high-permeability layer, as well as in the cracked zone. Three-phase relative permeabilities in the water-oil-gas system were obtained by the Stone model.

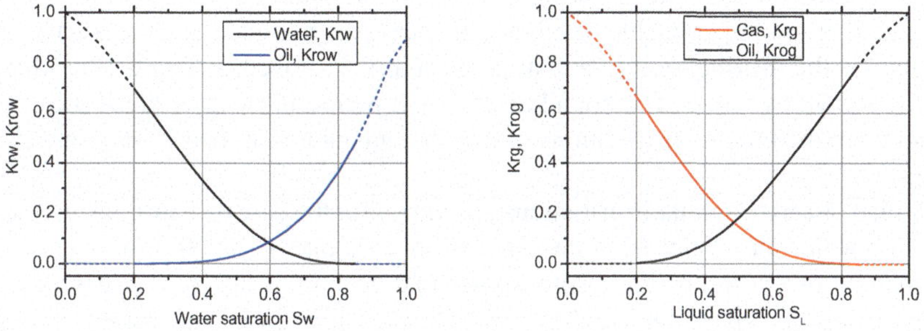

Fig. 1. Two-phase relative permeability curves

The initial saturations in the high-permeability layer were set to $S_w = 0.02$, $S_o = 0.98$, $S_g = 0$. In the initially impermeable rock matrix (i.e., in the closed pores) it was assumed that $S_w = 0$, $S_o = 1$, $S_g = 0$. Initial content of kerogen in the high-permeability layer was set to 5%, in the rock matrix 19.9%.

The initial porosity (i.e., fraction of volume accessible to fluids) in the high-permeability layer was set to 20% (which means that after complete burnout of kerogen the porosity becomes 20 + 5 = 25%), while in the impermeable rock matrix it was set to 0.1% (after complete conversion of kerogen it becomes 0.1 + 19.9 = 20%).

Simulations were carried out for the components listed in Table 1. The initial composition of native oil in the high-permeability layer was the following (molar fractions): Lites 50%, Sat 30%, Arom – 20%. This composition was chosen on the basis of experimental information that Bazhenov formation oil is of light type.

In the high-permeability layer, the horizontal absolute permeability was assumed equal to $K_r = 200$ mD, vertical $K_z = 20$ mD. The initial absolute permeability of rock matrix was set to zero. The minimum vertical permeability in the induced fracturing zone (necessary for the determination of the minimum irreversible crack opening e_0, see (9)) was taken equal to $K_0 = 15$ mD. For the horizontal permeability in the induced fracturing zone, the proportionality coefficient was $\psi = 0.1$.

The initial pressure in the impermeable rock matrix was set to 300 atm, which corresponds to anomalously high reservoir pressure of Bazhenov formation deposits. The initial pressure in the highly permeable layer was set to 200 atm (the same pressure was maintained on the right boundary of the permeable layer. This scheme corresponds to injection of air into the permeable layer after it has been developed at the natural flow regime, so that the reservoir pressure decreased with respect to its initial level.

4.2 Simulation Results

Consider first the results obtained for 2 m-thick permeable layer into which air is injected continuously at a constant volumetric rate of 20000 m^3/day (n.c.). In Fig. 2, distributions of temperature T, volumetric heat release rate Q_H, mole fraction of oxygen y_{O2}, as well as cracked zone boundary (shown as a variable CRACK taking the values of 0 and 1 in the impermeable and cracked rock) are presented at time 5 days. Also presented is the distribution of vertical permeability K_z, together with the oil filtration velocity shown by arrow. The boundaries of high-permeability layer are shown by the horizontal dashed lines. Distributions of the same quantities at time 17 days are shown in Fig. 3.

In Figs. 4 and 5, similar distributions are shown for the case of 1 m-thick permeable layer into which air is injected at the same volumetric rate of 20000 m^3/day (n.c.).

Analysis of the simulation results shows that heating of the rock matrix proceeds slowly, being limited by the conductivity of rock, while the in-situ combustion front travels away from the injection well, consuming the hydrocarbons (native oil and kerogen) contained in the highly permeable layer. As a result, conditions for fracturing of rock matrix (i.e., high enough fraction of kerogen converted to liquid and gaseous hydrocarbons, and high enough temperature necessary to maintain the pore pressure) are achieved only *behind* the in-situ combustion front, and not before it. Accordingly, rock fracturing with the formation of filtration channels also occurs behind the combustion front.

Another important observation is that after the rock matrix has been fractured, the hydrocarbons from the pore space (gas, synthetic oil) enter the fractured zone and move along the filtration channels towards the lower pressure zone (permeable layer). Due to anisotropy of permeability caused by predominantly vertical orientation of cracks, synthetic oil is moving downwards, see the velocity vectors presented in Figs. 2, 3, 4 and 5.

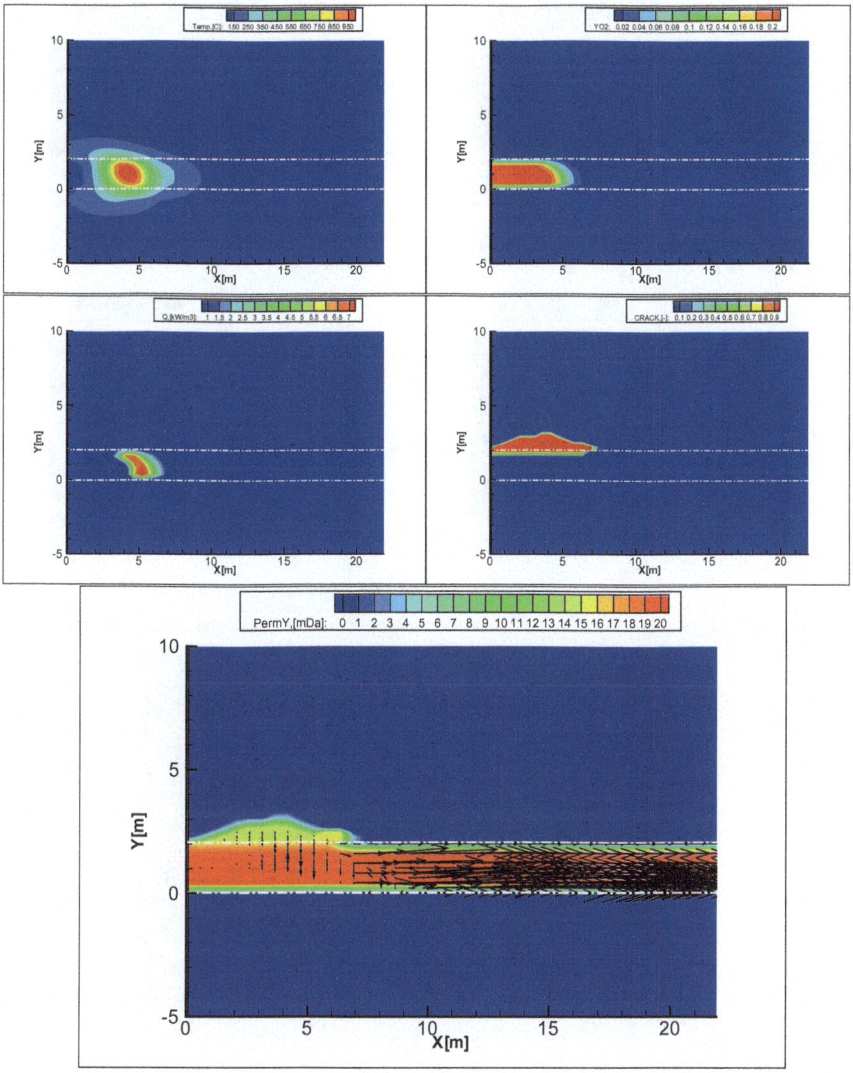

Fig. 2. Temperature, mole fraction of oxygen, heat release rate density, cracked zone boundary, vertical permeability and oil velocity (arrows) after 5 days (left-to-right, top-to-bottom), thickness of high-permeability layer 2 m.

Thus, for the technological scheme considered, simulations show that most of the synthetic oil formed by decomposition of kerogen will be transported towards the high-permeablity layer, reaching it behind the in-situ combustion front. In this zone, oxygen is present due to continuing air injection, and the temperature is high enough due to recent passage of the in-situ combustion front. Therefore, the hydrocarbons entering the

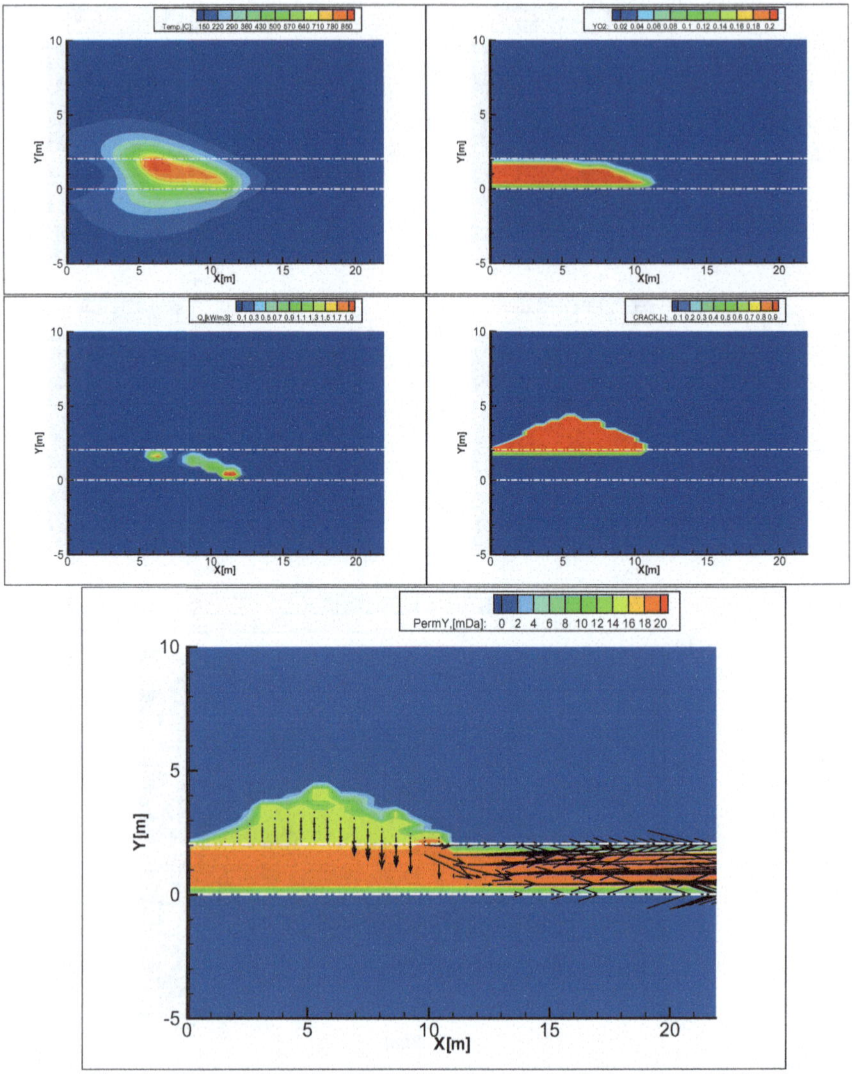

Fig. 3. Temperature, mole fraction of oxygen, heat release rate density, cracked zone boundary, vertical permeability and oil velocity (arrows) after 17 days (left-to-right, top-to-bottom), thickness of high-permeability layer 2 m.

permeable layer, will be reacting with the oxygen: this is clearly seen in the distributions of combustion power density shown in Figs. 2, 3, 4 and 5: the combustion front becomes inclined, stretching and moving towards the induced fracturing zone where concentration of oxygen also becomes reduced. It is important to note that continuous heating of the rock matrix leads to the increase in the size of cracked zone; the gases

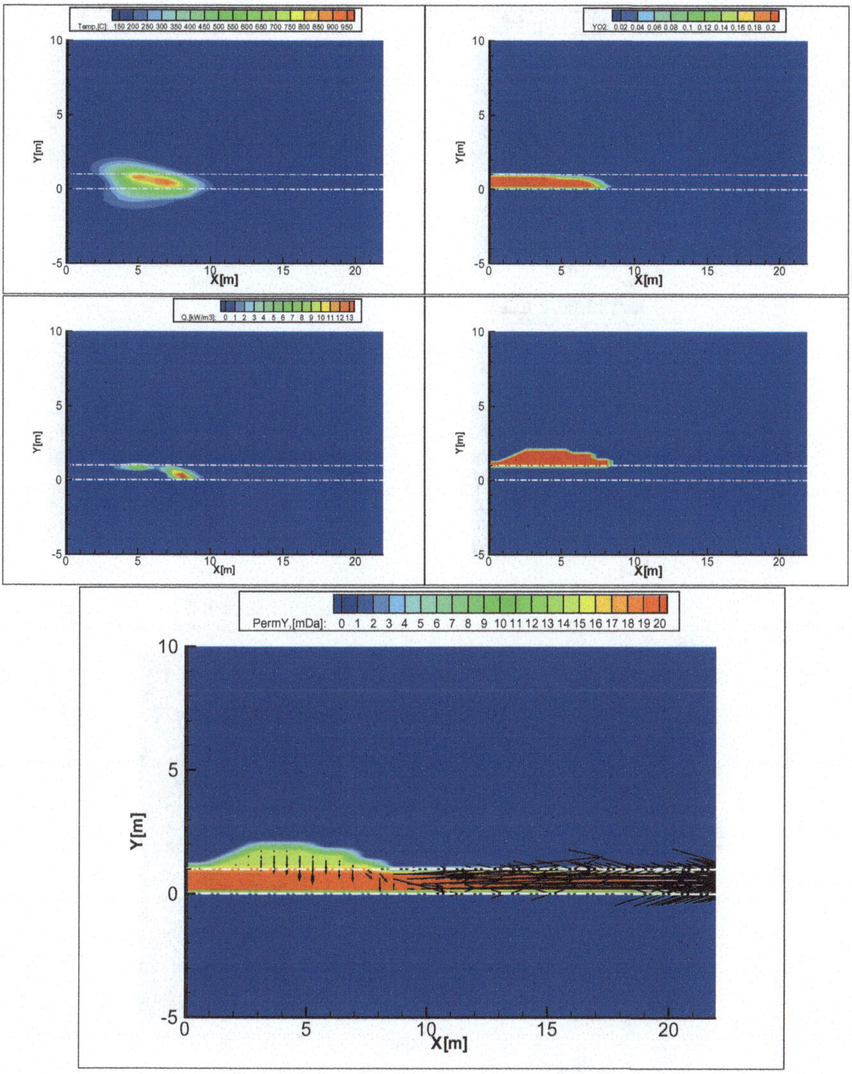

Fig. 4. Temperature, mole fraction of oxygen, heat release rate density, cracked zone boundary, vertical permeability and oil velocity (arrows) after 5 days (left-to-right, top-to-bottom), thickness of high-permeability layer 1 m.

and liquid hydrocarbons are released from the pore space into the fractured rock and propagate towards the thermal wave since the pressure remains high where pyrolysis of kerogen proceeds. This pressure distribution makes penetration of oxygen into the cracked zone impossible, and combustion of synthetic oil proceeds in the permeable layer.

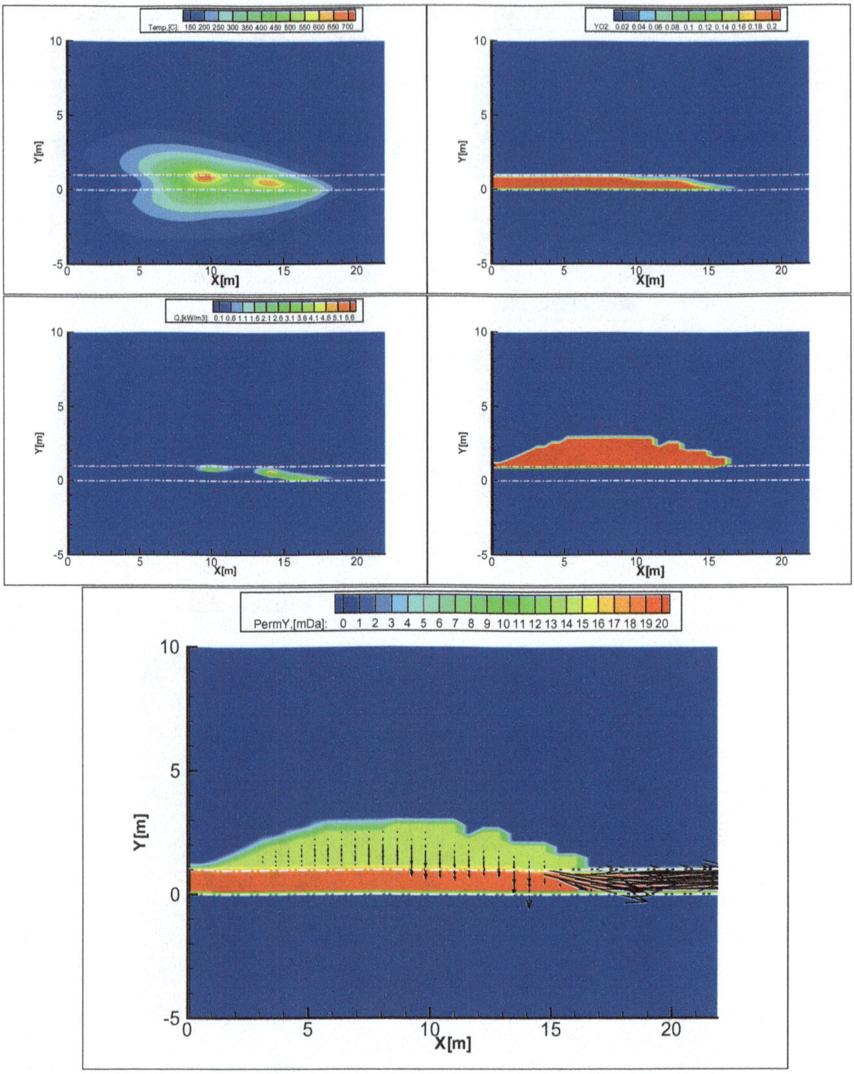

Fig. 5. Temperature, mole fraction of oxygen, heat release rate density, cracked zone boundary, vertical permeability and oil velocity (arrows) after 20 days (left-to-right, top-to-bottom), thickness of high-permeability layer 1 m.

5 Conclusions

Simulation results obtained in this work on the basis of a coupled compositional and geomechanical model clearly show that thermal gas treatment of initially impermeable kerogen-containing rocks of Bazhenov formation indeed provides a viable way towards development of vast hydrocarbon reserves contained in this deposit. In-situ combustion can be arranged by injection of air into highly permeable carbonate layers available in the practically impermeable rock matrix. The combustion heat released in the permeable layer can convert kerogen to synthetic oil and gas by thermal decomposition in the closed pores. The high internal pore pressure seems to be sufficient to cause fracturing of rock material and formation of a hydraulically connected system of cracks, or filtration channels.

However, an important conclusion drawn from the simulations is that these potential benefits can be turned real only if a proper technological scheme is developed. Straightforward injection of air into the permeable layer, in hope that more oil will be produced from it, is not viable for the reasons related to the specifics of physical and chemical processes involved. Slow heating of matrix rock, and faster propagation of in-situ combustion front result in the development of cracked zone and outflow of synthetic oil into the permeable layer behind the in-situ combustion front. Most of this oil will be burned in the air injected from the surface, and losses related to this may render the process unprofitable.

Thus, comprehensive studies of the coupled fluid dynamics, geomechanical, chemical processes are necessary in order to evaluate the viability of different technological schemes and approaches towards the development of Bazhenov formation. The mathematical model described in this paper provides a theoretical basis for such studies.

Acknowledgement. Support by Russian Ministry of Science and High Education (Project Reg. No. AAAA-A17-117021310385-6) is acknowledged.

References

1. Akhmadeishin, I.A., Kokorev, V.I., Darishchev, V.I., Shchekoldin, K.A.: Tight and unconventional hydrocarbon reserves: experience and forecasts. In: Proceedings of International Science Application Conference, Kazan, 3–4 September 2014, FEN, Kazan 2014, pp. 6–9 (2014)
2. Afanasiev, I.S., Gavrilova, E.V., Birun, E.M., Kalmykov, G.A., Balushkina, N.S.: Bazhenov formation. General review, unresolved problems. Vestnik OAO Rosneft, no. 4, pp. 20–25 (2010). ISSN 2074-2339
3. Bokserman, A.A., Vlasov, V.N., Ushakova, A.S., Kokorev, V.I., Chubanov, O.V.: Field exploration of the in situ oxidation process due to thermal-gas effect on the Bazhen's series deposits. Neftyanoe Khozyaistvo, no. 5, pp. 78–82 (2011)
4. Chen, Z., Huan, G., Ma, Y.: Computational Methods for Multiphase Flows in Porous Media. SIAM, Philadelphia (2006)
5. STARS. Advanced processes and thermal reservoir simulator. Version 2007. User's Guide. Computer Modelling Group Ltd., Canada (2007)
6. Kristensen, M.R.: Development of models and algorithms for the study of reactive porous media processes, Ph.D Thesis. Frydenberg, Copenhagen (2008)
7. Nesterov, I.A., Shapiro, A.A., Stenby, E.H.: Numerical analysis of a one-dimensional multicomponent model of the in-situ combustion process. J. Pet. Sci. Eng. **106**, 46–61 (2013)
8. Peng, D.Y., Robinson, D.B.: A new two constant equation of state. Ind. Eng. Chem. Fundamen. **15**, 59–64 (1976)

9. Brusilovskiy, A.I.: Phase Transitions in Oil and Gas Field Development. Graal', Moscow (2002)
10. Lin, H., Duan, Y.Y.: Empirical correction to the Peng-Robinson equation of state for the saturated region. Fluid Phase Equilib. **233**, 194–203 (2005)

Current Status of Oil Recovery from Bazhenov Formation: Efficiency Analysis of Existing Technologies and New Approach

A. M. Polishchuk[1], S. A. Vlasov[1], and S. E. Yakush[2(✉)]

[1] Research and Technology Organization ITIN, Moscow, Russia
itinmsk@gmail.com

[2] Ishlinskiy Institute for Problems in Mechanics RAS, Ave. Vernadskogo 101 Bldg 1, Moscow 119526, Russia
yakush@ipmnet.ru

Abstract. Bazhenov formation in West Siberia is featured by vast hydrocarbon reserves, however, its development is difficult due to the geological structure and properties of the deposit. The principal part of organic substances is contained in the impermeable rock matrix in the form of solid kerogen. The fundamental problem which must be solved for recovery of hydrocarbon reserves is to convert kerogen into fluids, and create a connected fracture system allowing flow of synthetic oil towards permeable reservoirs available. Existing technologies of enhanced oil recovery (including hydraulic fracturing and thermal gas treatment) are reviewed, and their applicability to Bazhenov formation conditions is discussed. It is shown that none of the technologies is capable of solving the problem of hydrocarbon recovery from the rock matrix where most of the kerogen reserves is concentrated. A new approach is proposed consisting in the combination of hydraulic fracturing and thermal methods, together with proper arrangement of hydrocarbon flow path in the induced crack system. Necessity for comprehensive mathematical modeling on the basis of coupled geomechanical, chemical, and filtration models is highlighted.

Keywords: Kerogen · Hydraulic fracturing · Thermal methods · Bazhenov formation

1 Introduction

Increasing demand for oil and simultaneous deterioration of the structure of petroleum reserves attract close attention to the problems arising in the development of tight and unconventional oil reservoirs. One such a reservoir is the Bazhenov formation (BF) in West Siberia, featured by vast hydrocarbon reserves. By the estimates of many specialists, the hydrocarbon reserves related to BF may reach a trillion tons, on the area of more than a million of square kilometers [1]. The potentially recoverable reserves are estimated to be in the range from 5 to 40 billion tons.

However, practical experience of the development of Bazhenov formation deposits gained over more than fifty past years have clearly indicated that the existing technologies do not allow this high potential to be used for commercial hydrocarbon

V. I. Karev et al. (Eds.): *Physical and Mathematical Modeling of Earth and Environment Processes (2018)*, SPEES, pp. 395–410, 2019.
https://doi.org/10.1007/978-3-030-11533-3_39

recovery. Over the whole period of reservoir development, oil recovery did not exceed 12–13 million tons [2].

The difficulty of Bazhenov formation development is mainly related to the geological structure of reservoirs and their properties. This oil deposit is featured, as a rule, by the presence of two different layers: a dense, practically impermeable rock matrix with permeability as low as 10^{-8}–10^{-5} μm^2, and relatively thin collectors (carbonate layers) having well-developed fracturing and possessing permeability of 10^{-5}–1 μm^2. The latter layers also include lens formations.

The oil contained in the layers of relatively high permeability can be recovered more or less easily with conventional vertical wells. It is by penetration into these reservoir regions that significant oil production rates were obtained, and, provided that these geological formations were large enough, commercial operation of wells could last for long enough time (of the order of years). For example, the wells drilled into the primary productive layer of Bazhenov-Abalak petroleum system of Salym deposit were operated for many years, providing debits of the order of 10 t/day [3].

However, such zones of high productivity are, as a rule, of limited size and distributed over the cross section very unevenly. Also, their thickness is relatively small, reaching just 1–10% of the total thickness of oil-saturated reservoir. The length and thickness of lens formations are also quite small, from few meters to few hundred meters, with the thickness as low as one meter. High permeability of the reservoir layers can provide, upon penetration, oil well production rate up to a hundred tons/day. But the volume of connected porous space in these layers is relatively small, and oil production rate of such wells falls down quite rapidly.

No regular pattern of distribution of high-permeability zones over the whole deposit has been found so far. Hydrodynamic connectivity between separate lenses is either insignificant, or is totally absent. As a result, the flowrates of closely spaced wells could be different by two or more orders of magnitude, and large number of wells turned out to be simply “dry” [1].

An important fundamental difference between BF and conventional oil reservoirs lies in the fact that transformation processes in the organic substance contained therein (kerogen[1]) into hydrocarbons have not yet been finished. This allows many geologists to consider BF as an oil-source bed. At the contemporary stage of metamorphism of these rocks, the organic substance is contained in two forms: (i) as a light oil, and (ii) as a solid rock-forming substance, kerogen. In the highly permeable zones, at the same time, most of hydrocarbon contents exist in the form of liquid low-viscosity hydrocarbons (native oil, or oil resulting from transformation of kerogen), while contents of solid kerogen there is only few percent. In the impermeable rock matrix, on the contrary, the picture is very different: the content of native oil there is about one percent, whereas the fraction of kerogen is as high as forty percent of the total rock volume, or even higher than that.

Due to the specific features mentioned above, the principal part of organic substances is contained in the impermeable matrix in the form of solid kerogen. Therefore, the main problem encountered by specialists concerned with the development of

[1] Kerogen is an organic substance scattered in sedimentary rocks, insolvable by organic solvents.

Bazhenov formation, is formulated in very simple terms: firstly, it is necessary to provide conditions necessary for transformation of kerogen into liquid and gaseous hydrocarbons (in other words, to speed up the natural processes dozens or hundreds thousand times), and, secondly, to transform the impermeable rock matrix into a permeable reservoir over which these hydrocarbon fluids will be able to move towards the production oil wells.

However, simplicity of this formulation is illusive, and no acceptable solution to the problem has been found so far. This paper is devoted to the analysis of pros and cons of existing technologies, and to discussion of possible approaches to the problem of Bazhenov formation development.

2 Existing Technical Approaches to Development of Bazhenov Formation

So far, a number of approaches towards the development of BF have been proposed. Generally, they can be divided into two groups.

2.1 Traditional Solutions

This group includes technologies providing recovery of the native (liquid) oil contained in the connected void space (cracks and caverns) of Bazhenov formation rocks. For the most part, these are traditional oil recovery technologies or their extensions. First of all, such technologies include oil reservoir development by vertical wells operated in the natural regime and aimed at recovery of the native oil only. Efficiency of this technology turned out to be quite low due to large number of "dry" wells and rapid decrease in the flow rate of oil production wells. This oil recovery scheme proved to be commercially rather unprofitable [4].

An enhanced version of this technology consists in drilling horizontal or controlled directional (inclined) wells, side holes etc., or in the application of known methods of bottom hole treatment (BHT), including hydrochloride acid treatment, acid-cut clay treatment etc., as well as directed unloading of reservoir rock.

The next stage of this technological direction is the hydraulic fracturing (HF)[2], of which there exist many versions, including the directional hydraulic fracturing, and its modern modification, multistage hydraulic fracturing (MSHF). It is this approach that is the most "fashionable" at the moment, and for many petroleum companies this is the mainstream technology [1, 4–7].

The reason for rapidly increasing interest to such technological approaches is quite straightforward: it is the so-called "shale revolution" thanks to which the USA reached the top place in the world with respect to natural gas production, and became one of the top oil-producing countries.

[2] As a result of hydraulic fracturing (HF), a well-developed system of filtration channels (cracks, induced permeability) is created in the reservoir rock enhancing the production well flowrate.

In the United States, the above-mentioned approach turned out to be very efficient for shale gas recovery. Inspired by the shale gas results obtained, American engineers massively unfolded similar works aimed at production of oil from tight and low-permeability reservoirs. Without HF, development of such oil reservoirs would be practically impossible.

Note that the USA hold a large number of drilling rigs, totaling to 60% of the number of rigs in the whole world. If in 2009 just as low as 20% of the US drilling rigs were used for oil production, in 2012 this number rose to as high as 80%. This initiated the full-scale production of shale oil in the United States [3].

It should be noted that in some sense the United States were lucky. Several large-scale formations are located on the US territory, including Bakken in the North, and Eagle Ford and Permian in the south of the USA. Application of MSHF allowed efficient development of these formations to be started in the USA, with up to 80% of oil being recovered from low-permeability reservoirs given by the shale plays Bakken and Eagle Ford. As a result, the total oil production in the USA increased dramatically.

In this respect, the central question to be considered is to what extent the MSHF technology is applicable to the development of Bazhenov formation?

Bakken Formation is Different from Bazhenov Formation. In order to answer this question, first of all, the geological characteristics of these two objects must be compared. This is very important because, on the basis of apparent similarity in the rock types of Bakken and Bazhenov formations, specialists of a number of leading Russian oil-producing companies try to enhance oil production from BF by using the HF technology as the principal tool [8].

Bakken formation, providing the main shale oil production in the USA, consists of three stratigraphic units [9, 10] presented in Fig. 1.

The lower shale member is formed by clayey-siliceous argillites, rich with organic substances and possessing the porosity of about 3.6% and permeability of up to 0.0001 10^{-7} m^2. The maximum thickness of this layer is 15 m.

The middle member is formed by sandstones, dolomite sandstones, dolomites, aleurolites, shale rock. The organics content there is up to 7%, with porosity of about 5% and permeability (0.4–1.0) 10^{-3} m^2; the maximum thickness is 40 m.

The upper shale member, with respect to its lithology, porosity, and permeability is similar to the lower shale unit; its maximum thickness is about 26 m.

The main resources of light oil are concentrated in the middle part of Bakken formed by low-permeability and relatively homogeneous rocks. These rocks contain dolomites with low clay contents, and carbonates. Due to tectonic processes, the reservoir already has a natural system of cracks, horizontal and vertical, where the main part of the light oil is contained. In accordance with definition of International Energy Agency, this oil falls into the category of *Tight oil*. This light oil can be recovered by the methods of multistage hydraulic fracturing, it is classified by EAI as *Conventional*, and, formally, it cannot qualify as “shale oil”.

According to EAI classification, *Shale oil* is comprised of light hydrocarbon fractions produced by thermal treatment of solid shale rocks with high contents of kerogen. It is this oil that is classified by EAI as *Unconventional*.

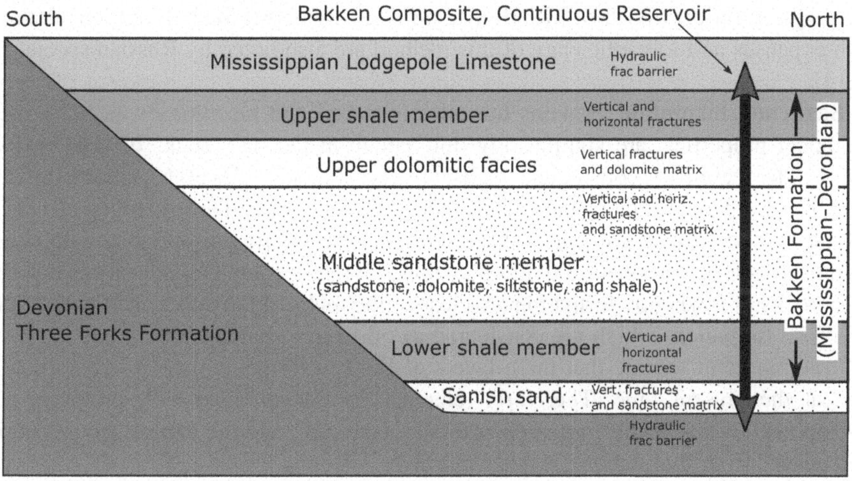

Fig. 1. Scheme of stratigraphic cross-section of Bakken formation with crack direction in the continuous collector (adapted from [9, 10])

For the above-mentioned conditions, HF technologies are quite applicable and efficient. By hydraulic fracturing, the initially discontinuous zones of natural fracturing can be connected to form a united interconnected high capacity system.

Horizontal wells are drilled to the distances of up to 3.5 km, and two-three dozens of hydraulic fracturing operations are performed. To maintain the oil production rate, it is necessary to drill new and new wells, and due to deterioration of petroleum reserve structure the annual drilling volume must be increasing. On Bakken formation, an average well becomes low-production in 7 years, while on Eagle Ford fields this occurs in 4 years. The number of wells drilled in Bakken exceeds 12 000, of them more than 4 thousand are horizontal wells where HF was performed [11].

Analysis of the development of American reservoirs shows that intensive drilling can provide rapid increase in the oil production rate, however, it then decreases quite rapidly. For example, it is estimated that peak oil production rate can reach about 2.8 million barrels per day on Bakken field, and 2.46 million barrels per day on Eagle Ford; however, rapid decrease in the production rates is forecasted. More comprehensive history of Bakken deposit development can be found in [12].

Another feature of the method of drilling horizontal wells and performing MSHF is its low (and, often, negative) profitability. Experience of oil well operation shows that over the first year after hydraulic fracturing the volume of recovered oil decreases more than twice, whereas in five years the well closes down because low production rate makes such a well non-profitable. On the contrary, during oil recovery from conventional reservoirs, the oil production rates decreases just by 7% per year, and the total amount of recovered oil turns out to be several times higher.

Moreover, despite the use of multilateral horizontal wells and multistage hydraulic fracturing, the oil recovery factor on Bakken and Eagle Ford plays remains very low, ranging from 3 to 10%.

A significant drawback of MSHF method is its high cost, totaling millions dollars. High expenses and low efficiency of this method are also noted by Russian specialists [7, 13].

Bazhenov formation contains the oil-source bed and reservoirs which, by their geological properties, are significantly different from the tight rocks of Bakken formation. The main rock constituents of BF are clay minerals (25–30%), silicon dioxide (35–60%), carbonate minerals (8–12%) and solid organic substance kerogen (10–20% and more). The relative content of these substances in the rock is variable, with either silicon dioxide or carbonate dioxide prevailing. By its geological and physical properties, rocks of Bazhenov formation are close to the rocks of the upper and lower layers if Bakken formation, which are considered as oil-source shale rocks.

It is important to note that these layers of Bakken formation *are not being developed* at the present time. Therefore, reference to Bakken field as a prototype for development of Bazhenov formation is totally irrelevant, and the experience gained in oil recovery from that deposit is not applicable.

The traditional technologies considered above possess one common feature. They all are aimed at the recovery of light native (i.e., already available in the reservoir) oil, contained in cracks and caverns of relatively high-permeability layers. Therefore, they are fundamentally incapable of solving the principal problem, to involve in the recovery process of the main fraction of organic substance contained in the impermeable rock matrix in the form of native oil and, for the most part, in the form of solid organic substance, kerogen.

2.2 Thermal Methods

The second group includes technological approaches aimed at in-situ transformation of kerogen to liquid and gaseous hydrocarbons, and their subsequent recovery.

Field data obtained by different authors and concerning the development of a whole range of fields in BF showed that such development parameters as porosity, initial production rate and cumulative oil production are functions of the reservoir temperature. The higher the temperature, the higher are the initial production rate and cumulative oil production. One of the reasons for this dependency is that conversion ratio of kerogen is different (with the increase in temperature, the conversion rate also increases).

On the basis of these research results, it can be stated with confidence that the most prospective method of efficiency enhancement with respect to development of Bazhenov formation consists in thermal treatment of oil reservoir [14–20].

Among such **thermal treatment projects** aimed at in-situ kerogen conversion and oil production, the following ones are relevant and deserve attention.

Pilot Project by Royal Dutch Shell – ICP (In-situ Conversion Project) [21, 22]. Thermal treatment is achieved by gradual heating of isolated regions of shale reservoirs over quite a long time (few years) by using electric heating of reservoir. Electric heaters are placed in 550 m deep wells, providing the increase in the reservoir temperature to 350 °C. This level was chosen on the basis of laboratory tests which showed that at this temperature the highest quantity of hydrogen molecules is released from kerogen to react with hydrocarbons and produce higher-quality oil. It takes, as a rule, from 2 to 4 years from the beginning of heating to start of oil production. Following the heating, light hydrocarbons are supplied through the injection well (diesel or jet fuel), which allows separation of light and heavy fractions, and obtain at the production well mouth a low-boiling hydrocarbon mixture (e.g., unstable gas condensate and oil-well gases).

Exxon Mobil ElectroFrac [23]. This technology is a symbiosis of hydraulic fracturing and thermal treatment methods. Initially, a system of cracks is created in the reservoir between two rows of horizontal wells; these cracks are then filled from the surface by an electrolyte. Upon application of voltage, the space filled with electrolyte is heated, and this heat is transferred to the surrounding rock. When some threshold temperature is reached, thermal decomposition (pyrolysis) reaction begins in the kerogen, leading to formation of hydrocarbons. The synthetic oil produced in this way is then recovered by traditional methods.

Chevron In-Situ Process [24, 25]. The technology involves oil recovery from the reservoir and partial in-situ thermal decomposition of kerogen achieved by injection of preheated carbon dioxide at high pressure exceeding the overburden load, i.e., hydraulic fracturing is performed. Energy efficiency is achieved by high pressure air recirculation through the depleted reservoir zones. In this technology, air plays the role of heat transfer agent, its heating in the depleted reservoir zones occurs due to burning of heavy hydrocarbon residues. The hot air is supplied into new reservoirs through the injection wells, due to its high temperature it stimulates kerogen decomposition. This method provides oil yield exceeding 90%, its field testing is performed in the Green River formation (USA).

In-Situ Gasification [26, 27]. Some versions of the technology developed by Chevron company are the in-situ gasification approaches developed by two companies, General Synfuel International (GSI) and Mountain West Energy (MWE).

Technology of GSI company involves injection into the oil-producing reservoir of hot air which, by heating the reservoir, causes pyrolysis of kerogen contained therein.

Technology of MWE company implies injection into the kerogen-containing reservoir of hot natural gas, it has been tested on Teapot Dome heavy oil deposit.

Both the above-mentioned technologies use a single well for injection and recovery of hydrocarbon vapor.

It should be noted that all the technologies considered above were developed and tested on shale deposits having relatively small depth, and they hardly can find direct application in the conditions of Bazhenov formation.

In-situ Catalytic Rethorting Technology (ICRT). Was suggested by Strategic Alliance GeoEn together with ZAO ResTech [10].

The physical basis of ICRT is high-temperature reservoir treatment by injection of the work agent obtained by fuel burning in the bottom hole assembly. Combustion products include water/vapor (unsaturated, saturated, or superheated), flue gas, CO_2, H_2, O_2, N_2, organic solvents and catalyst in the form of nano-sized (<100 nm) or, in some cases, ultra-dispersed (>100 nm) particles.

Formation of the work agent is suggested to be achieved by a specially developed bottom hole generator where the fuel supplied from the ground level is burned; this generator provides catalytic heterogeneous oxidation of hydrocarbon-containing liquid and gaseous fuel compositions.

In our view, practical implementation of this technology will be hindered by a number of obstacles difficult to overcome, namely:

- Permeability of BF matrix rock has is less than 10^{-5} μm^2, i.e., the matrix is practically impermeable; hence, the "penetration" velocity is going to be negligible. Therefore, in-situ retorting of kerogen can hardly be performed without pre-fracturing of the rock.
- The principal device for this technology is the bottom-hole steam generator. However, long-term practical experience gained in this direction, development of a generator working in deep (more than 1 km) wells is a tremendous scientific and technical problem, unsolved so far.
- One more important aspect of the problem must be kept in mind. Simultaneous supply of fuel and oxidizer to the bottom hole of the well, and running the combustion reaction in the bottom-hole steam and gas generator is technically difficult enough and unsafe. First of all, it should be taken into account that large quantity of gas is released upon combustion of the fuel supplied through the well. This increases the bottom hole pressure and decreases the well injection capacity. This, in turn, would entail the change in the steam generator output pressure and modify the fuel-oxidizer ratio at the well mouth. In other words, to ensure safe operation of the bottom-hole burner, a complex and reliable control system with efficient feedback must be implemented. However, taking into account the system lags caused by separation of bottom hole sensors and control equipment by more than two kilometers, development of such equipment is a quite difficult technical problem on its own. One must also keep in mind that one of the components (air) is in the gas state.
- And, finally, the most important point. All these difficulties are viewed as unnecessary for development of BF. The high enough reservoir temperature (above 100 °C) provides for self-initiation of low-temperature hydrocarbon oxidation reactions in which necessary heat is going to be released. The use of a interconnected surface and underground equipment, on one hand, complicates implementation of the technology, and, on the other hand, increases its cost, metal consumption and operational expenses.

2.3 Thermal Gas Treatment (TGT)

TGT is considered by many specialists as the most prospective technology for the development of Bazhenov formation, and field tests are now being carried out by AO RITEK.

The method of thermal gas treatment of oil reservoirs was developed in the USSR in 1971, as an approach to recovery of light oils from tight reservoirs by air injection [28]. In the USA, this technology was developed independently, it is known as HPAI (High Pressure Air Injection) [29]. The first project of this type was implemented in 1963 on Sloss field (USA) [30].

Field tests of TGT technology in application to BF conditions have been started by AO RITEK in 2009 on the test field of Sredny Nazym deposit, these tests are being carried out up to now [31–33].

TGT integrates the thermal and gas methods of enhanced oil recovery, it implies high-pressure injection of air and water into the reservoir, initiating the oxidation processes (combustion) with release of heat, CO_2, and broad fraction of light hydrocarbons. These products of oil combustion, together with nitrogen supplied with air, promote oil displacement towards the production wells.

To initiate the oxidation reaction in the reservoir, its temperature must be raised above some threshold level of 65–70 °C. (Note that the temperature in Bazhenov formation reservoirs is high enough for self-initiation of combustion, without any need for external igniters). Simultaneously, due to high temperature and pressure in the reaction zone, the matrix part of reservoir is heated up, and kerogen transformation into light oil and hydrocarbon gases begins there, resulting in induced fracturing that provides hydrodynamic connectivity between the matrix and permeable reservoir layers. Comprehensive review of this technology and analysis of the related literature can be found in [34].

Further development of in-situ combustion method was achieved in the horizontal well technology THAI [35, 36], catalytic THAI (CAPRI) [37], and lateral gravity in-situ combustion LGSC using one horizontal production well and one or several vertical injection wells located nearby [38].

Results of field tests with application of TGT, being performed by OAO RITEK, have not lead to any optimistic results so far. Therefore, a question arises on whether efficient application of TGT in the Bazhenov formation conditions is possible in its "classical" implementation, or some modification is necessary in view of the specific features of this deposit.

To answer this question, consider a typical scheme of the distribution of different zones over the reservoir under TGT, presented in [39], see Fig. 2.

In the TGT technology, air is supplied by the injection well into permeable layers of the reservoir having good hydrodynamic connectivity with the production well. The layers located below or above the permeable one (the matrix) are either impermeable for air, or contain lens reservoirs having no hydrodynamic connectivity to the production well. As a consequence, the oxidizer cannot penetrate the matrix and, therefore, no reaction between the oxygen and organic substance contained therein is possible. All chemical interaction of oxygen occurs solely in the high-permeability layer where exothermal reactions run in the oxidation zone (see Fig. 2); in these reactions, combustion products are generated which play the role of an effective displacing agent. The impermeable rock matrix (the overburden in Fig. 2) is heated only by conductive heat transfer.

Simulations of the thermal processes in the configurations relevant to Bazhenov formation show that the oxidation (in-situ combustion) front will be moving along the permeable layer much faster than the thermal front in the impermeable matrix (for more details on the mathematical model and simulation results, see [40, 41]). As a result, the kerogen pyrolysis zone, as well as the zone of induced fracturing in the matrix, will be lagging behind the zone of oxidation in the high-permeability layer. The synthetic oil, obtained in the course of kerogen decomposition, will then be supplied through the

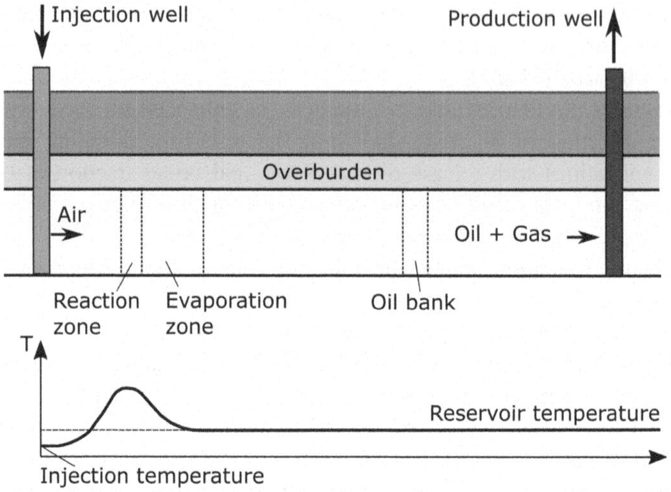

Fig. 2. Main zones in oil reservoir during thermal gas treatment (TGT) (adapted from [39])

system of predominantly vertical cracks developing due to thermal fracturing of matrix rock, reaching the well-heated zone of the permeable layer *behind* the in-situ combustion front. Reacting at high temperature with the oxygen of air supplied by the injection well, this oil will be undergoing high-temperature oxidation (combustion), with water and carbon dioxide as the ultimate products.

It can be seen that, upon this scheme of thermal gas treatment, all processes run in the carbonized high-permeability layer only. Practically all the effect amounts to enhanced oil recovery from this layer in comparison with native oil recovery from the same layer upon the natural well operation regime. Evidently, no significant oil recovery enhancement can be expected in this case. According to studies [42], capacity of carbonate reservoirs (including cracks and caverns) residing on the contact boundary between the Bazhenov and Abalak formations, which can form a hydraulically connected space, is as low as just 2–5%. The main potential hydrocarbon reserves, at the same time, are concentrated in the matrix which, upon implementation of the "classical" TGT technology, remains uninvolved in the Bazhenov formation development.

These considerations led the authors to the firm conclusion that, without fundamental modification, the currently existing TGT technology is capable of enhancing oil recovery only from the permeable reservoirs, and it cannot achieve the principal target, consisting in the *involvement of kerogen of the matrix rock* into development.

3 New Technical Approach

On a number of Bazhenov formation sites, existing highly-permeable carbonate layers are formed by vuggy and fractured reservoir rocks, tend to be located near the reservoir bottom (bordering Abalak formation). On the Bolshoi Salym deposit, of this type is reservoir KC_1 located near the boundary between Bazhenov and Abalak formations

[43]. The same point of view was supported by the group from ZAO MIMGO, lead by V.S. Slavkin. They came to the conclusion that the main oil reservoirs of Bazhenov horizon (Tutleim formation) of Galyanov and Sredny Nazym oil fields are tight carbonatized fractured layers located in the lower part of Bazhenov formation deposits (on the border with Abalak formation). The thickness of this layer is 1–4 m, and its permeability can reach several μm^2 [11, 43, 44].

The *principal goal* of the proposed new technical approach is the involvement in development process of the hydrocarbon reserves *contained in the impermeable matrix rock*, achieved by thermal conversion of kerogen into hydrocarbon fluids, together with the creation of a connected system of fracture channels for their evacuation.

To elucidate the essence of the proposed technical approach, consider a simplified model scheme for Bazhenov formation reservoir. Keeping in mind the typical rock data discussed above, consider two reservoir sublayers shown in Fig. 3.

To achieve the goal stated above, development of this object is performed in the following way. In the matrix rock (2), located above the permeable collector layer (3), a dead-ended horizontal filtration channel (1b) is arranged. For example, the role of such a channel can be played by a horizontally drilled well, or a system of wells.

Opposite to the vertical injection well bore (1a), a vertical well (9) is drilled, penetrating the highly-permeable reservoir layer (3). Then, multistage hydraulic fracturing is performed to create a system of vertical cracks (4) connecting the horizontal well bore (1b) with the reservoir layer (3). On one hand, this allows the oxidizer (e.g., air) injected through the well to penetrate into the rock; on the other hand, it provides the initial system of filtration channels necessary for the flow of low-temperature synthetic oil oxidation products.

After the initial crack system has been created, injection of oxidizer into the horizontal well begins. Upon interaction of oxygen and organic materials, oxidation of the matrix begins, accompanied by heat release in exothermic reactions (zone 6) which, in turn, promotes pyrolysis of kerogen (zone 7) contained in the closed pores, their thermal fracturing, and development of a secondary system of cracks (4a).

The process must be organized in such a way that the heat of oil oxidation be enough to raise the temperature of matrix rock to some 250–350 °C. Only at such temperatures, decomposition processes leading to the formation of liquid and gaseous hydrocarbons will be running in the matrix rock [45–47] (see also Sect. 2.2).

As was shown experimentally, at such temperatures significant (dozens to hundred times) increase in the permeability of matrix rock occurred, caused by the formation of cracks [48, 49]. When the kern was heated below 300 °C, the induced fracturing was insignificant, with only a few cracks formed, beginning near solid mineral particles. In this case, the kern permeability remained low. However, when the temperature exceeded 350 °C, generation of sub-vertical and sub-horizontal fractures was intensified, the cracks started to join forming a network of micro and macro cracks over the whole volume; this was the main reason for the dramatic increase in the rock permeability [49].

It is the availability of such a connected system of induced filtration channels that can provide propagation of the synthetic oil and gaseous hydrocarbons generated by pyrolysis of kerogen towards the production well (9).

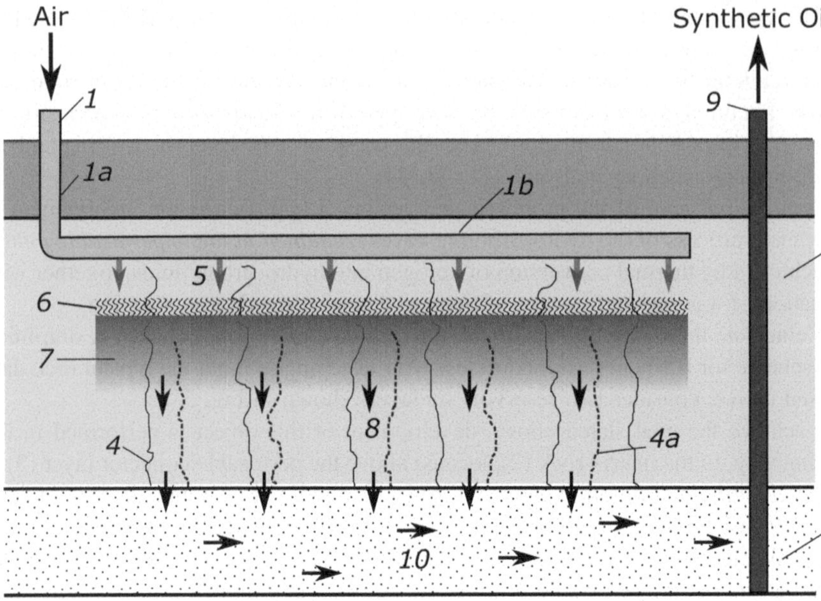

Fig. 3. Schematic geological model of Bazhenov formation reservoir and proposed approach to its development: 1 is injection well (1a and 1b are vertical and horizontal well bores); 2 is impermeable layer with high kerogen contents (matrix rock); 3 is highly carbonatized layer with high permeability (reservoir layer); 4 are cracks from hydraulic fracturing, 4a are secondary induced fractures developing due to thermal treatment; 5 is oxidizer (air) injected in the reservoir; 6 is the reaction (in-situ combustion) zone; 7 is the kerogen pyrolysis zone; 8 is synthetic oil flowing towards the permeable reservoir layer; 9 is the production well; 10 is oil flow direction in the permeable layer.

4 Discussion and Conclusions

Note that success of the proposed technology in its practical implementation is contingent upon deep theoretical studies of all its aspects, including geomechanics, fluid dynamics, chemistry and heat and mass transfer. This can be achieved by multivariate numerical simulation of the whole process, allowing the optimal conditions to be determined with respect to location of wells and regime of injection of agents into the reservoir.

Such simulations must be carried out comprehensively on the basis of a mathematical model including coupled solution of problems of chemical kinetics, heat and mass transfer, mechanics of solids, with account for rock porosity and permeability changes upon its heating. In the models, such highly important processes proceeding in the initially impermeable matrix rock must be taken into account as thermal and chemical processes in individual pores under conditions of kerogen pyrolysis, conditions for matrix rock fracturing (geomechanics of the processes), formation of induced cracks, geometrical properties of the crack system, including the principal orientation of the cracks, their relative location etc. This would allow the degree of hydrodynamic connectivity of the crack system, its porosity and permeability, affecting significantly the filtration direction and rate of the synthetic oil, to be evaluated.

Finally, of paramount importance for proper description of thermal treatment of Bazhenov formation rocks is the development and implementation of a coupled algorithm describing the interconnection and mutual effects of thermal, chemical, heat and mass transfer, and geomechanical processes in the permeable and impermeable layers of Bazhenov formation.

Acknowledgement. Support by Russian Ministry of Science and High Education (Project Reg. No. AAAA-A17-117021310385-6) is acknowledged.

References

1. Martsinkevish, B.: Bazhenov Formation News. http://geoenergetics.ru/2016/09/05/novosti-bazhenovskoj-svity. Accessed 16 Nov 2018
2. Akhmadeishin, I.A., Kokorev, V.I., Darishchev, V.I., Shchekoldin, K.A.: Tight and unconventional hydrocarbon reserves: experience and forecasts. In: Proceedings of International Science Application Conference, Kazan, 3–4 September 2014, pp. 6–9. FEN, Kazan (2014)
3. Afanasiev, I.S., Gavrilova, E.V., Birun, E.M., Kalmykov, G.A., Balushkina, N.S.: Bazhenov formation. general review, unresolved problems. Vestnik OAO Rosneft **4**, 20–25 (2010). ISSN 2074-2339
4. Yakovleva-Ustinova, T.: Gone to formation. Russian companies increasingly investing in shale. Oil Gas J. Russia, 23–27 (2014)
5. Ampilov, Yu.P.: Shale oil in Russia. Recovery prospective under sanctions. Oil Gas J. Russia, 24–30 (2015)
6. Sarancha, A.V., Garina, V.V., Mitrofanov, D.A., Levitina, E.E.: Results of test development of Bazhenov formation on Zapadno-Sakhalinskoye deposit. Fundamentalnie Issledovaniya **2** (14), 3052–3055 (2015)
7. Search and evaluation of prospective for commercial oil recovery from Bazhenov formation deposit on fields licensed to OOO Nord Imperial and OOO Aliansneftegaz (2014). http://rosgeoportal.ru/association/materials/SitePages/downloads/tomsk20082014_05.pdf
8. Markova, N.: Rosgeologiya drills Bazhen. http://neftianka.ru/tag/bazhenovskaya-svita. Accessed 16 Nov 2018
9. Pollastro, R.M., Roberts, L.N.R., Cook, T.A.: Geologic assessment of technically recoverable oil in the Devonian and Mississippian Bakken Formation. In: U.S. Geological Survey Williston Basin Province Assessment Team, Assessment of Undiscovered Oil and Gas Resources of the Williston Basin Province if North Dakota, Montana, and South Dakota, Chap. 5. U.S. Geological Survey Digital Data Series DDS-69-W, Reston VA (2010)
10. Prishchepa, O.M., Averianova, O.Yu., Visotskiy, V.I., Moriaru, D.: Bakken formation: geology, oil and gas capacity, and development history. Neftegazovaya Geologiya **8**(2), 1–28 (2013)
11. Technology of in-situ catalytic retorting. Strategic alliance GeoEn and operational company ZAO ResTech, Moscow, July 2013
12. Alekseev, A.A.: The Bazhenov formation: in search of big shale oil in Upper Salym. ROGTEC **34**, 14–27 (2013). https://rogtecmagazine.com/wp-content/uploads/2014/09/02_The-Bazhenov-Formation-in-Search-of-Big-Shale-Oil.pdf. Accessed 16 Nov 2018
13. Volkov, V.: Heading for Bazhenov formation. Neftegazovaya Vertical **12**, 26–29 (2013)

14. Bokserman, A.A.: Relevance of modern enhanced oil recovery methods as prerequisite for overcoming the decline in oil production. Neftyanoe Khozyaistvo **10**, 34–38 (2004)
15. Bokserman, A.A., Mishchenko, I.T: Overcoming the negative tendencies in the development of fuel and energy industry in Russia. Tekhnologii TEK **4**, 30–33 (2006)
16. Bokserman, A.A., Graifer, V.I., Kokorev, V.I., Chubanov, O.V.: Thermal gas treatment method of enhanced oil recovery. Interval **7**(114), 26–33 (2008)
17. Graifer, V.I., Bokserman, A.A., Nikolaev, N.M., Kokorev, V.I., Chubanov, O.V.: Integration of thermal and gas enhanced oil recovery methods as the fundamental technology for development of unconventional and tight oil deposits. In: International Forum on Nanotechnologies Rusnanotech, Moscow (2010)
18. Bokserman, A.A., Vlasov, V.N., Ushakova, A.S., Kokorev, V.I., Chubanov, O.V.: Field exploration of the in situ oxidation process due to thermal-gas effect on the Bazhen's series deposits. Neftyanoe Khozyaistvo **5**, 78–82 (2011)
19. Kokorev, V.I.: Fundamentals of controlling the thermal gas treatment of Bazhenov formation rocks in application to geological conditions of Sredny Nazym and Galyanov oil fields. Neftepromyslovoe Delo **6**, 29–32 (2010)
20. Gavura, A.V., Vlasov, S.A., Krasnopevtseva, N.V., Polishchuk, A.M., Salganik, R.L.: New approaches to the problem of enhanced oil recovery from deposits with carbonate reservoirs. In: Proceedings of IV International Symposium Theory and Practice of Enhanced Oil Recovery Methods, Moscow, 18–19 September 2013, vol. 1, pp. 67–74. OAO VNIIGAZ, Moscow (2013)
21. Birger, J.: Oil from stone. Fortune (2007)
22. Covell, J.: Environmental review of selected oil shale technologies. In: Proceedings of 28th Oil Shale Symposium, Golden, Colorado, pp. 126–134. Colorado School of Mines, Colorado (2008)
23. Symington, W.A., Olgaard, D.L., Otten, G.A., Phillips, T.C., Thomas, M.M., Yeakel, J.D.: ExxonMobil's Electrofrac Process for In Situ Oil Shale Conversion. AAAPG Annual Convention. American Association of Petroleum Geologists San Antonio
24. Environmental Assessment: Chevron Oil Shale Research, Development & Demonstration. U.S. Department of Interior. Bureau of Land Management. White River Field Office, November 2006. CO-110-2006-120-EA
25. Arutyunov, T.V., Arutyunov, A.A.: Methods and technologies of hydrocarbon recovery from shale deposits. Nauch. Trudi KubGTU **1**, 1–24 (2015). https://ntk.kubstu.ru/file/314. Accessed 17 Nov 2018
26. Kolomiychenko, O., Chernov, A.: Technologies for development of Bazhenov formation. The in-situ catalytic retorting technology. Oil Gas J. Russia **10**, 58–66 (2015)
27. Grushevenko, D., Grushevenko, E.: Oil of shale plays is a new challenge on energy market? Analytical review, INEI RAN, November 2011
28. Bokserman, A.A.: Results and prospective of application of thermal reservoir treatment methods. In: Proceedings of Branch Seminar on Thermal Methods, Ukhta, 5–8 October 1971, pp. 10–16. VNIIOENG, Moscow (1971)
29. Klinchev, V.A., Zatsepin, V.V., Ushakova, A.S., Telyshev, S.V.: Air injection method in oil recovery. Neftyanoe Khozyaistvo **8**, 76–80 (2014)
30. Das, S.S.: A study of oxidation reaction kinetics during an air injection process. MSc thesis, Australian School of Petroleum Faculty of Engineering, Computer and Mathematical Science, The University of Adelaide, Adelaide, Australia (2009). https://digital.library.adelaide.edu.au/dspace/bitstream/2440/60192/9/01front.pdf. Accessed 17 Nov 2018

31. Bokserman, A.A., Kokorev, V.I., Plynin V.V., Ushakova, A.S.: State-of-the-art and prospective of thermal gas treatment enhanced oil recovery method application to Bazhenov formation deposits. Int. Forum ENERCON. http://www.enercon-ng.ru/common/img/uploaded/exhibitions/enercon/doc/doklad-2010/Presentation-Plynin.ppt. Accessed 17 Nov 2018
32. Plynin, V.V.: Thermal gas Method and Bazhenov Formation. http://www.energyland.info/analitic-show-50375. Accessed 17 Nov 2018
33. Kokorev, V.I.: Technical and technological foundations of innovative development methods for deposits with tight and unconventional oil reserves. Doctor of Sciences thesis, Moscow (2010)
34. Asghari, K.: Review of Field Implementations of In-Situ Combustion and Air Injection Projects. Petroleum Technology Research Centre Incubation Project, University of Regina, pp. 1–98 (2009)
35. Xia, T.X., Greaves, M., Turta, A.T.: Main mechanism for stability of THAI – Toe-to-Heel Air Injection. J. Can. Pet. Technol. **44**(1), 42–48 (2005)
36. THAI in situ oil recovery requires minimal use of natural resources. Oil Gas Netw. **5**(2) (2004)
37. Alvarado, V., Manrique, E.J.: Enhanced oil recovery: an update review. Energies **3**, 1529–1575 (2010)
38. Zongzhan, X., Dengfa, H., Xiaoheng, W.: Methods study on channeling of in situ combustion development in developed heavy oil reservoir. Appl. Mech. Mater. **316–317**, 834–837 (2013)
39. Betelin, V.B., Yudin, V.A., Afanaskin, I.V., Volpin, S.G., Kats, R.M., Korolev, A.V.: Development of national thermohydrosimulator as an essential step towards development of unconventional hydrocarbon deposits in Russia. NIISI RAN, Moscow (2010)
40. Yakush, S.E., Galybin, A.N., Vlasov, S.A., Polishchuk, A.M., Akhmadeishin, I.A., Bugaev, K.A.: Mathematical modeling of induced fracturing in low-permeability reservoirs upon thermal gas treatment. In: Physical and Mathematical Modeling of Processes in Geomedia, pp. 241–243. IPMech RAS, Moscow (2017)
41. Yakush, S.E., Galybin, A.N., Vlasov, S.A., Polishchuk, A.M.: Modeling of thermal gas treatment of low-permeable reservoirs of Bazhenov formation. In: Physical and Mathematical Modeling of Processes in Geomedia, pp. 83–85. IPMech RAS, Moscow (2018)
42. Kuzmin, Yu.A., Sudat, N.V.: Specific features of geologic structure, estimation, and accounting in the state balance of hydrocarbon reserves in the Bazhenov formation deposits in Khanty-Mansiysky Autonomous Region-Yugra. Bestnik Nedropolzovaniya KHMAO 24 (2011)
43. Slavkin, V.S., Alekseev, A.D., Koloskov, V.N.: Some aspects of geological structure and oil recovery potential of Bazhenov formation in the west of Shirotnoye Priob'e. Neftyanoe Khozyaistvo **8**, 100–105 (2007)
44. Nemova, V.D.: Conditions for the formation of oil reservoirs in the deposits of Bazhenov horizon near the joint of Krasnoleninsky dome and Frolovskaya megadepression. Neftegazovaya Geologiya **7**(2), 1–14 (2012)
45. Kokorev, V.I., Sudobin, N.G., Vlasov, S.A., Polishchuk, A.M., Gorlov E.G.: Thermal decomposition of bituminous rocks of Tutleim (Bazhenov) formation from Krasnoleninsky refion deposits. In: Proceedings of II International Symposium Theory and Application of Enhanced Oil Recovery Methods, vol. 1. OAO VNIINEFT, Moscow (2009)
46. Kokorev, V.I., Vlasov, S.A., Polishchuk, A.M., Sudobin, N.G.: Study on thermal treatment of Bazhenov formation rock samples. Neftepromyslovoe Delo **3**, 12–19 (2010)

47. Kokorev, V.I., Sudobin, N.G., Vlasov, S.A., Polishchuk, A.M., Gorlov E.G.: Thermal decomposition of bituminous rocks of Galyanov oil field of Bazhenov formation. In: Proceedings of Conference Nanophenomena in Hydrocarbon Deposit Development: From Nanomineralogy and Nanochemistry to Nanotechnologies, pp. 261–266, Moscow (2008)
48. Mattson, E., Palmer, C., Johnson, E., Huang, H., Wood, T.: Permeability changes of fractured oil shale cores during retorting. In: Proceedings of 29th Oil Shale Symposium. Colorado School of Mines, Colorado (2009)
49. Kang, Z., Yang, D., Zhao, Y., Hu, Y.: Thermal cracking and corresponding permeability of Fushun oil shale. Oil Shale **28**(2), 273–283 (2011)

The Method of Joint Statistical Analysis of the Fields of Salinity Temperature and Density of Sea Waters and the Results of its Application in Studies of the Properties of these Fields

V. A. Sokolov[1(✉)], P. A. Egorov[1,2], and S. P. Apukhtina[2]

[1] State Oceanographic Institute named after N.N. Zubova, Moscow, Russia
vasokolov@inbox.ru
[2] Moscow Polytechnic University, Moscow, Russia

Abstract. A new method of joint statistical analysis of the fields of temperature - T, salinity - S, density of sea water-ρ is described, which is a generalization of the method of joint statistical t-S analysis. On the basis of the proposed method, the properties of joint statistics of fields T, S, ρ are investigated and the connection of the described properties with modern theories of self-organization of complex nonlinear systems (synergetic) and nonequilibrium processes in open thermodynamic systems is established. The concept of participation of water masses in redistribution of energy flows in sea waters from large-scale processes to smaller-scale processes is substantiated. From the described properties of the joint field statistics T, S, ρ follows the physical inconsistency of the existing methods for calculating the values of T, S, ρ in the areas of marine hydrological fronts. Recommendations to calculations of their physically reasonable values are given.

Keywords: Water masses · Marine hydrological fields · Joint statistics of hydrological fields · Synergetic · Thermodynamics of non-equilibrium processes

1 Introduction

In the waters of the World Ocean in the fields T, S ρ (hydrological fields) there are stable structural formations, called water masses (WM) [1, 2], with characteristic scales: horizontally from hundreds to thousands of km, vertically - from tens up to hundreds of meters. The lifetime of a WM depends on their spatial dimensions and the consistency of the hydrological fields forming them, which is characterized by their symmetry in the structure of the WM [3]. The modern understanding of the nature of marine WMs is associated with the movement of these formations in space from the sources of nucleation to the places of contact with other WMs, where in the process of

V. I. Karev et al. (Eds.): *Physical and Mathematical Modeling of Earth and Environment Processes (2018)*, SPEES, pp. 411–416, 2019.
https://doi.org/10.1007/978-3-030-11533-3_40

their interaction new WMs are formed [2]. In this description of WM there is no assessment of their role in the process of self-organization of the hydrological fields of the oceans, which exists in nature due to the stability of their global climate. Sea water is an open system with complex non-linear processes, in which there are their own patterns. The study of the processes occurring in complex nonlinear systems (hereinafter referred to as systems) is carried out by synergetic [4–6]. From the results obtained in it, it follows that in such systems there are processes of self-organization that ensure their stability. Self-organization of systems is carried out through the formation in them of symmetric vortex structures of intermediate scales. They transform energy in nonlinear systems from large-scale to small-scale processes more efficiently than turbulent processes. The paper studies the possibility of performing WMs in the sea waters of the functions of such vortex structures, since in the World Ocean there is a climatic stability of hydrological fields when energy enters them from the external environment. This is possible if mechanisms for ensuring such stability exist in an open nonlinear system of sea water.

The main conclusion of the modern theory of nonlinear processes occurring in open thermodynamic systems, including seawater, is to describe their temporal changes based on the principle of minimal production (changes) of the entropy of these systems [7]. The correspondence of the studied properties of joint statistics of marine hydrological fields to this principle is shown.

In a less explicit form than in the waters of the North Atlantic, the universality of the studied properties of joint statistics of hydrological fields was confirmed in studies of the hydrological fields of the Black Sea. The repeatability of the studied laws in different regions of the World Ocean and their consistency with modern physical theories confirms the objectivity of these laws.

2 Description of the Method and Its Application in the Study of Joint Statistics of Marine Hydrological Fields

The properties of the combined statistics of the T, S, ρ fields are analyzed on the basis of empirical histograms of the joint probability density functions T-S-ρ (JPDF T-S-ρ histograms). They are built according to T, S, ρ data located on fixed horizons of marine research areas. In the works [8–12] it was shown that on the histograms of JPDF T-S each WM has its own cluster. The histograms of JPDF T-S-ρ are obtained by applying to the constructed histograms of the JPDF T-S [8–12] isocline's ρ calculated from the values of T, S in the cell centers of these histograms taking into account the depth of the horizon under consideration. They characterize the probability of the occurrence of joint values of T, S, ρ, averaged at the scale of constructing the histograms of JPDF T-S in the area of marine research on the analyzed horizons. An example of such histograms constructed from data on the horizon of 100 m for the time intervals of the three phases of the Atlantic Multidecade Oscillations (AMO) climatic index [13] is presented below in Figure 1. Histograms were based on the climatic data T, S, ρ of the summer half, located in the western part North Atlantic on the area of a ten-degree trapezium with borders from 38^0 to 48^0 N, from 40^0 to 50^0 W, within which the branch area of the Gulf Stream is located. In Figure 1, the lines of isopycnus, presented in units of conditional density [2], indicate the distribution of the values of sea water density—ρ in histograms.

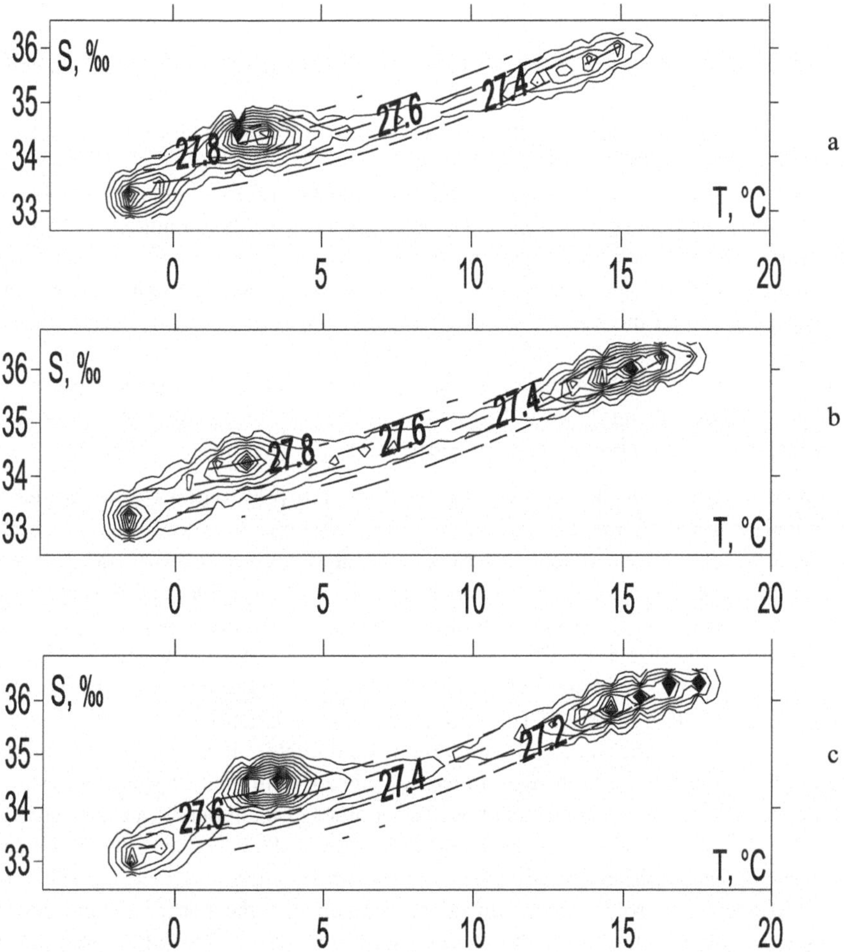

Fig. 1. Histograms of JPDF T-S-ρ, constructed for the three phases of the AMO climate index of the summer half year: 1928–1965: (a), 1988–1997, (b), 1998–2010, (c) according to data located in the western part of the North Atlantic within the boundaries of 38 to 48^0 N, from 40^0 to 50^0 W., on the horizon of 100 m, where the density—ρ is represented in as a parameter in arbitrary units in the form of isopyc lines.

3 Description of the Properties of the Joint Statistics of Marine Hydrological Fields with the Analysis of their Compliance with Modern Physical Theories

The results of the analysis of the histograms of Fig. 1 and previously published features of the joint statistics of the T, S fields [8–12] make it possible to formulate the following properties of the joint statistics of marine hydrological fields:

1. Each cluster in the histograms of JPDE T-S-ρ corresponds to its own water mass (WM);
2. The symmetry of statistical distributions of hydrological fields takes place only within a single WM;
3. In the zones of hydrological fronts, the joint statistics of T, S fields is bimodal, and T, S averages between clusters are located in areas of their unlikely values and, therefore, have no physical meaning;
4. Changes of T, S are interrelated - they have the same sign (an increase in T corresponds to an increase in S and vice versa) and is realized under conditions of minimal changes in the density of sea water—ρ;
5. The temporal stability of the positions of extremes of WM clusters and the constancy of their configuration in histograms at intervals of the three phases of AMO corresponds to the conservatism property of WM known in oceanography.

Similar constructions and analysis of the JPDF T-S-ρ histograms were carried out by the authors for the same intervals of the three AMO phases according to the T, S, ρ fields obtained in the western and eastern parts of the Black Sea at 0, 50, 100, 150 m for summer and winter semesters. In this paper, the results of these constructions are absent due to their large number (144 histograms). Here are only the general results of the analysis of these histograms. They are present, but in a less clearly defined form, all five of the above properties of joint statistics of marine hydrological fields. The WM clusters considered on the JPDF T-S-ρ histograms are symmetric formations similar to elliptical shapes. In the space of geographic coordinates, they correspond to their volume closed structures with their inherent symmetry, limited in space to the geographical coordinates by a sharp increase (by an order of magnitude) in the modules of the T, S gradients. The size of such WMs is intermediate between the large-scale processes of the considered marine regions, where the energy of external influences (the atmosphere and solar energy flows) is pumped into the marine waters and the processes of small scales to which this energy is transformed. The latter circumstance agrees with the main result of the theory of self-organization of complex nonlinear systems [4–6], which include sea waters. According to [4–6], the self-organization of processes in such systems is realized through the formation of symmetric vortex structures of intermediate scales in them. They provide in such systems a more efficient transformation of energy from processes large to small processes, compared with the mechanisms of turbulent processes. The structures of the WM on the histograms of SPPV T-S-ρ are suitable for participation in sea waters in such a mechanism of energy transformation from the external environment. They have symmetry, symmetry of the above-mentioned vortex structures and are constantly present in the studied marine area, ensuring the transformation of energy in marine waters from large-scale processes to shallow ones. These circumstances suggest the relationship properties 1, 2, 5 with the participation of WM in the processes of self-organization of marine hydrological fields, as a result of which their climatic characteristics are stable in time and space, which is consistent with the theory of synergy of nonlinear systems [4–6].

Property 3 is consistent with modern ideas about the importance of symmetry in the analysis of physical processes [3]. It follows from it that in the zones of hydrological fronts the average values of T, S are placed on the SPPV T-S histograms in the zones of their unlikely values, and therefore the physical meaning of such values is problematic. This conclusion casts doubt on the physical content of the calculations of T, S, ρ, in the zones of hydrological fronts, since their results are the mean values of these quantities in the scales of the steps of the calculated grids, which are unlikely to be realized in nature. The question arises about the actions that should be taken to obtain on the hydrological front probable values of T, S, ρ in the established confidence interval η%. The authors have obtained such a numerical solution based on the variation principle of minimizing the changes in the entropy of sea water in the calculated volume, carried out using empirical histograms of the SPPV T-S-ρ. Its description is made in the form of a separate article for open publication.

We show that the above properties 4, 5 correspond to a sufficient condition for performing the main conclusion of the theory of nonequilibrium processes in open thermodynamic systems. It follows from it that the evolution of such systems occurs under conditions of minimal production (changes) of their entropy [7]. The minimum entropy production of a thermodynamic system is determined by the formula: Min, where T is the system environment temperature in degrees Kelvin, δQ is the change in the total energy of the system in which Q is determined by the sum of: the internal energy of the system is U, its chemical potential and the work of external forces. In the conditions of sea water, its internal energy is determined by the temperature of sea water - T, the chemical potential is determined by the salinity of sea water, and the work of external forces is A = −P δV, where P is the pressure of sea water, δV is the change in its volume determined by the change in density of sea water ρ: A sufficient condition for the existence of a minimum of the total energy of sea water - δQ is the minimization of the cumulative changes in T, S, ρ, provided by the conditions 4, 5, which was to be shown.

4 Conclusions

A method of joint statistical analysis of temperature fields—T, salinity—S and density of sea water—ρ, representing a generalization of joint statistical T – S analysis [8–12], has been developed;

On the basis of the developed method, the properties of joint statistics of fields of temperature, salinity, density of sea water are described, their relationship with modern physical theories confirming the objectivity of these properties is shown;

The empirical substantiation of the new functionality of the water masses, related to their participation in the self-organization of thermodynamic processes in sea waters, was formulated and proposed.

Acknowledgement. The work was done in the framework of the State assignment № AAAA-A17-117021310371-9.

References

1. Dobrovolsky, A.D.: On the definition of water masses. Oceanology **1**, 12–24 (1961)
2. Mamaev, O.I.: Thermohaline analysis of the waters of the World Ocean, Gidrometeoizdat, 296 (1957) (in Russian)
3. Ovchinnikov, N.F.: Principles of conservation, symmetry, invariance structure. Nauka, 332 (1966) (in Russian)
4. Haken, G.: Synergetic. Mir. 404 (1980) (in Russian)
5. Olemskoj, A.I..: Synergetics of complex systems: phenomenology and statistical theory, Krasand, 374 p. (2009)
6. Seyidov, D.G.: Synergy of ocean processes. Gidrometeoizdat, 288 (1989) (in Russian)
7. Prigogine, I.R., Kondipudi, D.: Modern Thermodynamics, from heat engines to dissipative structures. World, 461 (2002)
8. Sokolov, V.A.: Application of joint statistical analysis of thermohaline and density fields for their filtration and presentation of marine hydrological observations data sets. In the collection, Messages on applied mathematics. EC them. Dorodnitsina RAS. Moscow. (2005) (in Russian)
9. Sokolov, V.A., Apukhtina, S.P.: A statistical model of the climatic conditions of the thermohalopicnic fields of the North Atlantic and the possibilities of its practical use in oceanographic research. Tr. Goin **212**, 48–63 (2009). (in Russian)
10. Sokolov, V.A., Gruzinov, V.M.: Problems of modeling hydrological fields in the Arctic seas. Arct. Ecol. Econ. **2**(14), 40–49 (2014)
11. Sokolov, V., Fomin, V.: The method of joint interpolation and analysis of probable values of thermohaline fields by the example of its use in the front zone of the Gulf Stream. Tr. Goin **216**, 48–59 (2015)
12. Sokolov, V.A.: Patterns and properties of the joint statistical distributions of the thermohaline fields of the North Atlantic and their use in solving oceanographic problems. Process. Geomedia **2016**(3), 267–277 (2016)
13. http://www.aoml.noaa.gov/phod/amo_faq.php

Thermohaline Structure of Water Above the Microbial Mats Site at Sevastopol Marine Area

A. A. Budnikov[1(✉)], I. N. Ivanova[1], T. V. Malakhova[2], and L. V. Malakhova[2]

[1] Faculty of Physics, M.V. Lomonosov Moscow State University, Moscow, Russia
aa.budnikov@physics.msu.ru, ivair@yandex.ru
[2] The A.O. Kovalevsky Institute of Marine Biological Research of RAS, Sevastopol, Russia
t.malakhova@imbr-ras.ru

Abstract. The paper studies the vertical profiles of the aquatic environment main parameters in the water column above the microbial community in the Sevastopol marine area. The profiles of temperature, salinity, turbidity, and dissolved oxygen were obtained directly above the microbial mats area. Presumably, the formation of abundant microbial mats and atypical profiles of salinity, turbidity and oxygen in the study area are caused by the same natural or anthropogenic factors.

Keywords: Microbial community · Dissolved oxygen · Sulfate reduction · Bubbly gas · Bottom sediments

1 Introduction

The one of the actual hydrobiological research topics is bacterial communities formation and their evolution on the seabed. The study of vital activity of microorganisms and their mutual influence to the environment is important problem, whose solution will allow to understand the formation mechanisms of these ecosystems.

The microbial mats are rather complex of microorganisms community, generally one or several types (of them) prevail. Depending on the microorganisms type and type of substrate the width of microbial film can vary from a few millimeters to a few centimeters [1–3]. A microbial fouling of surface sediments were found at several areas of the Black Sea: at the northwestern shelf near the lower edge of sulfide-enriched waters (110–200 m depth) and also in the sulfide zone of the Black Sea near the methane seep areas [4, 5]. In the last decade many microbial mats were detected near the Herakleian Peninsula (Crimean marine area) [6, 7]. Most often such fouling were located directly in the areas of methane gas emission, which are widely detected in coastal area of the Crimea peninsula (see Fig. 1) [8, 9].

V. I. Karev et al. (Eds.): *Physical and Mathematical Modeling of Earth and Environment Processes (2018)*, SPEES, pp. 417–423, 2019.
https://doi.org/10.1007/978-3-030-11533-3_41

The microbial mats were detected both on sandy sediments and in rocky recesses filled with detrital sediments. Microbial communities looked like flaky and finely dispersed formations or like filamentous fouling (see Fig. 2). The recent studies have shown that mats are usually precisely localized in space and, at the same time, have a seasonal character [7, 9]. For example, in Chersonesus Bay, the first signs of bacterial biofilms appearance were detected on the sand in July, and the maximum coverage was noted in September–October.

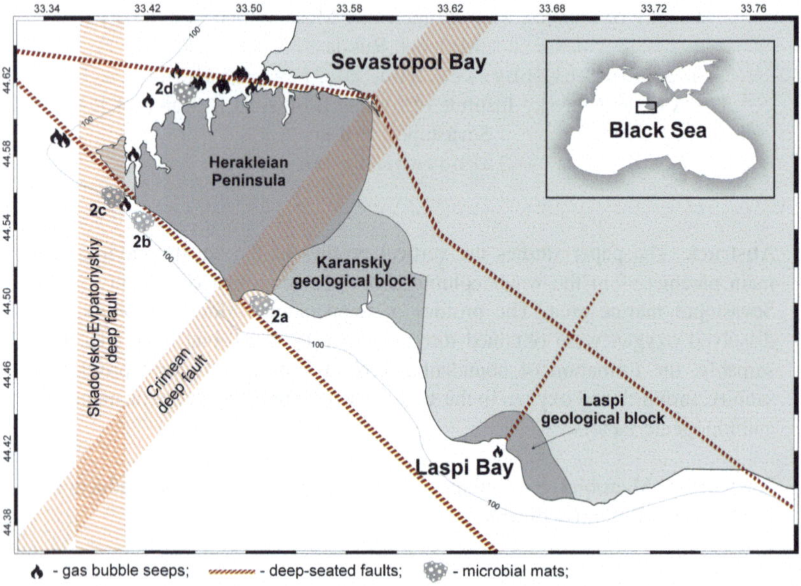

Fig. 1. The schematic map of research area. The deep faults, methane gas seeps and detected microbial mats are mapped. The structural and tectonic elements presented according to Shestopalov *et al.* [10] and Ivanov *et al.* [11].

Microbiological studies have shown that mats in different areas are represented by various communities [6, 7]. This fact can be caused either by different drivers of their expansion or by microbial succession. (Succession is the process of change in the species structure of a biological community over time as a result of the natural factors influence).

For the first time in the marine open area of Sevastopol, abundant white microbial mats were found in July 2018 at a distance of 100 m from the coastal strip, near Streletskaya bay (see Fig. 3). The total estimated area of the detected mats was several tens of square meters.

Fig. 2. Underwater photograph of a bacterial mats in the Sevastopol marine area (detected in 2017–2018). **a**—hollow contained detritus sediments upper layer being covered with white bacterial biofilm (Mramornaya Bay, Pimenov 2018); **b**—filamentous fouling on living thalli of algae; **c**—flaky microbial mat on sandy sediments (Chersonesus Bay, Bryukhanov 2018); **d**—free diver collects upper layer samples of microbial mat, discovered near Streletskaya Bay in July 2018

The purpose of this work was to determine in situ vertical profiles of main hydrological parameters of the water column above the microbial community: temperature, salinity, turbidity, dissolved oxygen.

Fig. 3. Underwater photograph of large-scale microbial mats, discovered near Streletskaya Bay in July 2018.

2 Object and Methods

Our study object was the microbial mats in the marine area of Sevastopol (44.611144 N, 33.461745 E) were found in July 2018. The white massive mats, which thickness reached to 2 cm, were located at 8 m water depth. The microbial community area was several tens of square meters. The underlying layer consisted of fine sand and a thin layer of finely dispersed dark gray silt. The bottom sediments of the adjacent areas were represented by coarse sand, pebbles, stones, overgrown with macro algae, mainly Cystoseira sp. On July 22 2018, the vertical profiles of the main hydrological parameters was determined in the water column above the microbial community from the board of a small-sized vessel.

The measurements were carried out using a RCM 9 LW multiparameter platform (Aanderaa) in the continuously operating mode. The temperature T, dissolved oxygen O_2, conductivity C and turbidity T_U were measured from the water surface to the bottom. The measurement accuracy was ±0.05 °C, 8 μM, 0.02 mS/cm and 0.4 NTU, respectively.

3 Received Data

The vertical profiles of the temperature T and turbidity TU in the water column directly above the microbial community are presented in Fig. 4a. The measurements showed that the water temperature from the surface layer to 3 m varied in a narrow range from 24.5 to 23.5 °C. Below the 3 m water depth a thermocline was observed: from 3 to 6 m, the temperature declined rapidly of more than 8 °C with a gradient of 2.9 °C/m. Near the bottom, in a layer from 7 to 8 m, the temperature changed insignificantly from 14.7 to 13.2 °C (Fig. 4a). Such a thermostructure in marine water is usually formed during the warm period of the year. The vertical stratification due to the presence of a seasonal thermocline determines the two-layer water structure and weak bottom layer ventilation under high surface heating. This, as a rule, leads to intensive depletion of dissolved oxygen in the water bottom layer due to various chemical and biological processes.

The turbidity graph showed a slight maximum at a depth of 5 m (see Fig. 4a).

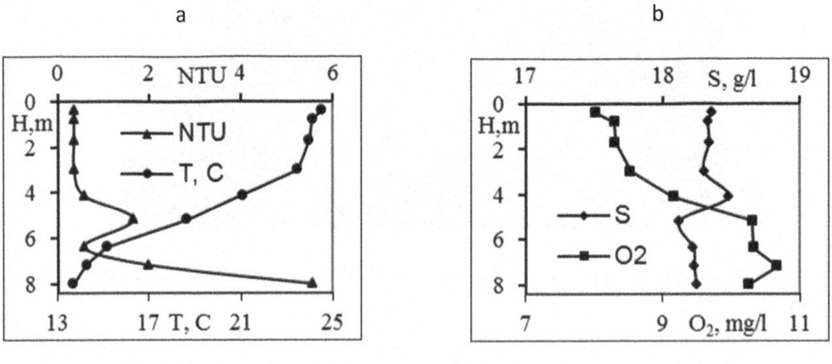

Fig. 4. The vertical profiles **a** of the temperature T (-●-) and turbidity T_U (-▲-), **b** salinity S (-♦-) and dissolved oxygen O_2 (-■-) in the water column directly above the microbial community (The marine open area of Sevastopol near Streletskaya Bay).

Besides, the vertical profiles of salinity S and dissolved oxygen O_2 in the water column directly above the microbial community are determined (Fig. 4b). The water salinity decreased with depth, its minimum value was obtained at a 5 m water depth, deeper it was slightly increased, but the values were less compared to the surface values (Fig. 4b). The oxygen content, as can be seen, increased with depth and was equal to 10.2 mg/l at the bottom. The bottom layers of water were generally less saline and more saturated with oxygen than near-surface ones.

In Fig. 5 we present also vertical profiles of the temperature, turbidity, water salinity and dissolved oxygen at station without microbial mats to compare thermohaline structure of water. The data were obtained in the Sevastopol Bay, July 2017 using the same instruments. The water temperature was almost evenly distributed in the column slightly decreasing from 28.3 °C at the surface to 24.3 °C near the bottom. Explicit thermocline was absent.

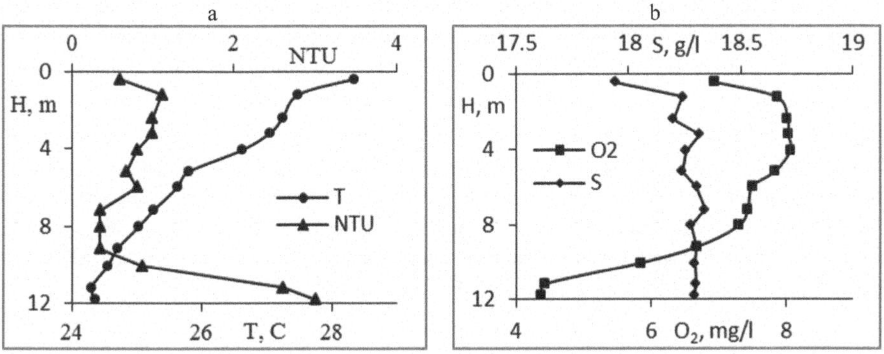

Fig. 5. The vertical profiles **a** of the temperature T (-●-) and turbidity T_U (-▲-), **b** salinity S (-♦-) and dissolved oxygen O_2 (-■-). The data obtained in the Sevastopol Bay, July 2017.

It can be seen that the salinity content from the water surface to the bottom is almost constant, the oxygen content near the bottom decreased to 4 mg/l, turbidity almost evenly increases from surface to bottom and has a value of 3 NTU near the bottom. The obtained thermohaline structure is rather typical for marine bays in the summer time and greatly differs from one obtained above the microbial mats in July 2018.

4 Discussion

The most common cause of the formation of fields of microbial mats in marine waters is the flow of reduced sulfur migrating to the surface of the sediments. At the water-bottom interface the sulfide oxidizing microbial communities can grow. The sulfide intake can be due to the metabolism of microorganisms inhabiting the upper layers of bottom sediments or to the discharge of groundwater enriched with reduced sulfur.

The study area where abundant microbial mats were discovered is located in the northwest of the Herakleian peninsula. The bottom relief of this area was flat and did not contribute to the local accumulation of detritus sediments. There were no any depression or rock cracks at the bottom of microbial mat location found, such as, for example, in the Mramornaya Bay. In this regard, a possible cause of the microbial mats formation here could be the local sulfide-enriched groundwater unloading—a phenomenon that has been observed previously on the shelf of the Black Sea and in other littoral zones of marine basins [12]. It is also can be cause the local dumping of organic waste of anthropogenic origin.

Increasing of dissolved oxygen concentration in the water column from the surface to the bottom layers is an atypical distribution of this parameter especially in the shallow coastal areas. Bottom communities of organic matter-rich sediments at this areas intensively consume oxygen, and it is also spent on the oxidation of hydrogen sulfide released by sulfur cycle bacteria during their metabolism. The consequence is a decrease in oxygen concentration with water depth up to anoxic conditions in the bottom layers. Taking into account the fact that most of the identified microbial mats communities are either obligate anaerobes or capable of growing under microaerophilic conditions, the presence of microbial communities we founded in such an oxidized environment indicates their great tolerance to high oxygen concentrations.

The resulting vertical distribution of salinity and oxygen in the water column above the mat indicates the correlation of these parameters. Given the openness of the coastline and the absence of stagnant zones in the study site, the revealed difference in the thermohaline structure (reduced salinity, increased oxygen content and a significant increase in turbidity) in the bottom layer and the connection of such a distribution of water parameters with the presence of bacterial mats requires further study. Presumably the reasons, which caused mats formation, also were influenced the atypical distribution of salinity, turbidity and oxygen in the study area.

5 Conclusion

1. Vertical distributions of hydrological parameters such as temperature, salinity, turbidity, dissolved oxygen in the water column above the microbial mat in the Sevastopol marine area were obtained.
2. Presumably formation of microbial mats and the atypical distribution of salinity, turbidity and oxygen in the study area caused by the one factor which could be whether natural or anthropogenic.

Acknowledgments. The work was carried out as part of the student practice of the Physics Faculty of the M.V. Lomonosov Moscow State University together with O. Kovalevsky Institute of Marine Biological Research of RAS on the topic of the state task IMBR of RAS "Molismological and biogeochemical principles of the homeostasis of marine ecosystems", state number registration AAAA-A18-118020890090-2.

References

1. Gromov, B.V.: Cyanobacteria in the biosphere. Soros Educ. J. **9**, 33–39 (1996)
2. Karatan, E., Watnick, P.: Signals, regulatory networks, and materials that build and break bacterial biofilms. Microbil. Mol. Biol. Rev. **73**, 310–347 (2009)

3. Jessen, G.L., et al.: Distribution and composition of thiotrophic mats in the hypoxic zone of the Black Sea (150–170 m water depth, Crimea margin). Front. Microbiol. **7**, 1011–1024 (2016)
4. Pimenov, N.V., et al.: Bacterial mats on coral-like structures at methane seeps in the Black Sea. Microbiology **66**(3), 354–360 (1997)
5. Treude, T., et al.: Consumption of methane and CO_2 by Methanotrophic microbial mats from gas seeps of the anoxic Black Sea. Appl. Environ. Microbiol. **73**, 2271–2283 (2007)
6. Bryukhanov, A.L., et al.: Microbiology **87**(3), 372–381 (2018)
7. Pimenov, N.V., et al.: Structure of microbial mats in the Mramornaya Bay (Crimea) coastal areas. Microbiology **87**(5), 681–691 (2018)
8. Egorov, V.N., et al.: Biogeochemical characteristics of methane distribution in the water and bottom sediments at the sites of jet gas emission in Sevastopol bays. Mor. Ekol. Zh. **11**(3), 41–52 (2012)
9. Malakhova, T.V., et al.: Microbial processes and genesis of methane gas jets in the coastal areas of the Crimean Peninsula. Microbiology **84**(6), 838–845 (2015)
10. Shestopalov, V.M., et al.: Modeling of ground_water resources in the southwestern part of mountaneous Crimea. Geol. Min. Res. Mir. Okean **4**, 5–28 (2008)
11. Ivanov, V.E., Lomakin, I.E., Topolyuk, A.S.: Geol. Min. Res. Mir. Okean **4**, 27–39 (2009)
12. Lein, A.Yu., et al.: Sulfide structures from the Black Sea bottom. Doklady Akad. Nauk **340**(5), 676–680 (1995)

Deformation and Destruction of Rocks on the True Triaxial Loading System with Continuous Acoustic Emission Registration

Shevtsov Nikolai[1(✉)], Zaitsev Alexey[2], and Panteleev Ivan[3]

[1] Institute for Problem in Mechanics of the Russian Academy of Sciences, Moscow, Russia
red3991@ya.ru

[2] Perm National Research Polytechnic University, Perm, Russia
a-zaitsev@mail.ru

[3] Institute of Continuous Media Mechanics UB RAS, Perm, Russia
pia@icmm.ru

Abstract. The necessity for involvement in the development of hydrocarbon fields with hard to recover reserves lying in complex geological conditions is the reason for the increased attention of producing and service companies to research in the fields of science related to the study of various properties of rocks. The paper presents experimental data on the measurement of acoustic emission simultaneously with laboratory modeling of the stress-strain state of rocks with unique experimental station test system for independent triaxial compression TSITC. The aim of the study was establishing the relationship between acoustic emission parameters and the stress-strain state of rocks when elastic deformation proceeds into plastic. In the course of biaxial experiments, it was demonstrated that the acoustic emission data obtained from sensors located on the sample free surface and compressive assembly of TSITC have the same patterns. This made it possible to carry out further triaxial experiments in which all acoustic emission sensors were located on the apparatus compressive assembly. The results of the experiments showed the presence of a legible connection between the acoustic emission total number of pulses and the deformation values of rock samples. It was also shown that the acoustic emission measurements allow predicting the experimental sample transition from elastic to plastic deformation (the beginning of sample destruction). The analysis of the acoustic emission pulses total number dependence on time during the experiment with the loading of rock sample with a uniform deformation increase demonstrated the stages of loading, when the greatest number of micro and macro defects in the sample occurs.

Keywords: Geomechanical model parameters · True triaxial loading · Elastic characteristics of rocks · Acoustic emission · Stress-strain state

V. I. Karev et al. (Eds.): *Physical and Mathematical Modeling of Earth and Environment Processes (2018)*, SPEES, pp. 424–432, 2019.
https://doi.org/10.1007/978-3-030-11533-3_42

1 Introduction

At the present time, the main energy resource is hydrocarbon raw materials, which extraction, as soon as easily accessible fields are exhausted, becomes a technologically difficult task and requires a scientific approach to its solution. For effective fields development, it is necessary to create their geomechanical model [1]. Wherefore determine a number of key characteristics of rocks that compose the reservoir of the field and mother rocks is required. First of all, these are data on the lithology, magnitudes and directions of the main initial stresses in the reservoir (natural stress state). Designing a field development demands the static elastic characteristics of rocks, such as Young's moduli, Poisson's ratios, shear moduli, and strength characteristics. This allows choosing the optimal mode of drilling and operation of wells, which minimizes the risks of sand production, wellbore collapse, and in some cases, allows increasing the production rate. Geophysical and laboratory methods both using for desired quantities measurements [2].

The main condition for the successful development of an oilfield geomechanical model is the correct determination of static elastic moduli of rocks. The principal problem of modern geophysical methods of deposits exploration is that they allow only dynamic modules to be determined, which can only be linked with static ones using empirical dependencies with large errors and a very limited application area. Static elastic moduli measurement is possible only in laboratory conditions. In addition to the characteristics mentioned above, developing a geomechanical model, strength characteristics of rocks composing the reservoir data are often required, which are also can be determined only by laboratory tests. From the science point of view, not only the strength characteristics are interesting, but also the mechanisms of rock destruction.

One of the methods for investigating these processes is to study the acoustic emission (AE) signals generated during irreversible or partially reversible changes in the structure of rocks. It is observed that the more heterogeneous materials have the more pronounced acoustic emission at its destruction [3]. Measurements of acoustic emission during experimental of the rock sample stress-strain state modeling will make it possible to specify data on the fracture mechanisms and the transition of elastic deformation to plastic in the rocks testing process [4]. From this point of view, the rock samples composing the reservoirs of oil and gas fields seems to be ideal candidates for this kind of research, since they often have a granular or layered structure and pronounced elastic characteristics anisotropy or transverse isotropy. At the moment, there is no established universal relationship between the acoustic emission and the stress-strain state of rocks, there are only experimental studies for specific samples. Also, the study of this type samples will complement the available results of AE studies during testing of the various rocks samples elastic-strength properties, for example in [5, 6].

2 Materials and Methods

Such an experiment was implemented in the Geomechanics Laboratory of IPMech RAS together with the Institute of Continuous Media Mechanics of the Ural Branch of the Russian Academy of Sciences and the Perm National Research

Polytechnic University on a unique test system for independent triaxial compression TSITC [7]. During this experiment, in parallel with deformations of the experimental sample measurements, acoustic emission data were received. This study was aimed at establishing the relationship between the stress-strain state of rock samples and acoustic emission parameters.

A true three-axis loading setup consists of three hydraulic presses, each transfers a force (up to 50 tons) using a pressure plate to an experimental sample of a cubic shape with a 40 mm rib length. The kinematic diagram of the pressure plates provides independent loading of the sample along each of the three axes according to any given program. Simultaneously with the loading of the experimental sample, its strain is measured on each axis. Sample loading control can be done both by force and by displacement. In the first case, the loading program is defined as the dependence of force on time along each of the axes, while in the second case, the program is the dependence of displacement on time for each axis.

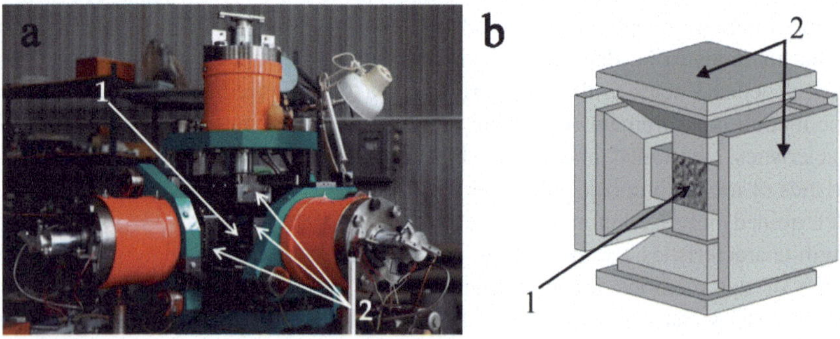

Fig. 1. Photograph of the TSITC setup (a) and schematic diagram of its compressive assembly (b). Experimental sample - 1, pressure plates - 2.

For conducting the presented experiments, samples of sandstone from the Perm Krai of the Russian Federation, Verkhnekamskiy (Bereznikovsko-Solikamsky) region were used. The pressure plates and the free face of the sample (Fig. 1) were fixed with 2SMEG-P broadband acoustic emission transducers (Deci, USA) operating in the 200–2000 kHz range and M-31 miniature transducers (Fujicera, Japan) with flat amplitude-frequency characteristic in the range of 300–800 kHz, respectively. The data were received and processed by the multichannel high-performance AMSY-6 (Vallen) system [8, 9].

The results presented in this paper were obtained during tests carried out according to the loading programs of an experimental sample corresponding to a generalized shift and uniaxial compression with lateral expansion. The program corresponding to the generalized shift consists of two stages: at the first stage, the sample is uniformly loaded along both axes, and at the second along one of the axes the load continues at the same speed as at the first stage, and the other is unloaded with the same speed. The program corresponding to uniaxial compression with lateral expansion is also implemented in two stages: first, the sample is uniformly compressed along both axes, after which the force value remains constant along one of the axes, while the sample continues to be loaded along the other axis, but loading is controlled by displacement.

3 Results

In the course of biaxial experiments, it was found that the acoustic emission data obtained from sensors mounted on the free surface of the sample and from the sensor attached to the TSITC pressure plate have qualitatively equivalent patterns of change. Later, this became a reason for a series of triaxial experiments with all three AE sensors were attached to TSITC pressure plates (Fig. 2).

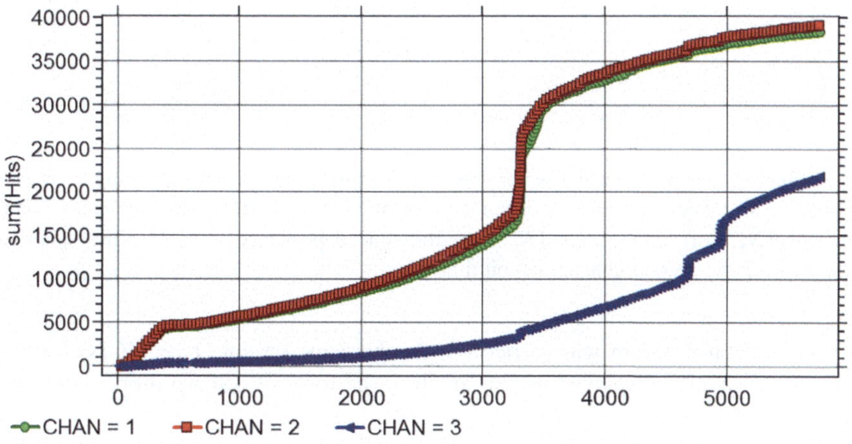

Fig. 2. Dependence of the AE impulses total number on time in three channels, where the third channel corresponds to a sensor located on the TSITC pressure plate [8].

This result was obtained during series of experiments carried out by loading a sample controlled by displacement, which means that in this experiment stress was studied value. At the beginning of the experiment, a uniform loading of the rock sample to a stress level of 11 MPa was carried out on both axes. Further along one of the axes (the second axis of the installation) a uniform increase in the value of displacement was ensured. And on the other, the force was constantly maintained at 11 MPa.

On the graph of the stress dependences along the second axis and the total number of pulses from the sensors located on the sample free surface (N_1^{total} and N_2^{total}) from time (Fig. 3, a, b), the moment of transition from elastic to plastic ($T \approx 2400\,\mathrm{s}$), which is visually reflected by a sudden decrease in the stress value, can be noticed. At this point, a uniform increase in the displacement value was manually suspended to avoid destruction of the sample, and resumed after 40 s. Later, another manual suspension of

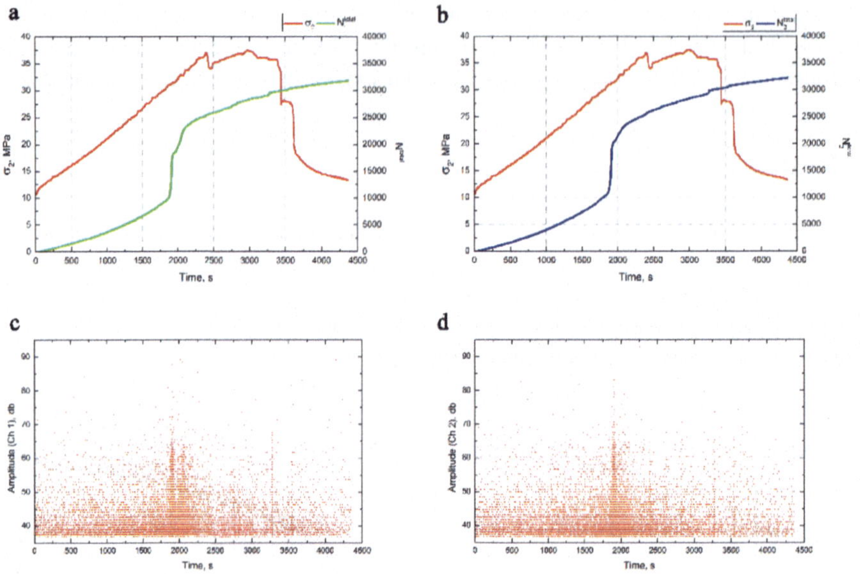

Fig. 3. Biaxial sandstone testing according to a loading program corresponding to uniaxial compression with lateral expansion: dependence of stress σ_2 and AE pulses total number on the first channel N_1^{total} (a) and the second N_2^{total} (b) from time; dependence of the AE pulses amplitude on the first (c) and second channel (d) on time.

the experimental program was carried out for the same reason ($T \approx 3360\,\mathrm{s}$) for 50 s. This moment is also displayed on the graph in the form of a sharp drop in the stress value.

A comparison of the results of acoustic emission and stress measurements showed that the maximum number of AE pulses recorded per second was revealed even before signs of destruction appeared on the stress-time dependence curve. This suggests that the critical accumulation of defects in the sample occurred much earlier than the elastic deformation was replaced by plastic. The general view of the total number of AE pulses dependence on both channels perfectly demonstrates that the most significant number of pulses occurs during a time in the vicinity of 1900 s, and neither has been registered before nor after such a number of pulses. Based on this observation, it was concluded that critical defects were formed at the point of time mentioned above. This is evidenced by the dependence of the AE pulses amplitude on time—it clearly shows that most of the pulses with amplitude above the average are also recorded in the time domain around the value of 1900 s. And in the time interval when the experimental sample was deformed plastically, the density of the pulses number is significantly reduced, although their average amplitude remains.

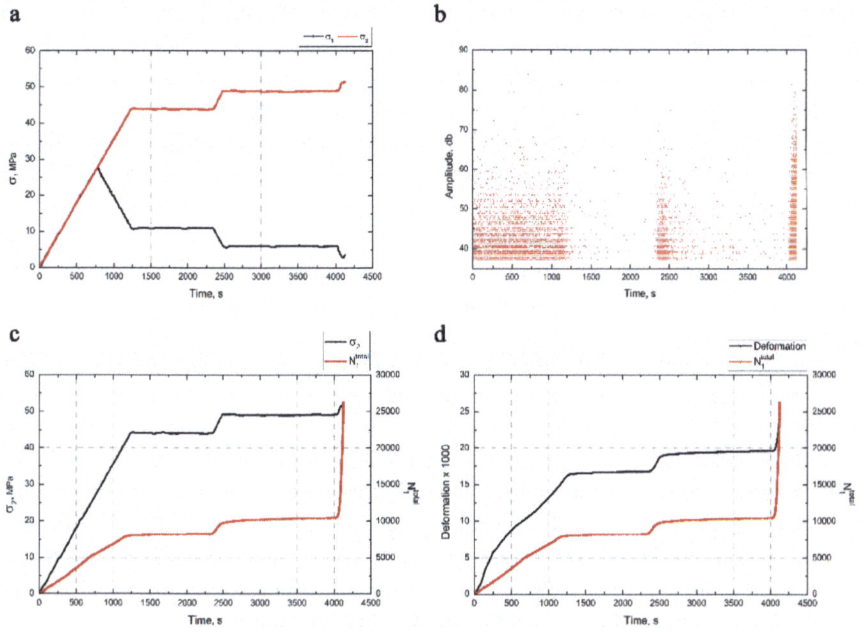

Fig. 4. Biaxial loading test according to generalized shift: loading program corresponding to the generalized shear (a), the dependence of the AE pulses amplitude on time (b), the dependence of stress σ_2 and the AE pulses total number N_1^{total} on time (c), the dependence of the strain on the second axis and the AE pulses total number N_1^{total} to time (d).

In contrast to the series of experiments described above, the second series was carried out according to a program corresponding to a generalized shift (Fig. 4a), controlled by force. The purpose of this experiment was to establish the relationship between the stress-strain state of the rock and the intensity of the acoustic emission pulses.

The first stage of the experiment consisted in the uniform compression of the sample along both axes to a stress value of 27.5 MPa. At the second stage, a uniform loading of the sample continued along one of the axes (the second axis of the installation), while a similar unloading occurred along the other axis (the first axis of the installation).

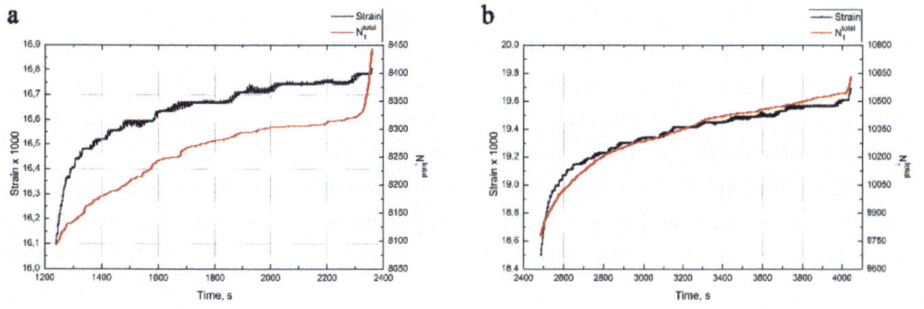

Fig. 5. Dependence of strain along the second axis and the total number of AE pulses N_1^{total} on time with the creep effect for stress levels $\sigma_2 = 44$ MPa (a) and $\sigma_2 = 49$ MPa (b).

When the stress on the second axis reached 44 MPa, the loading program was manually suspended, but the growth (and on the other axis a drop) of the strain did not stop. Then the experiment was continued, and a similar stop was made at a stress level of 49 MPa. Soon after the loading was resumed, the sample collapsed.

The effect, when the strain continues to grow at a constant stress, known as creep, has been studied in terms of acoustic emission. For each of the stress levels which the loading program was stopped, a graph of the strain and the AE impulses total value from time (Fig. 5) was plotted. In both cases, the curves of the deformation and the intensity of acoustic emission from time had similar patterns, which indicate a connection between the deformation of the rock sample and the emission of acoustic signals.

It should be noted that the intensity of AE activity, and the amplitude of acoustic pulses, has a maximum value at the moment of sample failure, which can be seen on the graph of the acoustic emission amplitude dependence on time (Fig. 4b). While during the observation of the creep effect, the intensity of the AE activity is minimal.

A similar result can be seen in the plots of stress and strain together with the total number of acoustic emission pulses as a function of time (Fig. 4c, d). The AE curve has the same shape as the strain and stress curves, as well as characteristic points identically located. This again confirms the assumption of the qualitative dependence of the flow of acoustic emission on the stress-strain state of a rock sample.

4 Discussion and Conclusions

The results of an experiments series showed the presence of a relationship between the stress-strain state of the sample and the intensity of the acoustic emission streaming. This is clearly seen in a series of experiments carried out with a program corresponding to a generalized shift, where the qualitative correspondence between the total number of AE pulses curve and the stress and strain curves takes place. There was a similar situation in the study of the creep effect, where the curves describing the change in the stress-strain state and the acoustic emission streaming had similar patterns. At the same time, the maximum intensity of the acoustic emission pulses appearance for these experiments was recorded at the end of the loading, simultaneously with the onset of the rock sample destruction.

In the course of an experiment series on the loading program corresponding to uniaxial compression with lateral expansion, which was carried out with displacement control, even at the stage of elastic deformation, there was a sharp increase in the number of AE pulses per unit time. These data demonstrate that the largest number of micro defects (which formation creates acoustic emission pulses) occurred in the sample significantly earlier than the process of destruction began. This is confirmed by the analysis of the AE amplitude dependences on time for all channels.

This fundamental difference in the behavior of samples that are so close in their properties can be explained by the fact that they were tested using completely different methods. If loading using “generalized shift” was implemented with control by force, and when approaching critical values, the experimental program was stopped for a considerable time to observe the creep effect, which additionally strengthened the rock, then loading according to the program "uniaxial compression with lateral expansion" was controlled by displacement. This test method was the reason that the occurrence of a large number of micro defects over a short period of time relative to the whole experiment did not lead the sample to abrupt destruction. This made it possible to continue the experiment and obtain acoustic emission data during plastic and behind limit deformation of the sample, which occurred for a considerable time, compared with experiments where the loading was controlled by force.

An analysis of the acoustic emission pulses amplitude dependences on time showed that in both considered experiments series, the highest density of AE signals streaming and the highest average amplitude were recorded at the moment of a sharp increase in the number of defects: for a generalized shift, this was destruction of the rock at the end of the experiment, and for uniaxial compression it occurred during elastic deformation. In both cases, the minimum density of pulses per unit of time was observed during stopping the loading of the sample simultaneously with the creep effect.

In conclusion, it can be said that the presented experiments revealed a qualitative connection between the stress-strain state of rock samples and acoustic emission parameters for several special cases. For two different methods of creating and controlling the stress-strain state in an experimental sample, analysis of AE data revealed loading stages, with an abrupt increase in the number of defects. All obtained results indicate the relevance of further research development on the presented theme.

Acknowledgments. The present work was supported by the Russian State Assignment under contract No. AAAA-A17-117021310371-9.

References

1. Nick, K., Xing, Z.: 3D reservoir geomechanical modeling in oil/gas field production. In: SPE Saudi Arabia Section Technical Symposium, 9–11 May, Al-Khobar, Saudi Arabia (2009)
2. Karev, V.I., Shevtsov, N.I.: Identification of geomechanical model parameters using geophysical and laboratory methods. Processes in Geomedia № 1(10), pp. 427–434 (2017). (in Russian)
3. Lavrov, A.V., Shkuratnik, V.L.: Acoustic emission during deformation and destruction of rocks. Acoust. J. **51**, 6–18 (2005). (in Russian)
4. Bashkov, O.V., Semashko, N.A.: Acoustic emission with changing deformation mechanisms of plastic structural materials. Phys. Mesomech. **6**, 59–62 (2004). (in Russian)
5. Shkuratnik, V.L., Filimonov, Yu.L.: On the interrelation of the acoustic emission parameters with the physicomechanical properties and the processes of salt rocks destruction. Interexpo Geo Siberia T.2 (2005). (in Russian)
6. Damaskinskaya, E., Frolov, D., Gafurova, D., Korost, D., Panteleev, I.: Criterion for fracture transition to critical stage. Interpretation **5**(4), SP1–SP8 (2017)
7. Karev, V., Kovalenko, Yu.: Triaxial loading system as a tool for solving problems of oil and gas production. In: The 12th International Congress on Rock Mechanics (Beijing, October 18 – 21, 2011) (2011)

8. Kovalenko, Y., et al.: Inelastic deformation, strain-softening and localized failure in sandstone media under triaxial quasistatic loading. In: Report on the 22nd European Conference on Fracture - ECF22, 26–31, August 2018
9. Damaskinskaya, E.E., Hilarov, V.L., Panteleev, I.A., Gafurova, D.R., Frolov, D.I.: Statistical regularities of formation of a main crack in a structurally inhomogeneous material under various deformation condition. Phys. Solid State **60**(9), 1821–1826 (2018)

Evolution of the Vorticity Contour in Vortex Energy Sources

Alexander Alekseyevich Solovyev[1]
and Dmitry Alexandrovich Solovyev[2(✉)]

[1] Faculty of Geography, M.V. Lomonosov Moscow State University, 1, Lenin Hills, 119991 Moscow, Russia
[2] P.P. Shirshov Institute of Oceanology RAS, 36 Nakhimovsky avenue, 117858 Moscow, Russia
solovev@ocean.ru

Abstract. This article utilizes the physical modeling of a differentially rotating shallow fluid to understand the role of restructuring the vorticity contour in realizing the effects of the evolution of intense atmospheric vortex formations. Also paper analyzes the results of laboratory experiments with a rotating fluid model with the conditions for the formation of the polygonal structure of the vorticity contour on the central part and transitions between the modes of the vortex structures at the periphery. Taking into account the diagram of eddy currents stability established for the model where various cases of instability development are visually indicated with changes in the barotropic velocity shift and angular velocity of the entire system, the selection rules for transition times between central polygonal structures and peripheral motion modes with different numbers of vortices are defined. Assuming the analogy and general principles of formation found during radio-location Doppler sounding of polygonal structures of the eye of hurricanes and accompanying peripheral vortices with a corresponding index of symmetry, illustrative calculations of the characteristic times of structural transformations of the configuration of the hurricane eye contour are presented until just before the destruction. Obtained results are important for the organization of the research and monitoring of tropical cyclones including space-based methods.

Keywords: Tropical cyclone · Contour of vorticity · Physical modeling · Polygonal eye · Tropical cyclone genesis · Hurricane monitoring indicator

1 Introduction

Intensive atmospheric vortices in the process of their evolution experience a structural reorganization of internal motions. The most distinct structural changes in the momentum and energy transfer processes are reflected in the transformation of motions in the vicinity of the region of maximum winds [1].

The transformations in the evolution of polygonal configurations of the central part of the trophic cyclones of the so-called wall of the eye, discovered in Doppler radar studies, and the corresponding set of mesoscale spiral vortex structures within the rain

V. I. Karev et al. (Eds.): *Physical and Mathematical Modeling of Earth and Environment Processes (2018)*, SPEES, pp. 433–440, 2019.
https://doi.org/10.1007/978-3-030-11533-3_43

zone stimulated a spectrum of numerous studies of the mechanisms of occurrence, evolution and destruction of hurricanes [2–6]. Attempts have been made to propose a method for diagnosing cyclogenesis at an early stage of nucleation using the merger index of mesoscale mid-level vortices [7, 8].

The use of the fusion index in mesoscale vortices as a precursor of the formation of hurricanes proved to be a rather suitable means of detecting cyclogenesis at an early stage. The later stages of the evolution of hurricanes up to this day remain a scientific challenge for forecasting and serve as an occasion to consider the processes of reformatting the eye and transitions between different modes of vortex structures behind the hurricanes' eye as a likely general process of the development of intense atmospheric vortices.

This paper proposes an idea to use the method of physical modeling to determine the predictor criteria for the evolution of the vorticity contour of intensive energy sources, such as tropical cyclones. This method despite the absence and impossibility of complete copying of the model from the original allows with the simulation of the main elements of the processes of momentum and energy transfer when simulating to achieve fairly accurate results for the forecast of the development of hurricanes. In this regard, as applied to the search for indicators of the intensification of tropical cyclones, for generalizing consideration, the previously obtained results of laboratory experiments with the model of a differentially rotating shallow liquid were used [9, 10].

In experiments with differentially rotating shallow water, polygonal configurations were observed in the central inner region of the flow [11]. This area corresponded to the processes in the zone of the hurricane's eye. With the change in the magnitude of the horizontal shear rate, the borders of the contour with constant vorticity changed. In the beginning, it was a circle, then an ellipse, triangle, and so on up to a hexagon. Along with the steady state of the polygonal contour, there was a picture of a periodic flip between motion modes with different values of the symmetry index or the mode number of peripheral vortices. Perturbations of the contour, along with the boundary of which the vorticity is constant, are interpreted as the result of the interference of waves developing in a velocity shift [12]. Similar situations are recorded in typhoons. For example, in the Winnie typhoon, radar sensing detected bounces between the four and six-angle "eye" configurations [13]. In model vortices and typhoons, as analysis shows, we are dealing with the same phenomenon, namely, with the inter-transformations between the modes of inertial oscillations [10, 14].

2 Analysis of Wave Contour Restructuring Processes

The results of observations of the temporal dynamics in vortex motions in a physical model of differentially rotating shallow water provide the basis for evaluating the natural transformations of the eye of an intense atmospheric vortex to use characteristic times that are spent on restructuring the vorticity contour. We will designate them as a symbol $\Delta\tau_{ij}$. The first index characterizes the oscillation frequency along the contour. The index value $i = 1$ will denote the state of the contour with low-frequency, long-wave oscillations, which modulated the movement along the contour, and the index

$i = 0$ - high-frequency, short-wave. The second index in the expression for the time of restructuring the contour of the "eye" means the index of symmetry of the polygonal configuration of the contour. It can vary from $j = 0$ (round "eye") to $j = 6$ (hexagonal eye).

Analysis of field observations and laboratory experiments leads to the following conclusion. The times $\Delta\tau_{ij}$ that are spent by differentially rotating shallow water to rebuild the shape of the wave contour are varying linearly with time. There are two types of eye states. They are conventionally designated by us in the form of a system of upper levels

$$\Delta\tau_{1j} = a_{0j} - \alpha t \tag{1}$$

and lower level systems

$$\Delta\tau_{0j} = a_{0j} - \beta t. \tag{2}$$

Here α, β - respectively, the parameters characterizing the rate of adjustment of the contour at the upper and lower levels, a_{0j}, a_{1j} - are the characteristic lifetimes of states with a different configuration of the wave contour.

The flip times at each level corresponding to the energy states of the rotating flow for movements of different scale. The upper row characterizes the amplitude of low-frequency averaged motions of the main flow. The bottom row corresponds to motion states with high-frequency components. In the process of evolution, a systematic transition of the energy of medium (low-frequency) movement into a pulsation (high-frequency) occurs. Then the energy goes partially into heat, and sometimes partially returns from pulsations to the average component. The transition from the upper to the lower levels determines the energy losses of the main flow and the selection of energy from this mode. In this case, a low-level mode is generated, during the development of which a part of the energy returns to the main current.

Consider the condition of transition from one mode to another in the mode of sequential oscillations on the example of typhoon Winnie. The Winnie typhoon in the system of the lower levels, two lines are distinguished $j = 4, j = 5$. For this lines

$$\Delta\tau_{04} = a_{04} + \beta t, \ \ \Delta\tau_{05} = a_{05} + \beta t \tag{3}$$

here $[\Delta\tau] = \text{min}$, $[a] = \text{min}$, $[t] = \text{min}$. Parameter values of a, β the can are found by analyzing the data in the article [13] by the method of least squares. In result: $a_{04} = 9{,}7$ min, $a_{05} = 9{,}4$ min, $\beta = 0{,}12$. Top levels for typhoon include one line with mode $j = 6$

$$\Delta\tau_{16} = a_{16} - \alpha t \tag{4}$$

And $a_{16} = 116\,\text{min}$, $\alpha = 0{,}2$.

Similar calculations were carried out with a laboratory model. It also confirms the linear law of type (1), (2). You can also talk about the system of upper and lower levels. The main difference between a typhoon and a model is in time scale values. The transition from full-scale to model values, as the comparison showed, should be carried out with the help of a large-scale time coefficient $\xi = 142$ min/s. This factor is obtained from the condition of the correspondence of the times determined by the intersection point of the corresponding lines of Eqs. (1), (2) for the model and the typhoon. Processing the experimental data for the model allowed us to formulate a rule of thumb by which the relative density of the level lines can be found

$$\frac{a_{1j}}{a_{1j+1}} = \frac{j+1}{j-1}. \tag{5}$$

In model experiments, for verification, calculations were made of the characteristic flip times between different states of a polygonal contour. In terms of tropical cyclones, equations of the type (1), (2) taking into account the coefficient ξ, are written as:

$$\Delta\tau_{12} = 77,2 - 0,098t$$

$$\Delta\tau_{13} = 74,3 - 0,098t$$

$$\Delta\tau_{04} = 20,4 + 0,056t.$$

Here all values are expressed in minutes. Table 1 presents a comparison of the calculated values of the flipping times from one contour configuration to another with certain values according to the experimental filming of the model's polygonal "eye" transformations. As can be seen, the calculated values are in good agreement with the experimentally determined values.

Table 1. The characteristic times of the vorticity contour reconstruction in the model system.

№	Mode number	$\Delta\tau$, s	t, s	$\Delta\tau$, min	t, min	$\Delta\tau$, min	t, min
		Experiment				Calculation	
1	3	0,37	0,38	53	53	53	53
	4	0,17	0,55	24	77	25	78
2	3	0,43	0,98	61	138	61	139
	4	0,25	1,23	36	174	30	170
3	3	0,38	1,60	53	227	53	223
	4	0,21	1,81	30	257	30	253
4	3	0,29	2,10	41	298	45	298
	4	0,29	2,39	41	339	40	338
5	3	0,26	2,65	37	376	38	376
	4	0,30	2,95	43	417	43	418
6	3	0,21	3,16	30	449	30	449
	2	0,32	3,48	45	494	47	495

Now let us consider the criteria that can be used to assess the necessary and sufficient conditions for the transfer between the polygonal "eye" configurations. As a result of the analysis of the experimental material obtained in the study of the model system, the following rules were established.

1. The transition from the upper levels to the lower ones is allowed provided the following:

$$\Delta\tau_{1j} > \Delta\tau_{0j} + \frac{\Delta\tau_{0j}}{1-\beta} \tag{6}$$

2. The transition from the lower levels to the upper levels is allowed provided the following:

$$\Delta\tau_{0j} < \Delta\tau_{1j} + \frac{\Delta\tau_{1j}}{1+\alpha} \tag{7}$$

The question of which line will be transferred from the upper levels to one of several lower levels, for which condition (6) is realized, must be solved separately by comparing their energy possibilities. If we compare the two lines j and $j+1$ the lower level, for which condition (6) is satisfied, then for this we should consider the inequality of the form:

$$\Delta\tau_{0j} \geq \frac{\Delta\tau_{0j+1}}{1-\beta} \tag{8}$$

If it is executed, then the transition to the line j rather than to the line $j+1$ is energetically more favourable. Similarly, when transferring from the lower levels to a series of upper lines is analyzed to determine which line j or the line $j-1$ transition will take, it is necessary to check the fulfilment of the next relation

$$\Delta\tau_{1j} \leq \frac{\Delta\tau_{1j-1}}{1+\alpha} \tag{9}$$

When the condition (9) is met, the transition occurs to the line $j-1$. In other cases, i.e. at

$$\Delta\tau_{1j} > \frac{\Delta\tau_{1j-1}}{1+\alpha}$$

the transition from lower levels to j the mode of the upper level. Conditions (8), (9) allow us to estimate the points in time and the critical values of the maximum permissible restructuring times of the typhoon "eye". We denote the limiting adjustment times for the mode j by the lower and upper levels, respectively: $\Delta\tau^*_{0j}$ and $\Delta\tau^*_{1j}$.

These times should be determined, as follows from the results of model experiments, the density of the nearest levels and their slope

$$\Delta\tau_{0j}^{*} = \frac{a_{0j} + a_{0j+1}}{\beta}; \ \Delta\tau_{1j}^{*} = \frac{a_{1j-1} + a_{1j}}{\alpha} \tag{10}$$

For Typhoon Winnie, the maximum allowable times for restructuring the polygonal eye of different degrees of symmetry along the lines of the lower level are presented in Table 2.

Table 2. Parameters for restructuring the contour of the eye for Winnie typhoon.

j	3	4	5	6	7
a_{0j}, min	16,4	9,7	6,0	3,8	2,6
$\Delta\tau_{0j}^{*}$, min	55,8	30,8	18,3	10,6	7,2

Figure 1 shows the data of actually observed values of periods $\Delta\tau$ of rearrangement of the vorticity contour of polygonal eye configurations in the Winnie typhoon for different points of time t of his life in comparison with the calculated lines of the corresponding levels of energy states of movements calculated in accordance with the proposed selection rules.

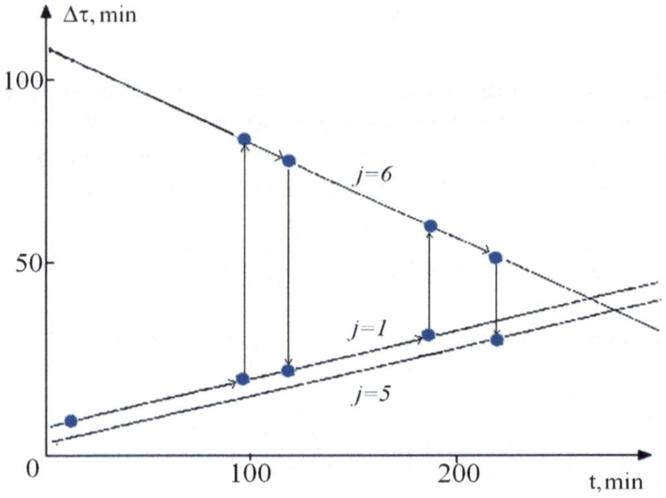

Fig. 1. Schematic representation of the dependence of the period of adjustment of the polygonal eye contour in Winnie's typhoon on the time of his life. Solid lines are calculated by formulas (1), (2). Circles are observational data [13]. The arrows indicate the direction of the structural adjustment of the vortex motions in the typhoon.

The results are in accordance with the observational data given in [2]. A general conclusion that should be made on the basis of the research conducted is as follows. The estimated algorithm for calculating the time of adjustment of the "eyes" of typhoons can be used for analytical verification in prognostic estimates of the evolution of various stages of development of hurricanes.

3 Conclusion

The submissions provide the basis for the following conclusion. As applied to the description of the processes of structural rearrangement of internal motions in tropical cyclones, we used the results of physical modeling of vortex formations in differentially rotating shallow water. Parallel changes in the polygonality of the vorticity contours in the model with the phenomenon of temporary transformation of the typhoon eye were made. Also, polygonal modifications of the eye are compared. In models and intense atmospheric vortices, the polygonality of the eye can be associated with the inter-conversion between the modes of gravitational-inertial oscillations. The characteristic lifetimes of states with different structures of the wave contour in the centre of the vortex and typhoon are found. Paper formulated the rules that determine the conditions for the transfer between the polygonal forms of the eye of the vortex. The proposed algorithm for calculating the restructuring times of the "eye" of intense atmospheric vortices can be used for analytical verification in prognostic estimates of the evolution of various stages of hurricane development.

Acknowledgment. This research work was carried out within the framework of the State task (№ 0149-2018-0001; № AAAA-A16-116032810088-8) and project № 14.W03.31.0006 of the Ministry of Science RF.

References

1. Makova, V.I., Pokazeev, K.V.: Calculation of wind and waves in tropical hurricanes. Meteorol. i Gidrol., no. 7. p. 9498 (1990)
2. Muramatsu, T.: The structure of polygonal eye of typhoon. J. Meteor. Jpn. **64**(6), 913–921 (1986)
3. Kossin, J.P., Schubert, W.H.: Mesovortices, polygonal flow patterns, and rapid pressure falls in hurricane-like vortices. J. Atmos. Sci. **58**(15), 2196–2209 (2001)
4. Lee, W., et al.: Doppler velocity signatures of idealized elliptical vortices. Terr. Atmos. Ocean. Sci. **17**(2), 429 (2006)
5. Martinez, Y., et al.: On the dynamics of concentric eyewall genesis: space–time empirical normal modes diagnosis. J. Atmos. Sci. **68**(3), 457–476 (2011)
6. Hendricks, E.A., et al.: Hurricane eyewall evolution in a forced shallow-water model. J. Atmos. Sci. **71**(5), 1623–1643 (2014)
7. Venkatesh, T.N., Mathew, J.: Prediction of tropical cyclone genesis using a vortex merger index. Geophys. Res. Lett. **31**(4) (2004)
8. Didlake Jr., A.C., Houze Jr., R.A.: Convective-scale variations in the inner-core rainbands of a tropical cyclone. J. Atmos. Sci. **70**(2), 504–523 (2013)

9. Inanov, V.N., Solovyev, A.A.: Laboratory modeling of elements of a tropical cyclone. In: Proceedings of the Fourth International Symposium on Tropical Meteorology, pp. 55–60. Gidrometeoizdat, Leningrad (1990)
10. Solovyev, A.A., Solovyev, D.A.: Evaluation of the temporal dynamics of oceanic eddies with initial peripheral rate shift. In: Karev, V., Klimov, D., Pokazeev, K. (eds.) PMMEEP 2017. SG, pp. 207–216. Springer, Cham (2018). https://doi.org/10.1007/978-3-319-77788-7_23
11. Solovyev, A.A., Solovyev, D.A.: The barotropicinstability of the oceanic jet currents. Hydrology **5**(5), 77 (2017)
12. Niino, H., Misawa, N.: An experimental and theoretical study of barotropic instability. J. Atmos. Sci. **41**(12), 1992–2011 (1984)
13. Hurricane Radar Data Sets. http://www.aoml.noaa.gov/hrd/data_sub/radar.html. Accessed 21 Aug 2018
14. Menelaou, K., Yau, M.K.: On the role of asymmetric convective bursts to the problem of hurricane intensification: radiation of vortex Rossby waves and wave–mean flow interactions. J. Atmos. Sci. **71**(6), 2057–2077 (2014)

Classifiers in Super-Resolution Problems

E. N. Terentiev[1(✉)] and N. E. Shilin-Terentiev[2]

[1] Department of Mathematical Modeling and Informatics, Faculty of Physics, M.V. Lomonosov Moscow State University, Moscow 119991, Russia
en.teren@physics.msu.ru
[2] Google, San Francisco, CA, USA
nikolay.terentyev@gmail.com

Abstract. Classification of varieties of super-resolution methods is based on the gap between the minimum value of the Modulation Transfer Function (MTF) M (O) (for discrete Apparatus Functions (AF) O) and zero. Classifiers underlie the concept of super-resolution. Conditionality is the main setting of AF O to increase resolution. The conditionality is numerically equal to the reciprocal of the minimum value |M(O)| or the magnitude of this gap.

Keywords: Super-resolution · Conditionality · Inevitability · Modulation transfer function · Convolution · Fourier transform

1 Introduction

An analogue of the conditionality parameter of the matrix is the conditionality parameter for a cyclic matrix or AF O.

The super-resolution problem begins with the construction of the set {O} of discrete AF O models using the continuous parO AF model with three parameters: the length of the definition domain Loc, the digitization step dx, and the conditionality parameter DIAP = 1/diap (from the word diapason) for values of MTF M(O) and M(R)).

Figure 1 shows examples of two digitization of a Gaussian with a length of the definition domain Loc = 6 with two different steps dx = 1 and dx = 1/2. The tasks of choosing a discrete model of AF O are considered on the set LO = {pO|[Loc, dx, DIAP]} reversible AF pR = pO − 1, with numerical inversion for MTF M(pR) = 1/M(pO).

If AF is DK - the Delta Kronecker symbol, then MTF (DK) == 1 is identically equal to one.

On the left graph below the unit to diap are the values of the MTF M(pO), above the unit to DIAP = 1/diap - the values M(pR) = 1/M(pO). The most conditioned AF DK, with DIAP = diap = 1. If DIAP increases, then conditionality falls.

The accuracy of the orthogonalization of two-dimensional Fourier harmonics is taken into account by the instrumental error (due to mantissa errors) Iz ~ 10–13. Hence, the minimum conditionality value is DIAP = 1/Iz.

We confine ourselves to AF O with the possibility of normalization by one $\sum O = 1$, this means that at zero the MTF M(O)(0,0) = 1.

V. I. Karev et al. (Eds.): *Physical and Mathematical Modeling of Earth and Environment Processes (2018)*, SPEES, pp. 441–455, 2019.
https://doi.org/10.1007/978-3-030-11533-3_44

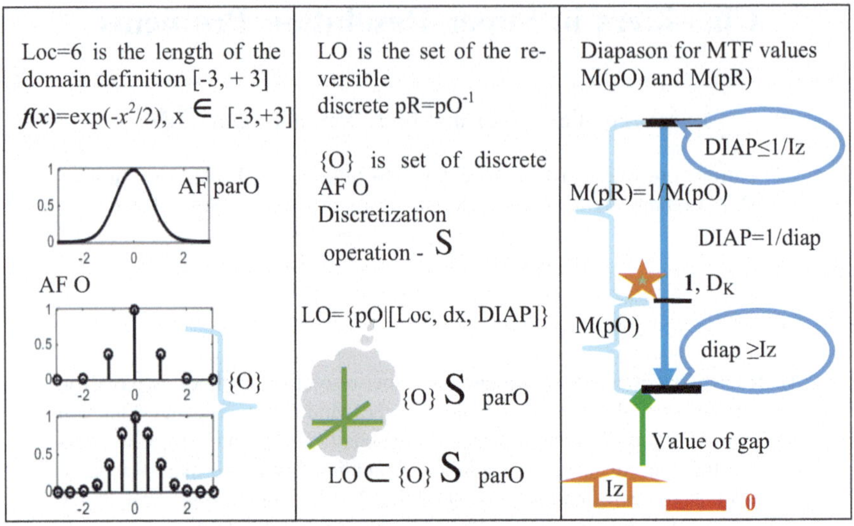

Fig. 1. Illustrations of basic concepts

We proceeded from the fact that low-precision images in one or two byte representation are available to us. Therefore, in the classifier of varieties of super-resolution methods, we distinguish five methods - cells depending on the length of the gap between the minimum eigenvalue md = min(|M(O)|) and zero and the accuracy of image representation.

Comment. In the constructions that claim to solve the inverse problem, there must be concepts such as "Conditionality" of the AF, matrix and, most importantly, "Measure or Indicator of Invertibility". AF pO and pR * O, which accompany the obtained results with super resolution imagers, should be presented.

Of course, there must be a construction about the quality of the solution, the inverse problem, which connects the result of the invertibility of the AF with the obtained errors of the type CAM AF O.

To compare the super resolution results, there must be a corresponding numerical value SR that estimates the obtained super resolution in imagers.

If these concepts are not present, then it is unclear what is at stake.

2 Configure AF PO by Setting the Conditionality

For a given two-dimensional discrete AF O in a linear convolution model, the eigen functions of an AF O are a set of two-dimensional Fourier harmonics H.

The basis of the calculation of two-dimensional MTF arrays Mij = M(O), i,j = 1: Loc + 1 is the eigenvalue problem O * H = Mij H, which is solved using the Discrete Direct Fourier Transform (DDFT) and even continuation operations [4].

The device with discrete AF O is present in the linear model of image registration with convolution $I_{\partial} = O_{\sigma} * I_x$. With SD (Standard Deviation) ∂ and σ, we note that the measured image I_{∂} and AF O_{σ} are known with errors.

In this paper, a simplified linear discrete model of image registration $I_{\partial} = O * I_x$ is assumed, i.e. with $\sigma = 0$.

Note that equally accurate with $\sigma = 0$ models of discrete AF O with different Loc and dx, obtained by digitizing with one continuous AF parO, are noticeably different in the resulting super-resolution.

Often there are situations when the original image with dx = 1, is required to interpolate (or to measure again) under another digitization step with reversible AF O. Denote md = min|M(O)| is not adjustable and mD = min|M(pO)| is adjustable gap between value MTF M(O) and zero. If md > 0, then $R = O^{-1}$ and $I_x = R * I_{\partial}$, i.e. a solution exists.

Suppose that we "removed" dangerous small MTF values from zero by distance mD > md with obtaining a two-dimensional MTF array pMij from Mij = M(O), i,j = 1:Loc + 1, i.e. we increased conditionality mD = 1/md AF pO.

Then, using the Discrete Inverse FT (DIFT) and the operation of even continuation, the inverse eigenvalue problem is solved: pO * H = pMij H in order to calculate the setted two-dimensional pO AF [4].

Since we then replace the AF O with a more conditioned AF pO, we estimate both the increased error Err (pO) and the reduced response to the noise Nor(pR) = ||pR||. ||pR|| is the norm of the resolving function of such a conditionality setting.

3 AF PO Errors with Conditionality Setting

The magnitude of the AF pO error was estimated by the value Err(pO) = SD(O − pO)/max(O), where

$$SD(O - pO) = \sqrt{\sum_{i,j=1}^{Loc+1} (O - pO)^2 / ((Loc + 1)^2 - 1)}.$$

A priori, the device should be "tuned" by the parameters [Loc, dx, mD] according to the initial continuous parO AF [2–5], so that it would be either natural md > 0, or interpolating the image under another digitization step, and increasing DIAP conditionality for AF O, mD > md > 0, see Fig. 2 and Fig. 4.

This means that in the Classifier we may find ourselves in methods **C, D** (measurements of low and high accuracy), but it would be better to immediately find ourselves in the ideal method **B** (usual inversion without conditionality settings), see Fig. 2.

The article [3] gives an example, when for AF O with Loc = image size the addition of one point (Loc increases by 1) leads to reversibility of AF O, the task becomes correct.

4 Classifiers

The super-resolution methods are classified according to the gap size md ($0 \le$ md ≤ 1) for AF O in the image registration model $I_ə = O * I_x$.

A	B	C	D	E	F, Finish
O=D_K	O~D_K	M(O),M(pO)	M(O),M(pO)	M(O),M(pO)	M(O)~D_K
md=1	**1≤md≤10^{-1}**	**10^{-1}≤md≤10^{-4}**	**10^{-4}≤md≤10^{-5}**	**Iz≤ md≤10^{-5}**	**0≤md≤Iz, Iz=10^{-13}**
M(D_K)==1	R=O^{-1}	md->mD pR=pO^{-1}	md->mD pR=pO^{-1}	md->mD pR=pO^{-1}	
Without inversion	*Usual inversion*	*Measurements low accuracy*	*Measurements high precision*	*Modeling experiments*	*Errors mantissa*

Fig. 2. Classifier of super-resolution methods. In the latter case, F cannot be implemented due to mantissa errors.

5 Initial Data

To illustrate the capabilities of the methods, three small fragments of the image of the surface of Mars with a stone will be taken. Note that the selected small fragments of images are low-contrast, low accuracy and all the brightness of the original represented by one byte, method **C** classifier (Fig. 3).

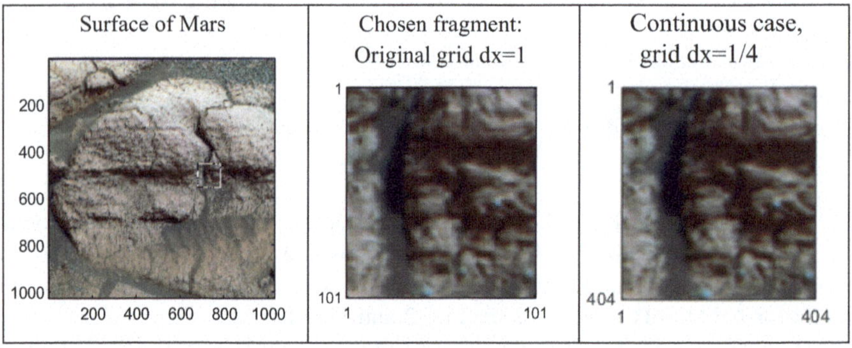

Fig. 3. Fragment of the image of the surface of Mars, presented on the grid with dx = 1 and the interpolated image using FDST in a continuous case on the grid with dx = 1/4.

6 Finite-Dimensional Sampling Theorems for Interpolation

FDST: Let three arrays of samples D, H and *H* be given. D is a two-dimensional array of samples (image on the source grid K, N) D = f (x0, y0). H is a two-dimensional orthonormal matrix of discrete Fourier harmonics H(x0) = {Hi(x0), i = 1:K, x0 = 0: K − 1} and H(y0) = {Hj(y0), j = 1:N, y0 = 0:N − 1}. *H* is the "continuous" Fourier harmonics *H*(x) = {*H*i(x), i = 1:K, x = 0:dx:K − dx}, *H*(y) = {*H*j(y), j = 1:K, y = 0: dy:K − dy}, dx < 1, dy < 1. All Fourier harmonics in the matrices H and *H* are numbered by the first index, arranged horizontally

Then a continuous function:

$$f(x, y) = \sum_{i,j=1}^{K,N} C_{i,j}\, H_i(x)\, H_j(y) \tag{1}$$

$$\begin{aligned} c_{i,j} &= (f(x0, y0), H_i(x0)\, H_j(y0)) \\ &= \sum_{x0,y0=1}^{K,N} f(x0, y0)\, H_i(x0)\, H_j(y0), \quad i = 1 : K, \; j = 1 : N \end{aligned} \tag{2}$$

passes through the sampling points f (x0, y0).

Scalar products (2) realize two-dimensional DDFT and Fourier series (1) gives us an interpolated two-dimensional result.

In the one-dimensional case of D = f(x0), the interpolation result is written in one line:

$$f(x) = (H(x0) \, * D')' * H(x) \tag{3}$$

The asterisk * in (3) denotes the multiplication rule "row*column with summation" when multiplying matrices and the stroke 'is transposition.

The first (left) asterisk implements a one-dimensional DDFT (2), and the second asterisk is the sum of a discrete Fourier series (1).

Note that the matrices of the digitized Fourier harmonics H (x0), H (y0) are unitary by construction and are intended for the realization of one-dimensional DFT.

Such interpolation (1) gives good results if there is no difference of samples on the border of images. Otherwise, it is necessary to apply special methods in order to protect from boundary effects [2, 3].

There are variants of FDST with the operations of integration and differentiation over arrays of samples in addition to interpolation [6].

7 Selection of an Adequate Model in Classifiers Methods C, D and E

The main task for choosing AF pO = pR^{-1} is set as a minimum task:

$$\min_{LO}\{||pR|| | Err(pO) \leq err\}, \quad LO = \{pO|[Loc,dx,mD]\} \tag{4}$$

with the construction of the Characteristics of the Adequacy of the Model (CAM) AF O [2, 3]

$$\{x = Nor(pR) = ||pR||, \ y = Err(pO), \ z = II(pR * O)\} \tag{5}$$

Adjustable gap is bounded from above. md < mD < 1, Iz is instrumental zero. In connection with this restriction, we consider the CAM AF O variant taking into account the Iz inversion of the two-dimensional MTF M (O) (in the simplified notation for formulas without indices):

$$M(zR) = \begin{cases} 1/M(O), \ \textit{for all} \ |M(O)| > Iz \\ M(O) \end{cases} \tag{6}$$

$$\{x = ||zR||, \ y = Err(zO), \ z = II(zR*O)\} \tag{7}$$

In the CAM AF O (5.7), the z-axis values of the Indicators of Inversion II(pR * O) = (pR * O)(0,0) and II(zR * O) = (zR * O)(0,0), i.e. convolution values of the adjustable pR and non-adjustable zR with the initial AF O at the point (0,0). In the reversible case II(zR * O) = 1, zR = R=O^{-1}. It may also happen that by changing the length of the domain of definition Loc, we get the reversible case II(pR * O) = 1, pR = O^{-1}.

In CAM AF O (5.7) the Indicators of Inversion are the basic concept for interpreting the result of super-resolution, which corresponds to the device with AF pR * O. For an ideal super-resolution pR = O^{-1} or pR * O = D_K, where D_K is the Delta-Kronecker symbol.

8 Estimation of Super-Resolution Value SR

Suppose there is a normalization for AF O: $\sum_{i,j=1}^{Loc+1} O_{ij} = 1$, then the MTF M(O) (0,0) = 1, in this case we will estimate the tunable Super-Resolution by value

$$SR = BW/\sum_{i,j=1}^{Loc+1} M_{ij}(pO) \leq \sum_{i,j=1}^{Loc+1} 1/\sum_{i,j=1}^{Loc+1} M_{ij}(pO) \geq 1, \tag{8}$$

$$BW = \sum_{i,j=1}^{Loc+1} M_{ij}(pR)M_{ij}(O), \tag{9}$$

where BW (9) is the BandWidth. If AF O = D_K, then SR = 1. It is clear that the problem of super-resolution arises for cases when AF O $\neq D_K$.

The BW is limited by the area of definition of AF O $BW \leq \sum_{i,j=1}^{Loc+1} 1 = (Loc + 1)^2$. The value of SR should not depend on the large size of the areas of definition of AF O and the size of the images.

Similarly to the adjustable SR (8) and BW (9), we can enter the maximum possible values of zSR and zBW with the minimum conditionality DIAP = 1/Iz. But note that this is realized with the maximum response to the noise Nor(zR) (7).

9 Examples CAM AF PO

CAM AF O (4) presents methods from **A** to **E**, marked by bright rectangles. The CAM of a discrete model with dx = 1 and the size of the domain of definition Loc = 12 corresponds to method **C**. The CAM is a continuous model with dx = 1/4, Loc = 12 closer to methods **D**, **E**, see Fig. 4.

If the CAM of the continuous model is depicted in one scale, then the CAM of the continuous model will conditionally take the place of a long rectangle marked with the expressions Err (pO) = 3.1 and dx = 1/4, see Fig. 4.

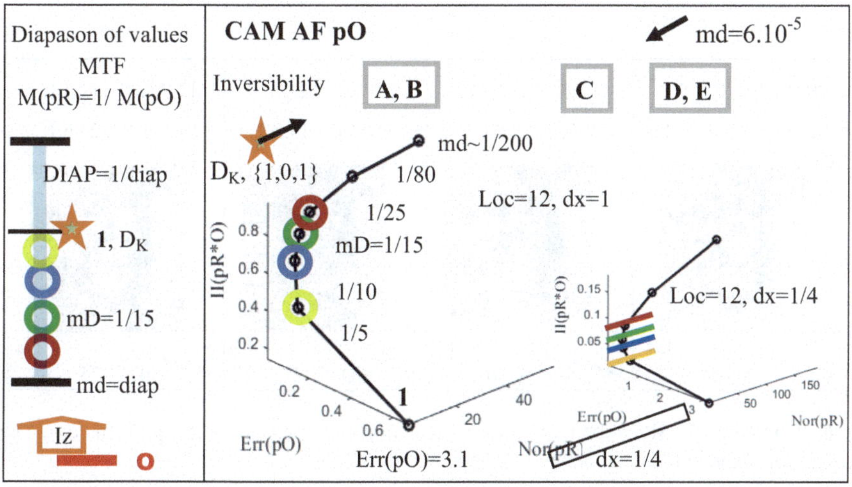

Fig. 4. The gaps md = min|M(O)| and mD = min|M(pO)| with CAM AF pO.

Mathematical focusing or adjusting the increase in image resolution begins with a decrease from the unit of the adjustable gap to the unregulated $1 \geq mD \geq md > 0$. If mD = 1, then, as a result of super-resolution the original image remains and the error Err (DK) has a maximum value.

With a decrease in mD (the conditionality of AF pO decreases), the values of the super-resolution SR and the responses to the noise Nor (pR) increase, the errors Err (pO) decrease to zero, see Figs. 4, 5 and 6.

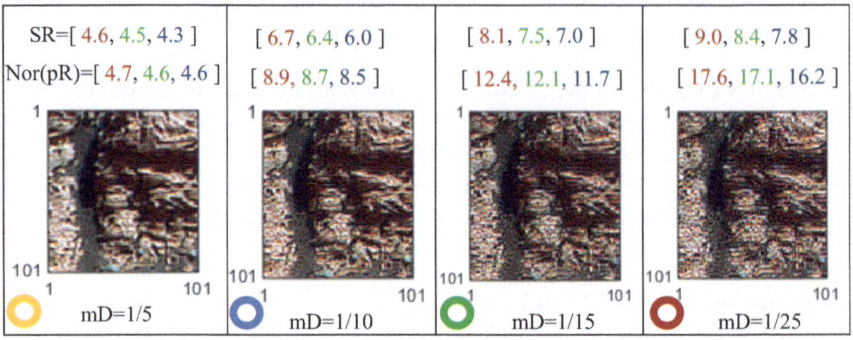

Fig. 5. Conditioned super-resolutions SR and noise response Nor (pR) in the initial grid dx = 1

For single-byte images (low accuracy method **C**) limitation (inadequacy) is realized near mD ~ 1/25, because there are changing ("knocking out" red) small details in the image (because of high response to noise - Nor (pR)), see Fig. 5.

Therefore, we assume that for single-byte images, as in our example (∂ ~ 3–5%), the values of SR ~ 9.5 are permissible. For two byte images, a super resolution of SR ~ 20–30 is expected, methods of type **D** in the classifier.

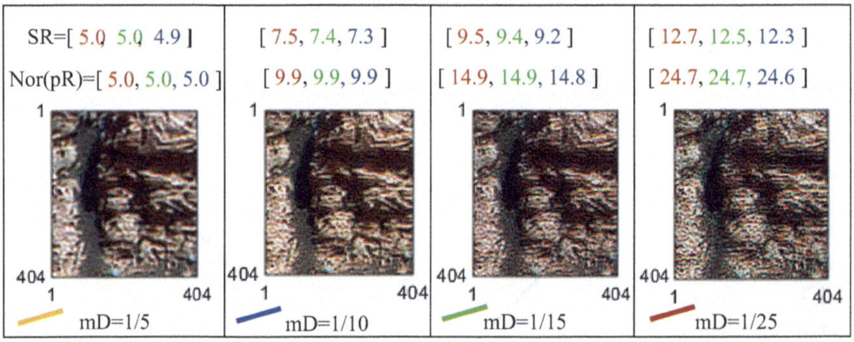

Fig. 6. Conditioned SR super-resolutions and noise responses Nor (pR) in case with interpolation dx = 1/4.

10 Gaps md and mD

Figures 6 and 7 show the gaps md = min|M(O)| - not adjustable and mD = min|M(pO)| - adjustable gap between MTF M (O) and zero (Fig. 8).

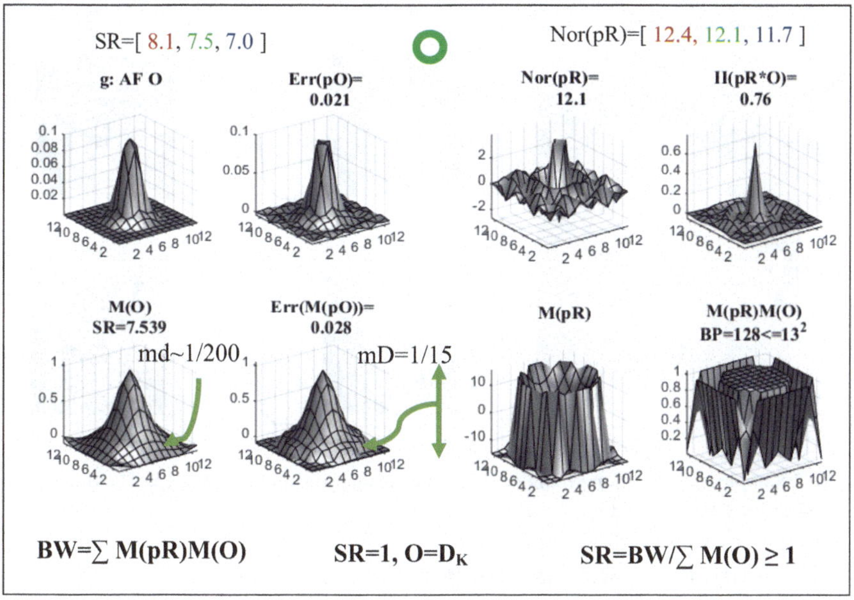

Fig. 7. Discrete model dx = 1 with gaps md ~ 1/200 and mD = 1/15.

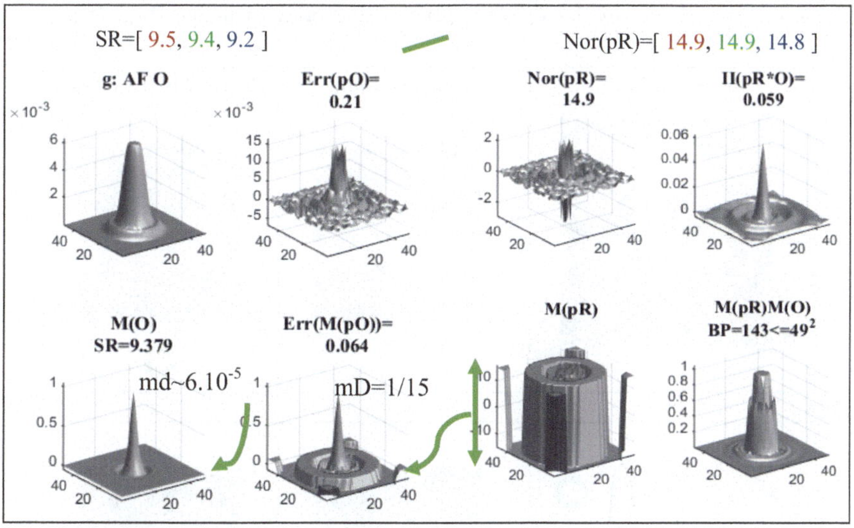

Fig. 8. The continuous model dx = 1/4 with gaps md ~ 6.10–5 and mD = 1/15.

11 Structure of Stone Similar to Fossilized Coral

In the second image, another area of the stone image is taken to demonstrate the conditioned super-resolution. In our opinion, the structure of this area is similar to petrified coral. We draw the attention of specialists to this result (Fig. 9).

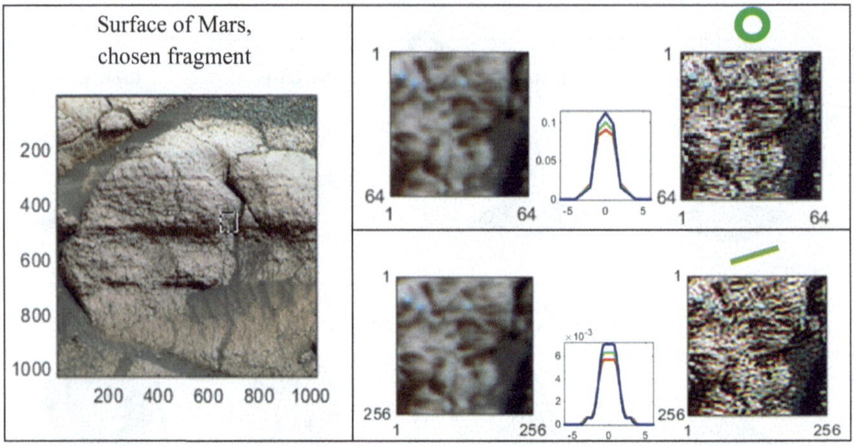

Fig. 9. Examples of conditioned mD = 1/15 super-resolutions in discrete dx = 1 and continuous dx = 1/4 cases with Loc = 12.

12 Resolution of Objects Smaller Than a Pixel

On the third selected fragment of the image of a stone, an area with suspicion of sand in the form of dunes was selected. In discrete and continuous cases, separate grains of sand are distinguished, smaller than the pixel of the original image (Fig. 10).

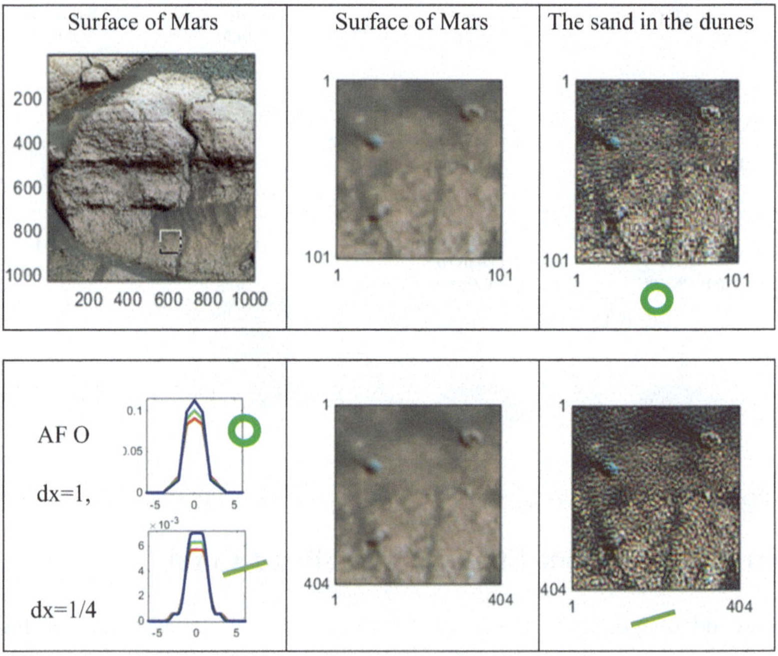

Fig. 10. The conditional mD = 1/15 super-resolution of individual sand grains is presented in the discrete dx = 1 and in the continuous cases dx = 1/4, Loc = 12.

13 On the Magnitude of the Super-Resolution SR

In reversible cases, SR, zSR (8, 9) as a function of three parameters [Loc, dx, mD] is a function with saturation by two parameters: the length of the domain of definition of Loc and the conditionality of mD. In the discrete case dx = 1, this is seen in Fig. 11 on the left. In the continuous case of dx = 1/4, this is evident in Fig. 11 on the right for large values of mD in the case of Iz inversions.

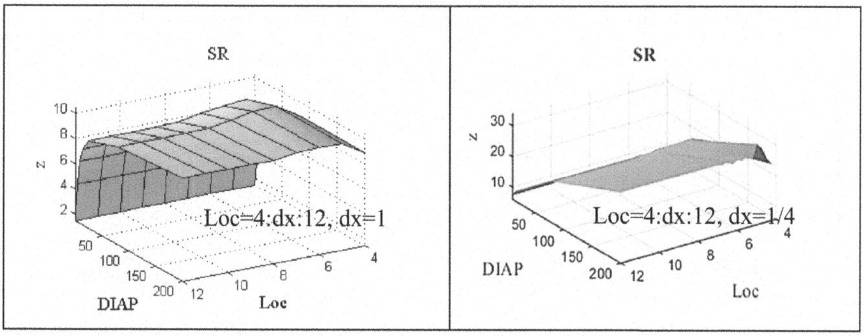

Fig. 11. Examples of dependences of the SR super-resolution values on parameters [Loc,dx, mD] in LO (4) that are tuned for local domains of definition of AF pO in discrete and continuous cases.

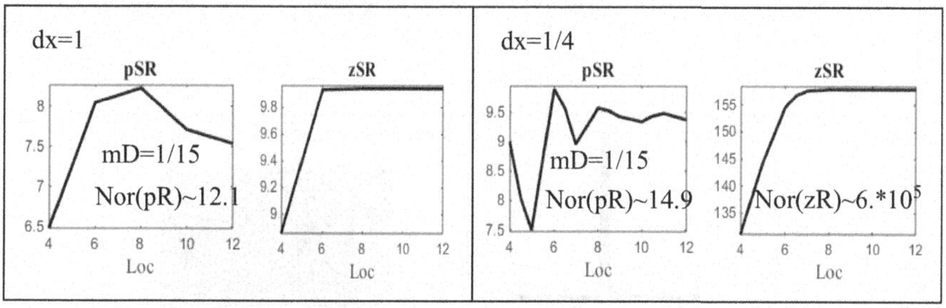

Fig. 12. Dependences of the values of super-resolutions pSR = SR and zSR on mD, Loc in the discrete dx = 1 and in the continuous dx = 1/4 cases during Iz inversions.

Moreover, since the image size, as a rule, is larger than the size of the Loc definition domain, the value (with saturation) of the conditioned super-resolution SR does not depend on the image size.

Iz inversion (6, 7) with zSR allows us to find the amount of super-resolution for very large mD up to Iz. If at mD = 1/15 we have SR ~ 9.5. If mD ~ Iz, then zSR will be of the order of ~ 156.

Such a large super-resolution can practically be achieved only in model problems, the E method, because this is accompanied by a huge response to the noise Nor (zR) $\sim 6.\ 10^5$, see Fig. 12.

14 Application of the Convolution Theorem

Usually, convolution is associated with the FT, but this can be done if certain conditions are met to prevent boundary effects. If AF O is even, then the image must be even too [1, 2]. Figure 13 shows how the "boundary effect" manifests itself.

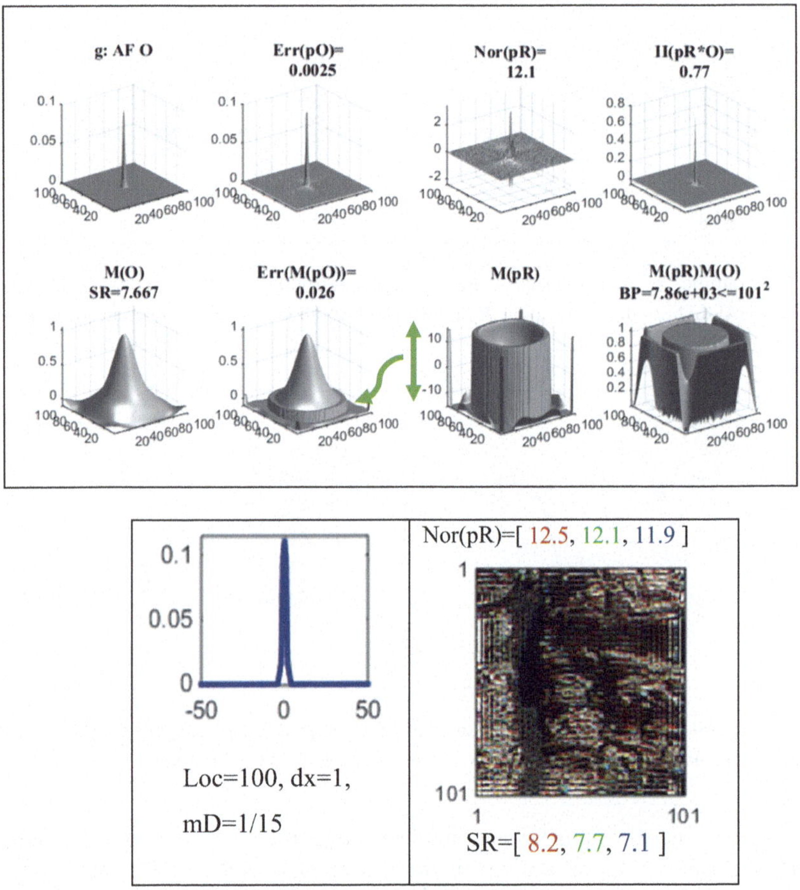

Fig. 13. The domain of definition of AF O is the entire image, the discrete case dx = 1, boundary effects.

To suppress boundary effects, the image should be "continued to torus" with adjustable sizes of neutralization areas [1, 2]. But in our experience, it is better not to use the FT, but necessary to continue the function beyond the boundary of the image "by mirror reflection" [1, 2]. This increases the amount of computation, and errors on the border of the image are not so noticeable, see Fig. 14.

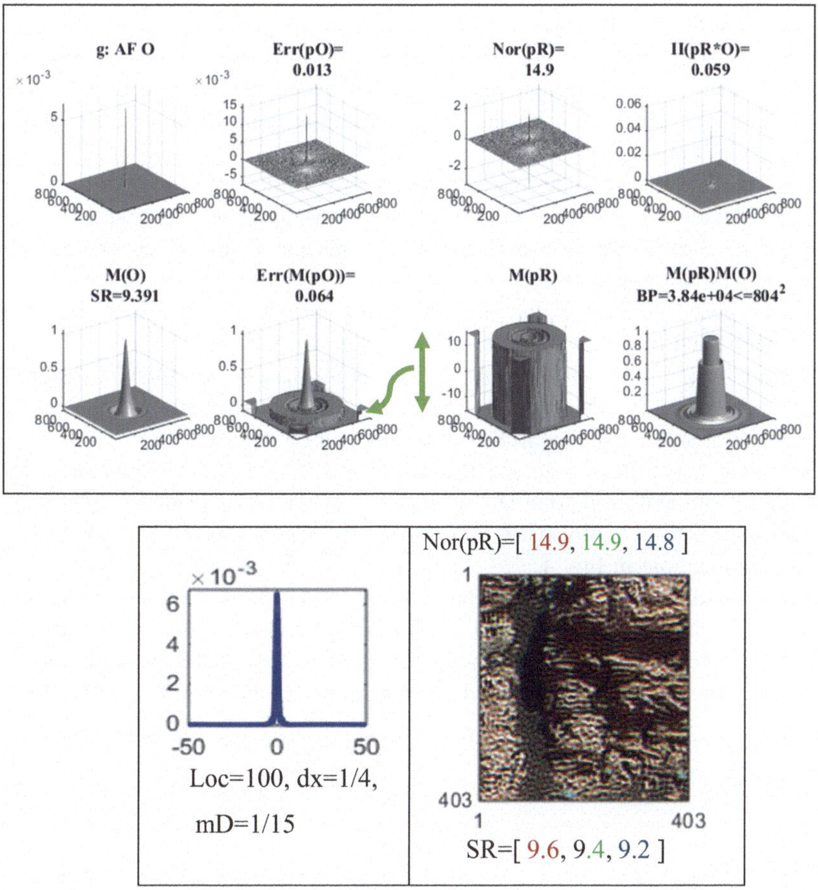

Fig. 14. The area of definition of AF O is the entire image, the continuous case of dx = 1/4, protected mode.

15 Discussion

The approach to the problem of conditional super-resolution with Classifiers orders the creation of models to achieve adequate super-resolution.

In our super-resolved images, jumps, differences in brightness, i.e. the solutions of the ill-posed inverse problem of compensating for distortion of AF O are not smooth.

Not the smoothness of solutions allows this method to be reasonably and safely applied, for example, in control loops of reactors, navigators, etc.

Interesting modifications of the methods can be implemented in new radar technologies (with a controlled antenna pattern), in synthetic aperture radars, etc. Results with higher SR ~ 17 have already been obtained in passive radio imaging [3]. This result corresponds to approximately two bytes accuracy of the measurement data. In this paper, low-contrast fragments of single-byte Martian images with a low relative accuracy of 3–5%, i.e. even the corresponding part of one byte.

Fine details of images are steadily distinguished in compared to the regularization method. If there is a need, then solutions (super-resolved images) can be "smoothed out" (of course, with loss of resolution) according to a priori information about the smoothness of the solutions.

It is possible to use super-resolution methods with interpolation to achieve the limiting adequate super-resolution in electron microscopy and tomography.

It is necessary to develop a Classifier for inaccurate image registration models $I_{\partial} = O_{\sigma} * I_x$, $\partial > 0$ and $\sigma > 0$.

The proposed concept of the super-resolution problem with a classifier is much more complicated than the classical regularization method [7].

The article [5] gives an example of a complex five-beam Antenna Patter (AP), when compensating for distortion of the AP, an ideal super-resolution with a zero error Err (pO) and a small response to the noise Nor (pR) is realized.

Of course, it is necessary to calculate in more detail all difficult situations with 3D CAM AF O with exit from point D_K, by segment, sector in the plane Err (pO) = 0, Nor (pR) > 1 then with exit into space 3D CAM with Err(pO) > 0 and Nor(pR) > 1, see a very simple example in Fig. 4.

An important property: when changing the adjustable gap mD to non-adjustable gap md > 0 (with decreasing AF pO conditionality) and when the error tends to zero ∂- > 0, the approximate super-resolution SR tends to the ideal super-resolution zSR. Almost real super-resolution is limited to the response to the noise Nor (pR), Nor (zR).

This is essentially an analogue of the asymptotic statement of A.N. Tikhonov for the regularization method [7].

References

1. Terentiev, E.N., Terentiev, N.E.: Izvestiya RAN. Ser. Phys. **79**(12), 1633–1637 (2015). (in Russian)
2. Terentiev, E.N., Terentiev, N. E.: Bulletin of the Russian Academy of Science. Physics, **79** (12), 1427–1431 (2015). https://doi.org/10.3103/s1062873815120229
3. Терентьев,Е.Н.: …//УЗФФ Московского Университета, 9 с., №6, 1761005 (2017). (in Russian)
4. Терентьев, Е.Н., Терентьев, Н.Е.: // 7-я межд. конф. АКУСТООПТИЧЕСКИЕ И РАДИОЛОКАЦИОННЫЕ МЕТОДЫ ИЗМЕРЕНИЙ И ОБРАБОТКИ ИНФОРМАЦИИ, 15–17 сен. 2014, Суздаль, Россия, Программа, с. 49–53 (in Russian)
5. Терентьев, Е.Н., Терентьев, Н.Е.: // там же, с. 53–57 (in Russian)

6. Terentiev, E.N., Terentiev, N.E., Farshakova, I.I.: PMMEEP 2017, Physical and Mathematical Modeling of Earth and Environment Processes, 19 Chap., Principles of Controlling the Apparatus Function for Achieving Super-Resolution in images, pp. 171–182, Number of pages 382, Springer International Publishing, https://doi.org/10.1007/978-3-319-77788-7_19
7. Tikhonov, A.N., Ufimtsev, M.V.: Statistical processing of experimental results. Publishing house of Moscow University (1988) (in Russian)

Bayes Estimates of Variations of the Duration of the Navigation Period Along the Northern Sea Route in the XXI Century from Simulations with Ensemble of Climate Models

O. V. Kibanova[1,2(✉)], A. V. Eliseev[1,2,3], I. I. Mokhov[1,2,4], and V. Ch. Khon[2]

[1] Faculty of Physics, Lomonosov Moscow State University, Moscow 119991, Russia
kibanova.o.v@gmail.com
[2] A.M. Obukhov Institute of Atmospheric Physics, Russian Academy of Sciences, Moscow 119017, Russia
[3] Kazan Federal University, Kazan 420008, Russia
[4] Moscow Institute of Physics and Technology, Dolgoprudnyi, Moscow Oblast 141701, Russia

Abstract. In the 21st century, a significant increase in the duration of the navigation period on the Northern Sea Route is expected. This is due to the stronger warming of the climate in the Arctic. Current climate models are characterized by considerable uncertainty in the sensitivity of the characteristics of sea ice in the Arctic Ocean to current climate changes. In order to take into account, the uncertainty of the results of numerical calculations for models, together with the uncertainty of the observational data, Bayesian analysis methods are used. This paper analyzes the changes in the duration of the navigation period (NPD) on the Northern Sea Route (NSR) for the Coupled Models Intercomparison Project, phase 5 (CMIP5) climate model ensemble under Representative Concentration Pathways (RCP) anthropogenic scenarios using Bayesian statistics. For various scenarios of anthropogenic impact, it was found that the expected duration of the navigation period on the Northern Sea Route will be 2–3 months by the middle of the XXI century and 3–6 months by its end.

Keywords: Climate changes · The arctic · Sea ice · North sea route · Satellite data · Climate models · Bayesian statistics · CMIP5

1 Introduction

Variations in the sea-surface temperature in the arctic latitudes in recent decades were lager then the global mean changes by a factor of two. At the same time, the area of sea ice in the Arctic is rapidly decreasing. The area of the Arctic sea ice in 1979–2014 in September it decreased at a rate of 1.3% per year [1]. With the continuation of modern climate warming in the 21st century, the Arctic Ocean can be free of sea ice in the

V. I. Karev et al. (Eds.): *Physical and Mathematical Modeling of Earth and Environment Processes (2018)*, SPEES, pp. 456–462, 2019.
https://doi.org/10.1007/978-3-030-11533-3_45

summer-autumn seasons. This tendency contributes to an increase in the duration of the navigation period on the Northern Sea Route [2–4].

Climate models, in particular the models included in the CMIP5 ensemble, are characterized by considerable uncertainty in the reproduction of the characteristics of sea ice in the Arctic Ocean compared to the observed climate changes, both from the results of numerical calculations of the climate reproduction of the XX century and calculations of climate change in the 21st century [5]. In order to obtain more reliable estimates of changes in the PNP on the SMP in the 21st century, it is necessary to use analysis methods that take into account the uncertainties in the results of climate model estimates and uncertainties in the observational data. For this, in particular, one can use Bayes averaging [6–8].

In our work, we analyze changes in the duration of the navigation period using calculations of the Coupled Models Intercomparison Project, phase 5 (CMIP5) climate model ensemble for Representative Concentration Pathways (RCP) 4.5 and 8.5 scenarios of anthropogenic impact using Bayesian statistical methods.

2 Data and Methods

For the analysis, model calculations of sea ice compactness in the Arctic Ocean of the CMIP5 ensemble for the period 1979–2100 were used. The initial year of this period was chosen as the beginning of satellite observations. We use numerical calculations of the "historical" scenario for the period 1979–2005. and the results of numerical calculations with the CMIP5 climate model ensemble under the scenarios of anthropogenic and natural impacts of the RCP family for the period 2006–2100. Of the four anthropogenic RCP scenarios, we select the moderate (RCP 4.5) and more "aggressive" (RCP 8.5) anthropogenic scenarios. To find the duration of the navigation path, we considered that the model cell was free of sea ice, if it had an average monthly ice concentration of no more than 15%. In addition, the duration of the navigation period was determined by taking into account how much Z of the route of the Northern Sea Route was ice-free (for Z, values of 80% and 90% were used).

We highlight some of the characteristics in order to take into account both the quality of the reproduction of the current climate models, and climate change at different time scales. These characteristics are (1) the average long-term NPD duration $T_{Z,m}$, (2) the linear trend for a series of NPD data $T_Z^{'}$, and (3) the standard deviation of the short-period interannual variability σ_Z. Bayesian weights are calculated as the likelihood functions of these quantities for the measured data. To obtain observational data, we use the Scanning Multichannel Microwave Radiometer instrument on the Nimbus-7 satellite (SMMR) for the period 1980–2014. We assume that all variables are normally distributed and their means and standard deviations can be found from observational data. Thus, weights are calculated as:

$$w_j^{(k)} = N\left(X_j^{(k)}, X_j^{(0)}, \sigma_j^{(0)}\right), \quad (1)$$

where $j = 1 - 3$, k is mmodel index, $N(x, x_0, \sigma)$ is the normal distribution of variable x with the mean value x_0 and standard deviation σ. The weights taking into account traits 1–3 and $X_j^{(k)} = T_{Z,m}$ if $j = 1$, $X_j^{(k)} = T_Z'$, if $j = 2$, $X_j^{(k)} = \sigma_Z$, if $j = 3$. Here, $X_j^{(0)}, \sigma_j^{(0)}$ indicates that these are the data of observations. In addition, we assume that each trait contributes to the result independently of the others, and the resulting contribution of this model will be the product of the weights:

$$w_4^{(k)} \sim w_1^{(k)} w_2^{(k)} w_3^{(k)}$$

An additional condition is required for all weights:

$$\sum w^{(k)} = 1$$

In addition, combinations of weighting factors are used that take into account all three time scales. Using the calculated weight factors, the ensemble averaging $E(T_Z|D)$ and calculation of the intermodel (intraensemble) standard deviation $\sigma(T_Z|D)$ is performed:

$$E(T_Z|D) = \sum T_Z^{(k)} w^{(k)}$$

$$\sigma(T_Z|D) = \left\{ \sum \left\{ \left[\sigma(T_Z|D)^2 + T_Z^{(k)2} w^{(k)} \right] - E(T_Z|D, T_Z^{(k)})^2 \right\} \right\}^{1/2}$$

Where $T_Z^{(k)}$ are array of variable T_Z of traits 1–3.

3 Results

Weights for scenarios of anthropogenic impacts RCP4.5 and RCP8.5 and Z = 80% are illustrated on Fig. 1. Weights w_1 (characterizing the average multiannual value of the NPD) are distributed relatively uniformly between the models. A more non-uniform intermodal distribution of weights is noted for w_2, w_3 and w_4. The degree of this non-uniformity can be characterized using the information entropy normalized to a value corresponding to the weights $w_0 = 1 / K$:

$$H_j = -\left(\sum w_j^{(k)} \log_2 w_j^{(k)}\right) / \log_2 K$$

The values of this entropy are: $H_1 = 0.98, H_2 = 0.93, H_3 = 0.72$ and $H_4 = 0.66$. They decrease with increasing weights numbers. The greatest contribution to the inter-model differences in estimating the change in the NSR is due to the reproduction of the short-period interannual variability of sea ice compactness on the NSR. In addition, the pairwise correlation coefficients between weights are not statistically significant. This allows us to use the combined weighting factor in the form w_4.

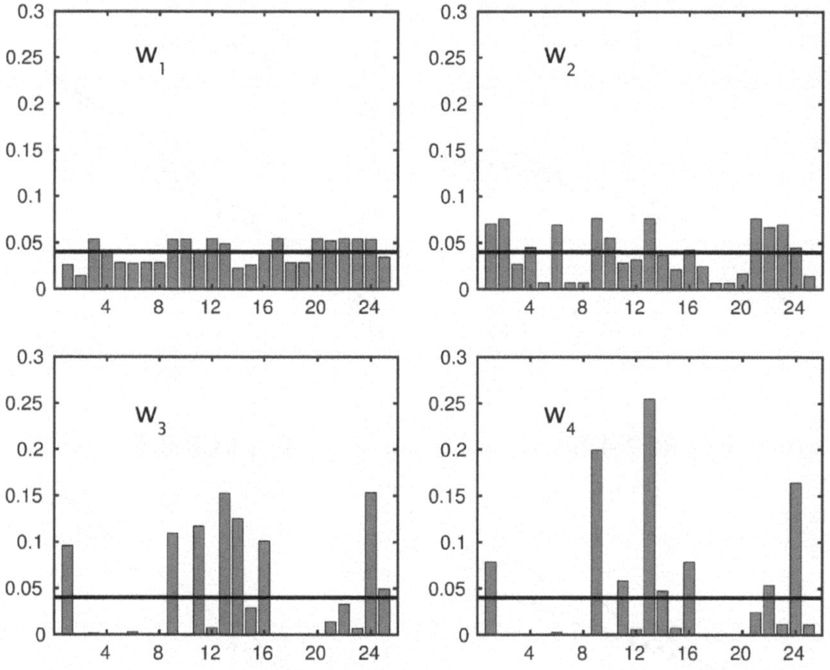

Fig. 1. Bayesian weight factors for the NPD on the Northern Sea Route (a case where 80% of the route is ice-free). For simplicity, an index indicating the climate model is omitted in the drawings. In determining the weights for 2006–2014. used the results of calculations with the script RCP 4.5. The number of the climatic model is indicated on the abscissa axis. The horizontal line shows the weight w (0) = 1 / K (K is the number of models in the ensemble).

It was found that the models of the CMIP5 ensemble are consistent with each other in an overall significant increase in the NPD at the NSR in the 21st century. This increase is more for the RCP 8.5 scenario than for the RCP 4.5, which is associated with a large increase in greenhouse gas concentrations in the atmosphere in the first scenario than in the second. In the case when 80% of the NSR route is free of sea ice, the change in the NPD under anthropogenic impact on the climate is slightly larger than in the case of 90%, the uncertainty intervals overlap for these values. Thus, the results of the work are not very sensitive to the choice of the value of Z (Fig. 2).

Models that best reproduce the current multi-year value of the NPD (models whose weights w_1 are greatest) are characterized by a small change in T_Z during warming caused by anthropogenic impact. In addition, the sensitivity of the NPD in the models that best reproduce the linear trend T_Z over the past decades (which is characterized by weights w_2) is maximum. The response of T_Z in models with the best reproduction σ_Z (i.e. with maximum weights w_Z) is greater than the response of models that best reproduce $T_{Z,m}$, but smaller than the corresponding response in models that best reproduce T_Z'. The response of models with the largest weights w_4 is similar response models with relatively large w_3 weights.

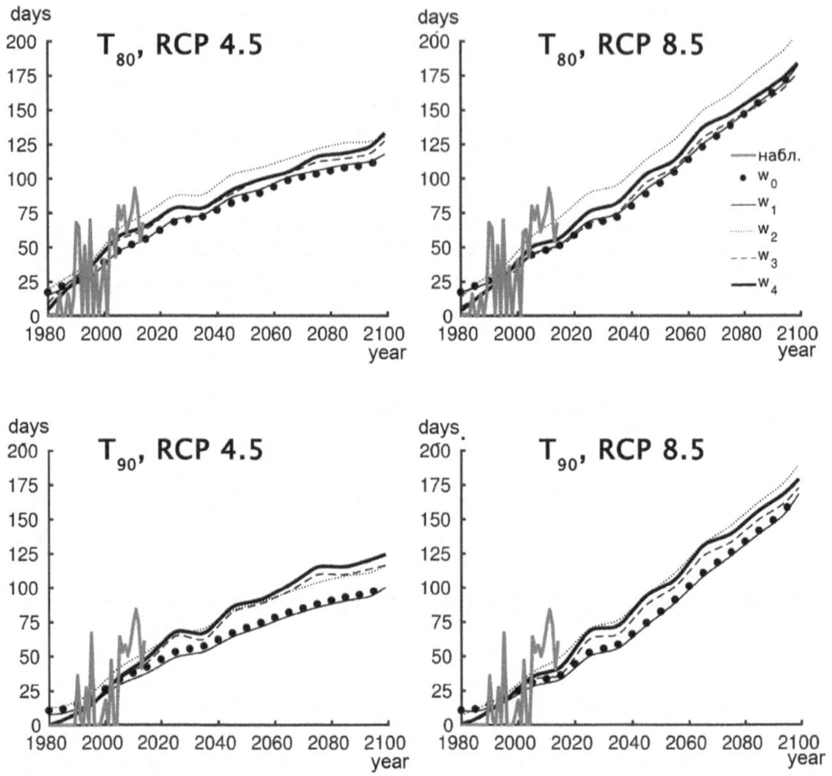

Fig. 2. The ensemble average value of the NPD on the Northern Sea Route for scenarios of anthropogenic impacts on the climate system RCP 4.5 and RCP 8.5, for different values of the Z parameter and for different options for calculating Bayesian weights.

The models that best reproduce the modern climate are not sensitive to anthropogenic warming. The models that best reproduce the linear trend over the past decades are the most sensitive to anthropogenic warming. Combining criteria in the form of a weight w_4 leads to an average response of the NPD to anthropogenic warming.

With all the choice of weighting factors used, the intermodel standard deviation $\sigma(T_Z|D)$ as a whole increases with time. As the weight number increases, the intermodel standard deviation decreases (Fig. 3). It can be observed that the intermodel variation for different scenarios of anthropogenic impact and Z values is very significant. But this does not mean that changes over time in the NPD are statistically insignificant. The change in $E(T_Z|D)$ for time variations should be compared not with $\sigma(T_Z|D)$, but with T_Z (standard deviation of interannual variability). The values of this standard deviation for different scenarios of anthropogenic influences, Z values and time points range from 1 to 2 weeks. Thus, the increase in the NPD on the Northern Sea Route becomes statistically significant even for the beginning of the XXI century.

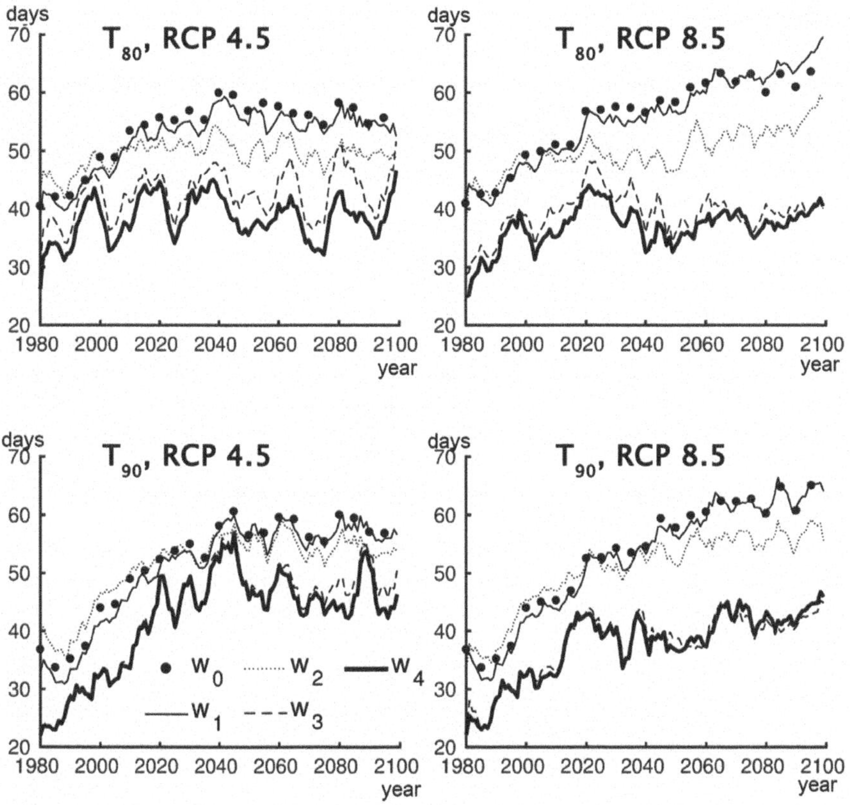

Fig. 3. The inter-model standard deviation of the NPD on the Northern Sea Route for scenarios of anthropogenic impacts on the climate system RCP 4.5 and RCP 8.5, for different values of the Z parameter and for various options for calculating Bayes weights.

Thus, the estimated duration of the navigation period on the Northern Sea Route for the scenarios of anthropogenic impacts of RCP is expected to be 2–3 months according to the ensemble model estimates. in the middle of the XXI century. At the end of the XXI century, the duration is expected to be 3–6 months.

Acknowledgments. This work was supported by the Ministry of Education and Science of the Russian Federation (Agreement no. 14.616.21.0082) using the results of the projects of the Russian Foundation for Basic Research and Programs of the Russian Academy of Sciences (Grants 17-05-01097, 18-05-60111). The development of the model weighting technics was supported by the Russian Science Foundation (grant 18-47-06203).

References

1. Mokhov, I.I.: Contemporary climate changes in the Arctic. Herald Russ. Acad. Sci. **85**(3), 265–271 (2015)
2. Mokhov, I.I., Khon, V.Ch.: The duration of the navigation period and changes for the Northern Sea Route: model estimates. Arct. Ecol. Econ. **2**(18), 88–95 (2015)
3. Mokhov, I.I., Khon, V.Ch., Prokof'eva, M.A.: New model estimates of changes in the duration of the navigation period for the Northern Sea Route in the 21st century. Dokl. Earth Sci. **468**(2), 641–645 (2016)
4. Khon, V.C., Mokhov, I.I., Semenov, V.A.: Transit navigation through Northern Sea Route from satellite data and CMIP5 simulations. Environ. Res. Lett. **12**(2), 024010 (2017)
5. Mokhov, I.I.: Assessment of the ability of contemporary climate models to assess adequately the risk of possible regional anomalies and trends. Dokl. Earth Sci. **479**(2), 482–485 (2018)
6. Eliseev, A.V., Semenov, V.A.: Arctic climate changes in the 21st century: ensemble model estimates accounting for realism in present-day climate simulation. Dokl. Earth Sci. **471**(1), 1183–1187 (2016)
7. Hoeting, J.A., Madigan, D., Raftery, A.E., et al.: Bayesian model averaging: a tutorial. Stat. Sci. **14**(4), 382–417 (1999)
8. Kibanova, O.V., Eliseev, A.V., Mokhov, I.I., Khon, V.C.: Variations in the duration of the navigation period along the Northern Sea Route in the 21st century based on simulations with an ensemble of climatic models: Bayesian estimates. Dokl. Earth Sci. **481**(1), 907–911 (2018)

Initial Strength Analysis of Anisotropic Concrete Supports for Spherical Mine Workings in a Sedimentary Rock Mass

A. V. Zaitsev(✉), A. A. Fukalov, and Yu. V. Sokolkin

Perm National Research Polytechnic University, Komsomolsky Av. 29, 614990 Perm, Russia
zav@pstu.ru

Abstract. The application of the method, i.e. component expansion of hoop and radial coordinate displacement vectors into trigonometric and generalized power series, helped obtain a new exact analytical solution for a heavy transversally-isotropic combined sphere in the equilibrium state of gravity forces. Uniformly distributed pressure was set on the internal surface, the external surface was considered to be rigidly fixed; stress and strain expressions are shown. The obtained new analytical solution helped analyze the influence of geometry on the distribution of independent invariants of stress tensor in the cross-sections of concrete monolithic supports for spherical mine workings and sedimentary rock mass surrounding mine; besides, a comprehensive multicriterion approach enabled the description of damage mechanisms.

Keywords: Thick-walled transversally-isotropic combined sphere · Gravity forces · Exact analytical solution · Concrete monolithic supports for spherical mine workings · Sedimentary rock mass surrounding mine · Mechanisms of initial damage state

1 Introduction

The building of spherical mine workings requires reinforcement as one of the main production processes; otherwise, their operation is impossible. The supports built for labor protection, the preservation of raw material and equipment placed inside are generally produced of concrete and are considered as bulky structures. Due to the fact, that concrete and rocks have an anisotropic nature caused by the presence of core reinforcing elements and mineral inclusions, we encounter an acute and significant task, i.e. to obtain new exact analytical solutions of boundary-value problem for the equilibrium state of thick-walled combined anisotropic elastic bodies with central symmetry under the action of internal and external pressures and gravity forces; hence, the obtained solutions will help work out engineering techniques of a proximate strength analysis of structure elements with similar geometry and boundary conditions. Moreover, these solutions can be used for the identification of constitutive constants and the material functions of anisotropic materials (concrete that is used for building

V. I. Karev et al. (Eds.): *Physical and Mathematical Modeling of Earth and Environment Processes (2018)*, SPEES, pp. 463–471, 2019.
https://doi.org/10.1007/978-3-030-11533-3_46

supports, underground structures and facilities are produced at the same time), for testing algorithms that implement numerical methods, and for the performance of experiment techniques with heavy bodies of elementary geometry.

2 Exact Analytical Solution

Let us examine the equilibrium state in gravity forces of a combined linearly-elastic anisotropic body that represents a structure consisting of two hollow thick-walled spheres with generic center rigidly fixed on the external surface of the radius ρ_2. Uniformly distributed pressure p was set on the internal surface of the radius ρ_1. Let us assume that the contact area, where the condition of full adhesion is performed, is located at the distance ρ_c from the body center. The material, that the combined body is produced of, is homogeneous spherically transversally-isotropic relative to any radius-vector drawn from the geometrical center. Let us use the spherical coordinate system r, θ and φ. All constants and functions that belong to the internal part of a combined sphere will be assigned the upper index 1, placed in brackets, and the external part – the index 2 correspondingly. Radial and hoop displacements ($u_r^{(i)}$ and $u_\theta^{(i)}$), radial ($\sigma_{rr}^{(i)}$ and $\varepsilon_{rr}^{(i)}$), hoop ($\sigma_{\theta\theta}^{(i)}$ and $\varepsilon_{\theta\theta}^{(i)}$), meridional ($\sigma_{\varphi\varphi}^{(i)}$ and $\varepsilon_{\varphi\varphi}^{(i)}$) stresses and strains, shear stress $\tau_{r\theta}^{(i)}$ and shear strains $\varepsilon_{r\theta}^{(i)}$ do not depend (due to body symmetry and external load) on the meridional coordinate φ, and satisfy the Cauchy's relation and equilibrium equations:

$$\begin{gathered} \varepsilon_{rr}^{(i)} = \partial_r u_r^{(i)}, \quad \varepsilon_{\theta\theta}^{(i)} = \frac{1}{r}\left(u_r^{(i)} + \partial_\theta u_\theta^{(i)}\right), \quad \varepsilon_{\varphi\varphi}^{(i)} = \frac{1}{r}\left(u_\theta^{(i)} \operatorname{ctg}\theta + u_r^{(i)}\right), \\ \gamma_{r\theta}^{(i)} = \partial_r u_\theta^{(i)} + \frac{1}{r}\left(\partial_\theta u_r^{(i)} - u_\theta^{(i)}\right), \quad i = 1 \text{ or } 2, \end{gathered} \tag{1}$$

$$\begin{gathered} \partial_r \sigma_{rr}^{(i)} + \frac{1}{r}\left(\partial_\theta \tau_{r\theta}^{(i)} + 2\sigma_{rr}^{(i)} - \sigma_{\varphi\varphi}^{(i)} - \sigma_{\theta\theta}^{(i)} + \tau_{r\theta}^{(i)} \operatorname{ctg}\theta\right) + F_r^{(i)} = 0, \\ \partial_r \tau_{r\theta}^{(i)} + \frac{1}{r}\left[\partial_\theta \sigma_{\theta\theta}^{(i)} + \left(\sigma_{\theta\theta}^{(i)} - \sigma_{\varphi\varphi}^{(i)}\right)\operatorname{ctg}\theta + 3\tau_{r\theta}^{(i)}\right] + F_\theta^{(i)} = 0. \end{gathered} \tag{2}$$

Here $F_r^{(i)} = -\gamma^{(i)} \cos\theta$ and $F_\theta^{(i)} = \gamma^{(i)} \sin\theta$ is the vector components of body forces, $\gamma^{(i)}$ is the specific gravity of the material, and $\partial_r(\ldots) \equiv \partial(\ldots)/\partial r$.

Constitutive equations for the spherically transversally-isotropic body:

$$\begin{gathered} \sigma_{rr}^{(i)} = A_{11}^{(i)}\varepsilon_{rr}^{(i)} + A_{12}^{(i)}\left(\varepsilon_{\varphi\varphi}^{(i)} + \varepsilon_{\theta\theta}^{(i)}\right), \quad \sigma_{\varphi\varphi}^{(i)} = A_{12}^{(i)}\varepsilon_{rr}^{(i)} + A_{22}^{(i)}\varepsilon_{\varphi\varphi}^{(i)} + A_{23}^{(i)}\varepsilon_{\theta\theta}^{(i)}, \\ \sigma_{\theta\theta}^{(i)} = A_{12}^{(i)}\varepsilon_{rr}^{(i)} + A_{23}^{(i)}\varepsilon_{\varphi\varphi}^{(i)} + A_{22}^{(i)}\varepsilon_{\theta\theta}^{(i)}, \quad \tau_{r\theta}^{(i)} = A_{44}^{(i)}\varepsilon_{r\theta}^{(i)} \end{gathered} \tag{3}$$

can be written with help of constants:

$$\begin{gathered} A_{11}^{(i)} = \frac{E_r^{(i)}}{\nu^{(i)}}\left(1 - \mu_{\theta\varphi}^{(i)}\right), \quad A_{22}^{(i)} = \frac{E_\theta^{(i)}}{\nu^{(i)} + \mu_{\theta\varphi}^{(i)}\nu^{(i)}}\left(1 - \mu_{r\theta}^{(i)2}\frac{E_\theta^{(i)}}{E_r^{(i)}}\right), \quad A_{12}^{(i)} = \frac{E_\theta^{(i)}}{\nu^{(i)}}\mu_{r\theta}^{(i)}, \\ A_{23}^{(i)} = \frac{E_\theta^{(i)}}{\nu^{(i)} + \mu_{\theta\varphi}^{(i)}\nu^{(i)}}\left(\mu_{\theta\varphi}^{(i)} + \mu_{r\theta}^{(i)2}\frac{E_\theta^{(i)}}{E_r^{(i)}}\right), \quad A_{44}^{(i)} = G_{r\theta}^{(i)}, \quad \nu^{(i)} = 1 - \mu_{\theta\varphi}^{(i)} - 2\mu_{r\theta}^{(i)2}\frac{E_\theta^{(i)}}{E_r^{(i)}}, \end{gathered}$$

defined by the Young's moduli $E_r^{(i)}$ and $E_\theta^{(i)}$ in the directions r and θ, the Poisson's ratios $\mu_{r\theta}^{(i)}$ and $\mu_{\theta\varphi}^{(i)}$, and the shear modulus $G_{r\theta}^{(i)}$ in diametral plane.

Given, that the heavy thick-walled combined spherical body with rigidly fixed external surface is in equilibrium state, then we receive the following formula:

$$\begin{gathered} u_r^{(2)}\big|_{r=\rho_2} = u_\theta^{(2)}\Big|_{r=\rho_2} = 0, \quad \tau_{r\theta}^{(1)}\Big|_{r=\rho_1} = 0, \quad \sigma_{rr}^{(1)}\big|_{r=\rho_1} = -p, \quad u_r^{(1)}\big|_{r=\rho_c} = u_r^{(2)}\big|_{r=\rho_c}, \\ u_\theta^{(1)}\Big|_{r=\rho_c} = u_\theta^{(2)}\Big|_{r=\rho_c}, \quad \tau_{r\theta}^{(1)}\Big|_{r=\rho_c} = \tau_{r\theta}^{(2)}\Big|_{r=\rho_c}, \quad \sigma_{rr}^{(1)}\big|_{r=\rho_c} = \sigma_{rr}^{(2)}\big|_{r=\rho_c}. \end{gathered} \tag{4}$$

The successive substitution of the relation (1) into (3), followed by the substitution of the gained result into the equilibrium equations (2) helps lay down the inhomogeneous systems of Lame's differential equations [1, 2]:

$$\begin{aligned} \gamma^{(i)}\cos\theta &= A_{11}^{(i)}\partial_r^2 u_r^{(i)} + \tfrac{1}{r}\left[2A_{11}^{(i)}\partial_r u_r^{(i)} + \left(A_{12}^{(i)} + A_{44}^{(i)}\right)\left(\partial_{r\theta}^2 u_\theta^{(i)} + \partial_r u_\theta^{(i)}\operatorname{ctg}\theta\right)\right] \\ &\quad + \tfrac{1}{r^2}\left[\left(A_{12}^{(i)} - A_{22}^{(i)} - A_{23}^{(i)} - A_{44}^{(i)}\right)\left(\partial_\theta u_\theta^{(i)} + u_\theta^{(i)}\operatorname{ctg}\theta\right)\right. \\ &\quad \left. + A_{44}^{(i)}\left(\partial_\theta^2 u_r^{(i)} + \partial_\theta u_r^{(i)}\operatorname{ctg}\theta\right) + 2\left(A_{12}^{(i)} - A_{22}^{(i)} - A_{23}^{(i)}\right)u_r^{(i)}\right], \\ -\gamma^{(i)}\sin\theta &= A_{44}^{(i)}\partial_\theta^2 u_\theta^{(i)} + \tfrac{1}{r}\left[2A_{44}^{(i)}\partial_r u_\theta^{(i)} + \left(A_{12}^{(i)} + A_{44}^{(i)}\right)\partial_{r\theta}^2 u_r^{(i)}\right] \\ &\quad + \tfrac{1}{r^2}\left[A_{22}^{(i)}\left(\partial_\theta^2 u_\theta^{(i)} + \partial_\theta u_\theta^{(i)}\operatorname{ctg}\theta\right) + \left(A_{22}^{(i)} + A_{23}^{(i)} + 2A_{44}^{(i)}\right)\partial_\theta u_r^{(i)}\right. \\ &\quad \left. - \left(A_{23}^{(i)} + 2A_{44}^{(i)} + A_{22}^{(i)}\operatorname{ctg}^2\theta\right)u_\theta^{(i)}\right], \end{aligned} \tag{5}$$

where by, due to the symmetry of the boundary-value problems the solutions can be found in the form of trigonometric sequences [3]:

$$u_r^{(i)} = \sum\nolimits_{n=0}^{\infty} u_{rn}^{(i)}(r)P_n, \quad u_\theta^{(i)} = -\sum\nolimits_{n=0}^{\infty} u_{\theta n}^{(i)}(r)d_\theta P_n. \tag{6}$$

Here $P_n = P_n(\cos\theta)$ is the Legendre polynomial, and $d_\theta(\ldots) \equiv d(\ldots)/d\theta$.

By the substitutions of the expressions (6) into (5) and (4) we obtain the infinite number of systems with ordinary differential equations as follows:

$$\begin{aligned} a_{1n}^{(i)}\left(u_{rn}^{\prime\prime(i)} + \tfrac{2}{r}u_{rn}^{\prime(i)}\right) + \tfrac{1}{r^2}\left(a_{2n}^{(i)}u_{rn}^{(i)} + ra_{3n}^{(i)}u_{\theta n}^{\prime(i)} + a_{4n}^{(i)}u_{\theta n}^{(i)}\right) &= A_n^{(i)}, \\ b_{1n}^{(i)}\left(u_{\theta n}^{\prime\prime(i)} + \tfrac{2}{r}u_{\theta n}^{\prime(i)}\right) + \tfrac{1}{r^2}\left(b_{2n}^{(i)}u_{\theta n}^{(i)} + rb_{3n}^{(i)}u_{rn}^{\prime(i)} + b_{4n}^{(i)}u_{rn}^{(i)}\right) &= B_n^{(i)} \end{aligned} \tag{7}$$

with the corresponding boundary conditions:

$$
\begin{gathered}
u_{rn}^{(2)}\Big|_{r=\rho_2} = u_{\theta n}^{(2)}\Big|_{r=\rho_2} = 0, \quad \left(u_{\theta n}^{\prime(1)} - \tfrac{1}{r}u_{\theta n}^{(1)} + \tfrac{1}{r}u_{rn}^{(1)}\right)\Big|_{r=\rho_1} = 0, \\
\Big\{\left(A_{11}^{(1)}u_{rn}^{\prime(1)} + A_{12}^{(1)}\tfrac{2}{r}u_{rn}^{(1)}\right)P_n(\cos\theta) \\
+ A_{12}^{(1)}\tfrac{1}{r}u_{\theta n}^{(1)}\left[\operatorname{ctg}\theta\, d_\theta P_n(\cos\theta) + d_\theta^2 P_n(\cos\theta)\right]\Big\}\Big|_{r=\rho_1} = \begin{cases} -p, & n=0; \\ 0, & n>0, \end{cases} \\
u_{rn}^{(1)}\Big|_{r=\rho_c} = u_{rn}^{(2)}\Big|_{r=\rho_c}, \quad u_{\theta n}^{(1)}\Big|_{r=\rho_c} = u_{\theta n}^{(2)}\Big|_{r=\rho_c}, \\
A_{44}^{(1)}\left[u_{\theta n}^{\prime(1)} - \tfrac{1}{r}\left(u_{\theta n}^{(1)} - u_{rn}^{(1)}\right)\right]\Big|_{r=\rho_c} = A_{44}^{(2)}\left[u_{\theta n}^{\prime(2)} - \tfrac{1}{r}\left(u_{\theta n}^{(2)} - u_{rn}^{(2)}\right)\right]\Big|_{r=\rho_c}, \\
\left(A_{11}^{(1)}u_{rn}^{\prime(1)} + A_{12}^{(1)}\tfrac{2}{r}u_{rn}^{(1)}\right)P_n + A_{12}^{(1)}\tfrac{1}{r}u_{\theta n}^{(1)}\left(\operatorname{ctg}\theta\, d_\theta P_n + d_\theta^2 P_n\right)\Big|_{r=\rho_c} \\
= \left(A_{11}^{(2)}u_{rn}^{\prime(2)} + A_{12}^{(2)}\tfrac{2}{r}u_{rn}^{(2)}\right)P_n + A_{12}^{(2)}\tfrac{1}{r}u_{\theta n}^{(2)}\left(\operatorname{ctg}\theta\, d_\theta P_n + d_\theta^2 P_n\right)\Big|_{r=\rho_c}.
\end{gathered} \tag{8}
$$

Here $u_{rn}^{\prime(i)} \equiv d\left(u_{rn}^{(i)}\right)/dr$ and

$$
\begin{gathered}
a_{1n}^{(i)} = A_{11}^{(i)}, \quad a_{2n}^{(i)} = 2\left(A_{12}^{(i)} - A_{22}^{(i)} - A_{23}^{(i)}\right) - A_{44}^{(i)}n(n+1), \\
a_{3n}^{(i)} = \left(A_{12}^{(i)} + A_{44}^{(i)}\right)n(n+1), \quad a_{4n}^{(i)} = \left(A_{12}^{(i)} - A_{22}^{(i)} - A_{23}^{(i)} - A_{44}^{(i)}\right)n(n+1), \\
b_{1n}^{(i)} = A_{44}^{(i)}, \quad -b_{2n}^{(i)} = A_{23}^{(i)} + 2A_{44}^{(i)} + A_{22}^{(i)}(n^2+n-1), \quad -b_{3n}^{(i)} = A_{12}^{(i)} + A_{44}^{(i)}, \\
-b_{4n}^{(i)} = A_{22}^{(i)} + A_{23}^{(i)} + 2A_{44}^{(i)}, \quad A_n^{(i)} = \begin{cases} \gamma^{(i)}, & n=1, \\ 0, & n=0, \quad n>1; \end{cases} \quad B_n^{(i)} = \begin{cases} -\gamma^{(i)}, & n=1, \\ 0, & n=0, \quad n>1. \end{cases}
\end{gathered}
$$

Therefore, the calculation of displacements, strains and stresses related to the solution of independent boundary-value problems n. Notably, for $n > 1$ the homogeneous systems of differential equations (7) are complemented with the homogeneous boundary conditions (8). Hence, all these systems have the trivial solution only: $u_{rn}^{(i)} = 0$ and $u_{\theta n}^{(i)} = 0$. This way, the boundary-value problem for a heavy transversally-isotropic combined sphere comes to the solution of the system (7) with the boundary conditions (8) by $n = 0$ and $n = 1$ [1–4].

By $n = 0$ the displacement performs in the meridional direction $u_{\theta 0}^{(i)} \equiv 0$, and the system of equations (7) significantly simplifies:

$$
A_{11}^{(i)}u_{r0}^{\prime\prime(i)} + 2A_{11}^{(i)}\frac{1}{r}u_{r0}^{\prime(i)} + 2\left(A_{12}^{(i)} - A_{22}^{(i)} - A_{23}^{(i)}\right)\frac{1}{r^2}u_{r0}^{(i)} = 0. \tag{9}
$$

The general solution of the Eq. (9) is as follows:

$$
u_{r0}^{(i)} = r^{-1/2-s^{(i)}}D_1^{(i)} + r^{-1/2+s^{(i)}}D_2^{(i)},
$$

where $s^{(i)} = \sqrt{1/4 + 2\left(A_{22}^{(i)} + A_{23}^{(i)} - A_{12}^{(i)}\right)/A_{11}^{(i)}}$, and the boundary conditions (8) help us determine the constants of integration:

$$\rho_2^{-s(2)} D_1^{(2)} + \rho_2^{s(2)} D_2^{(2)} = 0, \quad \rho_c^{-s(1)} D_1^{(1)} + \rho_c^{s(1)} D_2^{(1)} = \rho_c^{-s(2)} D_1^{(2)} + \rho_c^{s(2)} D_2^{(2)}$$
$$D_1^{(1)}\left[2A_{12}^{(1)} - A_{11}^{(1)}\left(\tfrac{1}{2} + s^{(1)}\right)\right]\rho_c^{-s^{(1)}} + D_2^{(1)}\left[A_{11}^{(1)}\left(s^{(1)} - \tfrac{1}{2}\right) + 2A_{12}^{(1)}\right]\rho_c^{s^{(1)}}$$
$$= D_1^{(2)}\left[2A_{12}^{(2)} - A_{11}^{(2)}\left(\tfrac{1}{2} + s^{(2)}\right)\right]\rho_c^{-s^{(2)}} + D_2^{(2)}\left[A_{11}^{(2)}\left(s^{(2)} - \tfrac{1}{2}\right) + 2A_{12}^{(2)}\right]\rho_c^{s^{(2)}}.$$

By $n = 1$ we observe an inhomogeneous system of the differential equations (7), so the solution can be found by the method of arbitrary-constant variation. Finally, let us lay down the displacements in the thick-walled transversally-isotropic sphere as follows [1–4]:

$$u_\theta^{(i)} = \left(C_1^{(i)} + C_2^{(i)}\tfrac{1}{r} + C_3^{(i)} r^{-1/2+t^{(i)}} + C_4^{(i)} r^{-1/2-t^{(i)}} + H_\theta^{(i)} r^2\right)\sin\theta,$$
$$u_r^{(i)} = D_1^{(i)} r^{-1/2-s^{(i)}} + D_2^{(i)} r^{-1/2+s^{(i)}} \tag{10}$$
$$+\left(d_2^{(i)} C_2^{(i)}\tfrac{1}{r} - C_1^{(i)} + d_3^{(i)} C_3^{(i)} r^{-1/2+t^{(i)}} + d_4^{(i)} C_4^{(i)} r^{-1/2-t^{(i)}} + H_r^{(i)} r^2\right)\cos\theta,$$

where

$$H_r^{(i)} H_1^{(i)} = \gamma^{(i)}\left(H^{(i)} - 2A_{12}^{(i)} - 2A_{44}^{(i)}\right), \quad H_\theta^{(i)} H_1^{(i)} = \gamma^{(i)}\left(2A_{11}^{(i)} - H^{(i)} - 2A_{44}^{(i)}\right),$$
$$H^{(i)} = A_{22}^{(i)} + A_{23}^{(i)}, \quad L^{(i)} = 2\left(A_{12}^{(i)} + A_{44}^{(i)}\right)^2 - A_{11}^{(i)}\left(H^{(i)} + 2A_{44}^{(i)}\right),$$
$$d_4^{(i)} L^{(i)} = Y^{(i)} - 2A_{44}^{(i)}\left(A_{12}^{(i)} + A_{44}^{(i)}\right)t^{(i)}, \quad Y^{(i)} = A_{44}^{(i)}\left(A_{12}^{(i)} - 2H^{(i)} - 3A_{44}^{(i)}\right),$$
$$t^{(i)} = \sqrt{\tfrac{9}{4} + \tfrac{1}{A_{11}^{(i)} A_{44}^{(i)}}\left[H^{(i)}\left(A_{11}^{(i)} + 2A_{44}^{(i)}\right) - 2A_{12}^{(i)}\left(A_{12}^{(i)} + 3A_{44}^{(i)}\right)\right]},$$
$$d_2^{(i)} = \frac{H^{(i)} + 2A_{44}^{(i)}}{A_{12}^{(i)} - H^{(i)} - A_{44}^{(i)}}, \quad d_3^{(i)} L^{(i)} = Y^{(i)} + 2A_{44}^{(i)}\left(A_{12}^{(i)} + A_{44}^{(i)}\right)t^{(i)},$$
$$H_1^{(i)} = 2\left[A_{11}^{(i)}\left(H^{(i)} - 4A_{44}^{(i)}\right) + 2A_{44}^{(i)}\left(H^{(i)} - 3A_{12}^{(i)}\right) - 2\left(A_{12}^{(i)}\right)^2\right],$$

The constants of integration $C_1^{(i)}$, $C_2^{(i)}$, $C_3^{(i)}$, and $C_4^{(i)}$, that enter into the formulae (10), are defined from solution of the system of linear algebraic equations (8) [3]:

$$-C_1^{(2)} + d_2^{(2)} C_2^{(2)} \frac{1}{\rho_2} + d_3^{(2)} C_3^{(2)} \rho_2^{-1/2+t^{(2)}} + d_4 C_4^{(2)} \rho_2^{-1/2-t^{(2)}} = -H_r^{(2)} \rho_2^2,$$

$$2\left[C_2^{(1)} A_{44}^{(1)}\left(d_2^{(1)} + 2\right) - C_2^{(2)} A_{44}^{(2)}\left(d_2^{(2)} + 2\right)\right] + C_3^{(1)} A_{44}^{(1)}\left(2d_3^{(1)} - 2t^{(1)} + 3\right)\rho_c^{1/2+t^{(1)}}$$
$$+ 2A_{44}^{(1)}\left(H_r^{(1)} - H_\theta^{(1)}\right)\rho_c^3 + A_{44}^{(1)} C_4^{(1)}\left(2d_4^{(1)} + 2t^{(1)} + 3\right)\rho_c^{1/2-t^{(1)}} = 2A_{44}^{(2)}\left(H_r^{(2)} - H_\theta^{(2)}\right)\rho_c^3$$
$$+ C_3^{(2)} A_{44}^{(2)}\left(2d_3^{(2)} - 2t^{(2)} + 3\right)\rho_c^{1/2+t^{(2)}} + A_{44}^{(2)} C_4^{(2)}\left(2d_4^{(2)} + 2t^{(2)} + 3\right)\rho_c^{1/2-t^{(2)}},$$

$$C_3^{(1)}\left[4A_{12}^{(1)}\left(d_3^{(1)}+1\right)+A_{11}^{(1)}d_3^{(1)}(2t^{(1)}-1)\right]\rho_c^{1/2+t^{(1)}}+2C_2^{(1)}\left[2A_{12}^{(1)}\left(d_2^{(1)}+1\right)-A_{11}^{(1)}d_2^{(1)}\right]$$
$$+C_4^{(1)}\left[4A_{12}^{(1)}\left(d_4^{(1)}+1\right)-A_{11}^{(1)}d_4^{(1)}(2t^{(1)}+1)\right]\rho_c^{1/2-t^{(1)}}+4\left[H_r^{(1)}\left(A_{11}^{(1)}+A_{12}^{(1)}\right)+H_\theta^{(1)}A_{12}^{(1)}\right]\rho_c^3$$
$$=2C_2^{(2)}\left[2A_{12}^{(2)}\left(d_2^{(2)}+1\right)-A_{11}^{(2)}d_2^{(2)}\right]+C_3^{(2)}\left[4A_{12}^{(2)}\left(d_3^{(2)}+1\right)+A_{11}^{(2)}d_3^{(2)}(2t^{(2)}-1)\right]\rho_c^{1/2+t^{(2)}}$$
$$+C_4^{(2)}\left[4A_{12}^{(2)}\left(d_4^{(2)}+1\right)-A_{11}^{(2)}d_4^{(2)}(2t^{(2)}+1)\right]\rho_c^{1/2-t^{(2)}}+4\left[H_r^{(2)}\left(A_{11}^{(2)}+A_{12}^{(2)}\right)+H_\theta^{(2)}A_{12}^{(2)}\right]\rho_c^3,$$

$$\left(C_1^{(2)}-C_1^{(1)}\right)\rho_c+d_2^{(1)}C_2^{(1)}-d_2^{(2)}C_2^{(2)}-\left(H_r^{(2)}-H_r^{(1)}\right)\rho_c^3$$
$$=\rho_c^{1/2}\left(d_3^{(2)}C_3^{(2)}\rho_c^{t^{(2)}}+d_4C_4^{(2)}\rho_c^{-t^{(2)}}-d_3^{(1)}C_3^{(1)}\rho_c^{t^{(1)}}-d_4C_4^{(1)}\rho_c^{-t^{(1)}}\right),$$

$$2C_2^{(1)}\left[2A_{12}^{(1)}\left(d_2^{(1)}+1\right)-A_{11}^{(1)}d_2^{(1)}\right]\rho_1^{-1/2+t^{(1)}}+C_3^{(1)}\left[4A_{12}^{(1)}\left(d_3^{(1)}+1\right)+A_{11}^{(1)}d_3^{(1)}(2t^{(1)}-1)\right]\rho_1^{2t^{(1)}}$$
$$+C_4^{(1)}\left[4A_{12}^{(1)}\left(d_4^{(1)}+1\right)-A_{11}^{(1)}d_4^{(1)}(2t^{(1)}+1)\right]=-4\left[H_r^{(1)}\left(A_{11}^{(1)}+A_{12}^{(1)}\right)+H_\theta^{(1)}A_{12}^{(1)}\right]\rho_1^{5/2+t^{(1)}},$$

$$\left(C_1^{(1)}-C_1^{(2)}\right)\rho_c+C_2^{(1)}-C_2^{(2)}+\left(H_\theta^{(1)}-H_\theta^{(2)}\right)\rho_c^3$$
$$=\rho_c^{1/2}\left(C_3^{(2)}\rho_c^{t^{(2)}}+C_4^{(2)}\rho_c^{-t^{(2)}}-C_3^{(1)}\rho_c^{t^{(1)}}-C_4^{(1)}\rho_c^{-t^{(1)}}\right),$$

$$C_1^{(2)}\rho_2+C_2^{(2)}+C_3^{(2)}\rho_2^{1/2+t^{(2)}}+C_4^{(2)}\rho_2^{1/2-t^{(2)}}=-H_\theta^{(2)}\rho_2^3,$$

$$2C_2^{(1)}\left(d_2^{(1)}+2\right)\rho_1^{-1/2+t^{(1)}}+C_3^{(1)}\left(2d_3^{(1)}-2t^{(1)}+3\right)\rho_1^{2t^{(1)}}$$
$$+C_4^{(1)}\left(2d_4^{(1)}+2t^{(1)}+3\right)=2\left(H_\theta^{(1)}-H_r^{(1)}\right)\rho_1^{5/2+t^{(1)}},$$

and are not demonstrated due to their cumbersome expressions.

The successive substitution of the Eq. (10) into the Cauchy's relation (1), and after that to defining constitutive equations (3) help write the expressions for the components of a stress tensor [3]:

$$\sigma_{rr}^{(i)}=\tfrac{1}{2}r^{-3/2-s^{(i)}}\left\{D_1^{(i)}\left[4A_{12}^{(i)}-A_{11}^{(i)}(1+2s^{(i)})\right]+D_2^{(i)}\left[4A_{12}^{(i)}-A_{11}^{(i)}(1-2s^{(i)})\right]r^{2s^{(i)}}\right\}$$
$$+\left\{2A_{12}^{(i)}\left[\left(1+d_2^{(i)}\right)C_2^{(i)}r^{-2}+\left(1+d_3^{(i)}\right)C_3^{(i)}r^{-3/2+t^{(i)}}+\left(1+d_4^{(i)}\right)C_4^{(i)}r^{-3/2-t^{(i)}}+\left(H_r^{(i)}+H_\theta^{(i)}\right)r\right]\right.$$
$$\left.-\tfrac{1}{2}A_{11}^{(i)}\left[2d_2^{(i)}C_2^{(i)}r^{-2}+(1-2t^{(i)})d_3^{(i)}C_3^{(i)}r^{-3/2+t^{(i)}}+(1+2t^{(i)})d_4^{(i)}C_4^{(i)}r^{-3/2-t^{(i)}}-4H_r^{(i)}r\right]\right\}\cos\theta,$$
$$2\sigma_{\varphi\varphi}^{(i)}=2\sigma_{\theta\theta}^{(i)}=r^{-3/2-s^{(i)}}\left\{D_1^{(i)}\left[2\left(A_{22}^{(i)}+A_{23}^{(i)}\right)-A_{12}^{(i)}(1+2s^{(i)})\right]\right.$$
$$\left.+D_2^{(i)}\left[2\left(A_{22}^{(i)}+A_{23}^{(i)}\right)-A_{12}^{(i)}(1-2s^{(i)})\right]r^{2s^{(i)}}\right\}+2\left\{\left(A_{22}^{(i)}+A_{23}^{(i)}\right)\left[\left(1+d_2^{(i)}\right)C_2^{(i)}r^{-2}\right.\right.$$
$$\left.+\left(1+d_3^{(i)}\right)C_3^{(i)}r^{-3/2+t^{(i)}}+\left(1+d_4^{(i)}\right)C_4^{(i)}r^{-3/2-t^{(i)}}+\left(H_r^{(i)}+H_\theta^{(i)}\right)r\right]-A_{12}^{(i)}\left[2d_2^{(i)}C_2^{(i)}r^{-2}\right.$$
$$\left.\left.+(1-2t^{(i)})d_3^{(i)}C_3^{(i)}r^{-3/2+t^{(i)}}+(1+2t^{(i)})d_4^{(i)}C_4^{(i)}r^{-3/2-t^{(i)}}-4H_r^{(i)}r\right]\right\}\cos\theta,$$
$$\tau_{r\theta}^{(i)}=-A_{44}^{(i)}\left[\left(2+d_2^{(i)}\right)C_2^{(i)}\frac{1}{r^2}+\left(\tfrac{3}{2}+d_3^{(i)}-t^{(i)}\right)C_3^{(i)}r^{-3/2+t^{(i)}}\right.$$
$$\left.+\left(\tfrac{3}{2}+d_4^{(i)}+t^{(i)}\right)C_4^{(i)}r^{-3/2-t^{(i)}}+\left(H_r^{(i)}-H_\theta^{(i)}\right)r\right]\sin\theta.$$

Based on the obtained solutions in the partial case, by the substitution of $E^{(i)} = E_r^{(i)} = E_\theta^{(i)}$, $\mu^{(i)} = \mu_{r\theta}^{(i)} = \mu_{\theta\varphi}^{(i)}$ and $G^{(i)} = G_{r\theta}^{(i)} = E^{(i)}/(1+2\mu^{(i)})$, we receive the expression for stresses, deformations and displacements in the points of hollow and combined heavy isotropic spheres with similar boundary conditions [5, 6].

3 Initial Strength

The found new exact analytical solution (6) permit the analysis of gravity forces' contribution into the stress state of concrete monolithic supports of spherical mine workings surrounded by sedimentary rock mass as well as perform an improved evaluation of initial strength and determine the domains of initial damage. Since rocks and concrete have a natural anisotropy, the underlying approach for the strength analysis of structure elements, produced of such materials, should be a multicriterion approach [7], that examines different mechanisms of partial or full loss of a bearing capacity typical for anisotropic media. In the work [8] independent values were introduced:

$$J^I = \sigma_{\theta\theta}, \quad J^{II} = \sigma_{rr}, \quad J^{III} = \sqrt{(\sigma_{\varphi\varphi} - \sigma_{\theta\theta})^2 + 4\tau_{\varphi\theta}^2}, \quad J^{IV} = \sqrt{\tau_{\varphi r}^2 + \tau_{\theta r}^2},$$

that are invariant to orthogonal transformations, which are acceptable under a transversally-isotropic body. These values help describe diverse damage mechanisms

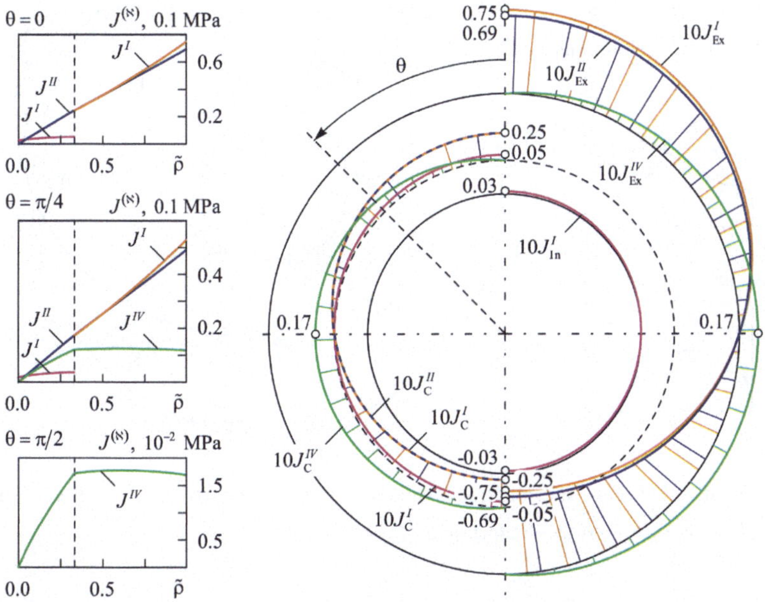

Fig. 1. The distribution of invariants of the stress tensor on the fixed external ($J_{Ex}^{(\aleph)}$), free of pressure internal ($J_{In}^{(\aleph)}$) surfaces and the contact area ($J_C^{(\aleph)}$)

caused by the tension or compression in the hoop or meridional and radial directions, or by the shear on isotropy surfaces and in the diametral plane correspondingly.

The Figs. 1 and 2 show the distributions of non-nil invariants of the stress tensor $J^{(\aleph)}$ in the concrete spherical support ($E_\theta = 40,0\,\text{GPa}$, $E_r = 25,0\,\text{GPa}$, $G_{r\theta} = 11,0\,\text{GPa}$, $\mu_{\theta\varphi} = 0,075$, $\mu_{r\theta} = 0,15$, and $\gamma = 40\,\text{kN/m}^3$) surrounded by sedimentary rock mass ($E_\theta = 55,0\,\text{GPa}$, $E_r = 23,0\,\text{GPa}$, $G_{r\theta} = 29,0\,\text{GPa}$, $\mu_{\theta\varphi} = 0,29$, $\mu_{r\theta} = 0,32$ and $\gamma = 27\,\text{kN/m}^3$) by $p = 0\,\text{MPa}$ along the meridional and dimensionless radial coordinate $\tilde{\rho} = (r - \rho_1)/(\rho_2 - \rho_1)$. The geometry parameters were selected as follows: $\rho_1 = 2,5\,\text{m}$, $\rho_c = 3,1\,\text{m}$, and $\rho_2 = 4,3\,\text{m}$ (see Fig. 1) or $\rho_2 = 9,0\,\text{m}$ (see Fig. 2).

As we see, on the external surface of the monolithic support free of pressure the first invariant J^I is non-nil only, that is nonlinearly distributed along the $\tilde{\rho}$, and has a jump in the contact area. Besides, under the set material constants the character of the distribution J^I is greatly influenced by the ratio of the radiuses of the considered body. For instance, the growth of thickness of a surrounding rock mass leads to the replacement of the first invariant sign in the support.

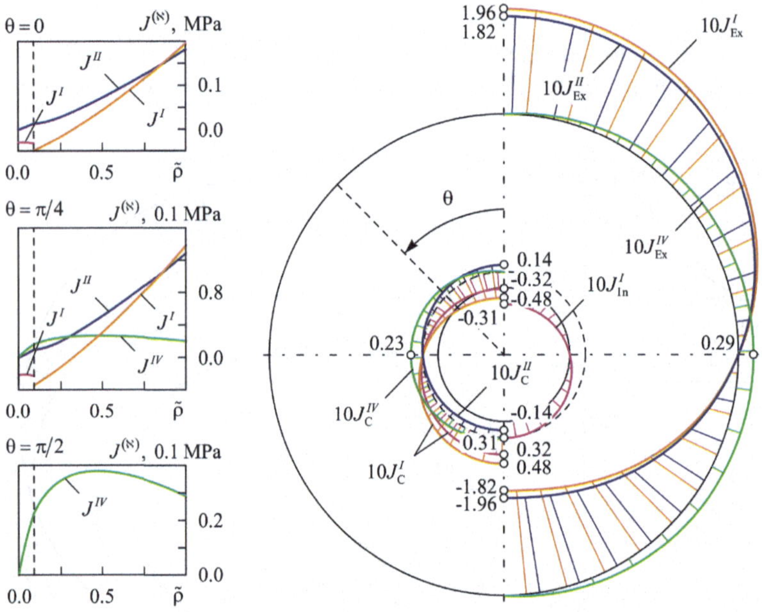

Fig. 2. The distribution of stress tensor invariants on the fixed external ($J^{(\aleph)}_{Ex}$), free of pressure internal ($J^{(\aleph)}_{In}$) surfaces and the contact area ($J^{(\aleph)}_{C}$)

By the change of the radial coordinate from the free of pressure internal surface to the fixed external surface, the second invariant J^{II} increases all around in the absolute value, at the same time, in the vault point it does not possess negative values, and in the lower half-sphere it does not possess positive values. Since J^{I} and J^{II} reach their highest absolute values in the points of the vertical axis that crosses the geometric center, these points are considered most dangerous as they might initiate support's damage caused by the tension or compression in the hoop and radial direction.

The third invariant J^{III} possess zero values in all the points of the concrete support and the surrounding sedimentary rock mass. Therefore, under the set load conditions and the type of an elastic symmetry damage mechanisms are not caused by the shear on the surface of isotropy.

In the points placed on the vertical central axis the forth invariant J^{IV} is reduced to zero and increases as the angle θ grows, and reaches its maximum values in case when $\theta = \pi/2$. Besides, in the radial direction from the fixed external surface of the monolithic support to the internal surface J^{IV} is decreasing all over and reaches zero values on the free of pressure internal surface. Hence, the points that cross the geometric center are deemed most dangerous in terms of potential support's damage that starts from the shear mechanism in the diametric plane.

Acknowledgements. The authors acknowledge the support of the Russian Foundation for Basic Research (Grant RFBR–Urals No 17–41–590148).

References

1. Zaitsev, A.V., Fukalov, A.A.: Elastic equilibrium state of thick-walled heavy transversally-isotropic spheres fixed on the interior surface. J. Samara State Tech. Univ. Ser. Phys. Math. Sci. **5**(21), 85–95 (2010)
2. Fukalov, A.A., Kutergin, A.V.: Exact analytical solutions to problems of the equilibrium state of elastic anisotropic heavy central and axial-symmetric bodies and their applications. Vestnik of Lobachevsky State Univ. of N. Novgorod, vol. 4, no. 4, pp. 1831–1833 (2011)
3. Zaitsev, A.V., Sokolkin, YuV, Fukalov, A.A.: Initial damage mechanisms of reinforced concrete monolithic supports for spherical mine workings located in sedimentary rock mass. PNRPU Mech. Bull. **4**, 59–74 (2013)
4. Zaitsev, A.V., Kutergin, A.V.: Equilibrium state of elastic thick-walled heavy horizontal orthotropic cylinder subjected to the action of nonuniform external lateral pressure. PNRPU Mech. Bull. **4**, 36–45 (2010)
5. Kozhevnikova, L.L., Kuznetsov, G.B., Matveenko, V.P., Shardakov, I.N.: Analytical investigation of the elastic equilibrium of a hollow sphere rigidly fixed along the outer contour. Strength Mater. **6**, 1057–1065 (1974)
6. Kuznetsov, G.B.: Elastic, Viscoelastic and Long-Term Strength of Cylindrical and Spherical Bodies. Nauka, Moscow (1979)
7. Wildemann, V.E., Sokolkin, YuV, Tashkinov, A.A.: Mechanics of Inelastic Deformation and Failure of Composite Materials. Nauka, Moscow (1997)
8. Pobedrya, B.E.: Mechanics of Composite Materials. Moscow State University Press, Moscow (1984)

On Influence of Inclusions and Cracks on Mechanical Properties of Rocks

K. B. Ustinov(✉) and E. V. Stepanova

A.Yu. Ishlinsky Institute for Problems in Mechanics RAS, Moscow, Russia
ustinov@ipmnet.ru

Abstract. The basic concepts of mechanics of heterogeneous media in relation to its application to rock mechanics are outlined. Brief evaluative review of approaches for obtaining effective characteristic is given. A special attention is devoted to specifying the ranges of applicability of the asymptotic and approximate expressions for effective elastic characteristics for particular cases of media with inclusions. Application of the general results of heterogeneous materials to rocks is described. Questions of co-relation between elastic properties of different nature is discussed.

Keywords: Heterogeneous media · Effective properties · Fracturing · Eshelby's tensor · Range of applicability · Cross-property correlation

1 Basic Concepts

Rocks are heterogeneous according to their nature. Heterogeneity is caused by both mineralogical composition and existence of defects, such as cracks, pores, inhomogeneities. This property is essential and have to be taken into account while describing mechanical properties of rocks.

One of the key concepts which is necessary for description mechanical processes in rocks and rock massifs in terms of continuum mechanics is the elementary volume concept. On the one hand, this volume has to be sufficiently small in comparison with dimensions of the domain under consideration and comparing to regions, within which the stresses changes significantly. And on the other hand the volume should be sufficiently large comparing to elements of structure. The peculiarity of rocks consists in the mutual contradiction of these requirements [1, 2], because typical sizes of inhomogeneities and defects may be comparable with the characteristic dimensions of the regions in question. In this relation problems of description of rock as a continuum possessing effective properties are of great importance.

Nowadays a lot of approaches have been developed allowing to describe effective (in particular, elastic) properties of heterogeneous media (e.g. [3]), starting from rather weak estimations based on "rule of mix", up to rigorous theoretical bounds, such as Hashin–Shtrikman bounds, restricting possible variation of effective moduli of mixes depending on properties of phases [4].

V. I. Karev et al. (Eds.): *Physical and Mathematical Modeling of Earth and Environment Processes (2018)*, SPEES, pp. 472–479, 2019.
https://doi.org/10.1007/978-3-030-11533-3_47

However these methods lose their attraction with increasing deviation of properties of phases: for the limit case of inclusions of zero modulus (voids) the bounds for moduli widen covering values varying from zero stiffness to the value of the intact material.

2 Inclusions and Cracks

To overcome the obstacles related to ambiguity in case of high difference in phases properties and to deal with other problems related to bodies with inclusions, the Eshelby method of an equivalent inclusion [5], (see also [6, 7]) seems the most appropriate. Nowadays on the basis of this approach solutions of rather general type are constructed yielding expressions for effective moduli of media with dilute concentration of inclusions having shapes of 3-axial ellipsoids [6]. The solution may be written as follows

$$\Lambda_{eff} = \Lambda_0 + (\Lambda_* - \Lambda_0)\left[I + S\Lambda_0^{-1}(\Lambda_* - \Lambda_0)\right]^{-1}\Omega \tag{1}$$

Here $\Lambda_0, \Lambda_*, \Lambda_{eff}$ are tensors of elasticity of matrix, inclusion and mixture, respectively; S is Eshelby's tensor [5], that depends on the inclusion shape; I is the unit tensor; Ω is volumetric concentration of the inclusions. A representation of the 4-th rank tensors involved in (1) as matrices may be applied for convenience.

Restriction imposed by considering inclusions elliptical is not drastic, because replacing inclusions of irregular shape with equivalent ellipsoids while modeling usually does not introduce strong additional error [8].

Since expressions for the effective characteristics (moduli) for the general case are rather awkward, solutions for particular cases are obtained, including the important cases of spherical inclusions, rigid disk-like inclusions, penny-shaped cracks, disk-like inclusions of material with not very high differences in elastic moduli from the matrix [9–11].

For media containing cracks, such as rocks, expressions for effective elastic moduli may be obtained not only by limit transitions of the general solution for media with inhomogeneities, but also by direct summation of deformation of the intact body and additional deformation caused by opening of the crack faces [10]. Contribution to increase in compliance (decrease of moduli) in this case is proportional to average value of the third power of the cracks sizes in their planes, rather than to volumetric fraction of defects (vanishing for ideal cracks having zero thickness). For effective elastic characteristics determination the fact mentioned above leads to inability of consideration of conventionally used values as characteristics of cracked media in geology and mining science.

3 Limiting Cases of Inclusions Properties

Degenerate cases, such as extremely thin and compliant/rigid inclusions, or needle-like rigid inclusions require careful treatment. More generally, such cases cover inclusion with very strong differences in dimensions, and which elastic properties strongly differ from the properties of the matrix. Examples of such media are media with cracks filled with fluid, debris or another compliant substance.

Existence of two small (or large if using the inverse values) parameters makes estimation of ranges of applicability not obvious for the particular simplified models.

The works [12–14] are devoted to determining ranges of applicability of the particular asymptotic models for soft/rigid and prolate/oblate spheroid inclusions. In these works the solutions for the particular cases are obtained by limit transitions or developing the general solution for spheroid inclusion with semi-axes $a_1,\ a_2 = a_3$ into series on the corresponding small parameters. Seven regions were distinguished for combinations of two parameters: aspect ratio $\rho = a_3/a_1$, and ratio of the shear moduli of the inclusion and the matrix $\eta = \mu_*/\mu_0$. For each region formulae for determining the effective moduli and elastic energy were written for both aligned inclusions and for inclusions distributed isotropically over orientations. For inclusions distributed isotropically over directions the formulae for effective compressive k^{eff} and shear μ^{eff} moduli for those seven cases are

A. extremely soft flat (crack-like) inclusions $\eta^{-1} \ll \rho^{-1} << 1$

$$k^{eff} = k^0\left[1 - \frac{4}{3\pi}\frac{(1-\nu_0^2)}{(1-2\nu_0)}\Omega\rho^{-1}\right],\quad \mu^{eff} = \mu^0\left[1 - \frac{8}{15\pi}\frac{(1-\nu_0)(5-\nu_0)}{(2-\nu_0)}\Omega\rho^{-1}\right] \tag{2}$$

B. extremely flat soft inclusions $\rho^{-1} \ll \eta^{-1} \ll 1$

$$k^{eff} = k^0\left[1 - \frac{3k^0}{3k^* + 4\mu^*}\Omega\right],\quad \mu^{eff} = \mu^0\left[1 - \frac{\mu^0(6k^* + 12\mu^*)}{5\mu^*(3k^* + 4\mu^*)}\Omega\right] \tag{3}$$

C. extremely flat rigid inclusions $\rho^{-1} \gg \eta \gg 1$

$$k^{eff} = k^0\left[1 - \frac{3k^0}{3k^* + 4\mu^*}\Omega\right],\quad \mu^{eff} = \mu^0\left[1 + \frac{\mu^*(9k^* + 8\mu^*)}{5\mu^0(3k^* + 4\mu^*)}\Omega\right] \tag{4}$$

D. extremely rigid flat inclusions $\eta \gg \rho^{-1} \gg 1$

$$\begin{aligned} k^{eff} &= k^0\left[1 + \frac{16}{3\pi}\frac{(1-\nu^0)(1-2\nu^0)}{(3-4\nu^0)(1+\nu^0)}\Omega\rho^{-1}\right],\\ \mu^{eff} &= \mu^0\left[1 + \frac{8}{15\pi}\frac{(1-\nu^0)(43-56\nu^0)}{(3-4\nu^0)(7-8\nu^0)}\Omega\rho^{-1}\right] \end{aligned} \tag{5}$$

E. extremely rigid needle-like inclusions $\eta \gg \rho \gg 1$

$$\begin{aligned} k^{eff} &= k^0\left[1 + \frac{2(1-\nu^0)(1-2\nu^0)}{3(1+\nu^0)[4(1-\nu^0)(\ln\eta - 1) - 1]}\Omega\eta^2\right],\\ \mu^{eff} &= \mu^0\left[1 + \frac{4(1-\nu^0)}{15[4(1-\nu^0)(\ln\eta - 1) - 1]}\Omega\eta^2\right] \end{aligned} \tag{6}$$

F. needle-like rigid inclusions $\rho \gg \eta \gg 1$

$$k^{eff} = k^0\left[1 + \frac{k^*\mu^*}{k^0(3k^* + \mu^*)}\,\Omega\right],$$
$$\mu^{eff} = \mu^0\left[1 + \frac{24k^0k^*\mu^*}{5\mu^0(k^* + \mu^*)(3k^0 + 7\mu^0)}\,\Omega\right] \tag{7}$$

G. needle-like soft inclusions $\rho \gg 1,\ \eta \gg 1$

$$k^{eff} = k^0\left[1 - \frac{k^0 + \mu^0}{\mu^0}\,\Omega\right],\ \mu^{eff} = \mu^0\left[1 - \frac{64k^0}{5(3k^0 + \mu^0)}\,\Omega\right] \tag{8}$$

Here ν_0 stands for Poisson's ratio of the matrix.

Diagram of the ranges of applicability of the asymptotics is given in Fig. 1. Thick and thin lines bounding ranges within which the corresponding asymptotics are valid with the accuracy of 3% and 10% respectively.

For all areas except *A, D* and *E* in Fig. 1 the changes in elastic moduli depend on volumetric concentration of the inclusions. For areas *A* and *D*, corresponding to extremely soft and extremely rigid disk-like inclusions the changes in elastic moduli depend on volumetric concentration divided by small parameter ρ, aspect ratio (which corresponds to the averaged third power of radius multiplied by the number of inclusion per volume unit). For area *E*, corresponding to extremely rigid needle-like inclusions the changes in elastic moduli depend on volumetric concentration multiplied by large parameter ρ, aspect ratio (which corresponds to the averaged third power of length multiplied by the number of inclusion per volume unit). Therefore for these three cases the impact is proportional to the third power of the maximal dimension of the inclusion.

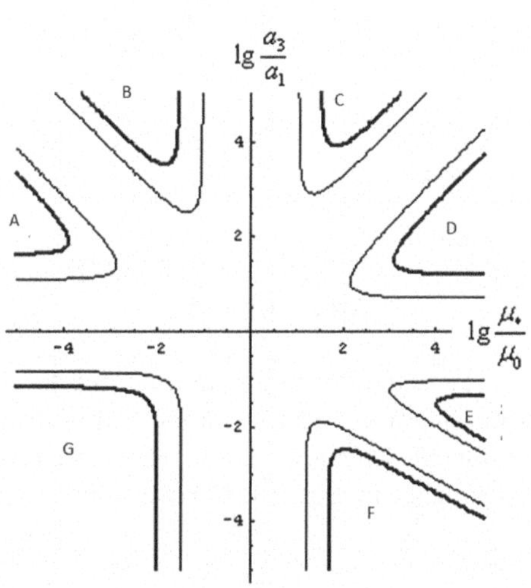

Fig. 1. Diagram of ranges of applicability of asymptotics for soft/rigid and penny-shaped/needle-like inclusions. *A* corresponds to $\eta^{-1} \ll \rho^{-1} \ll 1$; *B* - to $\rho^{-1} \ll \eta^{-1} \ll 1$; *C* - to $\rho^{-1} \gg \eta \gg 1$; *D* - to $\eta \gg \rho^{-1} \gg 1$; *E* - to $\eta \gg \rho \gg 1$; *F* - to $\rho \gg \eta \gg 1$; *G* - to $\rho \gg 1,\ \eta \gg 1$ (see (2)–(8)).

4 Approaches for Finite Concentrations

In case of finite concentration of inclusions they may not generally be considered as non-interacting. Generally, for media with finite concentration of inclusions there may not exist the universal solution, because the effective characteristics may strongly depend on particular realization of the distribution of the inclusions in space. However few powerful approximations are successfully used: method of effective field and differential method of effective media, also called differential scheme.

According to effective field method [15] each inclusion is considered as imbedded into the media with elastic properties of the matrix, influence of the other inclusions is accounted for in terms of effective field that is the sum of the externally applied field and additional field caused by action of all other inclusions. In the simplest variant, namely Mori-Tanaka method, this field is considered as uniform, which results to the following expressions for the effective moduli

$$\Lambda_{eff} = \Lambda_0 + (\Lambda_* - \Lambda_0)\left[(1 - \Omega)I + S\Lambda_0^{-1}(\Lambda_* - \Lambda_0)\right]^{-1}\Omega$$

Alternatively the effective media is used, according to which each inclusion is considered as imbedded into the media with elastic properties that coincides with the effective properties in question, and thus account for the presence of the other inclusion. The field acting on each inclusion is supposed to be the externally applied field [16]. In its original form, this method has been rightly criticized for unrealistic prediction for high concentration of inclusions. This inadequacy is eliminated in the differential form of the method, according to which, a body with a finite concentration of inclusions may be considered as the result of the process of successive imposing infinitesimal portions of inclusions [10], with the key assumptions of the method being supposed to be satisfied for the each increment. According to this modified method the effective elastic properties are found as the solution of the system of differential equations

$$(1 - \Omega)\frac{d\Lambda}{d\Omega} = (\Lambda_* - \Lambda)\left[I + S\Lambda^{-1}(\Lambda_* - \Lambda)\right]^{-1}$$

With the boundary condition

$$\Lambda(\Omega = 0) = \Lambda_0$$

Here coefficient $(1 - \Omega)$ at the derivative appears due to eliminating part of the volume at each increment already occupied with the inclusions [15, 17]

Apart from the considered schemes among others method of periodic lattices (e.g. [15]), and method of higher moments [18] have to be mentioned.

5 On Correlation of Properties of Various Nature

The problem of establishing correlations between different properties of heterogeneous materials in terms of structure attracted researches since may be work of [19]. Further the problem was studied in [20, 21].

In possible co-relations between properties of different natures, the key role is played by the tensor or non-tensor character, and ranks of the corresponding tensors. Examples of properties characterized by tensors of zero rank (scalars) are density, thermal capacity; examples of properties characterized by the 2-nd rank tensors are thermal expansion, thermal conductivity, electrical conductivity, relative permittivity, Biot's coefficient; examples of properties characterized by the 3-rd and 4-th rank tensors are piezoelectricity and elasticity, respectively. There are, also, non-scalar properties that do not express tensor nature, such as strength, fracture toughness. Co-relation between properties describing tensors with equal ranks may expected to be strong, however that is not necessarily, for example, if the inclusion-matrix contrast in the bulk moduli differ from the one in shear moduli [21]. One may expect much less co-relation between properties describing tensors with different ranks, although cor-relation is found for some cases, for example between elastic and conductive properties for bodies with cracks [22].

For properties, at least one of which is not described by a tensor, one could not expect noticeable co-relation. Consider, for example, influence of cracks on changes of elastic (4-th rank tensor) and strength (not a tensor value, e.g. [23, 24]) properties. Influence of cracks on the change in elastic moduli is proportional to the average of the third power of cracks radii, while the strength is determined by the size of the largest cracks. So the difference of the influence in these two cases is strongly depends on the distribution of cracks on sizes.

6 Discussion and Conclusion

Presence of cracks in rocks leads to one additional essential peculiarity of their mechanical behavior, namely, nonlinearity. For fractured rocks nonlinearity is observed even in elastic region of deformation. That effect is caused by gradual "switching off" due to closing of groups of cracks that occurs during compression [10]. Presence of cracks leads also to difference in elastic moduli during loading. In other case a share of the closed cracks may not open during reduction of compressive stress due to roughness of the crack faces causing elastic hysteresis.

Growth of multiple cracks and other defects during loading is the main reason for inelastic behavior of rocks. In order to describe such behavior the evolution law of cracks characteristics, or more general destruction, has to be specified.

Note, that since deformation and fracture processes are described by properties not having common tensor origin, the direct transform of tensor damage parameter describing deformation to address fracture may lead to strong discrepancy of the results of modeling [25].

Influence of distribution of crack over sizes is observed in measurements of elastic wave velocities. The reason is that effective elastic moduli depend strongly on the ratio of wave length and the characteristic size of the cracks. Thus relation between static and dynamic moduli, very important in mining science, is not obvious and determined not only by the difference between isothermic and adiabatic properties.

Acknowledgement. The work was done in the framework of the State assignment № AAAA-A17-117021310371-9.

References

1. Goodman, R.E.: Introduction to Rock Mechanics, 2nd edn. Wiley, New York and Chichester (1989)
2. Jaeger, J.C., Cook, N.G.W., Zimmerman, R.W.: Fundamentals of rock mechanics, 4th edn. Blackwell Publishing, Malden (2007)
3. Christensen, R.M.: Mechanics of composite materials. Lawrence Livermore Laboratory, University of California, Livermore. Wiley, New York (1979)
4. Hashin, Z., Shtrikman, S.: On some variational principles in anisotropic and nonhomogeneous elasticity. J. Mech. Phys. Solids **10**, 335–342 (1962)
5. Eshelby, J.D.: The determination of the elastic field of an ellipsoidal inclusion, and related problems. Proc. R. Soc. London. A **241**, 376–396 (1957)
6. Mura, T.: Micromechanics of Defects in Solids. Martinus Nijhoff Publishers, The Hague (1982)
7. Freidin, A.B.: Mechanics of fracture: Eshelby problem: textbook for higher educational institutions. St.-Petersburg State Polytechnic University Publishing house (2010). (in Russian)
8. Kachanov, M.: Effective elastic properties of cracked solids: critical review of some basic concepts. Appl. Mech. Rev. **45**, 304–335 (1992)
9. Walpole, L.J.: On the overall elastic moduli of composite materials. J. Mech. Phys. Solids **17** (4), 235–251 (1969)
10. Salganik, R.L.: Mechanics of bodies with a large number of cracks. Izv. USSR Ac. Sci. MTT. **4**, 149–158 (1973). (in Russian)
11. Kovalenko, Y.F., Salganik, R.L.: Fractured inhomogeneities and their influence on the effective mechanical characteristics. Izv. USSR Ac. Sci. MTT. **5**, 76–86 (1977)
12. Ustinov, K.B., Goldstein, R.V.: On application of classical Eshelby approach to calculating effective elastic moduli of dispersed composites. Int. J. Fracture. **147**(1–4), 55–66 (2007)
13. Ustinov, K.B.: Computation of the energy of an inhomogeneity: asymptotics and their scope. Mech. Solids **45**(2), 238–246 (2010)
14. Semenova, D.V., Ustinov, K.B.: Asymptotic computation of the inhomogeneity energy in a body in an external stress field. Mech. Solids **46**(3), 400–415 (2011)
15. Kanaun, S.K., Levin, V.M.: Self-Consistent Methods for Composites: Vol. 1 – Static Problems. Springer, Dordrecht (2007)
16. Budiansky, B.: On the elastic moduli of some heterogeneous materials. J. Mech. Phys. Solids **13**, 223–227 (1965)
17. Brinkman, H.C.: The viscosity of concentrated suspensions and solution. J. Chem. Phys. **20** (4), 571 (1952). https://doi.org/10.1063/1.1700493
18. Beran, M.: Statistical Continuum Theories. Springer, Vienna (1968)

19. Bristow, J.R.: Microcracks, and the static and dynamic elastic constants of annealed heavily cold-worked metals. British J. Appl. Phys. **11**, 81–85 (1960)
20. Milton, G.W.: The Theory of Composites. Cambridge University Press (2002). https://doi.org/10.1017/cbo9780511613357
21. Sevostianov, I., Kachanov, M.: Explicit cross-property correlations for anisotropic two-phase composite materials. J. Mech. Phys. Solids **50**, 253–282 (2002)
22. Sevostianov, I.: Explicit relations between elastic and conductive properties of a material containing annular cracks. Phil. Trans. Roy. Soc. L. A **361**, 987–999 (2003)
23. Singh, M., Samadhiya, N.K., Kumar, A., Kumar, V., Singh, B.: A nonlinear criterion for triaxial strength of inherently anisotropic rocks. Rock Mech. Rock Eng. **48**(4), 1387–1405 (2015)
24. Karev, V.I., Klimov, D.M., Kovalenko, Y.F., Ustinov, K.B.: Fracture model of anisotropic rocks under complex loading. Phys. Mesomech. **21**(3), 216–222 (2018)
25. Salganik, R.L., Gotlib, V.A.: Continuum versus discontinuum damage mechanics of creep caused by microcracking. Int. J. Fracture **101**, 181–201 (2000)

Scenario Planning in Arctic Offshore Oil and Natural Gas Resources Development: A Case of Russia

M. O. Morgunova[1(✉)] and E. A. Telegina[2]

[1] Joint Institute for High Temperatures of RAS, Moscow, Russia
maymorgunova@mail.ru
[2] Gubkin Russian State University of Oil and Gas (National Research University), Moscow, Russia

Abstract. Arctic offshore is considered to have high potential for new commercial discoveries of oil and natural gas. One of the most engaged countries in the offshore Arctic oil and natural gas resources development is Russia. However, due to the changing geopolitical and geoeconomic landscape Arctic offshore oil and natural resources development experience some challenges. The study focuses on evaluating Arctic offshore oil and natural gas resources development perspectives in Russia for a long-term period. This includes an algorithm and guidelines to identify the long-term strategic directions for the development these resources, and a scenario planning forecast of their development until 2050. The maximum annual projected production volumes of oil and natural gas combined from the offshore fields in the Russian Arctic compose approximately 1% of total production in Russia in 2016. Based on the obtained results, it is necessary to adjust the strategy for developing the offshore oil and natural gas resources in the Russian Arctic in accordance to the potential production volumes and changing macroeconomic situation.

Keywords: Arctic offshore · Russia · Scenario planning · Oil and natural gas

1 Introduction

There is a huge potential of new discoveries of oil and natural gas resources in the world, which includes remote and hard-to-reach resources offshore Arctic. The interest towards developing the oil and natural gas resources offshore Arctic is mainly motivated by its high oil and natural gas potential, the possibility of expanding the resource base for the energy companies, the importance these resources for some Arctic countries as a source of income, and the possibility of diversifying energy resources production [1].

Interest in oil and natural gas resources in the Arctic region appeared in Canada, the United States and Russia in the 1920s. Large-scale discoveries were made in the 1940s–50s, which activated oil and natural gas production at the fields around the Arctic Circle. In the late 1980s, significant discoveries took place in the Arctic seas, which led to a new round of activities in the Arctic region. In the period of stable and high oil prices in 2011–2014, there was a considerable interest in the development of

V. I. Karev et al. (Eds.): *Physical and Mathematical Modeling of Earth and Environment Processes (2018)*, SPEES, pp. 480–490, 2019.
https://doi.org/10.1007/978-3-030-11533-3_48

these resources from the Arctic states, national and international energy companies. At a later stage, when oil prices went down, many of the companies and governments of the Arctic countries revised their Arctic strategies in relation to the offshore oil and natural gas resources. In 2016 oil and natural gas production from the offshore Arctic fields globally was less than 1%, which is significantly lower than previously expected volumes [2]. In addition, oil and natural gas production in the Arctic is associated with a number of technological, climatic and logistical challenges, as well as the relatively high operating costs compared to other oil and natural gas provinces.

Oil and natural gas resources development expanded in the Russian Arctic and went further to offshore areas, which in general is considered as a very ambitious and important national mega project. The development pace of the Arctic offshore oil and natural gas resources in Russia is also changing due to the changing geopolitical and geoeconomic landscape associated primarily with the dramatic decrease of oil prices, as well as sanctions imposed on Russia by the European Union and the United States in 2014 [3]. It was planned to start production on 22 offshore Arctic fields by 2030 (according to the strategic goals of 2014). As a result of low oil price and sanctions the development of such giants as the Kharasaveyskoye, Kruzenshternskoye, Shtokmanovskoye fields was delayed.

The majority of challenges related to the offshore Arctic oil and natural gas resources development is studied by national and international researchers (for instance, see [2: 27–28]). However, the more detailed study of the oil and natural gas development prospects offshore Russian Arctic during the changing geopolitical and geoeconomic landscape from the long strategic planning point of view has not yet received a proper attention. The lack of medium- and long-term planning makes it difficult for the companies and government take decisions on the oil and natural gas resource development in the Arctic offshore areas in Russia. Imbalance is observed in rather short terms of strategic planning, lack of national macroeconomic forecasts, as well as inaccuracy in evaluations of oil and natural gas production and consumption volumes. For example, the government program documents are valid until 2020–2025, whereas the average time from discovery to the start of production of the offshore Arctic fields takes more than 20 years [2]. Currently, there is a need to revise the planning and forecasting mechanisms to re-assess the oil and natural gas development prospects offshore Russian Arctic from the long-term perspective. This also includes national strategic interests in the context of current macroeconomic trends.

The paper is structured as follows. Section 2 is focused on the methodological part of the study. Section 3 is presenting the results of the study in three subsections, followed by concluding remarks in Sect. 4.

2 Methods

The study is based on the in-depth literature review of international and national primary and secondary sources. Sources studied include publications of the International Energy Agency, the World Bank, the European Commission, the Arctic Council, the United Nations, the World Energy Council, as well as research publications of consultancy companies, analytical centres and scientific research institutes, scientific and industry periodicals, official statistical reports, open access materials of national and international energy companies.

In order to evaluate Arctic offshore oil and natural gas resources development perspectives in Russia for a long-term period during the changing geopolitical and geoeconomic landscape the following instruments were used.

First, the macroeconomic analytical pre-study was conducted with the purpose to create a consensus forecast. Extrapolation methods were used to adjust the indicators to the same flagship dates in the forecast.

Second, the risk assessment based on cross-measures of frequency and probability of risk occurrence were conducted in accordance with the World Energy Council [4].

Third, the scenario planning approaches of macroeconomic and industry type from the World Energy Council [5] were adapted to the study. Complex analysis methods were used to construct scenarios, including qualitative and quantitative tools (extrapolation and expert evaluation methods based on secondary sources). Quantitative macroeconomic parameters were extrapolated to 2050 based their average growth in the consensus forecast.

Forth, the industry specificity of scenarios is based on an assessment of the degree of development of oil and natural gas fields offshore Russian Arctic according the method suggested by Fadeev [6].

Finally, the algorithm and corresponding methodical recommendations are based on the principles of the Federal Law of the Russian Federation "On Strategic Planning in the Russian Federation" [7].

3 Results

Within the purpose to evaluate Arctic offshore oil and natural gas resources development perspectives in Russia for a long-term period during the changing geopolitical and geoeconomic landscape, the two tasks were defined:

- to develop an algorithm and propose methodical recommendations based on Russia's strategic priorities;
- to develop a forecast on the oil and natural gas resources development offshore Russian Arctic until 2050.

The results are presented in Subsects. 3.1 and 3.2 correspondingly, followed by forecasted production of oil and natural gas offshore Russian Arctic based on the developed scenarios in Subsect. 3.3.

3.1 Algorithm and Methodical Recommendations to Evaluate Arctic Offshore Oil and Natural Gas Resources Development Perspectives in Russia

The changing geopolitical and geoeconomic landscape creates a risky environment for the development of oil and natural gas resources offshore Arctic. Currently, Russia has limited technical and financial capabilities to independently ensure the full-scale development of these resources and require adjusting management approaches. High production costs, remoteness from the energy markets, as well as the current sanctions regime for technology transfer and restrictions on attracting capital, reduce the competitiveness of Russia's offshore Arctic oil and natural gas resources and postpone their development. Most of the long-term goals and objectives for the development of the Arctic region and these resources were not implemented due to the before mentioned changing geopolitical and geoeconomic landscape. The lack of medium and long-term planning complicates the decision-making process and planning for the development of the offshore Arctic resources. The imbalance is observed in short periods of strategic planning, the absence of a macroeconomic forecast, inaccuracy in the forecasting of oil and natural gas production volumes [8:84].

Based on the strategic importance of the Arctic region for Russia and its high oil and natural gas resource potential, it is necessary to revise the relevant strategic documents and corresponding macroeconomic indicators. The increased influence of external geopolitical and geoeconomic factors in the energy sector determines the need to develop a realistic strategy for the Arctic offshore oil and natural gas resources exploration, taking into account the macroeconomic situation. These require long-term and comprehensive planning.

In order to develop a methodical approach to evaluate Arctic offshore oil and natural gas resources development perspectives for a long-term period the paper suggests an algorithm (Fig. 1). This algorithm consists of five consecutive steps, takes into account macroeconomic changes and risks, and complies with the principles of the Federal Law of the Russian Federation "On Strategic Planning in the Russian Federation" [7]. It also includes the feasibility analysis and helps to determine the long-term prospects of oil and natural gas resources development offshore Russian Arctic based on national strategic priorities [8].

In addition to the algorithm, the study suggests supplementing recommendations, which are to provide guidelines for the stepwise and comprehensive evaluation of the Arctic offshore oil and natural gas resources development perspectives:

1. Analyse the key strategic priorities of Russia in the Arctic region for the development of offshore oil and natural gas resources, long-term priorities, goals and objectives of socioeconomic development based on national strategic documents;
2. Conduct a macroeconomic analysis based on the current transformation trends using national and international sources;
3. Analyse the current experience and specifics of the Arctic offshore oil and natural gas resources development and assess the level oil and natural gas fields' development in accordance with technological and economic criteria;
4. Assess the risks of developing the oil and natural gas resources offshore Arctic and rank them accordingly using qualitative and quantitative indicators;
5. Create a long-term forecast for the possible development of the oil and natural gas resources offshore Russian Arctic based on at least three macroeconomic scenarios using scenario planning tools and methods;

6. Estimate the potential volumes of oil and natural gas production offshore Russian Arctic;
7. Develop recommendations for adjusting the management approaches of the Arctic offshore oil and natural gas resources development;
8. Revise the forecast at least once in a three-year period, and at least once a year in a case of radical macroeconomic changes.

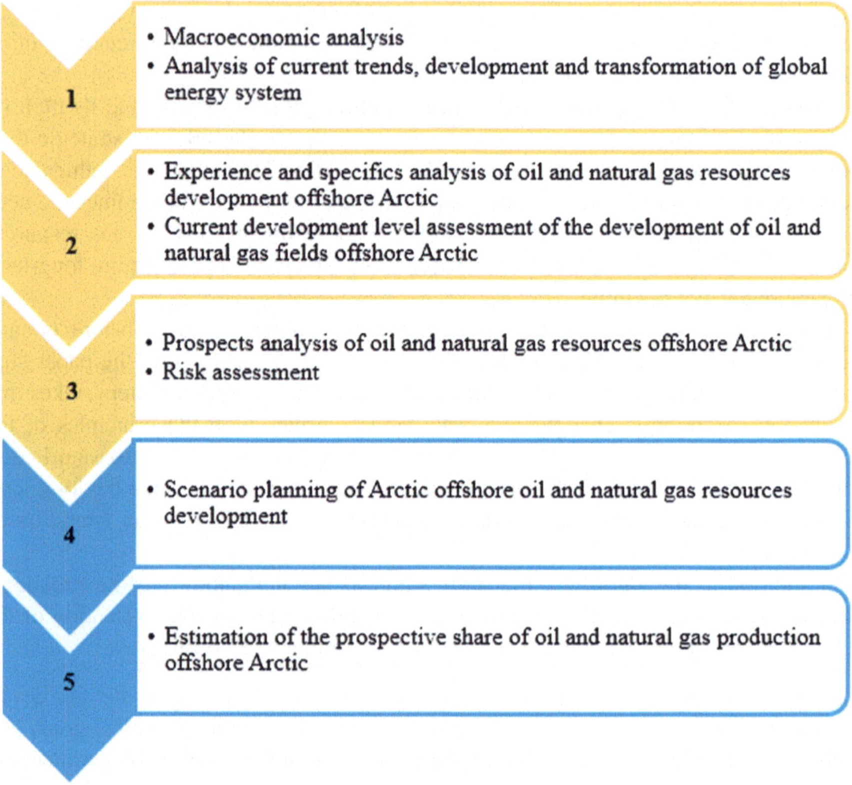

Fig. 1. The scenario planning algorithm [8].

The application of the algorithm and methodical guidelines can serve as a base for the evaluation of the Arctic offshore oil and natural gas resources development perspectives in Russia for a long-term period. Regularly applied, those instruments can contribute to the current existing planning approaches.

3.2 Scenario Forecast of Arctic Offshore Oil and Natural Gas Production in Russia

Estimation of the development prospects of oil and natural gas resources offshore Russian Arctic based on the existing forecasts is challenging, since these models are usually created for developed countries with stable institutional systems. Most of those forecasts are based on predefined trends in technological development (inertial forecasts).

In high-risk environments and in a changing macroeconomic situation (which are sanctions and oil price instability), scenario tools are often used. Scenario planning is a key instrument in forecasting broadly used in Russia and abroad. It is one of the most applicable options for the comprehensive forecasts development, which takes into account the diverse geopolitical and geoeconomic factors, as well as allows using interdisciplinary analytical tools. It has a certain 'degree of freedom' since allows to choose preferable input indicators [5], such as macroeconomic, technological, political, geopolitical, etc. For the long term, scenario forecasting of macroeconomic and industry type is applicable. Since long-term planning models are subject to limitations and changes, the common practice is constantly updating the forecasts.

This study considers three scenarios - alternative, moderate and active, which differ in macroeconomic indicators and trends in the development of the global energy market, as well as the intensity of development of the Arctic offshore fields. The period of scenario forecasting is extended to 2050, since the development of the Arctic offshore fields is an extremely long-term process.

Macroeconomic analysis in these scenarios is based on a current data from 17 forecasts of external analytical sources (the Energy Research Institute of the Russian Academy of Sciences, U.S. Energy Information Administration, International Energy Agency, the Organization of the Petroleum Exporting Countries, World Energy Council, companies BP, ExxonMobil) [8]. They were used as a basis for the consensus forecast with the purpose to take into account the different key trends considered by various analytical agencies. The quantitative parameters are population growth, the growth rate of the world economy, global oil and natural gas consumption and others (Table 1).

Based on the consensus forecast, the three scenarios of Arctic offshore oil and natural gas production in Russia were developed. The alternative scenario is based on the "green" energy forecasts of the global energy industry development, a moderate scenario is based on data from inertial forecasts (based on extrapolation), and an active scenario is based on forecasts of high economic growth.

The price of oil, as one of the key factors of economic and energy development, is one of the parameters in those scenarios. The oil price forecast is based on the principle of grouping the indicators according to similar types of scenarios. Only forecasts with relevant data available were utilized, where oil price forecast fell into the probabilistic corridor of predicted values (Fig. 2).

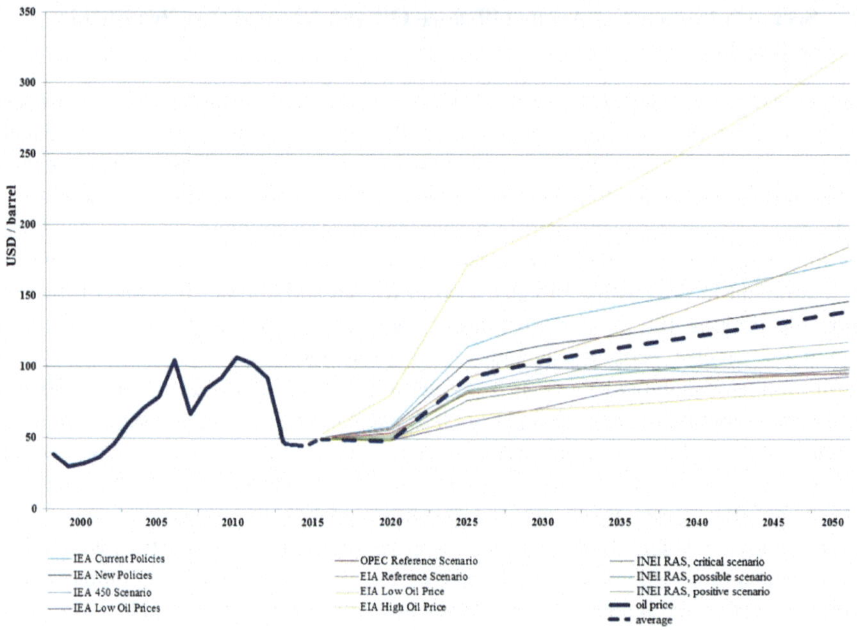

Fig. 2. Oil price forecasts summary [8].

The scenario parameters are presented in the following Table 1.

Table 1. Scenarios of Arctic offshore oil and natural gas development until 2050 [8].

Brief scenarios description/Indicators	Alternative moderate economic growth with a focus on sustainable development, climate and environmental issues	Moderate baseline scenario, current development trends, economic stabilization	Active world economy restoration and stable GDP growth; high oil prices, lack of significant macroeconomic risks
Population, billion people	9.84		
Annual growth, %	0.9		
Economic development, annual growth, %	3.2	3.2	3.5
Energy demand	A significant decrease (–16% to the baseline scenario)	Recovery of energy demand and its stabilization	High demand
Primary energy consumption, annual increase, %	0.6	1.2	1.4

(continued)

Table 1. (*continued*)

Brief scenarios description/Indicators	Alternative moderate economic growth with a focus on sustainable development, climate and environmental issues	Moderate baseline scenario, current development trends, economic stabilization	Active world economy restoration and stable GDP growth; high oil prices, lack of significant macroeconomic risks
Consumption of oil (liquid fuels), annual increase, %	–0.4	0.8	0.9
Natural gas consumption, annual increase, %	0.9	1.8	2.0
Share of oil and natural gas in the world energy balance, %	48	54	51
Demand for oil and natural gas	Significantly reduced (–28% oil and –21% natural gas to the baseline scenario)	Stable	Growing (3% oil and 4% natural gas to the baseline scenario)
Average price of oil by decades, USD/barrel (year)	56 (2020), 99 (2030), 97 (2040), 95 (2050)	55 (2020), 105 (2030), 123 (2040), 142 (2050)	55(2020), 104 (2030), 120 (2040), 132 (2050)
Realizable risks	Economic (price) and geoeconomic	Geoeconomic, geopolitical and technological	Geopolitical, geological and ecological
Intensity of the Arctic offshore development	Negligible	Moderate	Growing

One of the important elements of scenario planning is risk analysis [9]. The development of Arctic offshore fields is characterized by long lead investment and development time, as well as technological complexity. Thus, there are many related risks, which can potentially affect the development prospects. Each of the scenarios implements a certain set of risks (price, technological, geoeconomic, geopolitical, geological and environmental). In order to assess the risks in oil and natural gas resources development offshore Arctic, they have been studied, analysed and ranked based on the qualitative and quantitative evaluation. The risk assessment was based on cross-measures of frequency and probability of their occurrence [4]. The most urgent risks are price and technological risks, while the most critical are geoeconomic and environmental risks.

3.3 Future Production of Oil and Natural Gas Offshore Russian Arctic

Within the scenarios described above, the future production of oil and natural gas offshore Russian Arctic was evaluated. The evaluation is based on 24 examined Arctic offshore fields and 2 open structures in 8 offshore regions of the Russian Arctic [8:Appendix].

The three scenarios differ by the intensity of the offshore Arctic fields' development. In the alternative scenario oil and natural gas are produced only from already developed offshore fields (only the Prirazlomnoye field). In the moderate and active scenarios the possibility of developing certain offshore Arctic oil and natural gas fields was evaluated according the methodology by Fadeev [6]. This methodology suggests an integral coefficient for each of the identified offshore fields, which is based on the evaluation of their technical and economic characteristics. By rating the integral coefficient, the most and least promising for the future development offshore Arctic fields can be identified. Fields with an integral coefficient below the average (0.56) are not considered as promising for development. Thus, three gas (Ludlovskoye, Kamennomysskoye and Severo-Kamennomyskoye) and two gas condensate (Ledovoye, Shtokmanovskoye) fields are not advisable to be developed in the current technical and economic conditions. The gas condensate fields Leningradskoye, Rusanovskoye and Pomorskoye, the oil and gas condensate field Severo-Gulyaevskoye and the oil fields Medynskoye and Prirazlomnoye can be considered as ones having moderate development potential, since these fields are difficult to access or require high development costs. The most promising for development are the gas fields North-Kildinskoye and Murmanskoye and oil fields Varandey-Sea and Dolginskoye.

In moderate scenario the fields considered are the Dolginskoye, Murmanskoye, Varandey-Sea and Severo-Kildinskoye (with an integral coefficient above 0.6 [6]), and in the active scenario fields Pomorskoye, Severo-Gulyaevskoye, Medynskoye-Sea, Rusanovskoye and Leningradskoye (in addition to the previous ones, with a coefficient not lower than 0.56).

The forecasted volume of oil and natural gas production combined offshore Russian Arctic by three scenarios is presented in Fig. 3. The peak of production according to the alternative scenario will be passed in 2020–2023, according to the moderate scenario approximately in 2049, and according to the active scenario in 2048. The maximum annual projected production volumes are 4.8, 6.3 and 15.7 million tonnes of oil equivalent (mln t o.e.) respectively by each scenario, where those volumes compose approximately 1% of the total oil and natural gas production in Russia in 2016.

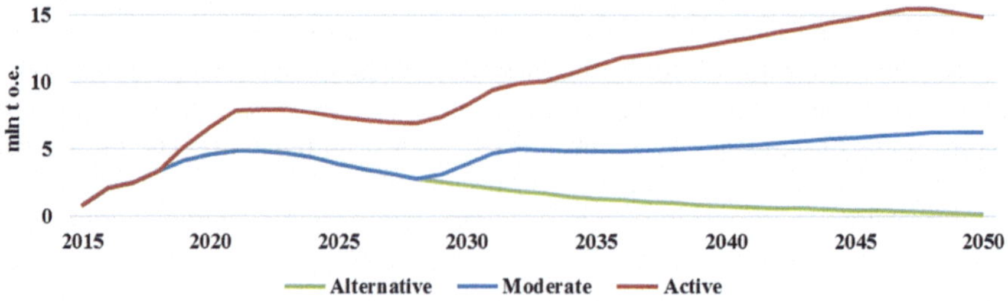

Fig. 3. The scenario based forecast of oil and natural gas production offshore Russian Arctic [8].

The "moderate" scenario is considered the most likely to occur, however, the role of geopolitical, geoeconomic factors and uncertainties can significantly influence the Arctic offshore oil and natural gas resources development.

It should be noted that a substantial part of the oil and natural gas potential offshore Russian Arctic is located at relatively shallow depths on the shelf of the Pechora and Kara Seas. These resources, as well as the resources offshore Yamal peninsula, are most likely to be exploited before the offshore fields. Moreover, more simple and less expensive technologies can be used for their development. Additionally, it should be noted that those calculations did not take into account the fields located partly offshore, such as Yurkharovskoye field. Production at the onshore part of the Yurkharovskoye field constituted 40.5 and 34.5 million tons of oil equivalent in 2014 and 2015 respectively [8].

Based on the obtained results, it can be concluded that it is necessary to adjust the strategy for developing the oil and natural gas resources of the Arctic offshore in Russia with accordance to the potential production volumes and changing macroeconomic situation.

4 Conclusion

Arctic offshore oil and natural gas resources, being highly interesting for the Arctic countries and energy companies, are being a challenging energy resource to develop. Russia is one of the most engaged countries, however, it is experiencing difficulties due to the challenging geopolitical and geoeconomic landscape.

With the aim to evaluate Arctic offshore oil and natural gas resources development perspectives in Russia for a long-term period, the algorithm, methodical recommendations and scenario forecast were developed. According to the scenario results, the future oil and natural gas production offshore Russian Arctic is going to be rather modest in all the three studied scenarios.

The study concludes with the necessity to adjust the strategy for developing the oil and natural gas resources of the Arctic offshore in Russia with accordance to the potential production volumes and changing macroeconomic situation, where the developed instruments can contribute to the current existing planning approaches.

References

1. Institute of Energy Strategy: Energy strategy of Russia for the period up to 2030 (Jenergeticheskaja strategija Rossii na period do 2030 goda). Energia (2010). (in Russian)
2. Morgunova, M.: Arctic Offshore Hydrocarbon Resource Development: Past, Present and Vision of the Future. Universitetsservice US-AB, Sweden (2015)
3. Morgunova, M., Wesphal, K.: Offshore Hydrocarbon Resources in the Arctic: From Cooperation to Confrontation in an Era of Geopolitical and Economic Turbulence? SWP Research Paper RP 3. Stiftung Wissenschaft und Politik German Institute for International and Security Affairs, Berlin (2016)
4. World Energy Council: World Energy Issues Monitor, Energy Price Volatility: The New Normal. United Kingdom (2015)

5. World Energy Council: World Energy Scenarios: Composing Energy Futures to 2050. London (2013)
6. Fadeev, A.M.: Assessment of the level of development of oil and gas fields in the Arctic as an important element of strategic management of the oil and gas complex (Ocenka urovnja razvitija neftegazovyh mestorozhdenij Arktiki kak vazhnejshij jelement strategicheskogo upravlenija neftegazovym kompleksom). In: Scientific and Technical Statements of SPbSPU, vol. 4, no. 223, pp. 81–90 (2015). (in Russian)
7. The Government of Russian Federation: On strategic planning in the Russian Federation (O strategicheskom planirovanii v Rossijskoj Federacii). In the version of federal laws of 23.06.2016 N 210-FZ, of 03.07.2016 N 277-FZ, dated 30.10.2017 N 299-FZ, dated December 31, 2017 g N 507-FZ (2014). (in Russian)
8. Morgunova, M.O.: Perspectives of development of oil and natural gas resources offshore Russian Arctic in the context of energy system transformation (Perspektivy osvoenija uglevodorodnyh resursov arkticheskogo shel'fa Rossii v uslovijah transformacii mirovoj jenergetiki). Thesis for the scientific degree, Ph.D. in Economic Sciences, Moscow (2017). (in Russian)
9. Zolotukhin, A.B., Gudmestad, O.T., Jarlsby, E.T.: Petroleum Resources with Emphasis on Offshore Fields. WIT Press, Boston (2010)

The manufacturer's authorised representative in the EU is Springer
Nature Customer Service Centre GmbH, Europaplatz 3, 69115 Heidelberg,
Germany. If you have any concerns regarding our products, please
contact ProductSafety@springernature.com

Printed and bound by CPI Group (UK) Ltd, Croydon, CR0 4YY
05/10/2026
02228145-0015